SPACE CHARGE DOMINATED BEAMS AND APPLICATIONS OF HIGH BRIGHTNESS BEAMS

SPACE CHARGE DOMINATED BEAMS AND APPLICATIONS OF HIGH BRIGHTNESS BEAMS

Bloomington, IN   October 10-13, 1995

EDITOR
S. Y. Lee
Indiana University

AIP CONFERENCE PROCEEDINGS 377

American Institute of Physics   Woodbury, New York

Authorization to photocopy items for internal or personal use, beyond the free copying permitted under the 1978 U.S. Copyright Law (see statement below), is granted by the American Institute of Physics for users registered with the Copyright Clearance Center (CCC) Transactional Reporting Service, provided that the base fee of $6.00 per copy is paid directly to CCC, 222 Rosewood Drive, Danvers, MA 01923. For those organizations that have been granted a photocopy license by CCC, a separate system of payment has been arranged. The fee code for users of the Transactional Reporting Service is: 1-56396-625-5/ 96 /$6.00.

© 1996 American Institute of Physics

Individual readers of this volume and nonprofit libraries, acting for them, are permitted to make fair use of the material in it, such as copying an article for use in teaching or research. Permission is granted to quote from this volume in scientific work with the customary acknowledgment of the source. To reprint a figure, table, or other excerpt requires the consent of one of the original authors and notification to AIP. Republication or systematic or multiple reproduction of any material in this volume is permitted only under license from AIP. Address inquiries to Office of Rights and Permissions, 500 Sunnyside Boulevard, Woodbury, NY 11797-2999; phone 516-576-2268; fax: 516-576-2499; e-mail: rights@aip.org.

L.C. Catalog Card No. 96-85165
ISBN 1-56396-625-5
DOE CONF- 9510361

Printed in the United States of America

CONTENTS

Preface ix
The Workshop Program xi
Workshop Participants List xv

TOPICS OF GENERAL INTEREST

Emittance Concept and Growth Mechanisms 3
    T. P. Wangler
Space Charge Tune Shift, Fast Resonance Traversal and Current
Limits in Circular Accelerators 23
    G. H. Rees
Space Charge Problems in High Intensity RFQs....................... 32
    M. Weiss
Critical Beam Intensity Issues in Hadron Colliders 42
    S. D. Holmes
A Summary of the 1995 International Symposium on Heavy Ion
Inertial Fusion.. 54
    T. J. Fessenden
Linear Accelerator for Tritium Production............................ 60
    R. W. Garnett, J. H. Billen, K. C. D. Chan, R. Genzlinger, E. R. Gray,
    S. Nath, B. Rusnak, D. L. Schrage, J. E. Stovall, H. Takeda, R. Wood,
    T. P. Wangler, and L. M. Young
Dealing with the High-Brightness Beam and Space Charge in CLIC 74
    G. Guignard

EXPERIMENTS WITH SPACE CHARGE DOMINATED BEAMS AND HALO FORMATION

Experimental Studies of Longitudinal Dynamics of Space-Charge
Dominated Beams.. 95
    J. G. Wang and M. Reiser
Maryland Transport Experiment: Comparison to Theory and Simulation 105
    I. Haber, D. A. Callahan, A. Friedman, D. P. Grote, D. Kehne, A. B. Langdon,
    M. Reiser, H. Rudd, H. Suk, D. X. Wang, and J. G. Wang
Critical Design Issues of LEBTs for High-Brightness Ion Beam
Transport and Focusing... 124
    S. K. Guharay
Space Charge Dominated Heavy Ion Beams in Electrostatic
Quadrupole (ESQ) Accelerators....................................... 134
    S. Yu, S. Eylon, E. Henestroza, and D. Grote
Performance and Measurements of the AGS and Booster Beams 145
    W. T. Weng
Space Charge Induced Non-Structure Resonances 160
    S. Machida and Y. Shoji

Halo Formation and Chaos in Space-Charge-Dominated Beams 169
    C. Chen
Overlapping Resonances and Chaos in Halo Formation 187
    A. Riabko and S. Y. Lee
Nonlinear Resonances, Chaos and Halo Formation in Space-Charge
Dominated Beams... 206
    J.-M. Lagniel
Lowest-Order Phase Space Structure of a Simplified Beam Halo
Hamiltonian.. 219
    D. L. Bruhwiler
The Envelope Dynamics of Intense Charge Particle Beams in a FODO
Channel 234
    S. Y. Lee and K. Y. Ng
PIC Simulation of Short Scale-Length Phenomena......................... 244
    I. Haber, D. A. Callahan, C. M. Celata, W. M. Fawley, A. Friedman,
    D. P. Grote, and A. B. Langdon
Injection into a Circular Machine with a KV Distribution 260
    E. Crosbie and K. Symon
Reducing Space Charge Tune Shift with a Barrier Cavity 283
    M. Blaskiewicz
On Space Charge Dominated Beam Transport without Emittance Growth 290
    Y. K. Batygin

STATISTICAL AND FOKKER PLANCK APPROACH

Generalized Free-Energy Principle and Emittance Growth.................. 309
    P. G. O'Shea
Conceptual Foundation of the Fokker-Planck Approach to Space
Charge Effects... 322
    C. L. Bohn
RF Linac Designs with Beams in Thermal Equilibrium..................... 329
    M. Reiser and N. Brown

NEUTRON SPALLATION SOURCES

Critical Beam Dynamical Issues in Neutron Spallation Sources.............. 343
    M. Pabst, K. Bongardt, and A. P. Letchford
Halo Containment in the ESS Linac to Ring Transfer Line 354
    K. Bongardt, M. Pabst, and A. P. Letchford
Space Charge Beam Dynamics Studies for a Pulsed Spallation Source
Accelerator 363
    Y. Cho and E. Lessner
Longitudinal Tracking Studies for a High Intensity Proton Synchrotron 375
    E. Lessner, Y. Cho, K. Harkay, and K. Symon
Longitudinal and Transverse Tracking Studies for ESS..................... 391
    C. R. Prior

HEAVY ION INERTIAL FUSION

Induction-Accelerator Heavy-Ion Fusion: Status and Beam-
Physics Issues 401
    A. Friedman
Heavy Ion Fusion Experiments at LLNL 427
    J. Barnard, M. D. Cable, D. A. Callahan, F. J. Deadrick, S. Eylon,
    T. J. Fessenden, A. Friedman, D. P. Grote, K. A. Holm, H. A. Hopkins,
    D. L. Judd, R. L. Hanks, S. A. Hawkins, H. C. Kirbie, B. G. Logan,
    S. M. Lund, L. A. Nattrass, D. Longinotti, M. B. Nelson, M. A. Newton,
    C. W. Ollis, T. C. Sangster, and W. M. Sharp
Longitudinal Dynamics and Stability in Beams for Heavy-Ion Fusion 434
    W. M. Sharp, D. A. Callahan, and D. P. Grote

HIGH BRIGHTNESS ELECTRON BEAMS

Collective Effects in Isochronous Storage Rings 451
    A. W. Chao and K. J. Kim
Dynamics of Microbeams from a Point Emitter and its Array 458
    S. Kawasaki, H. Ishizuka, M. Shiho, and K. Yokoo
Space Charge Effects in the Injector and Driver Accelerator
for a UV FEL at CEBAF 468
    H. Liu and D. Neuffer
High Voltage High Brightness Electron Accelerator with MITL
Voltage Adder Coupled to Foilless Diode 479
    M. G. Mazarakis, J. W. Poukey, D. Rovang, S. Cordova, P. Pankuch,
    R. Wavrik, D. L. Smith, J. E. Maenchen, L. Bennett, K. Shimp, and K. Law
On Minimum Emittance Lattices 495
    S. Y. Lee

PANEL DISCUSSIONS

Issues Presented at the 8th ICFA Beam Dynamics Workshop Panel
Discussions 505
    R. L. Gluckstern

Author Index 521

PREFACE

The last workshop on space charge dominated beams was in 1978 at the Los Alamos National Laboratory, and the last workshop on the application of high brightness beams was in 1991 at the University of Maryland. Therefore, it was considered timely to have a workshop dealing with the space charge problem in high brightness particle beams.

The workshop on space charge dominated beams was approved by the ICFA beam dynamics panel, and preparation on the workshop began in December 1994. The organization and program committee were quickly formed from experts in the US, Europe, and Japan and consisted of:

Ben B. Brabson (Indiana University)
Yanglai Cho (Argonne National Laboratory)
Alex Friedman (Lawrence Livermore National Laboratory)
Samar K. Guharay (University of Maryland)
Patricia Halstead (Indiana University, Workshop Secretary)
Ingo Hofmann (GSI)
Steve Holmes (Fermilab)
K. J. Kim (Lawrence Berkeley Laboratory)
S. Y. Lee (Indiana University)
Roger Miller (Stanford Linear Accelerator Center)
Y. Mori (KEK)
Gunther Plass (CERN)
Grahame Rees (Rutherford)
Martin Reiser (University of Maryland)
Thomas P. Wangler (Los Alamos National Laboratory)
W. T. Weng (Brookhaven National Laboratory)

On the workshop organization, Ms. Pat Halstead provided a great help. Planning the workshop's agenda was done mainly through e-mail, saving time and traveling expenses. As the program began to take shape, we realized that the original choise of a 2.5-day workshop was insufficient to accomplish our goal. Despite the lack of time, we worked hard and managed to complete our workshop in time.

The workshop took a closer look at major breakthroughs occurring in the past few years. The mechanism of halo formation, which has been actively studied in the past few years, was a major topic in this workshop. The thermalized beam with the Vlasov-Fokker-Planck approach for non-KV beams received major attention and became a joint session. The design of neutron spallation sources evolved into another session. Yet another working group focused on the space charge effects in high brightness electron beams near the source and quasi-isochronous storage rings.

The results garnered from this workshop are divided into the following sections:

- Topics of General Interest
- Experiments with Space Charge Dominated Beams and Halo Formation
- Statistical and Fokker-Planck Methods
- Neutron Spallation Sources
- Heavy Ion Inertial Fusion
- High Brightness Electron Beams.

The conclusions reached in the panel discussion are included at the end. These results will hopefully be useful for beam physics research in the coming years.

S. Y. Lee
Bloomington,
Indiana University

The workshop Program

Plenary talks:

Thomas Wangler (Wangler@lanl.gov)
  The emittance, concept and growth mechanisms
Grahame Rees (JVT@IB.RL.AC.UK)
  Space-charge tune shift, fast resonance traversal, and current limits in circular accelerators
Tom Fessenden (Tom_Fessenden@macmail3.lbl.gov)
  Summary Report of the 1995 Symposium on Heavy Ion Fusion held at the Princeton Plasma Physics Laboratory Sept. 6-9, 1995.
Jim Alessi (alessi@bnldag.ags.bnl.gov)
  State of the art in ion sources
Mario Weiss, CERN (Weiss@cernvm.cern.ch)
  Beam dynamics and performance of high intensity RFQ
Bob Garnett (rgarnett@lanl.gov)
  Linear Accelerator for tritium production (100 mA 1 GeV)
Michel Pabst (KFA) (ESS@kfa-juelich.de)
  Critical beam dynamical issues in neutron Spallation source
Alex Friedman (af@llnl.gov)
  Induction-accelerator HIF status and feasibility
  (Driver Configurations for Heavy Ion Fusion)
Steve Holmes (holmes@fnal.gov)
  Critical beam dynamics issues in hadron colliders
R. Sheffield (rsheffield@lanl.gov)
  Beam dynamics issues relevant to the fourth generation light source
Gilbert Guignard (guignard@cernvm.cern.ch)
  High-brightness beam emittance growth and preservation in CLIC
Robert Palmer (palmer@slac.stanford.edu)
  Status of mu-mu Collider Conceptual Design

Summary talks:

M. Reiser
  Space charge dominated beam Experiments
I. Hofmann
  Theory
K.J. Kim
  beam dynamics relevant to ultrashort pulse electron beams

Talks presented in the experiment working group:

Simon Yu (Simon_Yu@macmail.lbl.gov)
  LBL experiments
John Barnard(jjbarnard@llnl.gov)
  LLNL HIF Experiments
J.G. Wang (jgwang@lpf.umd.edu)
  Experimental studies on longitudinal dynamics of space charge dominated beams
Samar Guharay (guharay@plasma.umd.edu)
  Critical issues for developing efficient LEBT for intense, high brightness ion beams
Subrata Nath (snath@lanl.gov)
  Beam funneling in high intensity proton linacs
W.T. Weng (weng@bnldag.ags.bnl.gov)
  Experimental measurements of AGS booster and AGS beams
Mike Blaskiewicz (Blaskiewicz@BNLDAG.AGS.BNL.GOV)
  Reducing space charge tune shift with a barrier cavity
Ed Crosbie and K. Symon
  Injection into a circular machine with a KV distribution
Sergie Nagaitsev (IUCF)
  Ways to reduce space charge effects: beam heating
Irving Haber
  Maryland experiments
Simon Yu
  LBL experiments
Ingo Hofmann
  GSI experiments

Theoretical approaches in space charge dominated beams:

Irving Haber
  Maryland/NRL/LANL: PIC tracking
Rob Ryne
  Symplectic computation of Lyapunov exponent and application to halo formation
Chi-ping Chen
  Halo formation and chaos in space charge dominated beams
Yuri Batygin
  On space charge dominated beam transport without emittance growth and Nonlinear beam dynamics in heavy ion transport line

Jean-Michel Lagniel
  Nonlinear resonances in space-charge dominated beams
Alexander Riabko
  Overlapping resonances and chaos in halo formation
Dave Grote
  3-D particle-in-cell simulations of space-charge-dominated:
  Emittance growth due to energy spread, bends, and
  "energy effect" aberrations in ESQ's
Bill Sharp
  Longitudinal "resistive wall" and fluid/envelope models
  of space-charge dominated HIF beams
Alain-Claude Piquemal
  Theoretical and numerical studies of halo in intense ion beams
  emittance growth or routes to reversibility in charged particle beams
W.W. Lee
  Delta-f simulation of space charge effects
David Bruhwiler
  Lowest-order phase space structure of a simplified beam
  halo Hamiltonian
K. Bongardt (KFA) (ESS@kfa-juelich.de)
  Halo containment in the ESS linac to ring transfer line
G. Rees (RAL) (jvt@ib.rl.ac.uk)
  High Intensity Features of the spallation sources ESS and ISIS
C. Prior (RAL) (C.R.Prior@rl.ac.uk)
  Transverse and Longitudinal Tracking Studies for ESS and ISIS
Y.C. Chae (yc@anlaps.aps.anl.gov)
  Argonne: beam dynamics issues in the Argonne NSS proposal
Eliane Lessner (ANLAPS)
  Longitudinal tracking studies for ANLPSS

Talks presented in the electron beam working groups:

Michael G. Mazarakis
  High voltage and high brightness electron accelerator with foilless diode
Tom Katsouleas
  Beam Issues for Plasma Accelerators
Hongxiu Liu
  Space charge effects in recirculating UV FEL Driver Accelerator
Bruce Carlsten
  Emittance growth in bends

K.J. Kim
> *Coherent Radiation in bends*

Luca Serafini
> *Analytical studies of space charge dominated beams in RF photo-injectors*

Sunao Kawasaki
> *Behavior of high-brightness microbeam in a periodic focusing system*

Alexander Molodozhentsev
> *Problems with quasi-isochronous storage rings*

S.Y. Lee
> *Minimum emittance for the DBA lattice with combined function dipoles*

Joint session on Fokker-Planck and Statistical methods:

Shinji Machida & Y. Shoji
> *Space-charge Effect in Synchrotron (KEK Booster)*

Ingo Hofmann
> *Collective Theory of Equipartitioning*

Courtlandt L. Bohn
> *Conceptual foundation for Fokker-Planck approach to space charge effect*

J. Struckmeier
> *Fokker Planck approach to emittance and entropy growth*

Steve Lund
> *Thermal equilibration of beams*

M. Reiser
> *Proposed RF Linacs with beams in thermal equilibrium*

Patrick O'Shea
> *Thermodynamics Potential and Emittance growth*

K. J. Kim
> *Entropy and emittance in e-beams and radiation*

Workshop Participants List

Mei Bai	Dept. of Physics, IU Bloomington, IN 47405
John J. Barnard	POBox 808 LLNL L-440 Livermore, CA 94550
Yuri Batygin	Inst. of Physical and Chemical Research Hirosawa 2-1, Wako-shi Saitama, 351-01 JAPAN
Mari Berglund	IUCF, Bloomington, IN 47405
Michael Blaskiewicz	BNL 911B, Upton, NY 11973-5000
Courtlandt L. Bohn	CEBAF 12000 Jefferson Avenue Newport News, VA 23606
Klaus Bongardt	ESS Project Office KFA Forschungszentrum Postfach 1913 D-52425 Julich, Germany
B.B. Brabson	Physics Dept. SW 117, Indiana Univ. IN 47405
Nathan Brown	1201 W. Energy Research Inst. Univ. of Maryland College Park, MD 20742
David L. Bruhwiler	Northrop Grumman Corporation ATDC 4 Independence Way Princeton, NJ 08540-6620
Bruce Carlsten	LANL, AOT-9, MS H851, Los Alamos, NM 87545
Yong-Chul Chae	Bldg 360 ANL, Argonne, IL 60439
Alex W. Chao	MS 26, POBox 4349 SLAC, Stanford, CA 94309
Chiping Chen	MIT Plasma Fusion Center, Cambridge, MA 02139
Yu-Jiuan Chen	LLNL, POBox 808, L-440, Livermore, CA 94550
Weiren Chou	Fermilab MS323, POBox 500, Batavia, IL 60510
C.M. Chu	Physics Dept. IU, Bloomington, IN 47405
Pat Colestock	POBox 500 MS307 Fermilab, Batavia, IL 60510

Edwin Crosbie	Bldg 360 ANL, Argonne, IL 60439
T.J. Fessenden	MS 47-112, 1 Cyclotron Rd. LBL Berkeley, CA 94720
Alex Friedman	POBox 808 LLNL, L-440 Livermore, CA 94550
Robert W. Garnett	POBox 1663 AOT-1, MS H817 LANL Los Alamos, NM 87545
R.L. Gluckstern	Physics Department Univ. of Maryland College Park, Maryland 20742-4111
David P. Grote	POBox 808 LLNL, L-440 Livermore, CA 94550
S. Guharay	1201 W. Energy Research Inst. Univ. of Maryland, College Park, MD 20742
Gilbert Guignard	SL Div. CERN, 1211 Geneva 23 Switzerland
Irving Haber	NRL, Code 6790, Washington, DC 20375
Kathy Harkay	Bldg 360 ANL, Argonne, IL 60439
Kohji Hirata	KEK, 1-1 Oho, Tsukuba, Ibaraki 305 Japan
Ingo Hofmann	Gesellschaft fur Schwerionenforschung POBox 110552 D64220 Darmstadt, Germany
Steve Holmes	PO Box 500 MS 307 Fermilab, Batavia, IL 60510
Gerry Jackson	PO Box 500 MS 307 Fermilab, Batavia, IL 60510
X. Kang	Phys. Dept. IU, Bloomington, IN 47405
Tom Katsouleas	USC EE Dept. LA, CA90089-0484
Sunao Kawasaki	Saitama Univ. Faculty of Science 255 Shimo-Ohkubo Urawa 338 Japan
David Kehne	CEBAF MS85A 12000 Jefferson Ave. Newport News, VA 23606

K.J. Kim	MS 71-259, 1 Cyclotron Rd. LBL Berkeley, CA 94720
Shane Koscielniak	TRIUMF 4004 WESBROOK MALL VANCOUVER BC CANADA V6T 2A3
Thomas Kroc	Fermilab MS 301 Box 500 Batavia, IL60510
Sergey Kurennoy	Phys. Dept. Univ. maryland, College Park MD 20742
Jean-Michel Lagniel	CEN Saclay LNS 91191 Gif/Yvette Cedex France
S.Y. Lee	Dept. of Physics, IU Bloomington, IN 47405
W. W. Lee	Theory Div. Princeton Plasma Phys. Lab. P. O. Box 451, Princeton, NJ 08543
Eliane Lessner	Bldg 360 ANL, Argonne, IL 60439
Hongxiu Liu	CEBAF 12000 Jefferson Ave. Newport News, VA 23606
Steve M. Lund	LLNL, L-440 PO Box 808 Livermore, CA 94550
Shinju Machida	KEK, 1-1 Oho, Tsukuba, Ibaraki, 305 Japan
James MacLachlan	Fermilab MS345 Box 500 Batavia, IL 60510
Christa Markovits	Paul Scherrer Inst. Accelerator Div. CH-5232 Villegan Switzerland
Michael G. Mazarakis	Sandia National Labs. MS1193 POBox 5800 Alburquerque, NM 87185
Elliott McCrory	POBox 500 MS307, Fermilab, Batavia, IL 60510
A. Molodozhensev	Phys Dept, IU Bloomington, IN 47405
Sergei Nagaitsev	IUCF, IU, Bloomington, IN 47405
Subrata Nath	LANL, AT Div. H817 Los Alamos, NM 87545

David Neuffer	CEBAF 12000 Jefferson Avenue Newport News, VA 23606
King-Yuen Ng	Fermilab, MS345, POBox 500, Batavia, IL 60510
Michael Olivier	Commisariat energie Atomic DSM/DIR CEN Saclay, Gif sur Yvette, 91191 France
Miguel OLIVO	Paul Scherrer Institut (PSI) CH-5232 Villigen PSI Switzerland
Patrick G. O'Shea	Box 90318 Duke Univ. Durham, NC 27708-0319
Michael Ottinger	3378-B Lake Austin Blvd. Austin, TX 78703
Michael Pabst	KFA Forschungszentrum Julich Postfach 1913 D-52425 Julich, Germany
Robert Palmer	BNL, POBox 5000, Upton, N.Y. 11973
Alex Pei	IUCF, Bloomington, IN 47405
A.C. Piquemal	Centre d'Etudes de Bruyeres-le-Chatel BP 12, 91680 BRUYERES-LE-CHATEL FRANCE
Milorad Popovic	MS 307, POBOX 500, Fermilab Batavia, Il 60510
C.R. Prior	Rutherford Appleton Laboratory Bldg R2 Didcot OX11 OQX United Kingdom
Grahame Rees	Rutherford Appleton Laboratory Bldg R2 Didcot OX11 OQX United Kingdom
Martin P. Reiser	1201 W. Energy Research Inst. Univ. of Maryland College Park,MD 20742
Alexander Riabko	Phys. Dept. IU Bloomington, IN 47405
Robert Ryne	POBox 1663 AOT-1, MS H817 LANL Los Alamos, NM 87545
P. Schwandt	2401 Milo Sampson Lane, Bloomington, IN 47405

Luca Serafini	Dept. of Phys. Univ. of Milan Via Celoria 16 Milan, Italy 20133
William Sharp	Box 1663 AOT-1, MS H817 LANL Los Alamos, NM 87545
Richard Sheffield	MS H851, LANL, Los Alamos, NM 87545
Yoshikiko Shoji	KEK 1-1 Oho, Tsukuba, Ibaraki, 305 Japan
J. Struckmeier	Gesellschaft fur Schwerionenforschung POBox 110552 D64220 Darmstadt, Germany
David F. Sutter	1510 Blue Meadow Rd. U.S. DOE ER-224, GTN Potmac, MD 20854
Keith Symon	Bldg 360 ANL, Argonne, IL 60439
L. C. Teng	Bldg. 360, ANL, Argonne, IL 60439
Jian-Guang Wang	Inst. for Plasma Research Univ. of Maryland College Park, MD 20742
Thomas P. Wangler	POBox 1663 AOT-1, MS H817 LANL Los Alamos, NM 87545
Mario Weiss	CH-1211 Geneva 23 CERN, Switzerland
W.T. Weng	Bldg 911B AGS Department POBox 5000 BNL Upton, NY 11973
Xiaoyu Wu	South Shaw Lane Cyclotron Lab, East Lansing, MI 48824-1321
Richard C. York	South Shaw Lane Cyclotron Lab, East Lansing, MI 48824-1321
Simon Yu	MS 47-112, 1 Cyclotron Rd. LBL Berkeley, CA 94720
X. Zhao	Physics Dept., IU Bloomington, IN 47405

TOPICS OF GENERAL INTEREST

Emittance Concept and Growth Mechanisms

Thomas P. Wangler

*Accelerator Operations and Technology Division
Los Alamos National Laboratory
Los Alamos, New Mexico 87545*

Abstract. We present an introduction to the subjects of emittance and space-charge effects in charged-particle beams. This is followed by a discussion of three important topics that are at the frontier of this field. The first is a simple model, describing space-charge-induced emittance growth, which yields scaling formulas and some physical explanations for some of the surprising results. The second is a discussion of beam halo, an introduction to the particle-core model, and a brief summary of its results. The third topic is an introduction to the hypothesis of equipartitioning for collisionless particle beams.

EMITTANCE CONCEPTS AND DEFINITIONS

Beams are composed of collections of particles, which can be described by six normalized phase-space coordinates (x, p_x/mc, y, p_y/mc, z, p_z/mc), where in rf linacs, if z is the direction of the center of momentum of the beam, z and p_z/mc are usually defined relative to a synchronous or reference particle. In general we can talk about a 6-D phase space distribution that describes the beam. As the beam is transported and accelerated, this distribution obeys Liouville's theorem, provided that the proper conditions are satisfied [1]. These are that 1) the continuity equation is valid in 6-D phase-space, which means that no particles are created or lost, and 2) the forces are derivable from a potential function that depends only on the three spatial coordinates and time. The second condition implies that we are ignoring the individual binary particle collisions, and representing the sum of the interparticle forces as a smoothed Coulomb force, averaged over all the particles. In most circumstances these conditions are a very good approximation, and the smoothed Coulomb force is commonly known as the space-charge force. Subject to these requirements the theorem states that the density in 6-D phase space, measured along the trajectory of a particle, is invariant, or equivalently that the 6-D volume enclosed by any isodensity contour is invariant.

Beam emittances are defined as a measure of the projections of the 6-D phase-space volume on the 2-D areas x-p_x, y-p_y, and z-p_z [2]. In linear focusing systems these areas are invariant, although their shapes will generally change as the beam propagates. Because most focusing systems are linear for small displacements, the invariance or approximate invariance of the emittances is the underlying reason why emittances are such useful quantities.

An arbitrary phase-space distribution can be described by an rms ellipse. The second moments of the distribution are used to define the ellipse parameters and the rms emittances. For example the definition of rms emittance in the x-p_x plane can be given as

$$\varepsilon_n \equiv \sqrt{\overline{x^2 p_x^2} - \overline{xp_x}^2} \, / \, mc. \qquad (1)$$

Similar results apply for the y-p_y and (z-z_c)-(p_z-p_{zc}) planes, where z_c and p_{zc} are the coordinates of the center of momentum of the beam. For historical reasons the emittance, defined by Eq.(1) is generally called the normalized emittance to distinguish it from the unnormalized emittance, which is specified in a phase space defined by position x and divergence angle x' = p_x/p_z. In the paraxial approximation the unnormalized and normalized transverse emittances, respectively, are

$$\varepsilon = \sqrt{\overline{x^2 x'^2} - \overline{xx'}^2}, \qquad (2)$$

$$\varepsilon_n = \varepsilon\beta\gamma = \beta\gamma\sqrt{\overline{x^2 x'^2} - \overline{xx'}^2}, \qquad (3)$$

where β is the ratio of the beam velocity to the speed of light, and $\gamma = 1/\sqrt{1-\beta^2}$ is the usual relativistic mass factor.

Any discussion of emittances should be accompanied by a word of caution, because the emittance conventions are not unique. The above definition of rms emittance corresponds to the convention introduced by Sacherer [3]. Another commonly used convention for rms emittance, introduced by Lapostolle [4], contains a factor of 4 in front of the square root of the above expressions. The situation is further confused by the fact that some Laboratories introduce other factors than the 4. Another confusion has to do with a factor of π, which is often appended, again for historical reasons. When a π is attached to the numerical value, ε becomes the area of an equivalent ellipse. For consistency we prefer not to introduce a factor of π, because in the envelope equation, to be introduced shortly, the symbol ε generally represents an area divided by π, rather than an area. Until there is a consensus in the accelerator community about a convention for emittance, this author prefers to use the definitions of rms emittance, given by Eqs.(2) and (3), which have no additional factors of π or 4, or any other numerical factors.

The rms emittance is invariant, if only linear forces act on the beam. Nonlinear forces produce filamentation with accompanying change in the rms emittance. When Liouville's theorem is satisfied, and when there is no coupling between the three orthogonal directions of motion, the true phase-space areas are preserved, even with nonlinear forces. However, the filamentation results in the capture of empty phase space between beam filaments [5], and intuitively, an effective emittance of the beam, including the area of both the filaments and the captured space, increases. For this case, it is found that the rms emittance also increases, essentially taking this emittance dilution into account. Therefore, the rms emittance is generally regarded as an effective emittance. It is interesting that under some conditions with nonlinear forces, the rms emittance may decrease [6]. Finally, the evolution of the rms beam size depends on the rms emittance, through an envelope equation. For transverse motion of continuous beams with elliptical symmetry, but otherwise arbitrary particle distributions, the rms envelope equations [3,4], also known as the K-V envelope equations [7], are

$$\frac{d^2 a_x}{ds^2} + \kappa_x(s) a_x - \frac{\varepsilon_x^2}{a_x^3} - \frac{K}{2(a_x + a_y)} = 0,$$

$$\frac{d^2 a_y}{ds^2} + \kappa_y(s) a_y - \frac{\varepsilon_y^2}{a_y^3} - \frac{K}{2(a_x + a_y)} = 0,$$

(4)

where a_x and a_y are the rms projections on the x and y axes, k_x and k_y determine the strength of the focusing, ε_x and ε_y are unnormalized rms emittances, and K is called the generalized perveance, given in terms of the particle charge q, mass m, and the beam current I, by

$$K = \frac{qI}{2\pi\varepsilon_0 \gamma^3 \beta^3 mc^3}.$$

(5)

For a round beam in an ideal uniform focusing channel, the rms envelope equations become

$$\frac{d^2 a}{ds^2} + k_0^2 a - \frac{\varepsilon^2}{a^3} - \frac{K}{4a} = 0,$$

(6)

where a is the rms projection on either the x or y axis, and k_0 is the transverse phase advance per unit length of the harmonic motion for a single particle at zero current. The uniform channel is frequently used to represent a smooth approximation to the particle motion in a quadrupole focusing channel. The three force terms in Eqs.(4) and (6) represent the external focusing force, an outward pressure force associated with the emittance, and the space-charge defocusing force.

BEAMS WITH SPACE CHARGE

Understanding space-charge-dominated beams is generally important to isolate the effects of space-charge, and to define the limits of high-intensity performance. The ratio of the space-charge to the emittance terms in the envelope equations can be used to determine when space-charge effects will become important. We refer to emittance dominated and space-charge dominated beams, according to

$$\frac{Ka^2}{4\varepsilon^2} \ll 1, \text{ emittance dominated,}$$
$$\frac{Ka^2}{4\varepsilon^2} \gg 1, \text{ space-charge dominated.} \quad (7)$$

A space-charge-dominated beam may be compared with a cold plasma, where collective effects are more important, whereas an emittance-dominated beam is associated primarily with random or thermal effects. Many accelerators that are characterized as high-current machines are not necessarily in the space-charge-dominated regime, and are often associated with an intermediate regime, where $Ka^2/4\varepsilon^2 \approx 1$. Because the external focusing force plays the same role as a background charge of opposite sign in a neutral plasma, it is not surprising that the behavior of beams with space-charge can sometimes be described in terms of concepts familiar from plasma physics. For example, a mismatched beam in a uniform channel will oscillate at the beam-plasma frequency, which can be written, for an equivalent uniform-density beam, as $\omega_p = (K\beta^2c^2/2a^2)^{1/2}$. If an effective thermal energy is defined as $k_BT = mc^2\beta^2\varepsilon^2/a^2$, a Debye length can be written as $\lambda_D = (2\varepsilon^2/K)^{1/2}$, which characterizes shielding effects in real beams; thus the particles tend to redistribute to shield the interior from the external fields. Also, at the edges of the beam, the particle density exhibits a characteristic Debye-length fall off.

spreading of the curved filament into a finite phase-space area in x – x' space, as shown in Fig. 1.

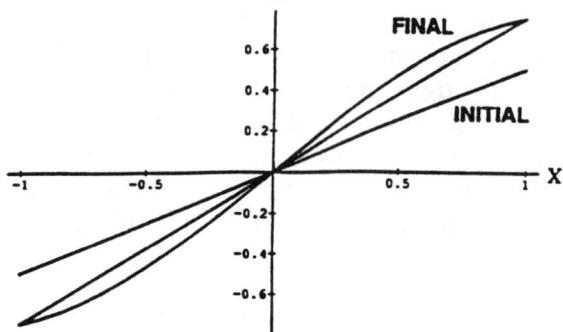

Figure 1. Effect from the model of the space-charge force on a parabolic-density beam with zero initial emittance. Initial and final phase-space distributions are shown.

The rms emittance grows linearly with respect to z, and for a circular beam, we find

$$\varepsilon_x(z) = \frac{Kz}{12\sqrt{5}} \qquad (8)$$

where K is the generalized perveance. The space-charge-induced emittance growth rate is independent of the rms beam size, which is the result of two compensating effects. As the beam size increases, the average space-charge force, and the transverse momentum impulses from space-charge decrease. But at fixed initial ε_x, the smaller momentum impulses are distributed over a larger beam, and the total phase-space area increase is independent of the beam size.

To obtain a prediction for the final emittance growth, it is necessary to take into account the saturation of the emittance growth that occurs after the transverse motion of the particles is included, and the beam relaxes into a matched density configuration. This stage of the dynamics is not easily adapted to a detailed dynamic model. However, we can appeal to the results from the numerical simulations, which tell us that the emittance growth of an initially rms-matched beam saturates

It is convenient to separate space-charge effects into two main categories, linear and nonlinear-force effects. The linear space-charge fields produce defocusing, resulting in increased beam size, but no emittance growth. The nonlinear fields also produce rms-emittance growth, and can be associated with significant halo formation at the periphery of the beam. Four categories of space-charge induced emittance growth can be identified [8], 1) charge redistribution [6,9], 2) kinetic-energy transfer [10,11,12], rms mismatch [13, 14], and 4) structure resonances [15,16,17]. Although not all of these processes are fully understood, certain general remarks can be made. Charge redistribution describes the process in which the phase-space distribution rearranges to provide shielding of the interior of the beam from external fields. It can occur for rms-matched beams, and develops within a very short time scale; the full emittance growth typically occurs within only one quarter of a beam-plasma period. The emittance growth is represented quantitatively by a nonlinear field-energy theory in which the excess field energy associated with the nonuniform initial distribution is transferred to thermal energy. Kinetic-energy transfer describes the tendency of highly space-charge dominate beams to reduce, by means of the space-charge interactions, any large thermal asymmetries between the different degrees of freedom. Rms mismatch refers to emittance growth associated with rms mismatched beams, in which the excess energy of the mismatch oscillations is converted into thermal energy. Understanding this mechanism is important for control of large-amplitude beam halos. The structure resonance is a parametric resonance, in which the particle amplitudes are driven by interaction with the periodic-focusing structure. The most serious case is known as the envelope resonance or envelope instability, which leads to growth of the transverse emittances and beam envelopes. The envelope instability can be avoided by limiting the zero-current phase advance per focusing period to $\sigma_0 < 90°$. The LBL beam-transport experiment [18] provides the best experimental study in a periodic quadrupole channel on emittance growth in space-charge dominated beams. The other significant experiment is the University of Maryland experiment in a periodic solenoid channel, which is reported in many published papers, most of which can be found in Reference [19].

SPACE-CHARGE-INDUCED EMITTANCE-GROWTH MODEL

Much of what has been learned about space-charge-induced emittance growth has been the result of computer-simulation studies. Although analytic approaches have been limited by the complex nature of the problem, more analytic work is necessary to provide better physical understanding. A simple analytic model of emittance-growth from charge redistribution provides some useful scaling rules for emittance growth.

In this model [20] we consider a continuous beam with a parabolic charge density, an elliptical transverse cross section, and zero initial emittance. We assume that as the beam travels over a short distance z, the space-charge impulse changes the transverse momentum of the particles, but the particle positions do not change appreciably; we assume that the parabolic density profile is constant. The model describes the initial stages of the charge redistribution process, and predicts how the emittance growth depends on the parameters. In $x-x'$ space there are two effects, 1) filamentation that changes an initial straight line in phase space (representing the zero-emittance initial beam) into a curved line, and 2) an x-y coupling through the space-charge term, which produces a

after the beam has propagated for one quarter plasma period [6, 21] The beam-plasma spatial period is related to the rms beam size a of the equivalent uniform beam by

$$\lambda_p = \sqrt{2\pi^2 a^2 / K}. \qquad (9)$$

Suppose we assume that the emittance grows linearly according to Eq.(8), while the beam propagates from z = 0 to $\lambda_p/4$. If the emittance growth rate is assumed to saturate at z = $\lambda_p/4$, as a result of the redistribution of the charges, we obtain for the final emittance

$$\varepsilon_x (z = \lambda_p / 4) = \frac{K\lambda_p}{48\sqrt{5}} = \frac{\pi}{48}\sqrt{\frac{2K}{5}} a. \qquad (10)$$

The result of Eq.(10) gives the same functional dependences, and is only about 30% lower in magnitude than the result for a parabolic beam from nonlinear field energy theory, a theory that is in excellent agreement with the numerical simulations [6]. Eq.(10) shows the result that the saturated emittance growth increases with increasing beam size, which is a surprising result for an effect that is caused by space-charge forces, which decrease with increasing beam size. The result can be explained by the fact that the larger beam size increases the plasma period, which means that the emittance grows for a longer time before saturation, at a rate which is independent of the beam size. Furthermore, the appearance of the generalized perveance K in these formulas shows that space-charge induced emittance growth is inherently a nonrelativistic effect, since $K \propto (\beta\gamma)-3$. The model can also be adapted to describe the transverse emittance growth of a bunch with the shape of a right circular cylinder. For a bunched beam the emittance growth in the x plane is affected by coupling to both the y and the z planes. Similar expressions are obtained in terms of a generalized perveance that has the form

$$K = \frac{qI_{ave}}{4\sqrt{3}\pi\varepsilon_0 mc^2 \beta^2 \gamma^3 bf}, \qquad (11)$$

where I_{ave} is the average current, b is the rms bunch length, and f is the bunch frequency. Eq.(11) predicts that to reduce κ and the emittance growth at fixed average current, one should choose a large bunch frequency to reduce the charge per bunch. This result implies that bunched-beam emittance growth is less for a higher frequency linac, and is a consequence of a smaller number of particles per bunch for a fixed average current at higher frequency.

BEAM HALO

The subject of beam halo has received much attention during the past several years [22-29]. The phenomenon of beam halo has practical importance because beam loss can result, if the halo is intercepted by the apertures of accelerator structures. Even a beam loss rate as small as 1 nA/m at about 1 GeV in a proton linac, can cause significant radioactivation corresponding to a dose rate of approximately 20 mrem/hr at a distance of 30 cm from the structure at 1 hour after beam is turned off. There are many unanswered questions about the origin and properties of the halo. It is known that the halo particles are those that possess higher than average energies of transverse motion. It is known from computer simulations that halos are produced from the space-charge-induced emittance growth processes, especially from the emittance growth that occurs in an initially rms-mismatched beam. This is not unexpected physically, because mismatched beams produce time-varying space-charge forces from the oscillations of the core of the beam that can, under certain conditions, transfer energy to a small fraction of particles, such as those that are initially just outside the main core. A precise definition of the halo does not exist; one simple definition is to include all particles that are outside an ellipse with the same aspect ratio as the rms ellipse, and with an emittance of 4 times the rms emittance for a continuous beam, and 5 times the rms emittance for a bunched beam. The factors of 4 and 5 correspond to the total emittance of the equivalent uniform beams, uniform beams that have the same first and second moments as the real beam.

Emittance growth is often compared with the classical filamentation process, where in one-dimensional motion a continuous nonlinear force produces oscillation frequencies that depend on the particle amplitudes. This results in a curvature, where an initial phase-space ellipse becomes distorted into an s shape, and eventually into a spiral structure in phase space. However, computer simulation studies show that after a few plasma periods, a round and initially rms-mismatched beam develops in $x-x'$ phase space into an inner core and an outer halo, instead of the classical spiral. To investigate the connection between the classical filamentation and the halo, it is revealing to study the evolution of an initial zero-emittance, radially mismatched beam, where all particles undergo radial oscillations. The beam oscillates in an azimuthally symmetric breathing mode, which is conveniently viewed in the radial $r-r'$ plane. One finds that filamentation is observed in $r-r'$ space, although the filamentary structure is more complicated than a simple spiral, because of the time dependence of the nonlinear space-charge force. The initial straight line in phase space rapidly evolves to a filament that is bent back and forth many times, forming a dense inner core, and an outer spiral-like structure. The spiral-like filament that is observed in $r-r'$ space projects into a diffuse halo in $x-x'$ space, because of the coupling of the x and y motions. Thus, the formation of a core and halo is found to result from the fundamental filamentation process, coupled with the projection onto the Cartesian $x-x'$ phase space. Similar spreading effects from coupling, as seen in the projected phase-space distribution, were described in the emittance-growth model of the previous section.

Recently, the particle-core model has been studied to provide a better understanding of the mechanism that transfers energy to particles, resulting in halo formation. In this model we consider a continuous beam in a uniform-focusing channel. A round, continuous, uniform-density distribution is used as a model for the core. The core is mismatched, so the radius of the core will oscillate in a breathing mode with an azimuthally symmetric core displacement . The halo particles are represented by test particles, which oscillate through the core, influenced by linear external focusing fields, and the nonlinear space-charge fields of the core. The model allows one to study the dynamics of the test particles. It consists of the envelope equation, and the single particle equation of motion, which are solved simultaneously. In dimensionless form these equations may be written as

$$\frac{d^2X}{ds^2} + X - \frac{\eta^2}{X^3} - \frac{I}{X} = 0, \qquad (12)$$

and

$$\frac{d^2x}{ds^2} + x - \begin{cases} Ix/X^2, & x < X \\ I/x, & x \geq X \end{cases}, \qquad (13)$$

where Eq.(12) describes X, the ratio of the core radius to the matched beam radius, s is the product of the axial distance times k_0, η is the tune-depression ratio of the core, or k/k_0, where k is the single particle transverse phase advance per unit length, k_0 is the same quantity without the space-charge force, and $I = 1 - \eta^2$. Eq.(13) is the equation of motion of a single particle, where the normalized displacement x is defined as the ratio of the displacement to the matched rms radius of the core. We note that the two equations contain a single dimensionless parameter η. The motion of a particle depends on the normalized core radius X, but the dynamics of the core is independent of the motion of the particle. This is a reasonable approximation, if the halo particles represent only a small fraction of the total beam.

The system is studied by systematically varying the initial state of the particle. A stroboscopic plot of the problem, obtained by taking snapshots of many independent particle trajectories, once per core-oscillation cycle at a specified phase of the core oscillation, reveals a separatrix defining three kinds of trajectories, 1) core dominated, 2) resonance dominated, and 3) focusing dominated. The plot is shown in Fig. 2. The core-dominated trajectories are the innermost trajectories on the stroboscopic plot. The focusing-dominated trajectories are the outermost trajectories, which describe a peanut-like shape in the plot. The resonance-dominated trajectories are most strongly affected by a parametric resonance, which occurs when the particle frequency is approximately half the core frequency. This resonance is responsible for the growth of the amplitudes that can form a halo. Particles that are initially nearest the core, and within the resonance-dominated region, are driven by the resonance, and acquire large amplitudes. The growth of the amplitudes is self limiting, because of the nonlinear decrease of the space-charge force outside the core. The model predicts an important result, that a maximum amplitude exists, which is the largest displacement on the separatrix, corresponding to a stroboscopic phase when the core has the minimum size. Resonance-dominated particles will cycle out and back as a result of the resonance, but because of the amplitude-dependence of the frequency, a significant phase mixing occurs, and there will usually be some particles with amplitudes near the maximum value. It is found that the model results are insensitive to the details of the charge-distribution of the core. This has been determined by study of a model where the uniform core distribution is replaced with a Gaussian distribution. Although the Gaussian distribution is not generally an equilibrium distribution in a linear focusing channel, this replacement exercise is still valid for testing the sensitivity of the results to the assumed core distribution.

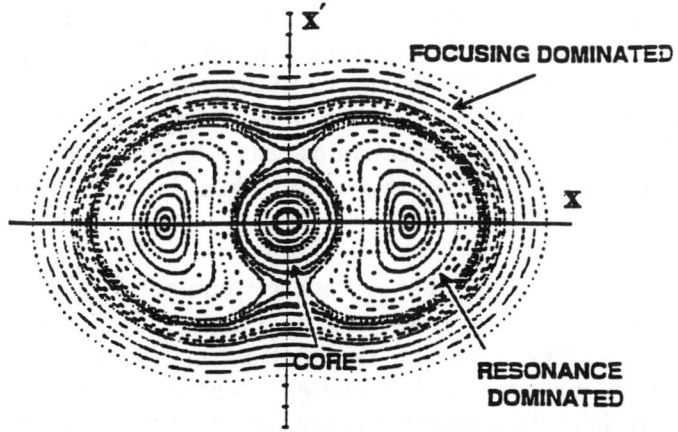

Figure 2. Stroboscopic plot obtained by taking snapshots of many independent particle trajectories, once per core-oscillation cycle at the phase of the core oscillation that gave minimum core radius. Initial coordinates were defined on the x and x' axes. The three regions shown in the figure as described in the text.

Evidence for chaos, associated with the breakup of the separatrix, is observed at low tune-depression ratios, when $k/k_0 < 0.4$. The growth time for halo in the resonance-dominated region can be estimated from the model, as the average time required for particles within the resonance-dominated region to go from the minimum to the maximum amplitude. It is found that this halo-growth time varies; the typical observed growth time is about 5 to 10 core cycles for space-charge dominated beams with $k/k_0 < 0.5$, and is about 30 to 40 cycles for $k/k_0 = 0.9$.

There are many questions remaining about halos. These include: 1) Do other modes of a mismatched beam contribute significantly to halo formation besides the breathing mode? 2) When other modes are included, is there a maximum halo amplitude? 3) How are the predictions of the particle-core model affected by periodic focusing? 4) What new effects are introduced from longitudinally mismatched beams?

EQUIPARTITIONING

A fundamental unanswered problem is that of predicting the evolution of a beam, whose initial state is given, when binary collisions, which would lead to thermal equilibrium and a Maxwell-Boltzmann distribution, are neglected. At present, we are able to predict the beam evolution only by direct computer simulation. Nevertheless, it has been suggested that the concepts of statistical mechanics and thermodynamics may be useful for describing beams, even when binary collisions are unimportant.

One such question is that of energy equipartitioning of beams, caused by collective space-charge effects. The space-charge-induced coupling between the motion in different planes provides a mechanism for energy transfer between planes. Does this lead to equipartitioning of the beam? To formulate the problem, suppose we consider the simple case of an rms-matched beam with uniform focusing in all three planes. In general the beam is not in thermal equilibrium. Suppose we define a temperature for each plane, by transforming to the center-of-momentum frame of the beam or beam frame, and defining temperatures in the beam frame according to

$$k_B T_x \equiv \frac{\overline{p_{bx}^2}}{m}, \quad k_B T_y \equiv \frac{\overline{p_{by}^2}}{m}, \quad k_B T_z \equiv \frac{\overline{p_{bz}^2}}{m}. \qquad (14)$$

Equipartitioning may be defined as the condition $k_B T_x = k_B T_y = k_B T_z$. By performing a Lorentz transformation back to the laboratory frame, and using the definition of rms normalized emittances, we can show that equipartitioning implies that

$$\frac{\varepsilon_{n,x}^2}{x_{rms}^2} = \frac{\varepsilon_{n,y}^2}{y_{rms}^2} = \frac{\varepsilon_{n,z}^2}{\gamma^2 z_{rms}^2}, \qquad (15)$$

If we consider the single particle equations of motion for an equivalent uniform beam, we can show that the rms beam sizes can be written as

$$x_{rms}^2 = \frac{\varepsilon_{n,x}}{\beta \gamma k_x}, \quad y_{rms}^2 = \frac{\varepsilon_{n,y}}{\beta \gamma k_y}, \quad z_{rms}^2 = \frac{\varepsilon_{n,z}}{\beta \gamma^3 k_z}, \qquad (16)$$

where k_x, k_y, and k_z are the phase-advances per unit length of the equivalent uniform beam including space charge, and γ is the relativistic mass factor for the center of momentum. Substituting Eqs.(15) into (16), we obtain for a convenient expression of the equipartitioning condition

$$\varepsilon_x k_x = \varepsilon_y k_y = \varepsilon_z k_z. \qquad (17)$$

However, equipartitioning is by no means an established principal for collisionless beams. Earlier work by Hofmann [11] identified a coherent anisotropy instability, which affects non-equipartitioned beams. Some questions regarding equipartitioning are posed by Hofmann at this conference. 1) How large an unbalance of Eq.(17) can exist through thermal anisotropy without exciting an instability that restores equipartitioning? 2) If an instability is induced by thermal anisotropy, will the beam evolve until Eq.(17) is established?

3) Will such a process produce beam halo, especially in the initially cold degree of freedom, which is heated by the energy-transfer process? Hofmann's paper gives some answers to these questions for a 2-D beam. Additional work should be carried out to answer these and similar questions for 3-D bunched beams in linacs.

Other questions arise of a very practical nature for an accelerator designer? Assuming that exact equipartitioning will occur with undesirable consequences for possible production of beam halo and associated beam losses, has led some to suggest that design procedures ought to give highest priority to injecting and maintaining equipartitioned beams in high intensity accelerators [30]. The question arises regarding what should be done, when numerical simulation studies show that thermal asymmetries develop and remain in rms mismatched beams. Should the focusing system be tailored to maintain equipartitioning, even if this means weakening the focusing and increasing the rms beam size relative to the aperture radius, which can by itself lead to enhanced beam loss at high energies [31]. It may be that the optimum design approach depends on the detailed requirements of the application. Thus, the design approach for a proton linac that must inject beam into a storage ring may be different than for a linac that must deliver a high-power beam directly to a target. Nevertheless, it is clear that these are important questions that will need further study.

CONCLUSION

Significant progress has been made in understanding space-charge-dominated beams, even since the 1991 University of Maryland High-Brightness Beam Workshop. Nevertheless there are still many unanswered questions about space-charge-induced emittance growth, equipartitioning, and halo formation. Many of these questions are motivated by practical questions stimulated by the design of a new generation of high-current accelerators.

ACKNOWLEDGEMENTS

The author thanks E. R. Gray and R. Garnett for some of the beam calculations discussed in the paper, and for helpful discussions, and M. Reiser and M. Pabst for discussions on equipartitioning.

REFERENCES

1. J. D. Lawson, "The Physics of Charged Particle Beams," Clarendon Press, Oxford, Second Edition (1988) 151-155.

   J. D. Lawson, "The Emittance Concept," AIP Conference Proceedings, (1991) 253.

2. C. Lejeune and J. Aubert, "Emittance and Brightness: Definitions and Measurements," Advanced Electronics and Electron Physics, Supplement 13A, Applied Charged Particle Optics, (1980) 184.

3. F. J. Sacherer, *IEEE Transactions on Nuclear Science,* **18** (3), 1105 (1971).

4. P. M. Lapostolle, *IEEE Transactions on Nuclear Science,* **18** (3), 1101 (1971).

5. J. D. Lawson, "The Physics of Charged Particle Beams," Clarendon Press, Oxford, Second Edition., 190 (1988).

6. T. P. Wangler, K. R. Crandall, R. S. Mills, and M. Reiser, *IEEE Transactions of Nuclear Science,* **32** (5), 2196 (1985).

7. I. M. Kapchinskij and V. V. Vladimirskij, "Limitations of Proton Beam Current in a Strong Focusing Linear Accelerator Associated with Beam Space Charge," *International Conference on High-Energy Accelerators and Instrumentation,* (CERN, Geneva, 1959), p.274.

8. For a historical review of emittance growth for different accelerators, including a more complete set of references to early work, see T. P. Wangler, "Frontiers of Particle Beams: Intensity Limitations," Lecture notes in Physics 400, M. Dienes, M. Month, and S. Turner (Eds.), Springer-Verlag, Berlin, 542-561 (1992).

9. J. Struckmeier, J. Klabunde, and M. Reiser, *Particle Accelerators* **15**, 47 (1984).

10. P. M. Lapostolle, *Proceedings of the Proton Linear Accelerator Conference*, Brookhaven National Laboratory Report BNL-50120 (1968).

11. I. Hoffman and I. Bozsik, *Proceedings of the lLnear Accelerator Conference*, Los Alamos National Laboratory document LA-9234-C (1982), p. 116.

12. R. A. Jameson, *Proceedings of the Linear Accelerator Conference*, Los Alamos National Laboratory document LA-9234-C (1982), p.125.

13. M. Reiser, *Proceedings of the Particle Accelerator Conference*, IEEE Catalog No. 91CH3038-7 (1991), p. 2497.

14. A. Cucchetti, M. Reiser, and T. P. Wangler, *Proceedings of the Particle Accelerator Conference*, IEEE Catalog No. 91CH3038-7 (1991), p. 251.

15. R. L. Gluckstern, *Proceedings of the Linear Accelerator Conference*, Fermi National Accelerator Laboratory (1970), p.811.

16. I. Hoffman, L. J. Laslett, L. Smith, and I. Haber, *Particle Accelerators* **13**, 145 (1983).

17. J. Stuckmeier and M. Reiser, *Particle Accelerators* **14**, 227 (1984).

18. M. G. Tiefenback and D. Keefe, *Transactions on Nuclear Science* **32** (5), 2483 (1985).

    M. G. Tiefenback, "Space Charge Limits on the Transport of Ion Beams in a Long Alternating Gradient Systems," Ph. D. thesis, Lawrence Berkeley Laboratory, LBL-22465 (November 1986).

19. M. Reiser, "Theory and Design of Charged Particle Beams," John Wiley and Sons, Inc., New York, 1994, pp. 482-491.

    M. Reiser, et. al., *Physical Review Letters* **61**, 2933 (1988).

    I. Haber, et. al., *Physical Review A* **44**, 5194 (1991).

20. T. P. Wangler, P. Lapostolle, and A. Lombardi, *Proceedings of the Particle Accelerator Conference,* Washington, DC, (1993), pp. 3606-3608.

21. O. A. Anderson, *Particle Accelerators* **21**, 197 (1987).

22. J. S. O'Connell, T. P. Wangler, R. S. Mills, and K. R. Crandall, *Proceedings of the Particle Accelerator Conference,* Washington, DC, (1993), p. 3657.

23. J. M. Lagniel, *Nuclear Instrument and Methods in Physics Research* **A345**, 46 (1994).

    J. M. Lagniel, *Nuclear Instrument Methods in Physics Research* **A345**, 405 (1994).

24. R. L. Gluckstern, *Physical Review Letters* **73**, 1247 (1994).

25. C. L. Bohn, *Physical Review Letters* **70**, 932 (1993).

    C. L. Bohn and J. R. Delayen, *Physical Review Letters* **50**, 1516 (1994).

26. C. Chen and R. C. Davidson, *Physical Review Letters* **72**, 2195 (1994).

Q. Qian and R. C. Davidson, *Physics of Plasmas* **2**(7), 2674 (1995).

Q. Qian, R. C. Davidson, and C. Chen, *Physics Review* **E51**(6), R5216 (1995).

27. C. Chen and R. A. Jameson, *Phusical Review* **E52**(3), 3074 (1995).

28. S. Y. Lee and A. Riabko, *Phusical Review* **E51**, 1609 (1995).

    A. Riabko, et. al., *Phusical Review* **E51**, 3529 (1995).

29. D.L. Bruhwiler, this conference.

30. M. Reiser, *Proceedings of the Accelerator-Driven Transmutation Technologies Conference,* Las Vegas, Nevada (1994) p. 364;

    also see M. Reiser, et. al., this conference.

31. R. W. Garnett, et. al. ,"Linear Accelerator for Tritium Production", this conference.

Space Charge Tune Shift, Fast Resonance Traversal and Current Limits in Circular Accelerators

G H Rees

Rutherford Appleton Laboratory, UK

INTRODUCTION

Space charge tune shifts, fast resonance traversals and current limits are important design issues for low energy, high power circular accelerators. Areas of interest are accumulator rings and fast cycling synchrotrons, and typical applications are for pulsed spallation neutron sources, heavy ion fusion storage ring drivers and booster injectors for high energy proton and ion facilities. Aspects of the three topics are discussed in the paper.

1. SPACE CHARGE TUNE SHIFT

Estimates of incoherent betatron space charge tune shifts and spreads are generally made in the following sequence:

1. Application of the Laslett tune shift formula,[1] using bunching factors derived for an assumed Hofmann-Pedersen 2-D elliptical longitudinal distribution.[2]

2. Finding solutions of the 2-D K-V envelope equation[3] for linear space charge, in the form of equivalent betatron and dispersion parameters at the bunch centre.

3. Estimating tune spreads for both 2-D parabolic and 2-D elliptical transverse beam distributions using K-V modified lattice functions.

4. Undertaking detailed tracking studies over the ring injection process,[4] to determine how the distributions and the tune shifts evolve.

Expressions for parabolic, elliptical and uniform 2-D transverse density distributions are, respectively:

$$\rho(x,y) = \frac{2\lambda}{\pi ab}\left[1 - \frac{x^2}{a^2} - \frac{y^2}{b^2}\right]$$

$$\rho(x,y) = \frac{3\lambda}{2\pi ab}\left[1 - \frac{x^2}{a^2} - \frac{y^2}{b^2}\right]^{\frac{1}{2}}$$

$$\rho(x,y) = \frac{\lambda}{\pi ab}$$

Here, λ is the line charge density and (a, b) are the horizontal and vertical beam semi-axes for an assumed elliptical transverse beam cross section. From these expressions, it may be seen that the density at the centre of the parabolic and elliptical distributions is 2 and 1.5 times respectively that for the uniform distribution. Hence, particles with x = 0 and y = 0 have tune shifts enhanced by these same factors. Such particles are usually avoided in the injection process. Horizontal space charge tune shifts for these distributions are:

$$\delta Q_{h(p)} = \delta Q_{h(parabolic)} = 2\delta Q_{h(u)}\left[1 - \frac{\int\left\{\frac{(2a+b)}{a^3(a+b)^2}\cdot\frac{\beta_h^2\varepsilon_h}{4} + \frac{1}{ab(a+b)^2}\cdot\frac{\beta_h\beta_v\varepsilon_v}{2}\right\}.ds}{\int\left\{\frac{\beta_h}{a(a+b)}\right\}.ds}\right]$$

$$\delta Q_{h(e)} = \delta Q_{h(elliptical)} = \frac{3}{2}\delta Q_{h(u)}\left[1 - \frac{\int\left\{\frac{(2a+b)}{a^3(a+b)^2}\cdot\frac{\beta_h^2\varepsilon_h}{8} + \frac{1}{ab(a+b)^2}\cdot\frac{\beta_h\beta_v\varepsilon_v}{4}\right\}.ds}{\int\left\{\frac{\beta_h}{a(a+b)}\right\}.ds}\right]$$

$$\delta Q_{h(u)} = \delta Q_{h(uniform)} = \frac{F_h}{\overline{B}}\cdot\frac{1}{4\pi}\int\frac{4\beta_h Nr_p}{\beta^2\gamma^3 a(a+b)}\cdot\frac{ds}{2\pi R}$$

where F_h is a scaling factor due to image charge effects,
β_h is the local value in the ring of the horizontal - β Twiss parameter,
N is the number of particles per bunch,
r_p is the classical radius of the proton (m),
s is the length measured along the central orbit in the ring (m),
R is the mean radius of the ring (m),
$\overline{B}$ is the average to peak intensity beam bunching factor, and
β, γ are the relativistic velocity and energy factors.

Here ε_h and ε_v refer to the trajectory emittance parameters for individual particles, so that $\delta Q_{h(p)}$ (or $\delta Q_{h(e)}$) include tune shift and tune spread terms. There is an equivalent expression for $\delta Q_{v(p)}$ in terms of $\delta Q_{v(u)}$ with a and b interchanged, as well as β_h and β_v, and also ε_h and ε_v. The required integrals for all three distributions may be evaluated by appropriate summation of terms derived in a modified magnet lattice program. When ε_h and $\varepsilon_v \neq 0$, tune shifts are reduced.

Image forces for long bunches may be found directly but, for short bunches, it is more accurate to calculate the forces with and without walls, after transforming from the laboratory to the beam frame (z to γz, with γ≈2), and then to transform back again to the laboratory frame. A beam debunching after a linac is best analysed by this procedure. Beam bunches in the line and if injected into a ring are not at equilibrium and have approximately cylindrical shapes, not ellipsoidal, as is often assumed. Image fields for dc beam current components may be found separately; incoherent image effects for transverse motion are usually small.

Cooling rings and heavy ion fusion storage rings are designed for large and rapid tune depressions, respectively. For other applications, typical ratios for the incoherent betatron tunes in the rings, with and without the peak space charge forces, are > 0.9, with the peak tune shifts < 0.4. For high beam power levels, the tune shifts are reduced, with values of 0.1 used in some spallation source ring designs; the transverse ring acceptances and beam emittances are increased to allow fewer foil interceptions in the H⁻ injection process, thus avoiding excessive stripping foil temperatures. Optimised injection painting[5] is a key design issue.

In the longitudinal plane, the ratios of the incoherent synchrotron tunes, with and without the peak space charge forces, are typically ≥ 0.8 so that both the transverse and longitudinal motion are in an emittance and not a space charge dominated regime. Longitudinal space charge forces are found in terms of a g parameter, with the longitudinal beam coupling impedance, Z/n, corresponding to $-jgZ_o/2\beta\gamma^2$ and with Z_o the free space characteristic impedance. For the case of a cylindrical beam in a cylindrical vacuum chamber, g is given in Figure 1 in terms of modified Bessel functions over a range of b/a, where b and a are here the aperture and beam radii respectively. At high angular frequencies, ω, the g parameter falls off, as indicated, but in a carefully designed low energy ring, the space charge impedance still remains the dominant impedance. Some reduction in the level of the longitudinal space charge forces may be made by profiling the vacuum chamber envelope to follow a constant value for b/a. This scheme has been used in the ISIS high intensity proton synchrotron.[6]

Fig. 1 Longitudinal Space Charge Parameter, g

[Tube to beam radius b/a = 2,3,....,10]

$$g \simeq \frac{4}{x^2}\left[\ 1 - 2I_1^2(x)\left(\ \frac{K_1(x)}{I_1(x)} + \frac{K_0(xb/a)}{I_0(xb/a)}\ \right)\ \right]$$

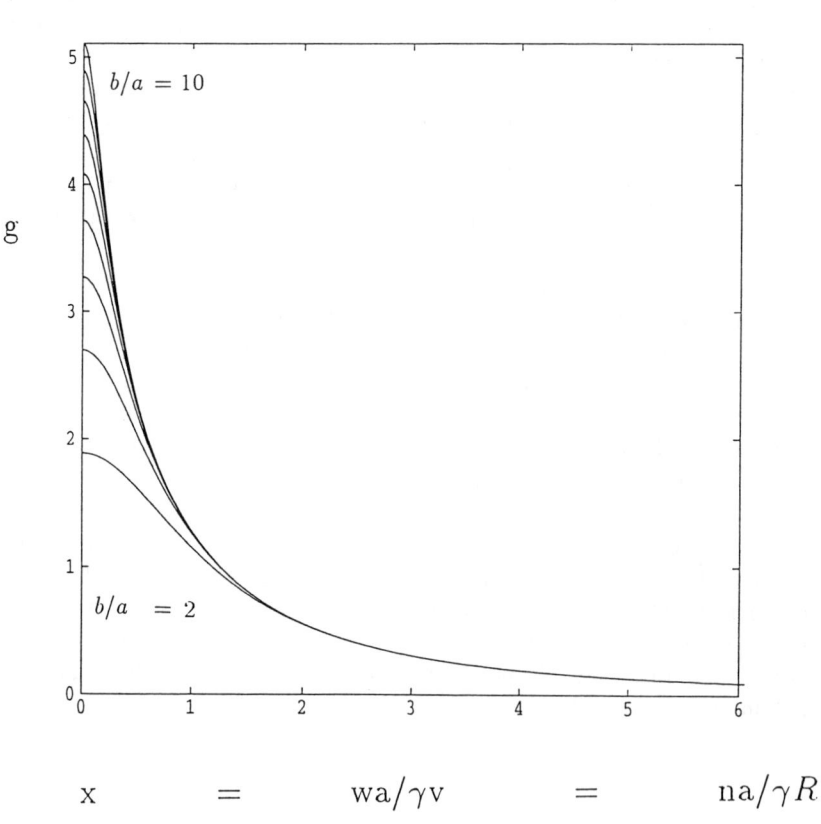

$x \quad = \quad wa/\gamma v \quad = \quad na/\gamma R$

2. FAST RESONANCE TRAVERSAL

Transverse space charge forces are directed away from the local centre of charge of a beam. In the absence of image forces, closed orbit excitations should not affect particle oscillations about the beam centre but only the motion of the beam centre itself, and then only if the coherent tune is close to an integer. In this view, it appears that fast traversals of one or more integer resonances under incoherent space charge forces are acceptable provided the image forces do not depress the coherent tune to approach or cross an integer value.

Coherent image effects arise, however, when a beam is not at the centre of its vacuum enclosure. The resulting forces are not uniform along an axis of displacement and this leads to the initial viewpoint being questioned. In the presence of closed orbit errors, there are additional image forces on some beam particles and these lead to enhanced motions during fast resonance traversals. Integer and half integer resonant motion may be excited, causing halo to form.

The effect has been observed on the ISIS proton synchrotron,[7] when the tune depressions are approximately 0.4. The vertical tune is below and the horizontal tune above the integer 4 and the tune shifts are sufficient to move the horizontal tunes of some particles, but not the coherent tune, across $Q_h = 4$. The performance of the ring is then very sensitive with respect to the fourth harmonic horizontal and vertical closed orbit deviations and corrections. Apertures in ISIS are rectangular and beam occupation is about 70% at the energies near 70 MeV. Image forces are found to vary significantly across an axis of displacement. Two effects occur, one when the coherent image coefficients vary along the axis of displacement, and one non-linear effect due to displacements orthogonal to the plane of particle motion. Examples from ISIS are given in reference 7, together with derivations of the approximate equations for incoherent betatron motion:

$$[x_j - x_c]'' + Q(Q - 2\delta Q_j)[x_j - x_c] = Q\delta Q [\xi_1 - \xi_e]/2 + kQ\delta Q[y_c/d]^2$$

Here, x_j and x_c are betatron motions of the jth and centre particle respectively,
x_j'' is the second derivative of x_j with respect to θ $(= s/r)$,
s is the path length along the orbit, R the mean ring radius,
Q is the unperturbed betatron oscillation tune value,
δQ_j is the incoherent tune shift, and δQ an average of the δQ_j values,
ξ_1 and ξ_e are coherent image coefficients at the centre and edge of the beam,
$[y_c/d]$ is the ratio of the displacement orthogonal to x to the half aperture, and
k is a non-linear orthogonal image coefficient.[7]

3. CURRENT LIMITS IN CIRCULAR ACCELERATORS

Discussion is restricted to low energy bunched beams, operating below the ring transition energy, without particle leakage from the bunch(es) into the bunch gap(s) during injection, storage or acceleration. Electrons produced in the residual gas are assumed to escape to the walls rapidly enough to avoid triggering an electron-proton (ion) instability. Of remaining concern is the excitation of longitudinal and transverse single bunch instabilities by the fields due to the space charge and the wall impedances. The space charge fields are dominant, both in the longitudinal and transverse planes, though coherent motion in the two planes assumes very different forms.

Coherent longitudinal motion of bunched beams is characterised by angular frequency components $n\omega_o$ and $n\omega_o \pm p$, where $n\omega_o$ is a harmonic of the beam angular revolution frequency and $p \simeq m\omega_s$, the product of the mode number and the synchrotron oscillation angular frequency. The modes are defined $m = 1$ for coherent dipole motion, $m = 2$ for coherent envelope motion, $m = 3$ for coherent sextupole motion and $m = 4, 5, 6....$ for higher order modes. The sidebands of mode m occur in pairs of equal or nearly equal amplitude and this has a strong influence on coherent stability when operating below transition energy. It may be shown that appreciable sideband excitation is required below transition to blow-up the longitudinal emittance of a bunched beam. Recommended is the use[8] of single sideband excitations eg $(n\omega_o - 4\omega_s)$ and $(n\omega_o - 8\omega_s)$, with n much larger than the rf harmonic number, h, and with the value for ω_s tracking an averaged value for the bunch particles.

A direct method for evaluating longitudinal stability for a bunched beam is the use of modulation response functions for the longitudinal beam coupling impedances. Excitations of the form $e^{j\omega t} e^{st}$ may be considered where $\omega = n\omega_o$ and s is the Laplace Transform Operator for the modulation. The combined response for a sideband pair is then obtained, allowing direct interpretation. Typical modulation response functions are shown in the Table 1 overleaf.

The space charge impedance is equivalent to a negative inductance of magnitude $- gZ_o/2\beta\gamma^2\omega_o$. Its modulation response has two components and, below transition, the first contributes to a coherent frequency shift, and the second (-Ls) to a damping of the coherent motion; of the other impedances shown, R gives neither damping nor antidamping, a cavity on resonance provides damping, while L, C or a cavity outside its bandwidth range all contribute antidamping. Since the longitudinal space charge impedance dominates, the coherent motion below transition energy is stable and damped, provided motion of the whole bunch is involved.

Component	Impedance	Response
Resistance	R	R
Inductance	$j\omega L$	$L(j\omega + s)$
Capacitance	$1/j\omega C$	$(s - j\omega)/C(\omega^2 + s^2)$
On Resonance	R	$R/(1 + \frac{2Qs}{\omega_o})$
Off Resonance	$R/(1 + \frac{2Q}{\omega_o} j\delta\omega)$	$R/(1 + \frac{2Q}{\omega_o}(j\delta\omega + s))$
Space Charge	$-j\omega L$	$-L(j\omega + s)$

Table 1. Modulation Response Functions for Longitudinal Impedances.

The damping and antidamping indicated by the response functions are reversed above transition when the longitudinal space charge and inductive wall fields change to become focusing and defocusing, respectively. A negative mass instability results if the space charge is dominant and if there is insufficient beam momentum spread for Landau Damping.

Energy loss due to a resistive impedance is compensated by the ring radio frequency (rf) system but no phase shift results for the coherent modulation frequencies. Antidamping due to one sideband component is exactly compensated by the damping due to its sideband partner. Barrier bucket type rf systems in a ring create a situation similar to that in an induction linac. Resistive impedances cause energy loss which is not compensated until the particles reach the region of focusing forces established for the ends of the bunch. Perturbations may grow during this interval.

An important question is whether the stability for local bunch disturbances is the same as for the entire bunch. The evidence from ISIS is that non-equilibrium distributions, created during non-adiabatic trapping, remain stable through acceleration even though $(Q_s/Q_{so})^2$ falls below 0.6 for a short interval initially. The distributions are smoothed under the defocusing action of the space charge. Interpretation may also be in terms of fast and slow waves, as envisioned for coasting beam coherent motion, with the perturbation damped over half a synchrotron period and antidamped in the subsequent half period; this is not observed at ISIS. Fast damping of the residue of the linac bunch structure is also observed at ISIS, even when operating an order of magnitude above instability thresholds, suggesting fast and slow waves may be coupled, in analogy with the sideband coupling for a bunched beam.

Coherent transverse motion for a bunched beam may assume various head-tail mode patterns, depending on the values of the bunch duration, machine chromaticities and gamma transition. For long bunches and natural chromaticities, the head-tail betatron phase shifts may be large, so leading to high head-tail mode numbers. Antidamping may not be analysed in the manner adopted for longitudinal motion as there are betatron sideband pairs with different amplitudes and widely spaced frequencies, and the beam coupling impedances may be very different at these frequencies. The resistive wall transverse impedance is one example of this, with its $\omega^{-1/2}$ dependence. To analyse the coherent motion, a dual complex number algebra may be employed, with one complex number representing the carrier frequency and a second, the modulation frequency.

The head-tail motion observed at the ISIS synchrotron has a single vertical displacement node at the centre of each of its two bunches and the mode m = 1 is believed to arise due to the resistive wall impedance. Over the instability interval, the head-tail chromatic phase shift is 2.8π, a value for which the mode m = 2 is expected and not m = 1. It is difficult to envisage how the traditional perturbation distribution assumed for head-tail motion may explain anti-damping for m = 1 motion for such a large phase shift. A modified distribution [9] has therefore been proposed in an attempt to explain the observations at ISIS, where a number of (n + Q) betatron sidebands are dominant.

In general, it may be observed that transverse stability is more difficult to achieve than longitudinal for most low energy, high intensity rings.

REFERENCES

1. L. J. Laslett, On Intensity Limitations Imposed by Transverse Space Charge Effects in Circular Particle Accelerators, Proceedings of 1963 Summer Study on Storage Rings, BNL-Report 7534, pp 324-367.

2. A. Hofmann and F. Pedersen, Bunches with Local Elliptic Energy Distributions, IEEE Trans. Nucl. Sci., Vol NS-26, No. 3, 1979, pp 3526-8.

3. I. M. Kapchinskij and V. V. Vladiminskij, Limitations of Proton Beam Current in a Strong Focusing Linear Accelerator Associated with the Beam Space Charge, Proceedings of the International Conference on High Energy Accelerators, CERN, 1959, pp 274-288.

4. C. R. Prior, Longitudinal and Transverse Tracking Studies for ESS and ISIS, Proceedings of this ICFA Workshop, 1995.

5. G. H. Rees, Injection, CERN Accelerator School General Course, CERN 94-01, 1994, pp 731-742.

6. G. H. Rees, Design for AUSTRON in the Light of Experience Gained at ISIS, Meeting on an Advanced Neutron Source, Vienna, May 1995.

7. G. H. Rees and C. R. Prior, Image Effects on Crossing an Integer Resonance, Particle Accelerators, Vol. 48, 1995, pp 251-257.

8. S. R. Koscielniak and G. H. Rees, Survey of Techniques for Longitudinal Emittance Blow-Up in the KAON Factory Collector Ring, Proceedings of the 2nd European Particle Accelerator Conference, EPAC 90, NICE, pp 1741-3.

9. G. H. Rees, Interpretation of the Higher Mode, Head-Tail Motion observed on ISIS, Particle Accelerators, Vol 39, 1992, pp 159 - 165.

Space Charge Problems in High Intensity RFQs

M. Weiss

*TERA Foundation, Via Puccini 11, Novara, Italy;
formerly at CERN-PS, 1211 Geneva 23, Switzerland*

Abstract. Measurements were made to check the performance of the CERN high intensity RFQs (RFQ2A and RFQ2B) and assess the validity of the design approach; the study of space charge effects was undertaken in this context. RFQ2A and RFQ2B are 200 mA, 750 keV proton accelerators, operating at 202.56 MHz. Since the beginning of 1993, RFQ2B serves as injector to the CERN 50 MeV Alvarez linac (Linac 2). In 1992, both RFQs were on the test stand to undergo a series of beam measurements, which were compared with computations. The studies concerning the RFQ2A were more detailed and they are reported in this paper.

INTRODUCTION

CERN has constructed two practically identical high intensity RFQs, RFQ2A and RFQ2B. RFQ2A was finished in 1990 and has delivered, from the start, the nominal current of 200 mA (1); unfortunately, it has been polluted by oil from a vacuum pump (2) and had to undergo a lengthy cleaning operation. At the beginning of 1992, RFQ2A was again on the test stand for extensive beam measurements. These measurements and the accompanying studies are reported in this paper. The measurements of RFQ2B, performed shortly afterwards, took considerably less time, giving practically the same results as RFQ2A. The RFQ2B was installed as injector to the CERN 50 MeV Alvarez linac (Linac 2) during the January-February 1993 shut down, as part of a general improvement program for the Large Hadron Collider (LHC).

DESIGN OPTIONS

The design philosophy of the RFQ2A has been described in Ref. (3). The usual linearized analytical formulae have been applied with a nominal "equivalent beam" (a beam having the same intensity and rms values as the real beam, but being of uniform space charge density distribution in real space) of 200 mA and transverse normalized rms emittances $\varepsilon_{rms,n}$ of 0.3 mm mrad. The chosen emittance value is on the lower side, as in this case it is more difficult to maintain a matched beam with an acceptable phase advance per structure period, σ_T. Essential for the design was the judicious choice of zero current phase advances σ_{0T} and σ_{0L} at the most critical position in the RFQ (downstream end in our case, where the synchronous phase angle $\phi_s = -35°$). The phase advances are a compromise between allowed surface fields E_s (usually $E_s \leq 2 E_k$, where E_k is the Kilpatrick breakdown limit) and acceptable tune depressions. After optimization, the values of σ_{0T} and σ_{0L} have been fixed as 35° and 31°, respectively. The resulting space charge parameters were $\mu_T \cong 0.92$ and $\mu_L \cong 0.65$, giving depressed tune values $\sigma_T \cong 10°$ and $\sigma_L \cong 18.5°$ (note that $\sigma^2 = \sigma_0^2 (1-\mu)$). The linear theory does not apply well to the longitudinal phase plane (except for small oscillation amplitudes) because of the strong nonlinearity and asymmetry of the restoring force; a lower μ_L had therefore to be chosen to obtain a good transmission efficiency, $\geq 90\%$, as confirmed by simulation programs.

To get another aspect of the importance of space charge, one can use the following equation in the transverse phase plane:

$$\sigma_{0T}^2 = \sigma_T^2(1 + \delta_{SC}) , \qquad (1)$$

where σ_{0T}^2 can be interpreted as the external focusing (multiplied by $\beta^2\lambda^2$) needed to obtain a phase advance σ_T in the presence of space charge represented by the space charge factor δ_{SC}. Comparing equation (1) with the usual formula

$$\sigma_T^2 = \sigma_{0T}^2(1-\mu_T) , \qquad (2)$$

one obtains:

$$\delta_{SC} = \frac{\mu_T}{1-\mu_T} . \qquad (3)$$

In our case, $\delta_{SC} > 10$, and the focusing is strongly dependent on space charge. In general:

$\delta_{SC} < 1$ *emittance dominated* regime;

$\delta_{SC} > 1$ *space-charge dominated* regime.

The main parameters of the RFQ2A are presented in Table 1 and the *four-vane* type structure can be seen in Fig. 1.

TABLE 1: Main Parameters of RFQ2A

RF frequency	202.56 MHz
Input energy	90 keV
Output energy	750 keV
Output current	200 mA
Beam pulse length	150 μs
Repetition rate	2 Hz
Trapping efficiency	~90 %
Vane voltage	178 kV
Final synchronous phase	-35 °
Modulation factor(max)	1.62
Mean aperture radius	7.87 mm
Cavity length	178.5 cm
Vane length	175.2 cm
Cavity diameter	35.4 cm
Number of cells	126

THE MEASUREMENT STAND

Figure 2 shows schematically the RFQ2A on its test stand. The following beam parameters are measured:

FIGURE 1. Four-vane structure of RFQ2A

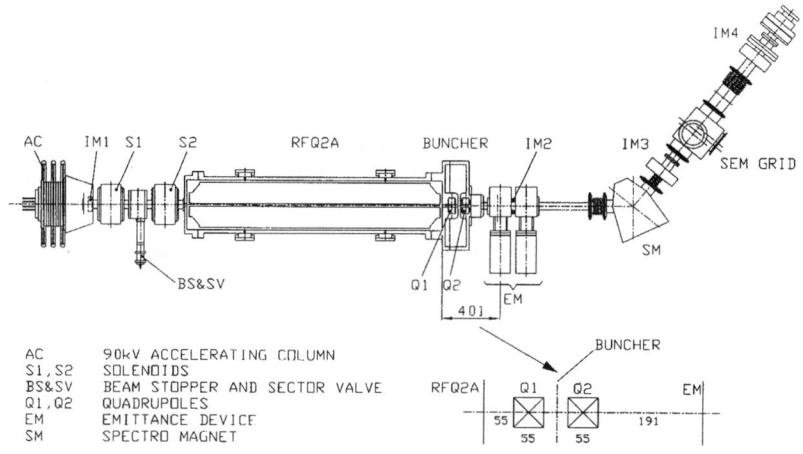

FIGURE 2. RFQ2A and its measurement stand

i. *beam intensity*: measured with beam transformers IM1 (input beam) and IM2 (output beam). Beam transformers IM3 and IM4 measure the intensity after the spectrometer magnet (SM);

ii. *beam emittance*: the emittance device (EM) measures the output emittances in the horizontal and vertical planes; it consists of a movable 0.2mm slit (in each plane) and a corresponding collector with 48 conducting strips each, placed on a ceramic board. The beam passing through the slit is stopped on the strips and the corresponding potentials across grounding resistors are measured. The slits are positively biased to about 1.5 kV to retain the secondary electrons (which would neutralize the beam), while the collector bias is about -40 V, to prevent secondary electrons from falling randomly back on the wires. The slit and collector are advanced synchronously through the beam in a given number of steps. The measurement at each step takes about 100 ns and the sample can be chosen anywhere in the beam pulse. The emittance is thus measured over many pulses, but at the same moment in each pulse;

iii. *energy spread*: the horizontal slit of the EM device followed by the SM (deflection angle 51⁰) and the secondary emission (SEM) grid are used to measure the energy spread in the beam. The edge focusing of the SM and the distances between the slit and the SEM grid are chosen so as to have the image of the slit on the SEM grid, with a magnification ratio of about 1. The SEM grid is biased to –50 V and its signals are sampled and measured in the same way as the EM signals;

iv. *bunch length*: the vertical slit of the EM device can be replaced by a *fast probe*, in fact a specially designed broadband coaxial line (4); the beam intercepted by the inner conductor (there is a hole in the outer one) is detected by a fast 1.2 GHz oscilloscope. The oscilloscope is precisely triggered by the RF reference signal, delayed so as to correspond to a given moment in the beam pulse.

Prior to the installation of RFQ2A on the test stand, we analyzed the 90 keV input beam. The EM device was placed at the position of RFQ2A and the characteristics of a matched input beam studied. The pressure being $\cong 10^{-5}$ mbar, there was a certain neutralization, but after about 40μs the beam was stable and the measured emittance did not change shape any more. The measured normalized rms emittance for input beams of about 220 to 230 mA was $\varepsilon_{rms,n} \cong 0.4$ mm mrad. With the solenoids S1 and S2, the input matched conditions could easily be satisfied.

BEAM MEASUREMENTS AND COMPUTATIONS

Transverse Phase Planes

As can be seen from Fig. 2, the beam is not measured directly at the exit of RFQ2A, but only after a distance of 401 mm. In this region are placed focusing elements, which are required when the beam is to be injected into the accelerator placed downstream (in our case Linac 2): the buncher acts as a "longitudinal" lens, the quadrupoles Q1 and Q2 serve to match the beam transversally (there are in fact two bunchers and four quadrupoles in total to match the beam to Linac 2). The RFQ2A emittance was studied with the buncher off and on. Figure 3 shows an example of emittance measurement output with the buncher off: the emittances, horizontal and vertical, are shown on the upper left side of each graph, whilst the 2-dimensional density profiles i(x,x') or i(y,y') are presented below. The values of

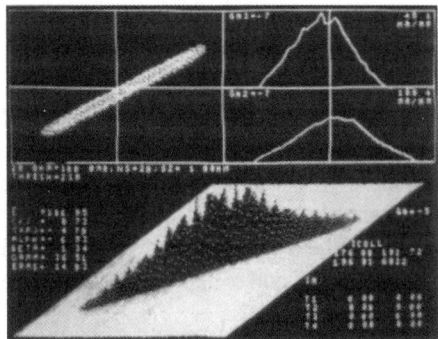 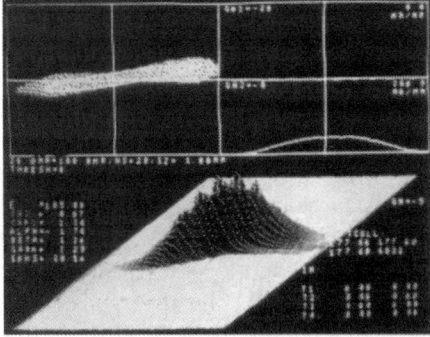

FIGURE 3. Emittance measurements: horizontal emittance (left) and vertical (right).

ε_{rms}, α and β, computed from the measurements, are also given on the graphs (unfortunately unreadable). Table 2 groups typical measurement results, with rms emittances normalized. It should be mentioned that the computed α and β can be quite imprecise, if the beam hits only a few strips on the collector of the EM device at each measurement step (5).

Comparing the rms emittances of Table 2 with the input value (0.4 mm mrad) one notes emittance growth factors of about 1.5 to 2.2. To understand this, one has to make some computations. We start by considering the situation in the RFQ2A; the beam simulation yields results which are presented in Table 3. The computed emittances show a growth factor of 1.36 in the (x,x') plane and 1.82 in the (y,y') plane. The reason for this difference in growth is the "time dependent" focusing at the RFQ2A output: particles with different phases are focused differently. This brings about an emittance growth that is bigger in the plane where the beam is larger, which in our case is the vertical plane. To eliminate the effect of "time dependent" focusing, one can represent the beam emittance at a "fixed time", instead of the usual "fixed position". The computed fixed time normalized rms emittances along the RFQ2A are shown in Fig. 4 (for convenience the two transverse emittances are summed up); one sees clearly that a transverse emittance growth occurs up to about cell 70, i. e. in the region where the longitudinal beam emittance is formed by non-linear processes. A density redistribution takes place, which usually brings about an emittance growth. After some beam loss (<10%), the final emittance growth is about 25%.

The beam simulation is continued between the exit of RFQ2A and the EM device; the details of this beam line are shown in the lower right part of Fig. 2. The evolution of beam emittances along this transport line is shown in Fig. 5. One observes again an emittance growth: the external focusing has changed and the beam density distribution adapts itself to the new situation. With the buncher off the growth factors are 1.25 (x, x' plane) and 1.15 (y,y' plane), while the corresponding factors for the buncher on are 1.35 and 1.25, respectively.

TABLE 2. Results of Beam Measurements with the Buncher off and on, respectively. (Effective Buncher Voltage $V_B T$=144kV; Units: β [m], $\varepsilon_{rms,n}$ [mm mrad]).

	α_x	β_x	$\varepsilon_{xrms,n}$	α_y	β_y	$\varepsilon_{yrms,n}$
Buncher off	-6.8	1.3	0.62	-1.8	1.2	0.82
Buncher on	-5.7	1.2	0.72	-1.3	1.1	0.89

TABLE 3. Results of Beam Simulation at Output of RFQ2A. (Units: Transverse Phase Planes: β [m], $\varepsilon_{rms,n}$ [mm mrad]; Longitudinal Plane: β [$^{\circ}$/ MeV], $\varepsilon_{rms,n}$ [$^{\circ}$ MeV]).

	(x,x') plane	(y,y') plane	($\Delta\phi,\Delta W$) plane
α	-2.80	2.80	-0.20
β	0.20	0.20	1590
$\varepsilon_{rms,n}$	0.55	0.73	0.125

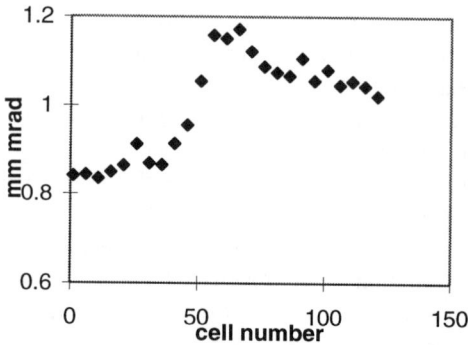

FIGURE 4. Computed fixed-time rms normalized emittances, $e_{xrms,n} + e_{yrms,n}$ along RFQ2A.

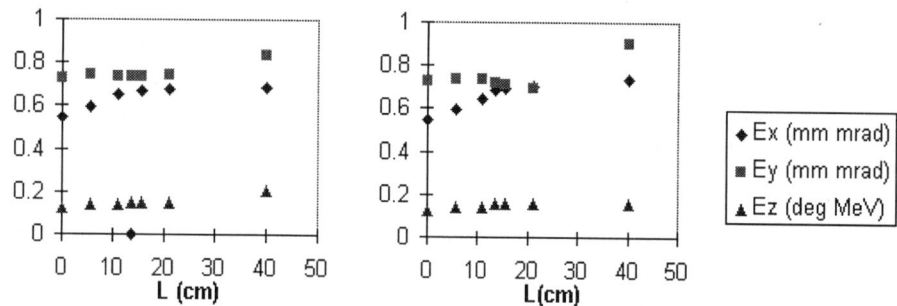

FIGURE 5. Evolution of rms emittances between the RFQ2A and the EM device; left: buncher off; right: buncher on.

The results of measurements and computations refering to the position of the EM device have been put in Table 4. There is a reasonable agreement between emittances, while the Twiss parameters, as expected, show some discrepances.

TABLE 4. Comparison between Measured and Computed Beam Parameters in the Transverse Phase Planes

	α_x	β_x	$\varepsilon_{xrms,n}$	α_y	β_y	$\varepsilon_{yrms,n}$
Measured						
Buncher off	−6.8	1.3	0.62	−1.8	1.2	0.82
Buncher on	−5.7	1.2	0.72	−1.3	1.1	0.89
Computed						
Buncher off	−5.0	0.9	0.68	−2.3	1.7	0.83
Buncher on	−7.0	1.6	0.74	−2.3	1.2	0.90

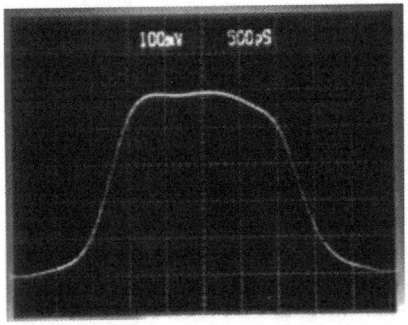 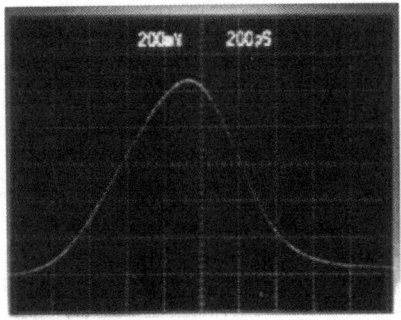

FIGURE 6. Measured phase spread $\Delta\phi$ without (left) and with buncher (right) at 144 kV.

Longitudinal Phase Plane

The measurements in the longitudinal plane were less precise than those in the transverse planes. The main interest was to estimate the energy spread ΔW and the phase spread $\Delta\phi$ when operating with and without the buncher. The precise output energy of the RFQ2A was of less concern, as with two bunchers in the transport line small energy corrections (if needed) could have been made. The measured $\Delta\phi$ without and with buncher (effective voltage $V_B T$ = 144 kV) is shown in Figure 6. Owing to the finite rise time of the oscilloscope, a correction has been applied when interpreting the results. The measured energy spread ΔW, again without and with buncher, is shown in Figure 7. Finally, the comparison with computed values is presented in Table 5. The agreement is relatively poor.

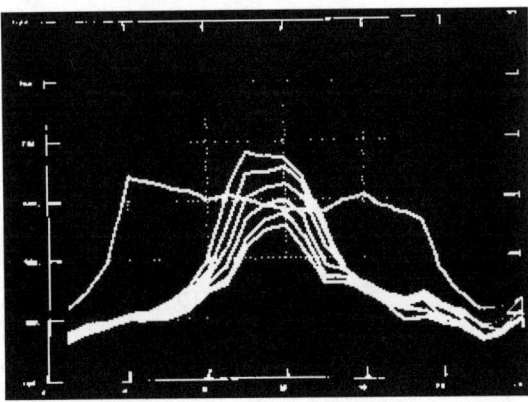

FIGURE 7. Measured energy spread ΔW without (broad curve) and with buncher; the third curve from top corresponds to an effective buncher voltage of 144 kV.

TABLE 5. Comparison between Measured (FWHM values) and Computed Values in the Longitudinal Phase Plane. (Units: $\Delta\phi$ [$^\circ$], ΔW [MeV]).

	Measured		Computed	
	$\Delta\phi_{FWHM}$	ΔW_{FWHM}	$\Delta\phi_{FWHM}$	ΔW_{FWHM}
Buncher off	185	0.100	202	0.160
Buncher on	48	0.037	40	0.060

CONCLUSION

The space charge studies of the RFQ2A complex were done primarily in order to assess the performance and reliability of this accelerator. The design parameters have been computed with respect to an unfavorable case of a small transverse beam emittance, $\varepsilon_{rms,n} = 0.3$ mm mrad (the aperture has, of course, been made big enough to accept emittances of about 0.5 mm mrad). However, the measurements showed an input emittance of $\varepsilon_{rms,n} \cong 0.4$ mm mrad and, in addition, an emittance growth factor of about 1.25 (fixed time emittance) has been detected in the RFQ2A. All this lowers the real space charge parameters, which become more favorable than those in the design study. The RFQ2A still operates in a space charge dominated regime with $\delta_{sc} \cong 3$. The operation is smooth, with a very low breakdown ratio (according to recent statistics, the down time of the RFQ2B in 1995 has been 4 minutes out of more than 6000 operational hours) in spite of the high electric surface fields, $E_s > 2 E_k$. The nominal beam intensity of 200 mA is easily reached and by increasing somewhat the input beam, output intensities of 225 mA are obtained without problems.

ACKNOWLEDGMENTS

Some of the measurements and analyses presented in this paper have been done with A. Lombardi, E. Tanke and M. Vretenar of CERN, with the valuable assistance of LANL visitors S. Nath and T. Wangler. A. Lombardi has helped, in addition, with the preparation of this paper. Many thanks go to them all.

REFERENCES

1. Vallet, J. L., Vretenar, M., and Weiss, M., "Field adjustment, tuning, and beam analysis of the CERN high-intensity RFQ," *Report CERN-PS/90–19 (HI)* 1990.
2. Tanke, E., Vretenar, M., and Weiss, M., "Performance of the CERN high-intensity RFQ", in *Proceedings of the Linear Accelerator Conference,* Albuquerque, NM, 1990, pp. 686-688.

3. Biscari, C., and Weiss, M., "Choice of parameters for the CERN high-intensity RFQ," in *Proceedings of the Linear Accelerator Conference,* Seeheim, Germany, 1984, pp. 106–108.
4. Knott, J., "Measurements of the instantaneous 200 MHz bunch currents at 750 keV for the CERN Linac 2," in *IEEE Transactions on Nuclear Science,* Vol. NS-30, No 4, 1983, pp. 2244–2246.
5. Guyard, J., and Weiss, M., "Use of beam emittance measurements in matching problems," in *Proceedings of the Proton Linear Accelerator Conference,* Chalk River, Canada, 1976, pp. 254–258.

CRITICAL BEAM INTENSITY ISSUES IN HADRON COLLIDERS

Stephen D. Holmes
Fermi National Accelerator Laboratory
Batavia, IL 60510

Introduction

I would like to discuss how some of the issues that have been talked about at this workshop (and some that haven't) are reflected in the performance of hadron colliders. Hadron colliders, be they proton-antiproton, proton-proton, or heavy ion, are typically supported by a half-dozen other accelerators each of which has its own set of performance characteristics and limitations. As a result, when designing, building, operating, or upgrading a hadron collider choices must be made that determine not only overall performance but also the ultimate configuration of the complex.

It is impossible to discuss here the full range of issues that one has to consider in projecting performance in a hadron collider. I will concentrate on a few and attempt to make some observations on how/when various effects relating to beam intensity are important. We will start with a short introduction that is intended to give the "lay of the land" in hadron colliders--what are the performance issues and what are the fundamental mechanisms that limit performance? We will then examine how choices in beam parameters can and have influenced performance, and how strategies are likely to change as we contemplate higher energy colliders. Finally, I will offer some opinions on what research directions are dictated for improving the luminosity delivered from hadron colliders.

Performance Issues in Hadron Colliders

The performance of any particle collider is characterized by two parameters--the center-of-mass energy and the luminosity. A discussion of energy limitations is beyond the scope of this presentation and, at least in hadron colliders built to date, unrelated to the beam intensity. The luminosity in a hadron collider is given by the expression:

$$L = \frac{fBN_1N_2}{2\pi(\sigma_1^2 + \sigma_2^2)} F(\sigma_z/\beta_L^*) = \frac{3\gamma fBN_1N_2}{\beta_L^*(\varepsilon_{N_1} + \varepsilon_{N_2})} F(\sigma_z/\beta_L^*) \quad (1)$$

where f is the revolution frequency of the accelerator, B is the number of bunches in each beam, N_1 and N_2 are the particle populations in each beam, σ_1 and σ_2 are the rms beam sizes (assumed round) in each beam, σ_z is the rms bunch length, β_L^* is the lattice function at the interaction point, γ is the standard relativistic factor (assumed $\gg 1$), ε_{N_1} and ε_{N_2} are the normalized beam emittances (assumed round), and $F(\sigma_z/\beta_L^*)$ is a form factor related to the ratio of the bunch length to the lattice function. The definition of emittance we use here is given in terms of the (observed) rms beam size:

$$\varepsilon_N = \frac{6\pi\beta\gamma}{\beta_L}\sigma^2 \quad (2)$$

Typical parameters leading to a luminosity of about 1.6×10^{31} cm^{-2}sec^{-1} at the Fermilab Tevatron are:

$N_p = 2.3 \times 10^{11}$	$\beta_L^* = 0.35$ m
$N_{\bar{p}} = 5.5 \times 10^{10}$	$\varepsilon_{N_p} = 23\pi$ mm-mr
B = 6	$\varepsilon_{N_{\bar{p}}} = 13\pi$ mm-mr
f = 47.7 kHz	$F(\sigma_z/\beta_L^*) = 0.6$
$\gamma = 959$	

As is clear from the luminosity formula the beam phase-space density, N/ε, is a critical element determining the luminosity performance of a hadron collider.

The Collider Complex

Any hadron collider is situated within an accelerator complex in which the beam (kinetic) energy typically swings through a range of more than six orders of magnitude. Fundamental limitations to be discussed here are related to: 1)space charge (including beam-beam); 2)synchrotron radiation; and 3)beam transfers. The relative importance of each of these effects depends on the energy regime of the beam.

Beam Intensity/Density Limitations

The fundamental fact of life in a hadron collider complex is that once the beam emittance is diluted its hard to recover. This is because no natural damping mechanism exists, as in electron synchrotrons, and efforts to cool the beam utilizing stochastic cooling at high energies have so far been unsuccessful. As a result preserving a high beam phase space density is at least as big a task as producing a high beam phase space density in a hadron collider complex. This fact is illustrated in Figure 1.

Figure 1 shows the measured proton beam vertical emittance at various stages of acceleration in the Tevatron collider complex. The data points represent an average over all proton fills between July 24, 1994 and July 23, 1995. The steps indicated on the figure are:

 1 Linac exit (400 MeV)
 2 Booster exit (8 GeV)
 3 Main Ring at 150 GeV after coalescing
 4 Tevatron at 150 GeV after proton injection
 5 Tevatron at 150 GeV after antiproton injection
 6 Tevatron at 900 GeV after squeeze
 7 Tevatron at 900 GeV in collision

The figure shows that the transverse emittance of approximately 7π mm-mr delivered from the linac at 400 MeV grows to typically 24π mm-mr by the time the protons are brought into collision with antiprotons at 900 GeV in the Tevatron. Clearly low emittance at the front end is a necessary-but-not-sufficient condition for low emittance in collision.

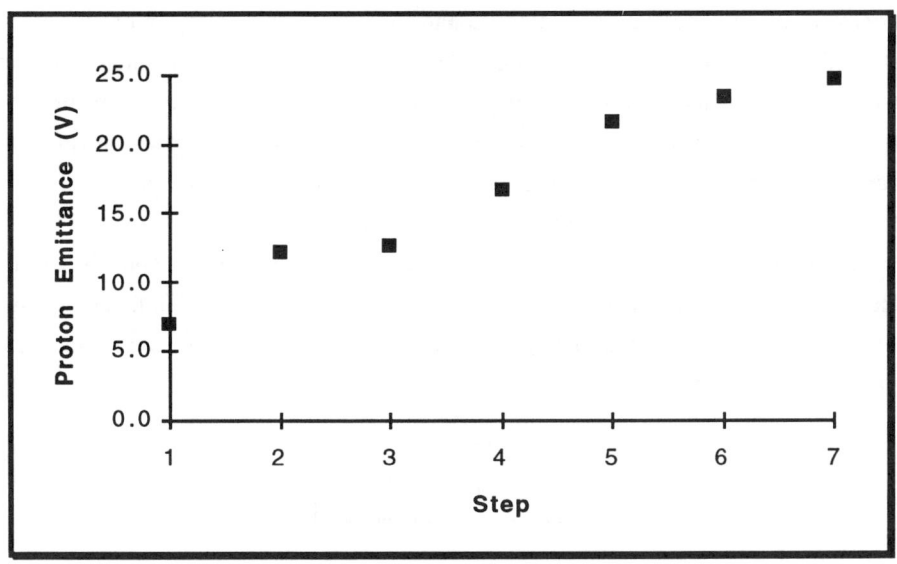

Figure 1: Proton beam vertical emittance at various stages in the accelerate-to-collision process in the Tevatron. Data points represent average of all store between 7/24-94-7/23/95.

Phase Space Density Limitations

The most fundamental effects limiting phase space density are due to macroscopic electromagnetic fields, generated by the beam, and applied to either the circulating beam itself or to its counter-rotating partner. The effect of a single beam's macroscopic fields on the individual particles making up the beam itself is called "space charge". Space charge is characterized by a net defocusing effect that is quantified in circular accelerators by the Laslett tune shift parameter:

$$\Delta v_{sc} = \frac{3r_o}{B_F} \frac{N_{TOT}}{\varepsilon_N} \frac{(1-\beta^2)}{\beta} \quad (3)$$

Here N_{TOT} is the total number of particles in the accelerator, B_F is the ratio of the average to peak circulating current, r_0 is the particle classical radius, ε_N is the beam emittance, and β is the beam relativistic velocity. As can be seen the size of the effect is proportional to the beam phase space density. The velocity dependence arises from the cancellation of electric and magnetic self-forces, and reduces the strength of the effect to zero as the beam becomes relativistic. For this reason space-charge is generally only a consideration in the lowest energy synchrotron in a hadron accelerator complex. This effect tends to limit the phase space density that can be achieved in low energy proton synchrotrons. Experience has shown that the value of Δv_{SC} that can be achieved in practice lies in the range 0.4-1.

In a colliding beam facility one beam also feels and responds to the electromagnetic field generating by its counter-circulating partner. In this case a net focusing force is generated that again is quantified as a tune shift. The beam-beam tune shift is given by:

$$\Delta v_2 = \frac{3r_o}{4} \frac{N_1}{\varepsilon_{N_1}} (1+\beta^2) \qquad (4)$$

Here Δv_2 is the tune shift (per crossing) of beam two caused by the particles in beam one. Again the size of the effect is proportional to the beam phase space density. The functional form of the beam-beam tune shift is similar to that of the space charge tune shift with the important exception that the electric and magnetic forces add. As a result the beam-beam tune shift can be, and often is, significant in high energy colliders and can limit luminosity performance. Experience to date has shown that a value of Δv summed over all beam-beam encounters of about .025 can be tolerated.

Space charge and beam-beam effects provide fundamental limitations in our ability to increase the phase space density of proton beams at the low and high energy ends of the acceleration/storage cycle. However, as is evident from Figure 1, there are other effects that can dilute the beam emittance at intermediate energies even if a large phase space density is achieved early in the acceleration chain. The most important of these relates to dilution arising from imperfect beam transfers. If the beam is transferred from one accelerator to another and is injected off the closed orbit by an amount ($\Delta x, \Delta x'$) the phase space will be diluted by an amount:

$$\Delta \varepsilon_N = \frac{3\pi\gamma}{\beta_L} \left[(\Delta x)^2 + (\alpha_L \Delta x + \beta_L \Delta x')^2 \right] \qquad (5)$$

where (β_L, α_L) are the lattice functions at the injection point. The relativistic factor gives this effect considerable importance in high energy beam transfers. Emittance preservation requires injecting onto the closed orbit to a high degree of accuracy (within <100 μm at multi-TeV colliders) and/or providing feedback systems that can damp beam motion in a period less that the decoherence time. Other potential transfer mismatches, including optical mismatches, are generally more benign.

The final effect that we will mention leading to beam emittance dilution in high energy colliders is intrabeam scattering. This effect can be significant at high energy and is very sensitive to both the transverse and longitudinal emittance. Since the growth rates scale as Z^4/A^2 this effect has a huge impact on the performance of heavy ion colliders such as RHIC where operations with Au^{+79} are being planned. Intrabeam scattering growth times measured in hours are expected both in RHIC and in the Tevatron as its luminosity approaches 1×10^{32} cm^{-2}sec^{-1}.

Beam Intensity Limitations

As discussed above a variety of effects act to limit the phase space density that can be achieved in a hadron collider. In some instances limitations apply to the total beam population rather than to the density. One example is beam instabilities arising from wakefields generated by the interaction of the beam with the surrounding environment. Such instabilities are typically independent of the transverse beam emittance and (coherent motion) can be controlled by active damping systems. More fundamental limits are related to the microwave instability, which is not susceptible to control by dampers.

As hadron colliders move into the multi-TeV range, as was planned for the SSC and is currently planned for the LHC, synchrotron radiation will play an increasingly important role in determining ultimate performance. The role of synchrotron radiation ranges from irrelevant at beam energies less than 1 TeV, to a major nuisance for energies in the range 1-30 TeV, to potential ally above 30 TeV. The major impact of synchrotron radiation is the heat load generated on the refrigeration system. The heat load (at the magnet operating temperature) is directly proportional to the total beam population, N_{TOT}:

$$\boxed{P(W) = \frac{6 \times 10^{-14} E^4(TeV) N_{TOT}}{\rho(km) R(km)}} \quad (6)$$

where ρ is the bend radius in the dipole magnets and R is the mean radius of the accelerator. The linear power density within the bending magnets is then given by,

$$\boxed{\frac{dP}{ds}(W/m) = \frac{6 \times 10^{-17} E^4 (TeV) N_{TOT}}{2\pi\rho^3 (km)}} \qquad (7)$$

Limiting this heat load to a level that can be extracted from a superconducting environment is a primary design criterion is multi-TeV hadron colliders and forces the designers to consider configurations in which N_{TOT} is minimized.

Strategies for Maximizing Luminosity in Hadron Colliders

A number of strategies and choices exist for ameliorating the effects described above and optimizing the performance of hadron colliders. The strategy followed in most instances depends upon what regime one is working in, i.e. whether limitations exist in the high or low energy accelerators and/or whether one is working in a regime in which the total beam population is a consideration.

Space Charge

At least two methods are available for minimizing the effects of space-charge in the low energy accelerators within a hadron collider complex. The most widely used technique involves the utilization of H⁻ injection into the lowest energy synchrotron. Multiple-turn H⁻ injection allows one to build up the beam density in the lowest energy synchrotron, thus removing the need for a very high current linac. Once this mode of operation is selected, the strategy for developing the highest phase space density possible is to inject at the highest energy possible into a synchrotron of the lowest circumference possible. Specifically, one attempts to maximize the ratio $\beta\gamma^2$/circumference (see equation 3) subject to financial and technological constraints. Figure 2 shows the impact on performance of the Fermilab 8 GeV Booster observed following an increase of the injection energy from 200 MeV to 400 MeV. The two sets of points show the dependence of the beam emittance delivered at 8 GeV on intensity for the two different injection energies. The two lines are drawn corresponding to a space-charge tune shift of about 0.4 for each of these cases. Clearly, raising the injection energy has had a highly beneficial effect on our ability to increase the phase space density delivered from this machine.

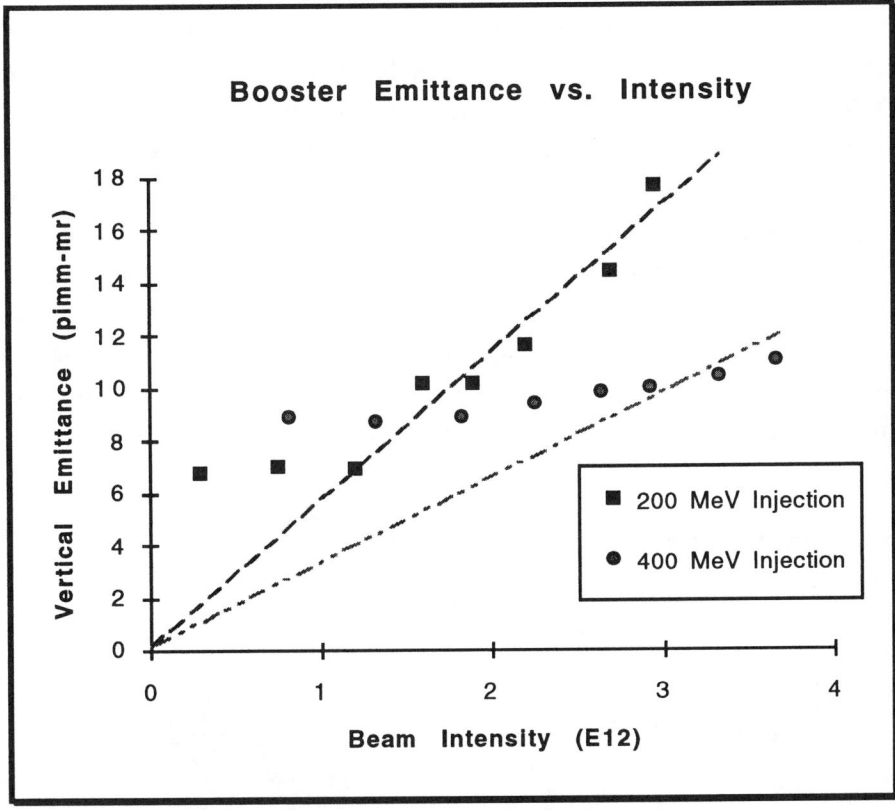

Figure 2: Vertical emittance delivered from the Fermilab Booster as a function of beam intensity for 200 MeV and 400 MeV injected beams. Contours of constant space-charge tune shift at the two injection energies are superimposed.

A second technique that is used at Fermilab to increase the transverse phase space density is called "bunch coalescing". This involves a manipulation of the beam in longitudinal phase space that combines several bunches into a single bunch. Coalescing is carried out at 150 GeV in the Main Ring at Fermilab and results in an approximate ten-fold increase in the transverse beam density. Of course, the total six-dimensional phase space must be preserved and so the longitudinal emittance of the coalesced bunch is somewhat greater than the sum of the longitudinal emittances of the (~ten) pre-coalesced bunches.

One could reasonably ask whether improvements in performance based on either of these consideration actually translate into improved performance in collision. The answer is yes. Figure 3 shows the achieved phase density of the 900 GeV proton beam in collision with antiprotons for 200 MeV linac operation and for 400 MeV linac operation. A 50-60% increase in phase space density is observed in the Tevatron for 400 MeV operations. This gain is attributed in approximately equal measure to the impact of 400 MeV linac operation and improvements in the Main Ring coalescing system. A secondary, but equally significant, impact of the improvement in beam phase space density delivered from the Booster has been to relieve some of the aperture problems present in the Main Ring and allow that machine to accelerate and deliver a significantly larger quantity of protons onto the antiproton production target. This has provided a 50% increase in the antiproton production rate and a corresponding contribution to increased luminosity.

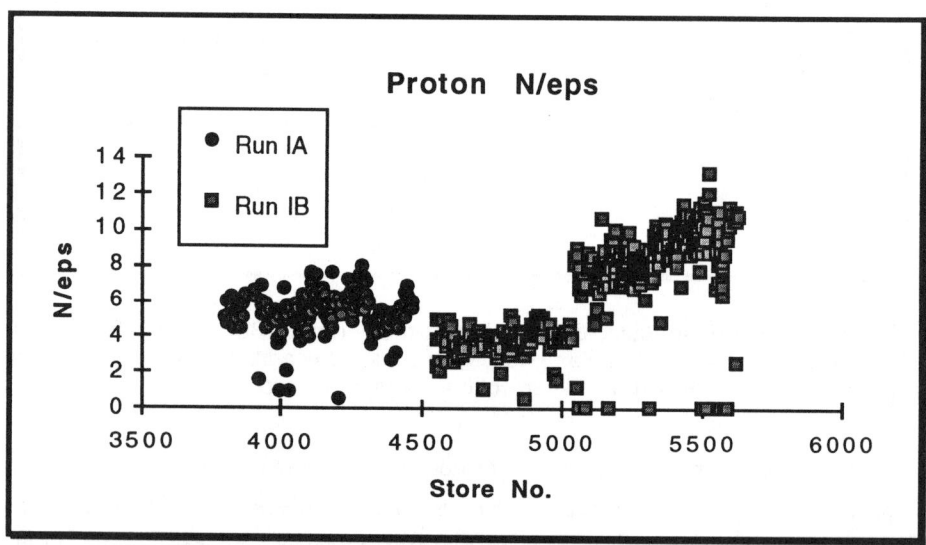

Figure 3: Proton phase space density as observed in the Tevatron before (Run IA) and after (Run IB) the 400 MeV linac upgrade. The discontinuity midway through Run IB is related to solution of a severe coupling problem in the Tevatron.

Beam-beam Effects

For a "weak-strong" scenario, such as exists in a proton-antiproton collisions, the luminosity formula can be recast as:

$$L = \frac{2\gamma f \Delta v_{HO}(BN_{\bar{p}})}{r_0 \beta^* N_{INT}(1+\frac{\varepsilon_{N\bar{p}}}{\varepsilon_{Np}})} F(\beta^*/\sigma_z) \qquad (8)$$

Here Δv_{HO} is the head-on tune shift, summed over N_{INT} encounters, seen by the antiproton bunches. One observes that the luminosity is proportional to the allowed tune shift, inversely proportional to the number of beam-beam encounters, and actually increases as the proton emittance is raised!

One sees immediately from this expression the value of creating separated orbits in a collider that is beam-beam limited. For B circulating bunches $N_{INT} = 2B$ if the two beams are circulating on a common closed orbit. This was the situation in the Tevatron prior to the introduction of electrostatic separators in 1991. During that period the Tevatron collider operated in a beam-beam limited mode and the proton beam was intentionally diluted to raise the luminosity as indicated by the formula. The introduction of electrostatic separators allowed a reduction of N_{INT} to two, while maintaining B at six, and was accompanied by an immediate factor of three increase in luminosity.

Current operations in the Tevatron are not beam-beam limited. However, it is expected that the Tevatron will return to this regime following commissioning of the Main Injector accelerator in late 1998.

Synchrotron Radiation

Future multi-TeV hadron colliders will operate in the regime in which the total beam intensity is limited by the allowed radiated power density. In this case the luminosity can be expressed as:

$$L \propto \frac{\rho^5 P^2}{\gamma^7 B \varepsilon_N} \propto \frac{P^2}{B_{MAG}^5 E^2 B \varepsilon_N} \qquad (9)$$

where P' is the linear power density and B_{MAG} is the magnetic dipole field. In this regime the luminosity is enhanced by: 1)lowering the emittance; 2)lowering the magnetic field (and increasing the circumference); and 3)minimizing the number of bunches (at a cost of more interactions/crossing and decreased beam lifetime due to interactions). This is the regime in which the SSC was being designed and led to a strategy based on creating and preserving a very low emittance proton beam. The LHC, now under design, will also operate in this regime.

Conclusions

The luminosity achievable in a hadron collider is inversely proportional to the beam transverse emittance, all other parameters remaining fixed. Consequently lower emittance as early as possible in the acceleration chain at a hadron collider is (almost) always desirable. However, in hadron facilities the problem of bringing low emittance beams into collision generally is not a reflection of limitations at the very low energy end of the chain. H⁻ injection and coalescing techniques allow an increase in the beam intensity and/or phase space density once the beam has been accelerated beyond $\beta \approx 0.5$. As a result emittance <u>preservation</u> is at least as important an issue as the creation of low emittance in a hadron collider complex.

Raising the injection energy of the lowest energy synchrotron in an accelerator complex has proven to be an effective approach to improving performance. Examples include the Brookhaven Booster and the Fermilab linac upgrade. However, our understanding of the role of space-charge in low energy synchrotrons is still rudimentary. Clearly, this is an area requiring continued study and attention.

Technological advances in a variety of areas will be critical to support continued improvements in the performance of high energy hadron colliders. Included are:

- Further development of beam feedback systems to minimize dilution during transfers

- Further investigation into the realization of bunched beam cooling systems capable of counteracting the effects of slow emittance growth at high energy, for example those due to intrabeam scattering, power supply and rf noise, and other mechanical motions.

- Development of medium energy electron cooling for the purposes of creating low emittance beams at a stage in the acceleration chain in which space-charge is not an important consideration.

The beam-beam interaction can limit performance in TeV scale proton-antiproton colliders. However, effects due to head-on encounters are unlikely to be important for multi-TeV proton-proton colliders because of the manner in which the parameters will tend to be chosen. What will still remain of importance however is the long-range beam-beam interaction. As bunches are spaced more closely together in an attempt to minimize the number of interactions per crossing the effects of parasitic crossings outboard of the collision point will become an important consideration.

Following the advent of hadron colliders operating with beam energies well beyond 1 TeV the premium placed on creating and preserving a small beam emittance will become much more critical. Keeping the radiation power density manageable while simultaneously generating a high luminosity will force designs to rely on a low beam emittance. In the more distant future designers of hadron colliders operating at tens of TeVs will find themselves in the happy position of enjoying the natural damping mechanism relied on in electron colliders. To gauge where we stand on that road I leave you with the following expression for the transverse damping time in a proton collider:

$$\boxed{\tau_x(hours) = \frac{1.5 \times 10^3 R(km)\rho(km)}{E^3(TeV)}}$$

A SUMMARY OF THE 1995 INTERNATIONAL SYMPOSIUM ON HEAVY ION INERTIAL FUSION

Thomas J. Fessenden
Lawrence Berkeley National Laboratory
Berkeley, California 94720

1.0 Introduction

The seventh International Symposium on Heavy Ion Inertial Fusion, HIF95, was held at the Princeton Plasma Physics Laboratory (PPPL) on September 6-9, 1995. The purpose of the symposium is to promote the international exchange of information in the field of Heavy Ion Inertial Fusion. The symposium was organized by the Lawrence Berkeley National Laboratory and the Princeton Plasma Physics Laboratory under the auspices of the American Physical Society, Division of Physics of Beams and chaired by R. Davidson of PPPL. The proceedings of the symposium will be published in the journal of Fusion Engineering and Design in the summer of 1996. The symposium was sponsored by the Office of Fusion Energy and the Office of Research and Inertial Fusion of the U.S. Department of Energy and by the Advanced Research Projects Agency.

The International Heavy Ion Fusion symposium is held approximately every two and a half years at locations throughout the world. The next HIF symposium is scheduled to be held in Germany-- most likely in the spring of 1998. Previous conferences were located at Frascati, Italy May 1993; Monterey, CA. Dec. 1990; Darmstadt, Germany Jul. 1988; Washington, DC May 1986; Tokyo, Japan Jan. 1984; Darmstadt, Germany Mar. 1982; Berkeley, CA. Nov. 1979; Argonne, IL. Sept. 1978; Upton (BNL), NY Oct. 1977; Berkeley, CA. Dec. 1976.

On March 22, 1995, Al Maschke passed away after an extended illness. More than any other individual, Al Maschke deserves to be called the father of heavy ion fusion. He realized in the early 1970's that the powerful accelerator technology developed for the high energy physics community could be extended to supply the energy and power required to implode an inertial fusion target with rep-rate and efficiency. HIF95 was dedicated to his memory.

The highlight of the symposium was the tour of the Tokamac Fusion Test Reactor and the presentation by D. Meade of PPPL of the very important magnetic fusion results recently obtained from TFTR. This was a unique opportunity, particularly for our foreign colleagues, to see and appreciate the largest U.S. tokamak experiment.

Proceedings of the 8th ICFA advanced beam dynamic workshop on Space Charge Dominated Beams and Applications of High Brightness Beams Oct.11-13, 1995

2.0 Major topics presented at HIF95

There was a total of 119 registered participants at HIF95. Of these 74 came from the U.S.; 18 from Germany; 8 from Russia; 6 from Japan; 5 from Italy; 4 from France; and one each from Switzerland, Spain, Poland, and Israel. During the symposium, 42 papers were presented orally, 64 papers were presented by poster, and 5 summary talks were presented at the end of the symposium.

2.1 Plenary Presentations

Professor H. Klein of the Johann-Wolfgang-Goethe Universitat in Frankfort, Germany in a plenary presentation discussed the technology of powerful particle accelerators. He pointed out that many accelerator systems throughout the world have, for many years, handled beams at parameters approaching those needed for inertial fusion. In the early 70's the Intersecting Ring at CERN stored 10^{15} protons carrying more than 4.5 MJ of at an energy near 30 GeV. The total power of this beam approached 1.2 TW. LAMPF at Los Alamos runs at nearly one MW of average beam power. The Fermilab beam in Batavia, Illinois contains approximately one megajoule of energy but at individual particle energies much too high to be considered for fusion. Klein also discussed some of the many proposals for expensive high energy accelerators that are under consideration. These include applications such as: spallation neutron sources; material irradiation; tritium production; transmutation of radioactive waste; reactor control with spallation neutrons; as well as heavy ion fusion. Klein pointed out that these accelerator systems will require extremely strong support from potential users if they are ever to be constructed.

Michael Campbell of the Lawrence Livermore National Laboratory described dramatic advances in solid state laser technology that support the design and development of the US National Ignition Facility (NIF) and the French Megajoule Project. Besides demonstrating the basic requirements for target ignition, these new facilities will provide data relevant to heavy-ion target design. Campbell also described technology that permits precise waveform control of the laser light on a sub-nanosecond time scale; highly efficient crystals for reducing the wavelength of the laser energy by a factor of three; and progress toward the development of laser diodes for pumping the solid state laser with high system efficiency--possibly even at repetition rates and economies appropriate for fusion.

2.2 Fusion Technology and Systems

There were several papers on the technology of ion beam fusion systems. Ralph Moir of LLNL described fusion chamber concepts that use a molten salt composed of fluorine, lithium and beryllium (FLIBE) to form a liquid wall that protects the chamber from the fusion blast. Moir projects that a reactor chamber fabricated of ordinary stainless steel should last 30 years if protected by a 0.5 m wall of FLIBE. He described the HYLIFE II system design that uses the heavy-ion recirculating induction accelerator to drive targets and projects a cost of electricity as low as 3.2 ¢/kwh. L. Taylor of Westinghouse suggested that a chamber first wall made of tantalum might be expected to withstand the pellet explosion with little to no surface erosion.

Plans for a heavy ion driven ignition facility study (HIDIF) were described by G. Plass of CERN. A coordinated effort between many European laboratories would study the many aspects of a 2 MJ, 300 TW driver system based on the rf accelerator-storage ring driver. This is the primary HIF approach being pursued in Europe. This system would use conventional accelerator technology of the type in use in Europe and other large high energy accelerator laboratories through out the world. Euratom funding is being sought for the study that could be complete in 1998 or 1999. Construction of the facility could then begin for completion about 2007 enabling ignition experiments in 2010.

2.3 Heavy Ion Fusion Programs

J. Quintenz of Sandia National Laboratory gave an overview of progress in programs in the U.S., Germany, Russia, and Japan working toward fusion energy using a light ion driver. The light-ion baseline approach is to use lithium ions at energies near 30 MeV and total beam energy of approximately 15 MJ at an intensity of 50-60 TW/cm^2 to produce a fusion yield of 600 MJ. Quintenz summarized the LIBRA SP light ion fusion reactor concept and reported progress by the Tri-Lab working group (LLNL, LBNL, SNL) that is looking to optimize the light and heavy ion approaches to ion driven fusion.

Aspects of the heavy ion fusion program and relevant experiments at GSI in Darmstadt, Germany were presented by I. Hofmann. He described experiments on heavy-ion storage rings, plasma focusing lenses, beam-plasma interactions that form the basis of the rf linac/storage ring approach to heavy ion fusion. Hofmann reported on the resistive instability studies on ESR facility at GSI. They find stable coasting beams at currents several times higher than Keil-Schnell current limit in good agreement with PIC simulations. GSI is one of the principal laboratories promoting the HIDIF study and its accelerator facilities are ideally suited for testing HIDIF physics concepts.

R. Bangerter of LBNL summarized the status and direction of the U.S. program for developing heavy ion fusion accelerators. The U.S. is pursuing the induction accelerator because it is well matched to the short intense pulses required by the target and because projections show a significant cost advantage over the rf-accelerator/storage-ring driver approach. The final driver choice must be coordinated with the NIF target physics results. Most of the U.S. HIF program is located at LBNL and LLNL and the laboratories are working in close collaboration. Plans for the realization of a series of experiments called the Induction Linac Systems Experiments (ILSE) were summarized by Bangerter and the first accelerator sub-section Elise was described in a paper presented by J. Kwan. ILSE would enable experiments and technology tests at a small fraction of the cost of a driver using beams at full driver size and intensity but at very much lower beam energy. LBNL has completed a 2-MeV, 0.8 A potassium injector that is intended for use with the Elise/ILSE accelerator. S. Yu of LBNL described the development and experiments inprogress using this injector. Papers on Elise physics and Engineering were presented by E. Lee and L. Reginato, respectively.

A. Friedman of LLNL summarized progress toward a recirculating induction accelerator that is being studied theoretically and experimentally as part of the U.S. program. Although the physics of this accelerator is more complicated, projections suggest that the induction recirculator may be approximately half the cost of an induction linac at driver scale. Papers by J. Barnard discussed modeling driver recirculator physics and M. Newton described the mechanical and electrical design of a small recirculator experiment being assembled at LLNL.

2.4 Accelerator Physics

Approximately 22 papers were presented on the physics of intense ion beam transport and manipulation. M. Reiser of the University of Maryland presented an overview of the physics of space charge dominated ion beams including a definition of "Free Energy" available for longitudinal or transverse emittance growth. R. Jameson (LANL) and R. Davidson (PPPL) discussed the evolution of beam halos and the onset of chaos during the transport and manipulation of intense ion beams using magnetic quadrupoles. I. Haber (NRL) reported a longitudinal instability observed in PIC simulations. J.G. Wang described experiments at Maryland that use intense electron beams to elucidate the physics of intense particle beam transport. Experimental and theoretical studies of ion storage ring instabilities were presented in papers by I. Hofmann and Hasse, (GSI) and Venturini (Maryland). Space charge effects at the ends of particle beam bunches were studied by W. Sharp and D. Grote at LLNL and by J.G. Wang and M. Reiser at Maryland. Designs of quadrupole magnets for transporting intense particle beams were presented by W. Fawley (LBNL), S. Lund (LLNL) and B. Wollnik (Giessen). C. Celata and O. Anderson presented papers on the transverse combining of intense ion beams and J. Madlung considered the funneling of beams from an rf accelerator. T. Fessenden (LBNL) reported experimental results from an intense ion beam transport using permanent magnetic quadrupoles. E. Henestroza (LBNL) and M. Reiser (Maryland) presented designs for beam matching sections.

2.5 Chamber propagation & neutralized transport

Eight papers were presented on the physics of beam propagation in the target chamber. D. Callahan of LLNL presented an overview of the physics issues. She concludes that partial beam neutralization will minimize the impact of beam stripping in the chamber and permit a tight controllable focus at the target. A. Tauschwitz (LBNL) presented initial results from a plasma focusing experiment that uses potassium beams from the LBNL 2 MV injector. Self-pinched transport concepts were discussed by D. Welch (Sandia) and by K. Hahn (LBNL). C. Olsen described IPROP simulations and charge neutralization transport experiments on the SABRE facility at Sandia.

2.6 Beam Target Interaction

There is considerable interest, particularly in Europe, in the basic deposition physics of ion beams interacting with matter. Experiments so far are exclusively with beams at much lower intensity than required for inertial fusion. Fifteen papers were presented on this subject. M. Chabot (IPN Orsay) described the basic physics of ion beam-plasma interactions and experiments performed on the Orsay Tandem accelerator. B. Sharkov presented plans for a Russian ion beam-plasma interaction program at ITEP that will generate Plasmas up to $10^{21}/cm^3$ and 3 eV using explosives; the projectile will be 1 MeV protons. M. Stetter presented initial experiments that study the interaction of heavy ion beams with rare gas cryogenic targets. The target is 1.0x1.0x0.7 cm crystal of krypton at 20 °K; the beam is Ne^{10+} at 300 MeV/u from the SIS accelerator at GSI. Results from this experiment will be used to test hydro codes. C. Deutsch summarized experiments performed with the Bruyères Van de Graaf that investigated the stopping of heavy ion beams in dense laser ablated plasmas. This experiment has provided a quantitative check of the standard stopping model.

2.7 Target Physics and Fabrication

Nineteen papers were presented on the physics and two on the fabrication of inertial fusion targets. There is general agreement that a good understanding of target physics has been achieved through the combination of simulation and laser experiments. However, although solutions at much lower drive energy are believed possible, no one suggested an indirect-drive heavy ion target that was predicted by sophisticated codes to operate at less than 10 MJ of drive.

John Lindl of LLNL described the target physics basis of indirect drive. For heavy ion drive, the drive pressure will be constrained by the focusability of the ion beam whereas laser driver pressures are constrained by plasma instabilities. Lindl summarized the main conclusions from the Nova experiments, discussed NIF target expectations, and showed how results obtained with a laser driver are applied to heavy ion targets. Atzeni described target designs for the European HIDIF initiative. Meyer-ter-Vehn suggested that the albedos of the holraum walls could be used to shape the radiation field and pointed out that the Asterix laser is scheduled to be shut down in 1996. D. Ho of LLNL presented his calculations of HI targets indirectly driven by two ion beams. He obtains a yield of 400 MJ from 8 MJ of heavy ions but restricts the motion of the internal shields in the target. If the shields are allowed to move, more than 10 MJ of drive are needed. R. Olsen of Sandia reported on target concepts for light ion fusion that use layers of oxygen that burn through to provide pulse shaping. A yield of 600 MJ was obtained from 14 MJ of 25 MeV lithium beams. Sandia finds that 12 to 24 ion beams are required to generate the foot and drive pulse with adequate symmetry for ignition and gain. Two sets of integrated calculations were reported by Romanov and Vatulin of the Russian Federal Nuclear Center in Arzamas. They looked at a target design suggested by the group under Mahrun at Frankfort University and concluded that shield motion would greatly limit gain. In 3-d calculations, they needed a drive of 10 MJ to produce a yield of 40 MJ.

The present state of inertial fusion target fabrication was discussed in a paper by K. Shultz (presented by L. Stewart) of General Atomic in San Diego. GA has the responsibility of providing the experimental targets for most U.S. laser-target experiments. The community has developed methods of controlling the thickness of cryogenic D-T on the pellet surface by using the heat of decay of tritium (beta-layering). Temperatures near 20 °K must be controlled within 0.2° to achieve the required surface precision. Experimental targets typically cost from $500 to $2000 because of the pedigree each requires. Studies indicate that these costs can be reduced to the 0.3$ /target required by inertial fusion energy but that a proof-of-principle development program is needed.

R. Petzoldt of LLNL discussed methods of injecting and tracking targets into a working inertial fusion reactor chamber. He suggested using a Gatling/BB gun to inject five targets per second at a velocity of 100 m/s. His analysis shows that the cryogenic targets can withstand the acceleration and in-flight warming associated with the injection technology. Petzoldt described method for optically tracking and steering the targets to the ignition point in the chamber. He also feels that a proof-of-principle development program is needed.

2.8 Miscellaneous

M. Basko of the Institute for Theoretical and Experimental Physics in Basko described their concept of a heavy ion driven inertial fusion facility that uses four isotopes of platinum at both positive and negative charge states in an rf-accelerator storage-ring concept. A factor of two in particle intensity is achieved by filling both the positive and negative buckets of the rf accelerator. In an accompanying paper D. Koshkarev calculated limiting currents in the accelerator and estimated the lifetimes of the ions in the storage rings and transport lines.

C. Deutsch of the University of Paris at Orsay presented progress on the production of cluster beams for inertial fusion. These included enhanced stopping of the clusters following fragmentation and discussions of cluster ablation and impact collision models.

There were three papers on sources of ions for HIF. W. Mróz of the Institute of Plasma Physics in Warsaw Poland summarized experimental production of multicharged ions from laser-produced plasmas. They have produced ions from mass 1-200 and charge state 1-50 at temperatures of 100's to several MeV using their 20 J, 1 ns neodymium-glass laser and a 50 J, 350 ps iodine laser at the Institute of Physics in the Czech Republic. Sources that produce high-charge-state heavy ions at very low emittance were described by S. Eylon working with Duly Research Inc. Alkali ion beams from a low emittance zeolite source are stripped in a gas cell and then separated. Stripped beams of unique and higher charge state permit more efficient acceleration at higher current.

Finally, T. Godlove of FMT is preparing a bibliography for heavy ion inertial fusion. The database runs on either IBM or Mackintosh personal computers and at present contains 630 records in 38 subtopics beginning with the first publications of Maschke in 1975.

Linear Accelerator for Tritium Production

R. W. Garnett, J. H. Billen, K. C. D. Chan, R. Genzlinger*,
E. R. Gray, S. Nath, B. Rusnak, D. L. Schrage, J. E. Stovall,
H. Takeda, R. Wood, T. P. Wangler, and L. M. Young

Accelerator Operations and Technology Division,
Los Alamos National Laboratory,
Los Alamos, New Mexico 87545

Abstract. For many years now, Los Alamos National Laboratory has been working to develop a conceptual design of a facility for accelerator production of tritium (APT) [1]. The APT accelerator will produce high energy protons which will bombard a heavy metal target, resulting in the production of large numbers of spallation neutrons. These neutrons will be captured by a low-Z target to produce tritium. This paper describes the latest design of a room-temperature, 1.0 GeV, 100 mA, cw proton accelerator for tritium production. The potential advantages of using superconducting cavities in the high-energy section of the linac are also discussed and a comparison is made with the baseline room-temperature accelerator.

DESIGN PHILOSOPHY

The main design objectives for the APT accelerator are to provide high beam transmission and low beam loss during all phases of linac operation. Advances of the past decade have made it possible to design a machine of this type by providing the framework required for emittance and beam halo control of high-current proton beams. The use of high-frequency accelerating structures reduces the amount of charge per bunch which minimizes the effects of space-charge forces on the beam. The use of strong transverse focusing reduces transverse emittance growth. Large transitions are avoided in the design, which provide for a well-matched beam to reduce halo formation from mismatch.

Beam losses in the APT accelerator must be extremely low (0.1-1.0 nA/m) so as to minimize structure activation and allow hands-on maintenance. In order to meet this design criterion we have required large apertures, both transverse and longitudinal, in order to contain all the beam. The transverse apertures in the accelerator have been chosen to provide a large aperture radius to rms beam size ratio. This ratio has been used as a figure of merit for our designs. The additional

---

*Engineering Sciences and Applications Division

requirement to have reasonable shunt-impedance values in order to reduce rf structure-power loss, was also considered when making the choices of transverse apertures. The synchronous phase profile, which was chosen for the accelerator, allows adequate longitudinal acceptance of the beam throughout all phases of acceleration with sufficient margin to avoid the nonlinear region of the acceptance. Precision beam diagnostics and rf system control will be required to monitor and correct deviations from nominal operation.

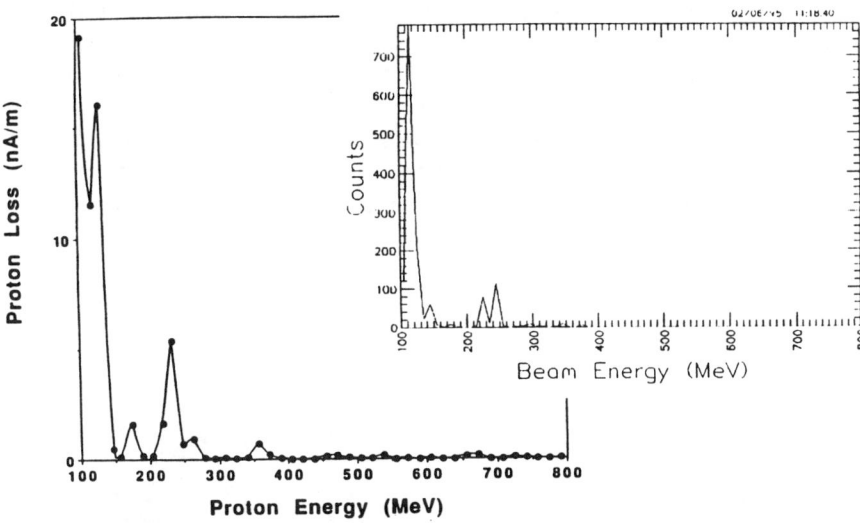

FIGURE 1. Beam loss in the LANSCE side-coupled linac from activation measurements as a function of beam energy. The inset in the upper right-hand corner is the simulation result. Measured losses are 0.1% and simulated losses are 1.2%.

LAMPF SIMULATIONS AND OPERATION

Comparisons have been made between simulations and beam measurements of the LANSCE accelerator (formerly LAMPF) which have allowed us to benchmark the simulation codes [2]. The rms beam properties, predicted by the simulation codes, agree well with measurements. However, to accurately predict the magnitude of the beam losses, a detailed knowledge of the tails of the beam distribution is necessary. Figure 1 shows both simulation results and results from activation measurements for the LANSCE side-coupled linac (SCL) from 100 to 800 MeV. As can be seen, the locations of large beam loss are well predicted by the codes, but, the fraction of total beam lost is overpredicted by approximately an order of magnitude. The simulations indicate that the primary cause of beam loss is due to poor longitudinal capture and poor longitudinal matching. No longitudinal matching is done to compensate for the acceptance change due to the 201.25 MHz to 805 MHz frequency transition from the drift-tube linac to the SCL. Inefficient beam bunching at injection leads to low-energy losses, which is observed at LANSCE. Bunching of the beam by a radiofrequency quadrupole (RFQ) is far superior to a conventional multi-buncher system. This has been demonstrated in simulation studies for possible upgrades to LANSCE and is shown in Fig.2. The longitudinal beam distribution at 100 MeV is seen to have an extremely long tail due to poor bunching prior to injection into the drift-tube linac at 750 keV. The strong longitudinal focusing and bunching of a RFQ can eliminate this tail as is seen in the figure. Simulations using both the RFQ beam and longitudinal matching at 100 MeV resulted in approximately two orders of magnitude reduction in beam losses. We believe that the RFQ is a critical, required component, if a high-quality, low-loss beam is to be produced in high-current applications. Additionally, all of these results indicate the importance of examining the effects of longitudinal beam dynamics, including bunching and matching, to avoid beam halo and beam losses.

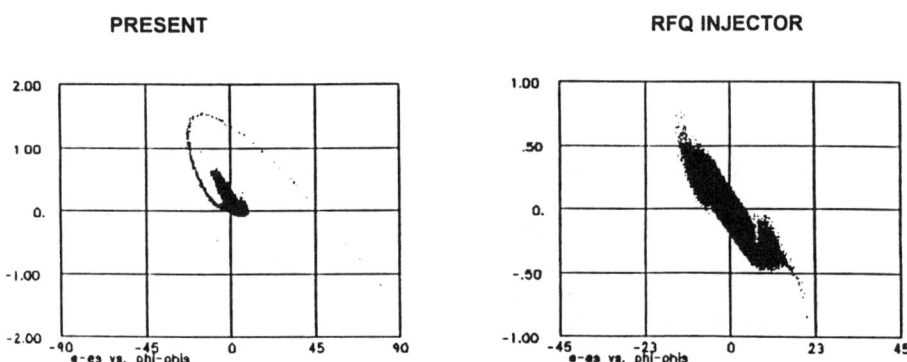

FIGURE 2. A comparison of the longitudinal phase space distributions at 100 MeV from simulations for the present LANSCE accelerator and a modified LANSCE using a 5.4 MeV RFQ as an injector.

ROOM-TEMPERATURE BASELINE DESIGN

A schematic diagram of the room-temperature baseline design is shown in Fig. 3. Table 1 lists some parameters for the baseline design. A 100 mA, cw, H^+ beam is injected at 75 keV into a 350-MHz RFQ, which accelerates the beam to 6.7 MeV [3]. An engineering drawing of this high-energy RFQ is shown in Fig. 4. The RFQ consists of four resonantly-coupled 2-meter segments with each segment made up of two 1-meter sections. Each section is fabricated from four vane-cavity quadrants which are brazed together. The 6.7-MeV RFQ output energy was chosen to provide a transition to the coupled-cavity drift-tube linac (CCDTL) structure [4] without requiring focusing quadrupoles in the drift-tubes.

The CCDTL structure, operating at a frequency of 700 MHz, will be used to accelerate the beam to 100 MeV. A generic 3-gap CCDTL structure is shown schematically in Fig. 5. The CCDTL structure is a hybrid structure combining the features of a conventional drift-tube linac and that of a coupled-cavity linac, such as the SCL. The CCDTL provides the field stability of a $\pi/2$-mode periodic structure with the high transit-time factor required to efficiently accelerate low-β ion beams. One of the main advantages of the CCDTL compared with the conventional DTL is that all quadrupole-focusing lenses are external to the accelerating cavities rather than installed within drift tubes. The CCDTL provides good efficiency for large bore radii in a high-frequency structure, and assures better mechanical stability for the quadrupoles. In addition, the drift-tube alignment requirements are relaxed, and accurate magnet alignment is achievable. Three-gap CCDTL cavities will be used to accelerate the beam to 20 MeV. From 20 to 100 MeV, 2-gap cavities will be used to maintain the high average shunt-impedance of ≥ 25 MΩ/m. At 100 MeV, a transition is made to a conventional, 700-MHz SCL, which accelerates the beam to the final energy of 1 GeV. SCL tanks with 6 cells/tank will be used from 100 to 154 MeV. Tanks having 7 cells/tank are used thereafter. The choices to use CCDTL and SCL structures allow the majority of the accelerator to operate at 700 MHz and in the $\pi/2$-structure mode. Because the RFQ operates at 350 MHz, only every other longitudinal bucket in the CCDTL and SCL, which both operate at 700 MHz, is filled. Therefore, a possible upgrade path, which would double the output beam current by filling all buckets at the same peak current per bunch, could be to funnel [5] another beam at some energy above 6.7 MeV.

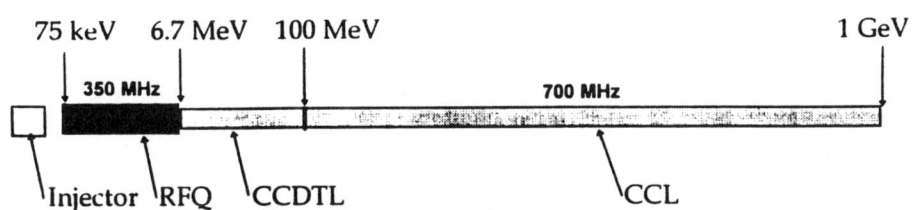

FIGURE 3. Schematic view of the proposed accelerator for tritium production.

Although transitions from one type of accelerating structure to another exist, there are no abrupt changes in the focusing period, and the beam experiences no change in the average transverse and longitudinal focusing forces at these locations. This is achieved by maintaining constant focusing through the transitions and allowing only very adiabatic changes throughout the linac. To do so, a FODO magnetic-quadrupole-focusing lattice with an 8-$\beta\lambda$ period, and magnets of constant gradient times length are used from 6.7-1000 MeV. At the

RFQ/CCDTL transition, the RFQ transverse zero-current phase advance per unit length (σ_0/L) was chosen to be 1.96°/cm. The CCDTL transverse-focusing-lattice parameters were chosen to give the same σ_0/L. The rf phases and amplitudes in the cavities were chosen to provide adequate longitudinal focusing, and efficient acceleration. It has been found in previous studies [6], that by maintaining

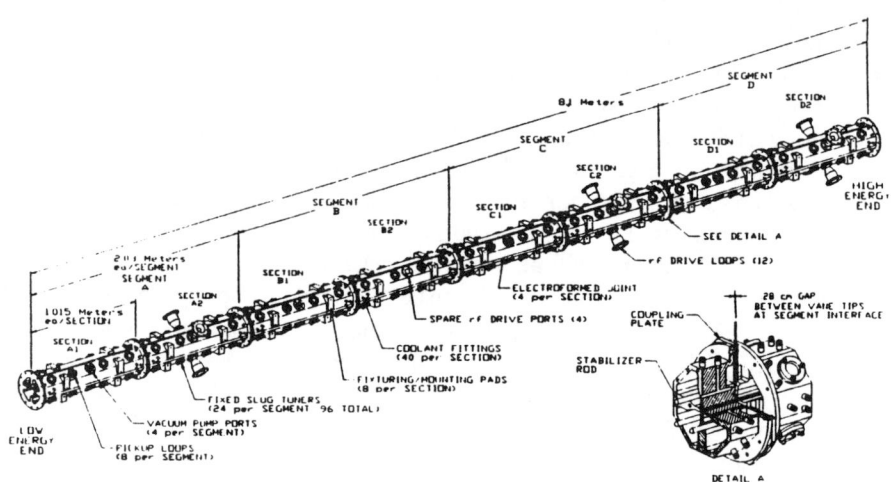

FIGURE 4. Engineering drawing of the 6.7-MeV RFQ.

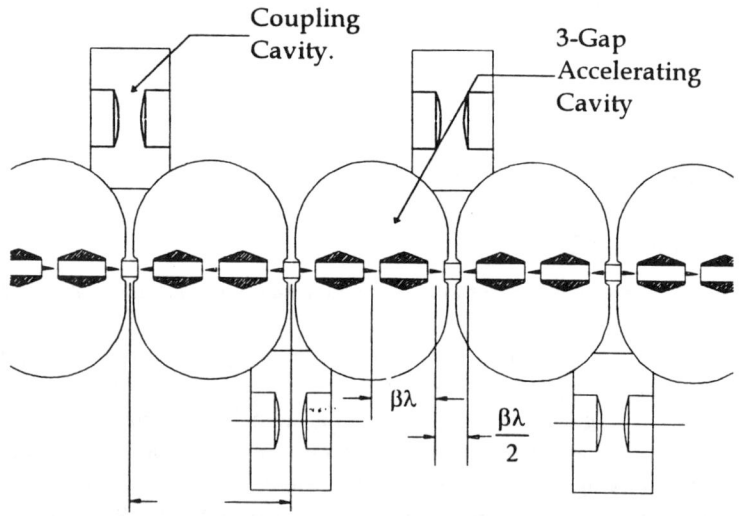

FIGURE 5. Schematic diagram of a generic 3-gap CCDTL.

constant focusing across transitions, it is possible to design an accelerator, which is relatively insensitive to beam current. The RFQ/CCDTL transition region has been designed so that no separate matching section is required while preserving the current-independence. This has been verified using multiparticle beam-dynamics simulations.

Table 1. Baseline Room-Temperature Accelerator Parameters.

Parameter	RFQ	CCDTL 3-Gap	CCDTL 2-Gap	CCL 6-cell, 7-cell
Energy (MeV)	0.075-6.7	6.7-20	20-100	100-1000
Frequency (MHz)	350	700	700	700
Beam Current (mA)	100	100	100	100
Aperture Radius (cm)	0.23-0.34	0.75-1.0	1.25-1.75	1.75-2.5
E_0T (MV/m)	1.38	0.85-1.87	1.50-1.32	1.35-1.71
Real-Estate E_0T (MV/m)	0.81	0.53-1.17	1.17-1.0	1.0-1.5
Synchronous Phase (deg)	-90 to -60	-60 to -40	-40 to -35	-35 to -30
Shunt Impedance (MΩ/m)	—	30-38	30-26	26-41
Transverse Phase Advance/Period (deg)	20	80	80	80-40
Quadrupole Lattice	FODO	FODO	FODO	FODO
Quadrupole Length (cm)	—	3.0	3.0	3.0
Quadrupole Gradient (T/m)		87.5	87.5	87.5
Transverse Emittance* (cm-mrad)	0.022	0.022-0.024	0.024	0.024
Longitudinal Emittance* (cm-mrad)	0.044	0.042	0.048	0.068
Ratio of Aperture Radius to RMS Beam Size	—	5-6	6-13	13-25

*Emittances are rms, normalized, area/π values.

To guarantee high availability and reliability of the accelerator beyond the RFQ, rf modules will consist of seven 1-MW klystrons powering approximately 200 accelerating cells. Structure studies indicate that modules containing 200 accelerating cells or less should have very stable field distributions. In this scheme, the accelerating structure itself acts as a manifold to accept rf power from multiple drives. Therefore, the structure locks the phase and amplitude of the rf field, which assures the proper longitudinal dynamics. This simplifies beam tuning and rf control. Such a module requires only 6 klystrons to provide adequate power, however, fault analyses have shown that by adding a seventh redundant klystron, the rf module reliability is significantly increased. All seven klystrons would be operated at lower power to increase tube lifetime. In the event of a single klystron failure, all klystrons would be brought up to high-power operation.

Figure 6 shows the transverse and longitudinal tune-depression ratios (σ/σ_0, where σ is the phase advance per period at full beam current) as a function of proton beam energy. As can be seen, although the average beam current is quite high (100 mA), the linac parameters have been chosen to avoid low depression ratios, which can lead to beam halo production when there is beam mismatch [7]. Figure 7 shows the equipartition ratio as a function of proton beam energy, defined as the ratio of $\sigma\varepsilon$ values for the transverse to longitudinal planes, where ε is the rms normalized emittance. In the first 20 MeV of the accelerator, where it is economical to do so, equipartitioning is maintained on average. As a result, there is very little transverse or longitudinal emittance growth up to 20 MeV as can be

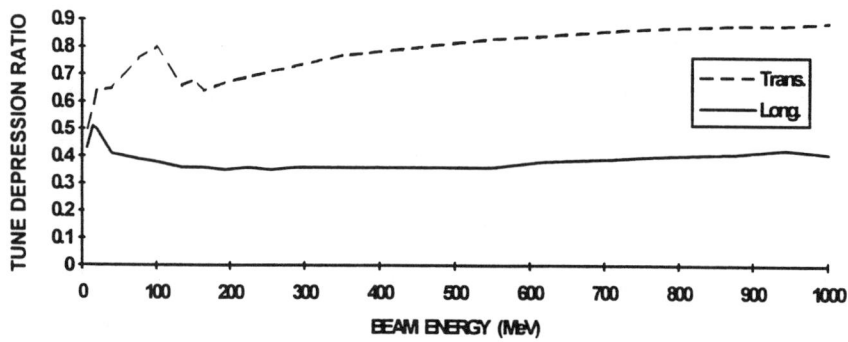

FIGURE 6. Tune depression ratios as a function of proton beam energy.

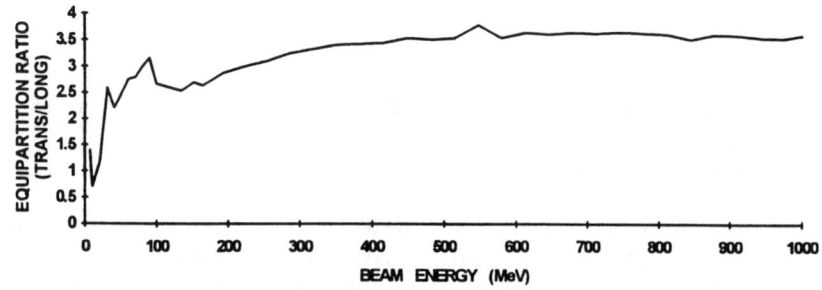

FIGURE 7. Equipartition ratio as a function proton beam energy.

seen in Fig. 8. Strong transverse focusing allows the transverse emittance to be held constant throughout the accelerator up to 1 GeV. Because the real estate gradient in the accelerator is nearly constant at 1.5 MV/m above 100 MeV, the longitudinal focusing grows weaker as a function of beam energy. As a consequence, the longitudinal emittance is seen to grow. Because there are no specific beam brightness or luminosity requirements for this application, this longitudinal-emittance growth is tolerable, as long as it does not affect the beam loss along the machine. The simulation results indicate that a large transverse aperture radius to rms beam-size ratio is maintained throughout the high-energy section of the accelerator, where structure activation from beam loss is of the greatest concern. The values of this ratio are also included in Table 1. Our studies show that reducing the transverse focusing strength above 100 MeV to bring the equipartition ratio closer to unity, as proposed elsewhere [8], has little effect on the emittance growths, but increases the transverse rms beam size, which increases the risk of beam loss.

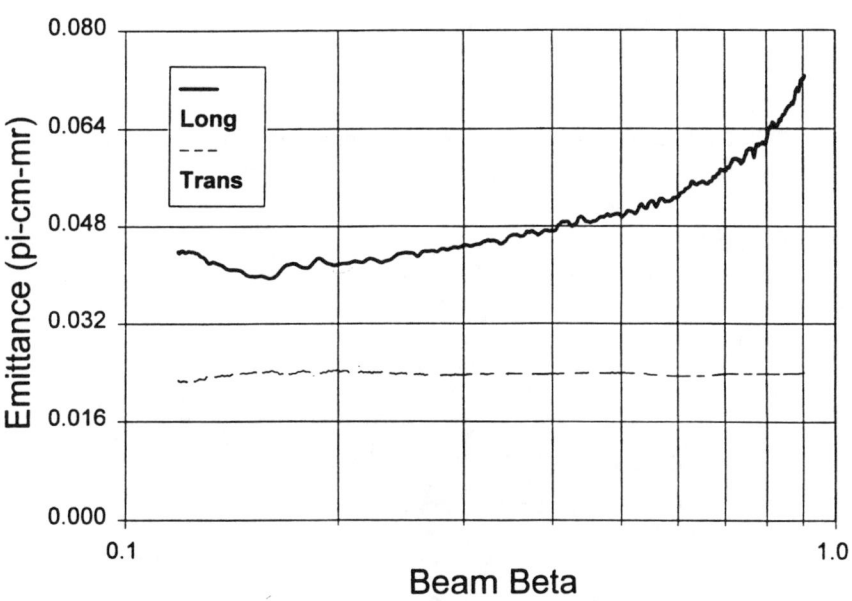

FIGURE 8. Normalized rms emittances as a function of beam velocity, β.

HIGH-ENERGY SUPERCONDUCTING OPTION

Recently, we have examined the feasibility of using superconducting rf cavities to accelerate the proton beam in the high-energy section of the linac from 100 to 1000 MeV. Cost studies indicate that as much as twenty million dollars per year could be saved in operating costs by using this technology. However, this technology is viewed by some as high risk. The purpose of our study was to develop a point design to a sufficient level so that technical feasibility, technology development requirements, risk factors, and cost could be determined.

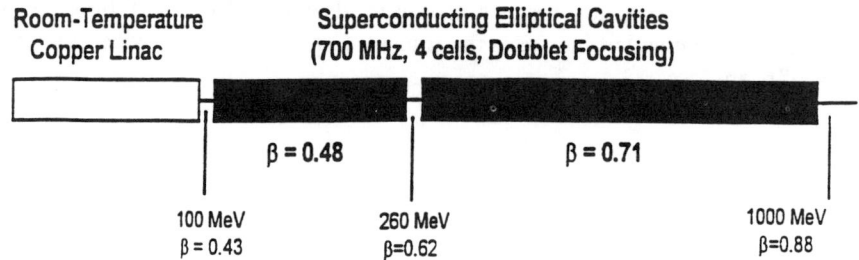

FIGURE 9. Schematic view of the high-energy superconducting option.

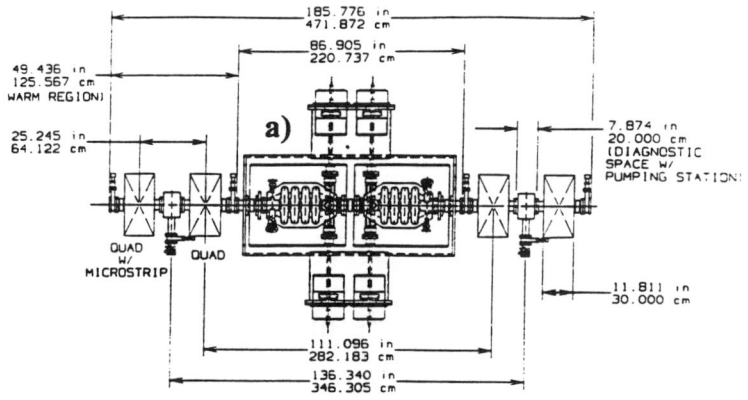

FIGURE 10. Engineering layout of a medium-β cryomodule. The distance from the midpoint between a pair of quadrupoles to the next midpoint is equal to one FDO period.

The superconducting rf linac (SCRFL) presented here is unoptimized. We have chosen conservative requirements for the various system components, most of which have already been demonstrated in existing accelerators or laboratory tests.

Figure 9 shows a schematic diagram of the proposed SCRFL, and Table 2 gives some of the accelerator parameters. The SCRFL consists of two sections called the medium-β and high-β sections. Each section is composed of identical 4-cell elliptical cavities, where each cell length is βλ/2. The medium-β section has cavities that are optimized at β=0.48 and will accelerate the beam from 100 to 261 MeV. The cavities in the high-β section are optimized at β=0.71, and will accelerate the beam up to 1000 MeV. A cryostat containing two cavities forms a cryomodule. Figure 10 shows an engineering drawing of a medium-β cryomodule, and one period of the magnetic quadrupole focusing lattice. In this example, transverse focusing is provided by quadrupole doublets between each cryomodule. The power from each klystron will be split among four cavities. RF power will be fed to each cavity using two antenna-type power couplers capable of handling 105 kW, each.

The use of short, multi-cell rf cavities has the advantage that a relatively high transit-time factor can be maintained, while having a large velocity acceptance. Figure 11 shows the transit-time factor for various multi-cell cavities as a function of the ratio of beam β to cavity design β. We have chosen 4-cell cavities in this example as a compromise to minimize the number of required system components, while maintaining a large velocity acceptance.

Table 2. High-Energy Superconducting Accelerator Parameters.

Parameter	
Energy Range (MeV)	100 - 1000
Frequency (MHz)	700
Beam Current (mA)	100
No. of β Sections	2
No. of Cavities	488
No. of Cryostats	244
No. of Klystrons	122
Cavities/Cryostat	2
Cavities/Klystron	4
Cells/Cavity	4
RF Couplers/Cavity	2
RF Power/Klystron (MW)	0.67 (med. β), 1.0 (high β)
RF Power/Coupler (kW)	72 (med. β), 105 (high β)
Accelerating Field, E_a (MV/m)	4.2-5.3
Average Phase (deg)	-35
Aperture Radius (cm)	5.0 (med. β) and 7.2 (high β)

The large velocity acceptance of these cavities allows operational flexibility. In normal operation, the multi-cell cavities will be operated for a specific energy gain per cavity (medium β ΔW=1.44 MeV, high β ΔW=2.1 MeV) with an average synchronous phase of -35°. An iterative procedure, done with a computer, is used to determine the required rf field amplitude and injection phase for each cavity, such that the desired energy gain per cavity and average synchronous phase is achieved. Therefore, the cavity rf amplitudes vary as a function of beam energy. Simulation results show emittance growths between 100 to 1000 MeV of 25% and 8%, respectively, for the transverse and longitudinal degrees of freedom. The ratio of transverse aperture radius to rms beam size in the superconducting sections ranges from 19 to 26.

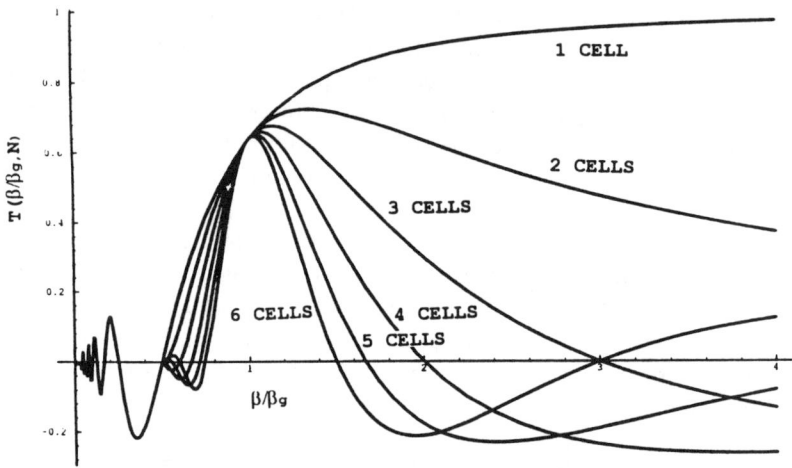

FIGURE 11. Transit-time factor for various multi-cell cavities as a function of the ratio of beam β to cavity design β.

To investigate alternative operating schemes that use the inherent flexibility of a linac built from independently-phased resonators, three examples were simulated. The simulation results are given in Table 3, below. Scheme 1 assumes that all cavities will be operated at a constant accelerating field of E_a=5.3 MV/m. This is the maximum field under normal operating conditions. In this scheme, the energy gain per cavity is no longer fixed. We have assumed a cavity average synchronous phase of -35°. As can be seen, the beam output energy is raised by 99 MeV. The change in output beam emittances and ratio of transverse aperture to rms beam size is small. Also shown in Table 3 is the minimum required beam current to produce 100 MW output beam power at 1099 MeV. This example demonstrates an alternative operating scheme which could be used in the event of source output current degradation. In Scheme 2, the average synchronous phase has been reduced to -25°. As is expected, the output energy is further increased to 1179 MeV. In Scheme 3, the cavity fields have been increased by 33%. This scheme demonstrates a possible upgrade path, which would require significantly increased power-coupler capabilities and klystron output to produce 130 MW of beam power, without requiring additional accelerating cavities. In the last two schemes, there is a slight degradation in the ratio of transverse aperture to rms beam size. Transverse emittance growth is observed in all cases, which is comparable to the 25% observed for the nominal operating mode. The effects of emittance growth on beam uniformity at the neutron production target have not been studied.

Table 3. Alternative operating schemes for the high-energy superconducting option. Required beam current is the beam current required to produce a 100-MW beam power.

Case	Output Energy (MeV)	Trans. Emittance Growth	Long. Emittance Growth	Required Beam Current	Aperture Ratio, Med.-β	Aperture Ratio, High-β
1	1099	17%	-5%	91 mA	18	21
2	1179	32%	98%	85 mA	18	20
3	1297	19%	-4%	77 mA	17	20

Experience at operating superconducting accelerator facilities has shown that often there is a large variation in the maximum accelerating gradients achieved in identical multi-cell accelerating cavities. Typically these are $\beta=1$ cavities used to accelerate electron beams such as at CEBAF [9]. If cavities fail or perform at lower than expected accelerating gradients, the gradients and rf phases in the other cavities are adjusted to compensate and provide the required additional energy gain. A possible solution to increase the availability of the proposed superconducting linac is to provide additional accelerating cavities, thus anticipating some fraction of cavity failures. In order to examine these effects, we simulated the case where 5% of all cavities are not operating (every 20th cavity off). To restore the beam energy, we added 5% additional cavities to the high-β section. Simulation results, using a simple algorithm for setting the cavity phases, showed a transmission of 100% with a reduced output beam energy of 993.4 MeV for this case. Small adjustments of the phases should restore the correct final beam energy. The transverse and longitudinal emittances were observed to grow by factors of 2.9 and 6.8, respectively; however, only small changes were observed in the aperture to rms values.

Additional simulation studies of various error, alignment, and failure conditions have also been completed for this point design. These studies indicate that the alignment tolerances required for the system proposed are greatly relaxed compared to the room-temperature design. Relatively large rf phase and rf amplitude errors are tolerable (5° and 3%). A maximum of two consecutive quadrupole doublets can fail without beam loss although there is a substantial reduction in the ratio of transverse aperture to rms beam size. Our initial studies indicate that the superconducting linac is tolerable to many imperfections and fault conditions.

SUMMARY

The details of our baseline design for a room-temperature linear accelerator for tritium production have been discussed. The main new features of this design are the use of a high-energy RFQ to accelerate the beam to 6.7 MeV and a new medium-β structure from 6.7 to 100 MeV, the CCDTL. The design presented here is conservative and will meet the design requirement for low beam loss and hands-on maintenance.

We have also discussed the option of using superconducting cavities in the high-energy section of the accelerator from 100 to 1000 MeV. The two main advantages of the superconducting option may be increased operational flexibility and substantial power savings. Table 4 shows a comparison between the performance of the proposed room-temperature, SCL and the superconducting option. Although the beam evolution in the two linacs is quite different, both will meet the requirements for accelerator production of tritium.

Table 4. Simulation results. A comparison of the performance of the room-temperature baseline design and the superconducting option for 100-1000 MeV.

	SCRF Linac	Room-Temperature Baseline
Transverse Emittance Growth	25%	None
Longitudinal Emittance Growth	8%	41%
Transverse Aperture To RMS Ratio	19 (100 MeV) 24 (260 MeV) 26 (1000 MeV)	13 (100 MeV) 21 (260 MeV) 25 (1000 MeV)
Transverse Tune Depression Ratio σ_t / σ_{ot}	0.50 (100 MeV) 0.81 (260 MeV) 0.98 (1000 MeV)	0.80 (100 MeV) 0.71 (260 MeV) 0.89 (1000 MeV)
Longitudinal Tune Depression Ratio σ_l / σ_{ol}	0.53 (100 MeV) 0.71 (260 MeV) 0.94 (1000 MeV)	0.38 (100 MeV) 0.35 (260 MeV) 0.42 (1000 MeV)

REFERENCES

1. T. P. Wangler et al., "Linear Accelerator for Production of Tritium: Physics Design Challenges," Proceedings of the 1990 Linear Accelerator Conference, Albuquerque, New Mexico, September 10-14, 1990.
2. R. W. Garnett, E. R. Gray, L. J. Rybarcyk, and T. P. Wangler, "Simulation Studies of the LAMPF Proton Linac," Proceedings of the 1995 Particle Accelerator Conference and International Conference on High-Energy Accelerators, Dallas, Texas, May 1-5, 1995.
3. L. M. Young, "An 8-Meter-Long Coupled Cavity RFQ Linac," Proceedings of the 17th International Linac Conference, Tsukuba, Ibaraki, Japan, August 21-26, 1994.
4. J. H. Billen, "A New Structure for Intermediate-Velocity Particles," Proceedings of the 17th International Linac Conference, Tsukuba, Ibaraki, Japan, August 21-26, 1994.
5. S. Nath, "Funneling in LANL High-Intensity Linac Designs," Proceedings of the Accelerator-Driven Transmutation Technologies Conference, Las Vegas, Nevada, July 25-29, 1994.
6. R. S. Mills, K. R. Crandall, and J. A. Farrell, "Designing Self-Matching Linacs," Proceedings of the 1984 Linear Accelerator Conference, Seeheim, Germany, May 7-11, 1984.
R. W. Garnett and P. Smith, "Design of a Current-Independent Matching Section for APDF," Proceedings of the 17th International Linear Accelerator Conference, Tsukuba, Ibaraki, Japan, August 21-26, 1994.
7. J. S. O'Connell, T. P. Wangler, R. S. Mills, and K. R. Crandall, "Beam Halo Formation From Space-Charge Dominated Beams in Uniform Focusing Channels," Proceedings of the 1993 Particle Accelerator Conference, Washington D.C., May 17-20, 1993.
8. M. Reiser, "Beam Physics Design Strategy for a High-Current RF Linac," Proceedings of the Accelerator-Driven Transmutation Technologies Conference, Las Vegas, Nevada, July 25-29, 1994; Also see M. Reiser et al., this conference.
9. C. Reece et al., "Performance Experience with the CEBAF SRF Cavities," Proceedings of the 1995 Particle Accelerator Conference, Dallas, Texas, May 1-5, 1995. (To be published).

Dealing with the High-Brightness Beam and Space Charge in CLIC

Gilbert Guignard
CERN, 1211 Geneva 23, Switzerland

INTRODUCTION

In the Compact Linear Collider (CLIC) complex [1] under study at CERN, questions related to space-charge forces and high-brightness beams are both pertinent. The former appear mainly in the CLIC test facility (CTF) [2] which runs at low energy (maximum around 73 MeV/c) with a high charge per bunch (up to 35 nC in a single bunch) while the latter are mostly relevant in the design of the CLIC main linac where a very small vertical emittance is required for achieving a high luminosity. Space-charge effects may also be present either in the pre-acceleration or after the deceleration of the drive beam (train of intense bunchlets) which generates the 30 GHz power needed to accelerate the main beam [3]. In addition, at this high RF frequency, the strong action of the wakefields produced by the beam is superimposed on the two physical phenomena just mentioned.

Hence, space-charge effects have to be included in the model-making of the beam dynamics in the CTF. Moreover, the presence of 30 GHz accelerating sections in this facility justifies the need for numerical tools to simulate the beam behavior in the simultaneous presence of high space-charge and strong wakefields. Therefore, in the first part of this report, there is a brief description of how we deal with these two effects in a new code, which allows the tracking of particles over relatively large distances. There is also a short summary of the simulations and experiments made with the CTF bunch compressor as an example of the space-charge influence on the bunch length. The second part of the report is then devoted to the main topic, which is the preservation of the small transverse emittances of a high-brightness single bunch in the main linac of CLIC. The challenge is to counteract the instabilities generated by strong wakefields over the whole length (measured in kilometers) of the linac.

We describe the blow-up mechanisms as well as the correction methods, which are proposed for and have been adapted to CLIC, and which incorporate developments and investigations made for other linear collider studies.

SPACE-CHARGE EFFECTS IN CLIC AND CTF
Simulations with Wakefields and Space-Charge Forces

There are several places in CLIC and in the CTF (which tends to reproduce at low energy the conditions of the two-beam scheme), where the effects of the space charge and of the wakefields are superimposed on each other. The CTF is essentially made of an RF gun that contains a photocathode, an RF booster increasing the beam energy to 13 MeV/c, a bunch compressor, an RF accelerating structure bringing the beam to 73 MeV/c (all these RF elements functioning at 3 GHz), and two 30 GHz CLIC structures (separated by a double bend of 180°), used as transfer and accelerating structure, respectively. In this range of energy, space-charge forces are important for the beam current foreseen [2] and wakefields are present in all RF components, in particular in the 30 GHz sections. Similarly, in the low-energy parts of the driver of the CLIC two-beam scheme [3], like the pre-acceleration below 100 MeV say, the magnetic switch-yard and the end of the drive linac where the beam might be strongly decelerated after energy transfer, space-charge forces and wakefields cohabit. In all cases, the beam is a train of high-charge (20–30 nC) bunches. These examples illustrate the need we had for a code capable of tracking particles in a multibunch high-current beam through RF structures, at low-energies and over relatively long distances.

There exist of course powerful programs like MAGIC and MAFIA [4] which include self-consistent field calculations in their 'particles-in-cell' mode. They have, however, some limitations associated with the mesh-size and with the computing time, when bunches are as small as in CLIC or CTF, and the distances to be tracked over are long. We therefore sought an alternative and simplified method of tracking particles in these conditions and also with strong space-charge as well as wakefields. On this point, we suggested the incorporation of the treatment of wakefields that is used in our tracking programs MTRACK [5] and DTRACK [6] into the code PARMELA [7] which has the capacity of tracking particles according to a variety of external fields and in the presence of space charge, at low energy ($v/c < 1$); PARMELA has in addition the advantage of dealing with 3D distributions of particles and of being well tested, but includes the propagation of a single bunch only. MTRACK and DTRACK codes consider a simplified model of wakefields and fully relativistic beams ($v/c = 1$, valid for high-energy electron beams); they have no radial forces and cannot therefore simulate space-charge forces but offer the possibility to track multibunch trains over long distances (for linacs of 3 or 6 km). With these considerations in mind, we thought that the easiest way to proceed was to insert the Green's function model of the wakefields, used in MTRACK codes, into PARMELA and to apply it to all macroparticles. There are several advantages to this approach: 1) space-charge and wakefield forces for specific RF conditions are given as input from outside the program, 2) the complexity of the wakefield model can be adjusted to the problem treated, 3) some effects can be switched on and off for study purposes, 4) tracking steps can be different and variable for the two main effects considered, 5) the 3D distribution of the macroparticles makes possible the numerical observation of bunch lengthening associated with finite transverse emittances, since the independent variable is 'time' (or RF phase and frequency), which allows a longitudinal mixing of the non-relativistic particles ($\beta < 1$). The disadvantage of this new code, called PARMTRACK [8], is that it is not self-sufficient with self-consistent field calculations, but it requires a separate calculation of wakefields with programs like KN7C [9], TRANSVRS [10], MAFIA [4] or ABCI [11], and some approximations are unavoidable.

The approximations depend on the model, which is based on solutions of Maxwell's equations either in the frequency domain or the time domain. In the frequency domain, the usual model description can be used for the delta-function wake potentials of the longitudinal and dipole fields, respectively,

$$W_L^\delta(t) = 2 \sum_{n=1}^{\infty} K_{0n} \cos \omega_{0n} t$$
$$W_T^\delta(t) = 2 \sum_{n=1}^{\infty} K_{1n} \frac{c}{\omega_{1n} a^2} \sin \omega_{1n} t \qquad (1)$$

where K_n and ω_n are the loss factors and frequencies of the longitudinal (index 0) and transverse (index 1) modes, while a is the cell iris radius. Up to now, quadrupole modes have not been included, but could be added later, since CLIC contains microwave quadrupoles [Eq. (11)]. The short-range wakefields can be estimated by using Eqs. (1) with a limited number of modes and by adding the contribution of an analytical extension covering the high-frequency part of the spectrum and using the optical resonator model [12]. Moreover, since the time steps in PARMTRACK correspond to different distances when β varies, spline functions are prepared in advance for further interpolations. As for the long-range wakefields, the high-frequency term contribution is small for distances large with respect to the bunch length σ_z and they are well-approximated by their first modes only.

In time domain, the calculation provides a global solution, integrated over the whole bunch. The problem is then to extract the delta-functions of the wake potential, calculated with bunches of finite length, small with respect to the cavity size but not infinitely small (the number of mesh points describing the geometry gives the actual limit). An elegant way to get the asymptotic behavior of the delta-potentials (for σ_z tending to zero) has been proposed by A. Mosnier [13] and used for PARMTRACK. The total loss factor $K(\sigma_z)$ is developed in a power series as a function of σ_z

$$K(\sigma_z) = \int W(s)\rho(s)ds = \sum_n A_n \sigma_z^{\alpha_n} \qquad (2)$$

with

$$W(s) = \int W^\delta(s^*)\rho(s - s^*)ds^* \ .$$

The same development but with different coefficients is then assumed for the delta wake potential as a function of s

$$W^\delta(s) = \sum_n a_n s^{\alpha_n} \ . \qquad (3)$$

Using all these definitions, the author of Ref. [13] finds a simple relation between the coefficients a_n and A_n, which are proportional with a factor $F(\alpha_n)$ that only depends on the power α_n,

$$a_n = A_n 2\pi \sigma_z^{(\alpha_n+2)} \left[\iint x^{\alpha_n} \exp\left(-\frac{(y-x)^2}{2\sigma_z^2}\right) \exp\left(-\frac{y^2}{2\sigma_z^2}\right) dx\, dy \right]^{-1} . \qquad (4)$$

The loss factors and the coefficients A_n are calculated in the time domain for a Gaussian distribution of particles and different values of σ_z. Then, relations (3) and (4) give an estimate of the asymptotic values of the wakefield Green's functions. The interest of this method originates from the fact that only very few terms in Eq. (3) are necessary for a good description.

Note that, in both approximations, a possible attenuation of the modes (behind the bunch) has been initially neglected. This excludes at the moment accelerating structures with damping.

The application of the wakefield kicks in PARMTRACK proceeds from the same method used in MTRACK for the slices that are making a bunch. For each 'test' macroparticle k of any bunch, the change in energy dE and the transverse kick θ are computed by adding the contributions, i.e. the wakefield Green's function amplitudes, of all the preceding macroparticles (index j), in a very simple way

$$dE_W(k) = -\sum_j W^\delta_{L_{jk}}(z_j) q_j \, c \, dt_k$$

$$\theta_y(k) = \frac{dp_x(k)}{E(k)} = \frac{1}{E(k)} \sum_j W^\delta_{T_{jk}}(z_j) q_j \, y_j \, c \, dt_k \qquad (5)$$

where q_j is the charge of the macroparticles, y_j their transverse displacements (either horizontally or vertically), E the total energy of the test particle and z_j the distance between the particles k and j. The accuracy will depend on the time interval (or step) used for computing the dynamics. The space-charge effects and low energies ($\beta < 1$) induce time variations of the distances z_j which thus do not remain a constant multiple of a given step Δz. This explains why interpolation of the Green's functions is required.

All these features have been implemented [8] in PARMTRACK. Additional complexity came from including a multibunch train, the possibility to have a group velocity of the energy flow different from c and from the existence of different zones of wakefields (boundary problems). Examples of results [8] are given for two Gaussian bunches with $\sigma_z = 0.7$ mm, separated by 1 cm and travelling through a 30 GHz CLIC accelerating structure, 10 cm long, followed by a drift of 5 cm. The initial energy is 60 MeV and the charge per bunch 20 nC, as in the CTF. The space-charge effect is switched off and both bunches are off-centered by 1 mm at injection. Initial emittance and energy-dispersion are zero, and longitudinal particle distribution is Gaussian. Figure 1 represents the divergence (or the kicks) of the two bunches at the end of the channel and their energy variation due to longitudinal wakefields. These figures reproduce the expected familiar distribution of the single-bunch wakes and the additional shift for bunch 2 due to long-range forces. Figure 2 indicates that the beam blow-up is already large with respect to the aperture radius (2 mm) of the 30 GHz structure.

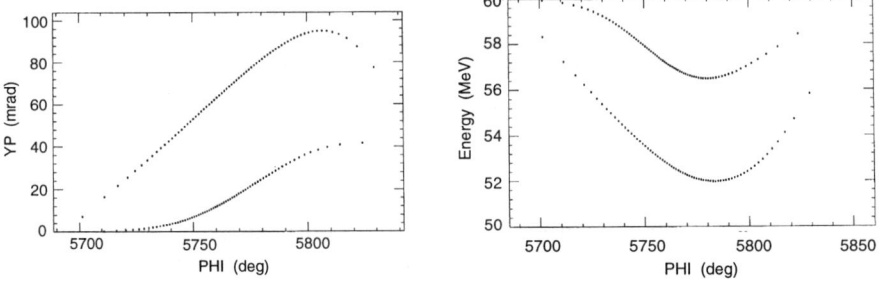

Figure 1: Transverse divergence (left) and energy variation (right) within two bunches, due to wakefields.

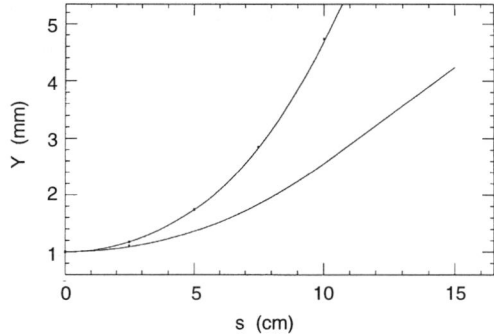

Figure 2: Transverse envelope of two bunches travelling through a CLIC structure (for 90% and 100% of the particles).

Space Charge in the CTF Bunch Compressor

To reproduce the drive-beam conditions in the CTF, the charge per bunch has to be as high as possible and the bunch length short with respect to the 30 GHz wavelength, i.e. σ_z of the order of 1 mm or 3.3 ps in time units. Pushing these two parameters results in a rise of the extracted power by the transfer structure. For minimizing the achievable σ_z, a bunch compressor [14], made of three rectangular bending magnets installed side by side to form an achromatic chicane, has been added in the CTF line, after the 3 GHz, 4-cell energy booster. Since the beam momentum at this position is about 13 MeV/c, space-charge forces become important when the charge increases from, say, 2 nC to 20 nC.

For a limited current of 2 nC, i.e. weak space-charge effects, and after optimization of the booster RF-phase, in order to obtain a good energy correlation and a small momentum spread, the compression rate is of the order of 6, as shown in Fig. 3. For a higher current, simulations done by PARMELA show that the space-charge forces smear out the longitudinal emittance and reduce the compression rate. Raising the charge up to 17 nC per bunch means that, for bunch lengths of the order of 10 ps, the beam is space-charge-dominated, with a large spread of the particles in the phase space. Increasing the initial bunch length σ_z from about 8 ps to 16 ps, the effect of the compressor is still observable but the compression does not exceed a factor 2 approximately (Fig. 4).

Bunch-compressor studies have been carried out and experiments performed in the CTF [15], in order to make extensive comparisons of PARMELA results with a real beam. Bunch lengths were measured after the bunch compressor for different charges and a variable deflection angle (or excitation current) taken in the first dipole of the chicane. Measurements were done with a streak camera that has a resolution of 4 ps, so that one has to correct for this error when observing bunches as short as 2 ps (e.g., at low charge).

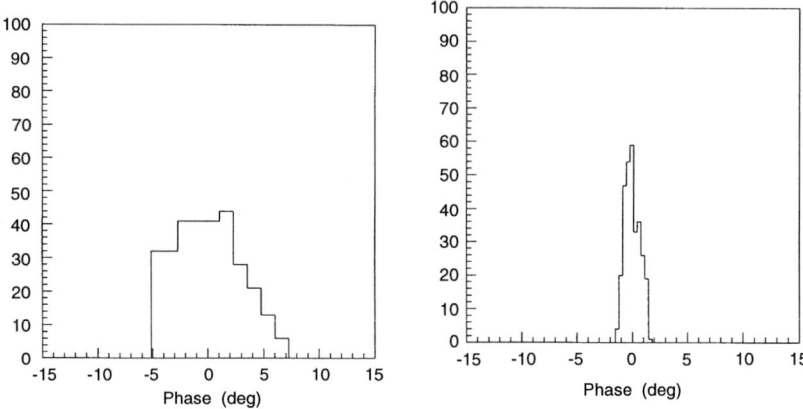

Figure 3: Calculated particle distribution of a 2 nC bunch, before (left) and after (right) compression.

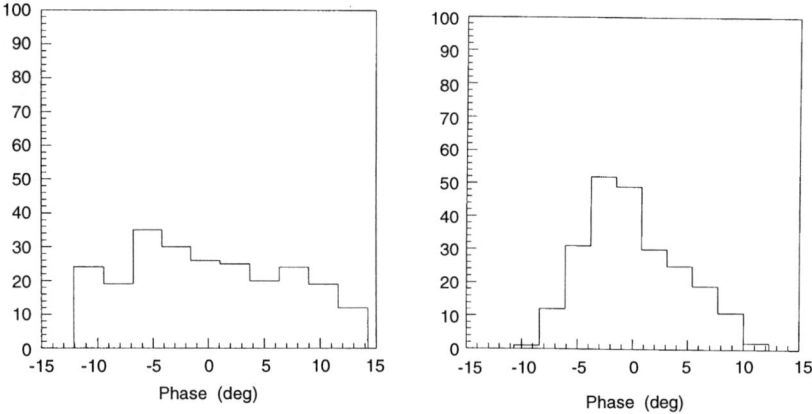

Figure 4: Calculated particle distribution of a 17 nC bunch, before (left) and after (right) compression.

With a charge of 2 nC per bunch, the measured bunch length goes through a minimum at the expected deflection angle of 17.5 degrees and the value corrected for the camera resolution agrees rather well with the calculated value (Fig. 5). To make an experimental comparison with a space-charge dominated bunch of 17 nC, long bunches were created by using a double laser pulse (on the photocathode) generating two 8 ps pulses separated by 9 ps, which give an equivalent bunch length (full width, half-height, i.e. FWHH) of 16.5 ps. In these conditions, the minimum of σ_z as a function of the compressor angle is flat and corresponds to a FWHH length of 9 ps (i.e. $\sigma_z = 2.6$ ps).

The agreement between simulations and measurements is good again (Fig. 6), taking into account that the energy correlation was probably not optimum and there were some losses of particles in the experimental case.

In general, the results given by PARMELA for the bunch length dependence on the space charge have been confirmed. Measurements are being carried out to make the same kind of comparison for the two betatron emittances.

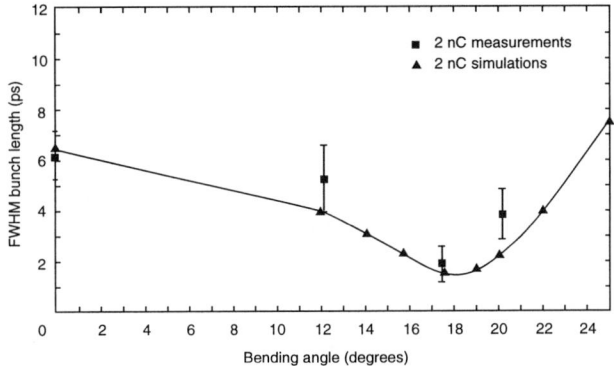

Figure 5: Simulated and measured bunch length at 2 nC, as a function of the bending angle of the (first) compressor magnet.

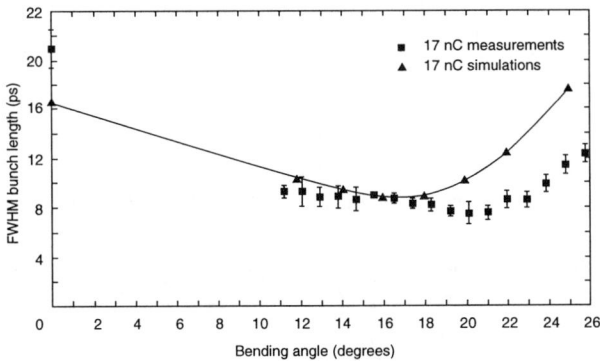

Figure 6: Simulated and measured bunch length at 17 nC, as a function of the bending angle of the (first) compressor magnet.

HIGH-BRIGHTNESS BEAM IN THE CLIC MAIN LINAC

Emittance and Brightness Requirements

When the beams collide head-on at the interaction point of a linear collider, the luminosity obtained is proportional to: the square of the bunch population N_b, the repetition rate $f_{\rm rep}$, the enhancement factors H_x, H_y due to beam disruption (or reduction factors of the beam transverse dimensions due to the pinch effect at collision), and to the number of bunches k_b. But it is inversely proportional to the product of the beam sizes $\sigma_x \sigma_y = R\sigma_y^2$ (if R is the beam aspect ratio σ_x/σ_y), i.e.,

$$L = \frac{k_b \, N_b^2 \, f_{\rm rep} \, H_x H_y}{4\pi R \sigma_y^2}. \qquad (6)$$

Equation (6) immediately shows the advantage of having a vertical emittance as low as possible to achieve a small σ_y. The total emittance must also remain limited and in these conditions the brightness of the beam, defined here as the density of particles in a 3D space,

$$B = \frac{N_b \, e}{\Delta t \, \varepsilon_x \varepsilon_y} = \frac{\widehat{I_b}}{\varepsilon_x \varepsilon_y} \qquad (7)$$

is high. For CLIC nominal parameters, B has a value close to 5×10^{26} A/m^2 at 250 GeV, which would correspond to the magnitude of the brightness of an unusually bright source.

Given the total emittance, the aspect ratio R can be selected — not necessarily to optimize the luminosity, but to reduce all the beam–beam backgrounds and improve the resolution on the center-of-mass energy of the collision. These effects can be characterized by the beamstrahlung parameter Υ (fractional critical photon energy), the fractional energy loss due to radiation δ_B that is a function of the enhancement factor H_Υ for quantum effects, and the detector occupancy Ω (number of 2γ events per bunch crossing), which all depend on the parameters entering Eq. (6) — in particular on R,

$$\begin{aligned} \Upsilon &\propto \frac{\gamma N_b H_y}{\sigma_y \sigma_z} \frac{1}{(1+R)} \qquad H_\Upsilon = \left(\frac{1}{1+1.33\Upsilon^{2/3}}\right)^2 \\ \delta_B &\propto \frac{\gamma N_b^2 H_y^2}{\sigma_z \sigma_y^2} \frac{H_\Upsilon}{(1+R)^2} \qquad \Omega \propto \frac{R N_b^2}{f_{\rm rep}(1+R)^2}. \end{aligned} \qquad (8)$$

Consequently, high R-values lessen the key parameters of Eqs. (8), as required for the experiments. Therefore, the CLIC main linac accelerates a beam which has an aspect ratio between 30 at injection (9 GeV) and 20 at 250 GeV.

The challenge is to keep such a small emittance ($\gamma \varepsilon_{tot} = 2.65 \times 10^{-6}$ rad·m) and the corresponding high brightness during the acceleration in the main linac, this in the presence of many sources of imperfections. Concerning static or slowly drifting imperfections, one can mention misalignments of linac components, the possible tilts of these components, the wakefields due to the 30 GHz cavities, and the energy dispersion of the optics. These issues and the associated corrections are reviewed for CLIC in the next sections, considering a single bunch.

BNS Damping and Energy Scaling

The presence of strong dipole wakefields implies large kicks originating from misalignments of the cavities and off-centered trajectories. In a single bunch, short-range wakes generated by the leading particles act downstream, distorting the bunch. The well-known technique for counteracting this effect consists in obtaining a coherent motion by imposing the same oscillation period (or focusing strength k^2) on all particles of the bunch. The condition, called autophasing by its author [16], can be written as:

$$k^2(z,s) = k_0^2 + \frac{r_e}{\gamma} \int_{-\infty}^{z} \rho W_T^\delta(z - z^*) dz^* , \qquad (9)$$

where the point-charge wake W_T^δ is integrated over the bunch of density $\rho(z)$.

The linearized version of this criterion, known as the BNS damping condition [17], writes

$$\frac{\partial k^2}{\partial z} = \frac{N_b r_e}{\gamma} \frac{\partial W_T}{\partial z} . \qquad (10)$$

In Eqs. (9) and (10), r_e is the classical radius of the electron defined as $e^2/m_0 c^2/4\pi\varepsilon_0$.

Solving Eq. (10) for the energy variation $\Delta p/\Delta z$ along the bunch by using finite differences, gives the energy spread (δ_{BNS}) required in order to achieve the right variation of k^2 with z, via a z–p correlation. In CLIC, δ_{BNS} is too large and hence in conflict with the minimization of the bunch energy-spread needed for the final-focus system. The solution retained is therefore to modulate the focusing strength k^2 in the bunch by generating a small fraction of the transverse focusing directly from quadrupole RF fields oscillating at the frequency of the accelerating fields. While the static average focusing is produced by magnetic quadrupoles arranged in a FODO lattice, the additional modulation comes from microwave quadrupoles (RQF), i.e., oval-bodied cavities with circular iris of 4 mm aperture [18], that provide an equivalent gradient whose maximum amplitude is given by

$$G_{\rm RF} = 0.80 \frac{\pi}{c\lambda_{\rm RF}} \widehat{E}({\rm V/m}) \ . \tag{11}$$

Since these RQFs are there only to modulate k^2 ($\lesssim 20\%$ total), they can run with a quadrupole phase close to zero (in fact 12°, as for the accelerating cavities) which corresponds to the maximum (in fact, the optimum for the energy spread) of the acceleration. Hence, this solution does not cost much physical space in the linac.

With the condition of smooth focusing and the equivalence between k^2 and $1/\beta^2$, conditions (9) and (10) indicate that keeping a constant stability margin along the linac implies scaling the β-function with $\sqrt{\gamma}$. With a constrained phase advance, this in turn means that both the focal distance f and the separation L_c of the lattice quadrupoles should be scaled with $\sqrt{\gamma}$. However, studies of discrete focusing lattice show that β-function and chromaticity $d\mu/d\delta$ (μ = phase advance, δ = energy deviation) are independent attributes with different sensitivities to a given phase μ. Taking this into account, the new practical form of the BNS damping criterion can be written as [19]

$$\frac{r_e}{2} \frac{\langle \beta \rangle}{\langle d^2\mu/d\delta ds \rangle} \frac{N_b \langle W_T \rangle_{\rm beam}}{\gamma \langle \delta \rangle_{\rm beam}} \cong 1 \ , \tag{12}$$

and the $\sqrt{\gamma}$-scaling of β set aside. In order to balance better the effects of energy dispersion and strong wakefields, and to achieve approximately the phase-advance chromaticity demanded by (12), CLIC has introduced a different scaling with energy that involves a variation of μ along the linac. This is done by independent and different scalings of L_c and f, according to Ref. [20]:

$$\frac{L_c(s)}{L_0} = \left(\frac{\gamma(s)}{\gamma_0}\right)^{\alpha_a} \quad \frac{f(s)}{f_0} = \left(\frac{\gamma(s)}{\gamma_0}\right)^{1-\alpha_q} \ . \tag{13}$$

With $\alpha_a = 0.3$ and $\alpha_q = 0.6$, this gave a 33% gain on the vertical emittance blow-up with respect to standard $\sqrt{\gamma}$ scaling. On this basis, the practical implementation of the elements has been studied [21] assuming that the linac focusing was arranged in sectors and the quadrupoles could not encompass the accelerating sections. In each sector, the focusing and lattice layout are constant. Both are adjusted to approximate as well as possible the continuous scaling (Fig. 7) defined by Eq. (13), by an initial phase advance of 95° and by an initial quadrupole separation of 3.5 m. There are six sectors of different dimensions (135 m to 1480 m) with phase advances decreasing from 90° to 60° and a quadrupole size rising from 0.26 to 0.94 m. The required gradient never exceeds 100 T/m, which is acceptable for an aperture diameter of say 20 mm, that makes possible misalignments and quadrupole displacements for trajectory correction. The layout periodicity is given by the nominal supporting girder (1.41 m long) and the quadrupoles always sit at the head of a girder. Near every magnetic quadrupole, one finds microwave quadrupoles of approximately the same total length. The girders which do not contain quadrupoles support four (30 GHz) accelerating cavities. While the focusing strength is constant in a sector, the quadrupole gradients are all different because of the acceleration. Between sectors, the betatron functions change abruptly and matching insertions are required. They are made of one cell from a sector end and one from the following sector (total of five quadrupoles powered independently). Such a layout makes the real linac longer than assumed initially but does not modify significantly the emittance blow-up since the additional elements (or drifts) do not generate wakefields as strong as those of the cavities.

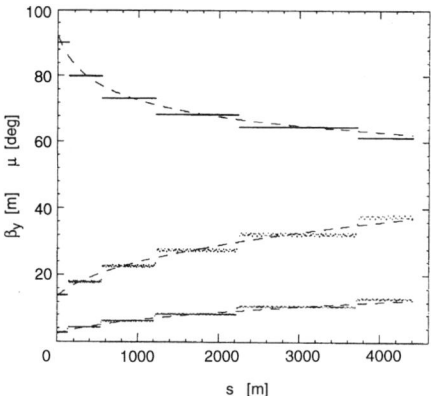

Figure 7: Energy scaling of the main linac Twiss parameters through constant lattice sectors.

With flat beams, betatron coupling plays an important role in the vertical beam-size growth by exchange of transverse 'energy' (square of the amplitudes). When the lattice is identical in the two planes ($\mu_x = \mu_y$), the effect is enhanced by resonance. Therefore, different phase advances should be selected in each plane using two families of quadrupoles with different focal distances. This was tried in CLIC [21] with $(\mu_x - \mu_y) = 10°$, which then allows an r.m.s. tilt of 1 μrad for the quadrupoles, whilst keeping the emittance target values ($3.0 \times 10^{-6} \times 1.5 \times 10^{-7}$ rad·m). The β-beating in each linac sector (Fig. 8) shows coupling. To reduce it and relax the tolerance, one can envisage coupling correction stations, using quadrupole rotations and emittance diagnostics. There must be sufficient stations to restrict the filamentation caused by betatron mismatch.

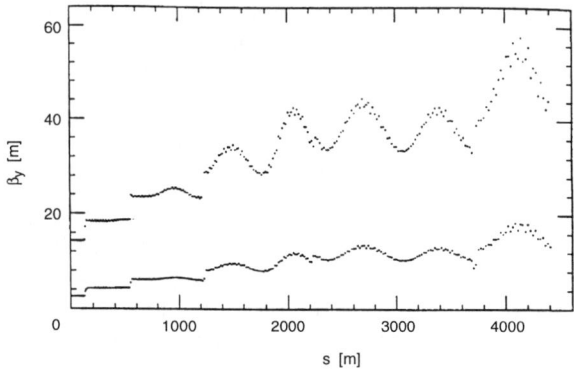

Figure 8: Betatron beating due to residual betatron coupling.

Correction of the Trajectories

Although autophasing and appropriate energy scaling of the focusing are indispensable in CLIC, it is just as vital to maintain small beam excursions in the elements of the linac. The final position tolerances to prevent emittance dilution are therefore tight. To keep them reasonable, however, trajectory corrections must be implemented. In addition, if the correction algorithms, which are based on beam measurements, are capable of re-aligning the linac components via active elements, the initial pre-alignment requirements can be relaxed with respect to the final tolerances needed.

For CLIC the accelerator will have to be pre-aligned so that the beam passes through the available aperture along a fraction of the linac. It is foreseen to do this by using signals from a stretched-wire system: the relative positions of two adjacent support girders (1.4 m long) will thus be maintained within a few microns transversely, while excursions from a straight line of about 0.2 mm will be allowed over longer distances of say 100 m. The quadrupoles will be supported and activated independently of the string of girders and the beam-position monitor (BPM) signals will be used to optimize their positions. By first using a one-to-few algorithm with approximately three monitors for one quadrupole, correction will be effected by moving the quadrupoles which are thus re-centered toward the ideal line within the position tolerances of the monitors (i.e. 10 μm). This algorithm, that centers the beam in several BPMs sitting in front of the cavity supporting girders, can be characterized by the function to be minimized [20]

$$\phi_1 = \sum_{i<j<i+1} \left[\langle x \rangle_j - \frac{R12_{ij}}{f_i} dx_i \right]^2 \qquad (14)$$

where j is the index of the BPMs which are located between the quadrupoles i and $i+1$, x_j the measured positions, $R12_{ij}$ the transfer matrix element from the quadrupole i to the BPM j, f_i the focal length of quad i and dx_i its displacement required for correction. The solution of Eq. (14) can be written [20], after including the measurement error ξ_m and the displacement inaccuracy ξ_d,

$$dx_i = \frac{\sum_j (\langle x \rangle_j + \xi_{m,j}) R12_{ij}/f_i}{\sum_j (R12_{ij}/f_i)^2} + \xi_{d,i} . \qquad (15)$$

Assuming that the initial prealignments of the quadrupoles are within 50 μm r.m.s. (but this value can easily be doubled to make the survey easier), Fig. 9 illustrates how this first correction indeed re-centers the quadrupoles.

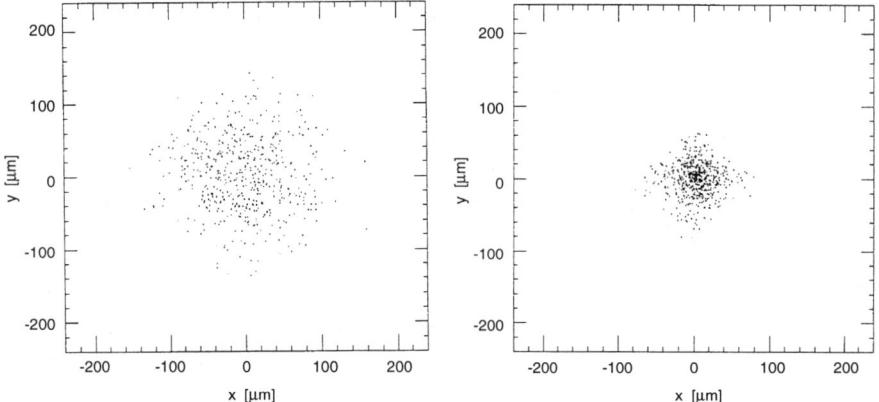

Figure 9: Transverse quadrupole positions after pre-alignment (left) and after corrections based on measurements (right).

Up to now in CLIC, the possibility of using high-order mode measurements or other diagnostics in order to move groups of four structures on girders (that are linked to form a chain) has not yet been considered and will be the subject of investigations. Before this is done, accelerating cavities are not supposed to be displaced during the first phase of corrections and, for them, pre-alignment requirements coincide with the final position tolerances (10 μm).

Next will come corrections based on trajectory difference measurements and on the observation of the emittance at 10 or 12 positions along the linac.

These may be either the dispersion-free (DF) and wake-free (WF) algorithms [22] or a correction method which uses difference measurements while simultaneously varying the bunch intensity as well as the quadrupole setting in order to minimize the wakefield effect and the dispersion [23]. The high resolution of the 30 GHz resonant cavity BPM (0.1 μm) makes these methods very efficient.

They all involve trajectory corrections over a fraction of the linac, using a certain number of quadrupoles which are moved in order to minimize at all the BPMs sitting in the segment defined by these quadrupoles, a function of the type

$$\phi_2 = \sum_j \frac{(y_j + Y_j)^2}{\sigma_r^2 + \sigma_a^2} + w\frac{(\Delta y_j + \Delta Y_j)^2}{2\sigma_r^2} \qquad (16)$$

where j is the index of the BPMs, σ_r and σ_a their r.m.s resolution or misalignment, y_j the trajectory deviation at the BPMs and Δy_j the difference of two trajectories measured in different conditions of quadrupole setting or beam current. The quantities termed Y_j and ΔY_j are equivalent to y_j and Δy_j, but they refer to the trajectory deviations or differences resulting from a unit-displacement of the quads (response of a corrector). The parameter w is just there to indicate that one can adjust the relative weight of the two terms in Eq. (16), beside the logical weighting with the errors. The different methods then only differ by the kind of trajectory difference that is included in ϕ_2:

1. When Δy and ΔY are measured while all quadrupole strengths are simultaneously increased or decreased (by 3.5% say), the change corresponds to a measurement of $\Delta y/\Delta p$ or of the dispersion. Minimizing ϕ_2 is therefore called dispersion-free (DF) correction [22].

2. When Δy and ΔY are measured by increasing the QF strengths while decreasing the QD strengths (by 3.5% say) or vice versa, the kicks all have the same sign on one side of the central line and the change mimics the wakefield effect. Minimizing ϕ_2 is then called wake-free (WF) correction [22].

3. When wakefields are as strong as in CLIC, taking the differences Δy and ΔY when changing the charge per bunch gives direct information about the dominant effect [23, 24]. This means

$$\begin{aligned}\Delta y &= y(N_b) - y(N_b/10) \\ \Delta Y &= Y(N_b) - Y(\text{model})\end{aligned} \qquad (17)$$

where N_b is the nominal charge (8×10^9) and $Y(\text{model})$ means that the response of the quads can be calculated from the optics model instead of measured at low charge. This correction is sometimes called 'measured-wakefield' (MW) correction [23].

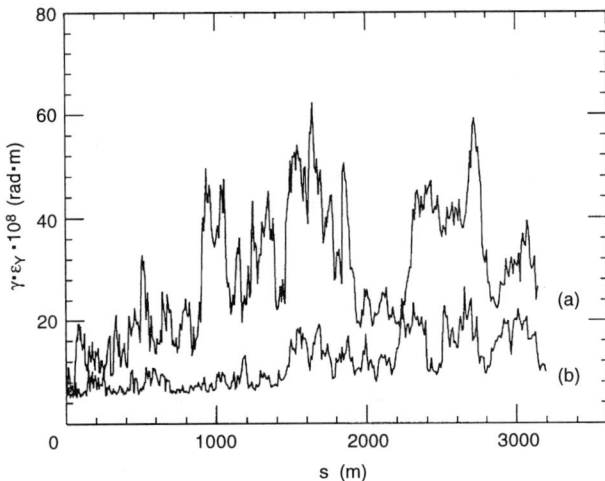

Figure 10: Vertical emittance growth in the main linac, after DF/WF correction (a) and DW (beam-based) correction (b).

4. If at low energy the wakefield effects dominate, dispersion begins to play a role anyway toward the end of the linac. It consequently seemed judicious to generalize the beam-based correction of method 3 by including information about the dispersion [23]. In this case, the differences considered are

$$\begin{aligned}\Delta y &= y(N_b/10, p_0 \pm \delta p) - y(N_b, p_0) \\ \Delta Y &= Y(N_b/10, p_0 \pm \delta p) - Y(N_b, p_0) \,.\end{aligned} \quad (18)$$

Since both the charge per bunch and the momentum (or quadrupole setting) are changed, this method was called by its author 'dispersive-wakefield' (DW) correction [23].

In order to illustrate how important it is in the main CLIC linac to have sophisticated correction algorithms based on direct measurements at variable beam current, results are presented after DF/WF corrections [25] [Fig. 10(a)] and DW corrections [23] [Fig. 10(b)]. They give the blow-up of the vertical (normalized) emittance in a 3.2 km long linac (no drifts, max. energy of 250 GeV), considering that after pre-alignment and first corrections the so-called final position tolerances are equal to 10 μm r.m.s. for all components. Simulations were done with $N_b = 8 \times 10^9$, $\sigma_z = 0.2$ mm, an injection energy of 9 GeV, and an energy spread of 0.23% [26]. Figure 10(b) shows the clear benefits of applying a beam-based correction.

CONCLUDING REMARKS

Space-charge effects have to be considered in parts of the CLIC design and, in particular, in the CLIC test facility which reproduces the two-beam scheme conditions at low energy. They are sufficiently important to justify systematic measurements of the bunch length and the bunch emittance in CTF and to run numerical simulations with the code PARMELA and its offspring PARMTRACK which includes longitudinal and dipole wakefields.

The most critical issue in CLIC remains the preservation in the main linac of the small vertical emittance of a single bunch. A lot of effort has therefore been put into the study of the most appropriate focusing scheme and elaborate trajectory-correction algorithms. Prototypes of microwave quadrupoles, necessary to damp single-bunch instability, have actually been built. Given the importance of accurate alignment, development work is being carried out on micron-displacement systems, submicron resolution BPMs and active pre-alignment schemes. The work is based on the principle of mounting four accelerator structures on a support girder, the positions of which can be adjusted by micro-movers. These movers with a resolution of 0.1 μm and an absolute accuracy of 1 μm over ± 4 mm provide both large displacements for initial alignment and micron movements for correction. The structures are fixed to the girder via V-supports with a precision of 3 μm. The ends of two adjacent girders are on a common platform which assures continuity of position between units. This system maintains the tolerances but allows the movement of groups of four structures in a continuous path. The quadrupoles are supported and activated independently. The test bench which has been built shows that the position tolerances which result from beam-dynamics studies can be satisfied and the corrections via quadrupole micro-displacements can be applied. The resulting transverse emittances at the end of the linac are adequate for physics experiments.

Acknowledgments

The author is thankful to F. Chautard, C. Fischer and A. Riche who have made available their most recent results.

References

[1] H. Braun et al., Proc. Part. Accel. Conf., Dallas, Texas, 1995.

[2] R. Bossart et al., Proc. Part. Accel. Conf., Dallas, Texas, 1995.

[3] B. Autin et al., Proc. European Part. Accel. Conf., London, 1994, p. 43.

[4] R. Klatt et al., Proc. LINAC 86, SLAC 303.

[5] G. Guignard, CERN Report, CERN SL/91–19 (AP), 1991.

[6] G. Guignard, Proc. XVth Int. Conf. on High Energy Accel., Hamburg, 1992, p. 894.

[7] B. Mouton, Report LAL/SERA/93–455, Orsay, 1993.

[8] A.J. Riche, Proc. Part. Accel. Conf., Dallas, Texas, 1995.

[9] E. Keil, Nucl. Instrum. Methods 100, p. 419, 1972.

[10] K. Bane and B. Zotter, Proc. XIth Int. Conf. on High Energy Accel., Geneva, 1980, p. 581.

[11] O. Napoly, Y.H. Chin and B. Zotter, CERN Report, CERN SL/AP 93–1, 1993.

[12] D. Brandt and B. Zotter, CERN Report, CERN/LEP TH/85–27, 1985.

[13] A. Mosnier, Report DAPNIA–SEA–92–06, CENS Saclay, 1992.

[14] F. Chautard and L. Rinolfi, CERN Report, CERN/PS 94–30 (LP), 1994.

[15] F. Chautard, Private communication.

[16] V. Balakin, Inst. Nucl. Phys., Novosibirsk, Preprint 88–100, 1988.

[17] V. Balakin, A. Novokhatsky and V. Smirnov, Proc. XIIth Int. Conf. on High Energy Accel., Fermilab, Batavia, 1983, p. 119.

[18] I. Wilson and W. Schnell, Proc. Part. Accel. Conf., San Francisco, 1991, p. 3237.

[19] W. Spence, ECFA Workshop Emittance 93, KEK, Japan, 1993.

[20] G. Guignard, Proc. Part. Accel. Conf., Washington, 1993, p. 3600.

[21] G. Guignard and G. Parisi, Proc. European Part. Accel. Conf., London, 1994, p. 893.

[22] T. Raubenheimer, SLAC Report 387, Stanford, 1991.

[23] C. Fischer, Proc. Part. Accel. Conf., Dallas, Texas, 1995.

[24] T. Raubenheimer, ECFA Workshop Linear Collider 93, SLAC, 1993.

[25] C. Fischer, Proc. European Part. Accel. Conf., London, 1994, p. 683.

[26] G. Guignard, Proc. Part. Accel. Conf., Dallas, Texas, 1995.

EXPERIMENTS WITH SPACE CHARGE DOMINATED BEAMS AND HALO FORMATION

EXPERIMENTAL STUDIES OF LONGITUDINAL DYNAMICS OF SPACE-CHARGE DOMINATED BEAMS

J. G. Wang and M. Reiser
Institute for Plasma Research, University of Maryland, College Park, MD 20742

Abstract

We review the experimental results of our research on longitudinal dynamics of space-charge dominated beams. The topics include three recent experiments. In the longitudinal compression experiments, we investigated the longitudinal density profiles of bunched beams. In the space-charge wave experiments, we studied the behavior of localized perturbations as fast and slow waves in beams. The resistive-wall instability experiments were designed to measure the growth of perturbations in a dissipative environment, and the damages to beam quality. On-going and future work is also discussed.

I. INTRODUCTION

Intense charged particle beams in advanced particle accelerators have many applications such as in experimental particle physics, nuclear waste transmutation and energy production, heavy ion inertial fusion, and free electron lasers.[1-3] The physics of these beams is often governed or significantly affected by strong self-forces due to space charge. It is very crucial to fully understand the dynamics of space-charge dominated beams in order to build reliable machines for various applications.

We have an active research program at the University of Maryland to study the physics of space-charge dominated beams. Previous work was concentrated on the investigation of the transverse dynamics of space-charge dominated electron beams. This included beamlet merging experiments, emittance growth and halo formation due to mismatch. A combination of experimental results, analytical work and computer simulation led us to important discoveries in the behavior of space-charge dominated beams, such as image and halo formation.[4,5]

During recent years, our main attention has been focused on the longitudinal dynamics of space-charge dominated beams. The research topics include the longitudinal compression experiments, the space-charge wave experiments, the resistive-wall instability experiments, and the longitudinal energy spread experiments. These experiments investigated the longitudinal distribution functions (line charge density profiles) of bunched beams, the behavior of localized perturbations in beams, the growth of perturbations in a dissipative environment, and the longitudinal energy spread of space-charge dominated beams from a thermionic electron gun.

In this paper we briefly review these experiments, and discuss on-going and future work.

© 1996 American Institute of Physics

II. LONGITUDINAL COMPRESSION EXPERIMENTS

The longitudinal compression experiments were designed to study the longitudinal distributions of space-charge dominated bunches and to explore the techniques of manipulating and handling such bunches. By using experimental data on beam profile evolution, we have showed how the longitudinal profile is maintained by a self-consistent parabolic distribution, while the rectangular distribution erodes at the head and tail. In addition, we have demonstrated a novel technique of restoring a rectangular bunch after its erosion to a cusp shape.

The experimental setup is shown in Fig. 1, which consisted of an electron beam injector [6] and 36 short-solenoid-focused transport channel. The key device was a gridded electron gun which was able to generate different longitudinal distributions of electron bunches. The diagnostics along the channel included five very fast wall-current monitors and three time-resolved energy analyzers. Typical beam parameters in the experiments were: energy of 2.5 - 5 keV, current of 25 - 70 mA, pulse length of 20 - 70 ns. The beams were fully space-charge-dominated in both the transverse and longitudinal directions.

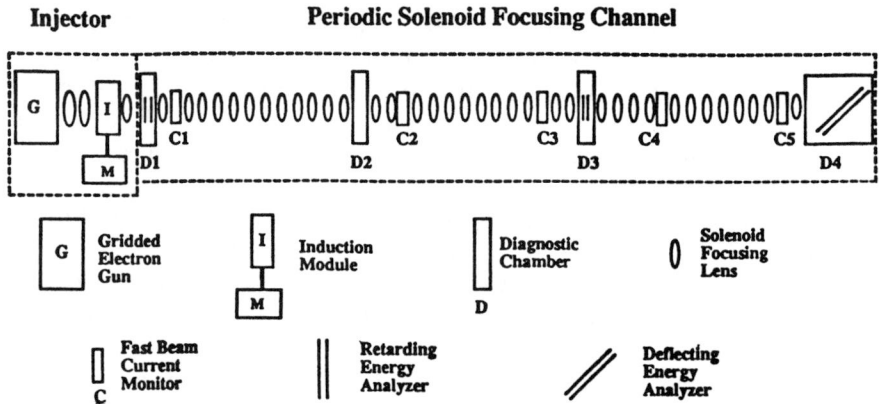

Fig. 1. Set-up of longitudinal compression and space-charge wave experiments.

1. Parabolic bunch experiment

Electron bunches with parabolic profiles were produced in this experiment by manipulating the waveforms of the grid-cathode pulse. They were further matched into the periodically focused channel for drift expansion or linear compression. The evolution of the beam current profiles and velocity distribution was studied with the wall-current monitors and the energy analyzers.

The experiment showed that the velocity distribution of particles and the space-charge self-force along the bunch is linear, that the parabolic profiles of the bunches are preserved during drift expansion and linear compression, and that the dynamics of parabolic bunches is governed by the longitudinal envelope equation.[7] The parabolic bunch profiles at five channel locations during drift expansion are shown in Fig. 2. A typical velocity distribution along the bunch is shown in Fig. 3. The detailed results can be found in references 8 and 9.

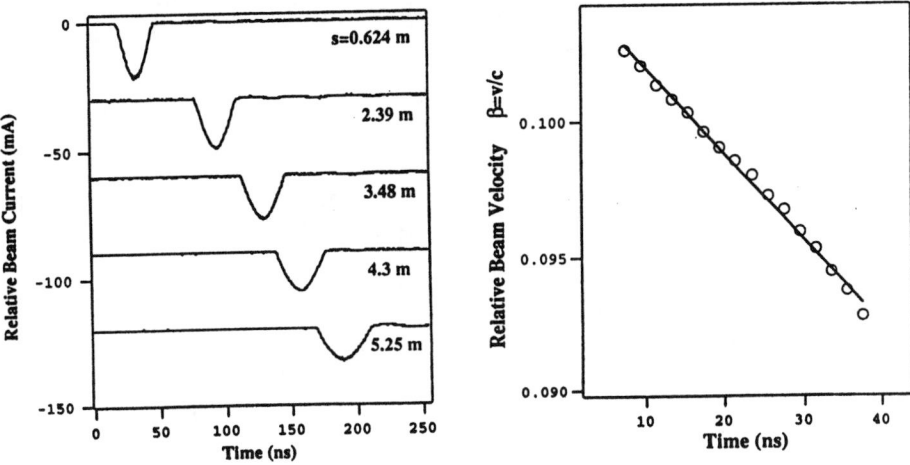

Fig. 2 Current profiles of parabolic bunch.   Fig. 3 Velocity distribution.

2. Rectangular bunch experiment

Rectangular bunches in the duration of about 50 ns and with a rise time of less than 1 ns were produced and matched into the transport channel. The evolution of such bunches during drift expansion and linear compression was studied. This was an extension of the previous work on rectangular bunches, conducted at LBL.[10]

The experiments showed that the velocity distribution of particles and the space-charge self-force along the bunch is nonlinear, that the rectangular profiles of the bunches are not preserved during drift expansion and linear compression, and that the dynamics of such bunches is governed by the solution of 1-D nonlinear fluid equations by the method of characteristics. In practice so called ear fields are essential for controlling the longitudinal profiles of rectangular bunches. Two typical experimental examples are shown in Figs. 4 and 5. The detailed results can be found in reference 11.

3. Restoration of rectangular bunches

The experiment on rectangular bunches also showed that the velocity distribution along the entire bunch is linear at the cusp point where the two simple waves traveling inward from both ends just meet each other at the center of an eroded, initially rectangular bunch. This feature led us to explore a simple mechanism to restore the rectangular shape of an eroded bunch, by employing a linear force on the bunch at the cusp point.[12] Due to the time-reverserability of 1-D fluid equation, a rectangular bunch will be regenerated downstream. An experiment to demonstrate this principle was performed. The result is depicted in Fig. 6, where (a) shows free expansion of a rectangular bunch, while (b) shows the restoration process in which a linear velocity tilt was imparted to the bunch at the cusp point (t=0) by an induction acceleration module in the beam injector.

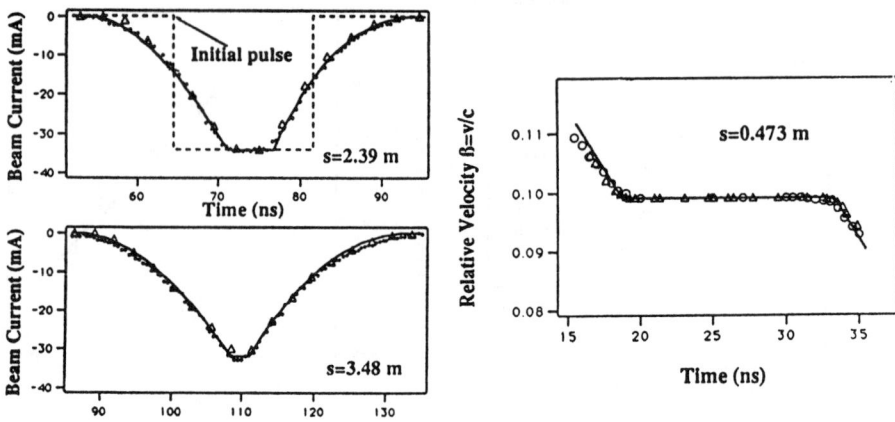

Fig. 4 Current profiles of rectangular bunch. Fig. 5 Velocity distribution.

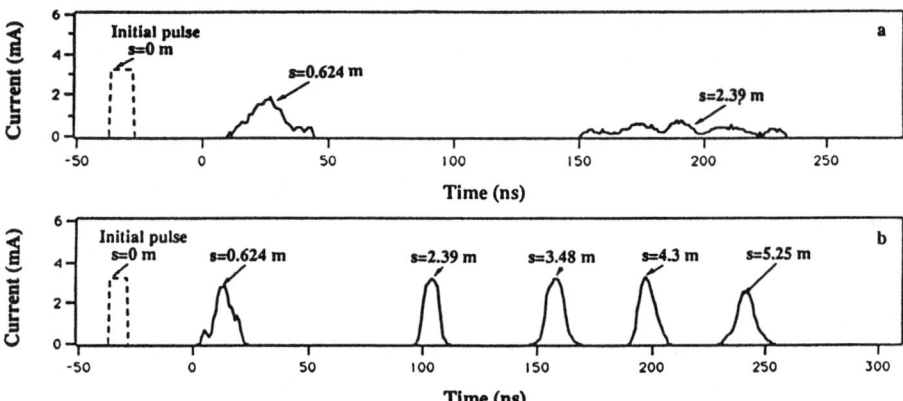

Fig. 6 Restoration of rectangular bunch after edge erosion.

III. SPACE-CHARGE WAVE EXPERIMENTS

The space-charge wave experiments were designed to generate localized perturbations in electron beams for the study of longitudinal instabilities, and to look into the problem of possible reflection of space-charge waves at bunched beam ends. During the experiments we found that single space-charge waves can be generated. In addition, we employed the technique of localized space-charge waves to measure the geometry factor and to determine experimentally the correct g-factor formula for space-charge dominated beams.

1. Generation of space-charge waves

Conventionally, space-charge waves are generated in velocity modulation devices always in pairs, i.e. both slow wave and fast wave with almost the same amplitudes. We have found that with the introduction of a current perturbation and strong enough space-charge effect, either a single slow wave or a single fast wave can be produced experimentally.[13] For combinations of the two waves different amplitude and polarity relations can exist. We also developed an analytical theory in the time domain for generating localized space-charge waves. The analysis shows that a single fast wave is generated if the ratio of the initial velocity perturbation to the space-charge wave velocity equals the ratio of the initial line-charge density perturbation to the beam line-charge density. If these two ratios are equal but with opposite signs, a single slow wave is generated.

In the experiment we perturbed the beam by a velocity kick on the grid-cathode pulse of the electron gun. This initial velocity perturbation in turn produced density and current perturbations in the electron beam. Depending on whether the gun was under the temperature limited operation or the space-charge limited operation, a single fast wave, a single slow wave, or two waves were possible. Figure 7 shows the localized velocity waves, evolving from a fast wave to a slow wave under various conditions.

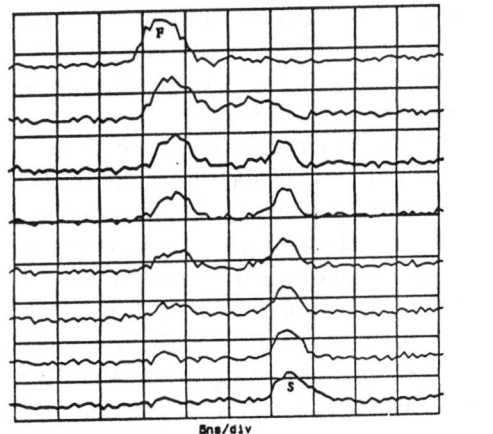
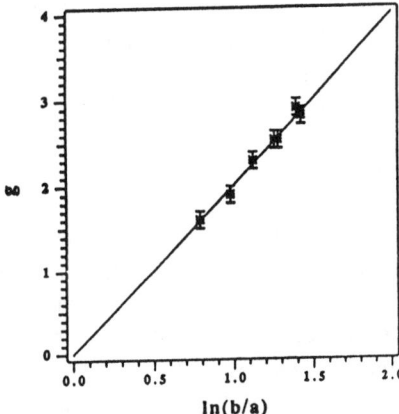

Fig. 7 Various velocity waves.   Fig. 8 Measurement of g-factor.

2. Measurement of the geometry factor

The geometry factor g is an important parameter in longitudinal dynamics to determine the perturbed axial field in a long beam. There have been two different theoretical models, yielding different g-factor formulas. One model [14] assumes a constant radius in a perturbed beam, yielding a formula of $g = 2 \ln(b/a) + 1-(r/a)^2$ where b is the pipe radius, a the beam radius, and r is the radial position within the beam. Another model [15] is based on the assumption of a constant volume density in a perturbed beam, yielding a formula of $g = 2 \ln(b/a)$. Associated with this difference, there was also an argument whether a surface wave or body wave could explain these formulas.

We performed an experiment to measure the g-factor and to determine experimentally the correct g-factor formula in a space-charge dominated electron beam.[16] The technique involved the use of localized space-charge waves. By launching two space-charge waves and measuring their separation as a function of the channel distance, the space-charge wave velocity was determined. This velocity, driven by the longitudinal electrical field, is a function of the g-factor. Thus, the g-factor was calculated from the experimental results. At the same time, we had independent measurements of the beam radius by a phosphor screen. This resulted in an experimental g-factor formula, which is $g = 2 \ln(b/a)$ as shown in Fig. 8. Further analysis shows that the assumption of a constant volume density in a perturbed space-charge dominated beam is correct, but the surface wave concept is rather misleading to interpret this phenomenon.

3. Bunch end effect on space-charge waves

The bunch end effect concerns the analysis of longitudinal instabilities. There have been theoretical investigations and computer simulations on this topic.[17-20] In the previous work the bunch end is referred as the vanishing density point. We performed an experiment to study the reflection and transmission of space-charge waves at bunched beam ends.[21] In our study, the bunch end was explicitly defined as the shoulder of an eroded rectangular bunch.

Figure 9(a) shows a single fast wave before reaching the beam front end where the top trace is the beam current without perturbation, the middle trace is the beam current with perturbation, and the bottom trace is the net perturbation signal, i.e. the difference between the top two signals. Downstream, this fast wave is reaching the bunch end and splits into two: one is the reflected wave and the other is the transmitted wave on the beam edge, as shown in Fig. 9(b). This is a clear demonstration of wave reflection at the shoulder of an eroded rectangular bunch. The speeds of the reflected wave and the transmitted wave were measured. Within the 5-m long channel distance, these speeds are the same as the speed of space-charge waves. We also developed an analytical theory to interpret the observed phenomenon. It turns out that a full reflection occurs if the beam edge length is much smaller than the perturbation wavelength. On the other hand, if the beam edge length is much longer than the perturbation wavelength, there is no reflection. In general, a partial reflection and partial transmission happens when a space-charge wave reaches the shoulder of an eroded rectangular bunch.

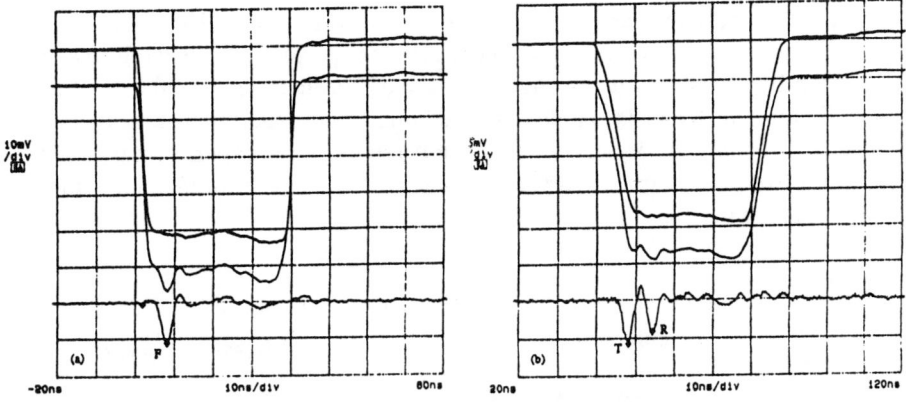

Fig. 9(a) Single fast wave close to front end. (b) Reflected and transmitted waves.

IV. RESISTIVE-WALL INSTABILITY EXPERIMENTS

Conventionally, the longitudinal instability has been studied with sinusoidal waves both in theoretical and early experimental work.[22] We believe that a time-domain approach with localized perturbations is more practical since perturbations to beams are usually in the form of localized pulses. The experimental setup, as shown in Fig. 10, consisted of the short-pulse electron beam injector as used for the space-charge wave experiments, a glass tube with a length of 0.96 m and a coated resistance of 5.4 kΩ, a 1.4-m long focusing solenoid, and diagnostics. In the experiment, electron beams of 3-8 keV, 30-80 mA, and about 100 ns in duration were passed through the resistive-wall tube, and perturbations to generate slow or fast waves were launched at the center of these beam pulses. Two current monitors at the entrance and the exit of the resistive tube were used to measure the growth and decay of the perturbation.[23]

Figure 11 shows the growth of a single slow wave where 1(a) is the current waveform with a slow wave at the entrance of the resistive channel, 1(b) is the net slow wave, 2(a) is the current waveform with the slow wave at the exit of the channel, and 2(b) is the net slow wave at the exit. It is clear that the amplitude of the slow wave is increased. Similarly, Fig. 12 shows the decay of a fast wave after passing through of the resistive-wall channel.

With localized perturbations in our experiment, the frequency spectrum of space-charge waves covers a wide range. Though the long wavelength limit condition is mainly satisfied, the dominance of space-charge wave impedance over the wall resistance breaks down for the very low frequency components of the perturbation pulses. The growth rate for these frequencies would be smaller.[24]

Overall, the amplitude increase of localized slow waves would be smaller than what usually expected. In the experiment different perturbation waveforms from a Gaussion-like shape to a rectangular one have been tested for the frequency effect. Fourier analysis is employed to compare the experiment and theory.[25] Fig. 13 and 14 show the results where the dots represent the experimental data, the crosses are from the Fourier analysis, and the triangles are from the conventional growth rate formula.

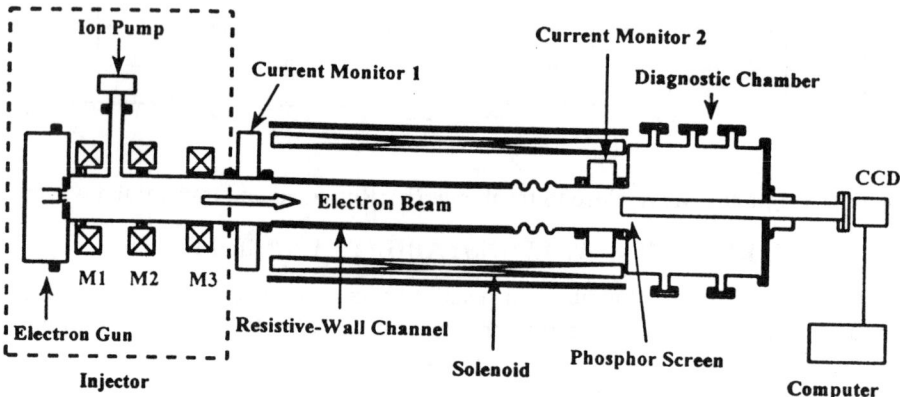

Fig. 10 Setup of resistive-wall instability experiment.

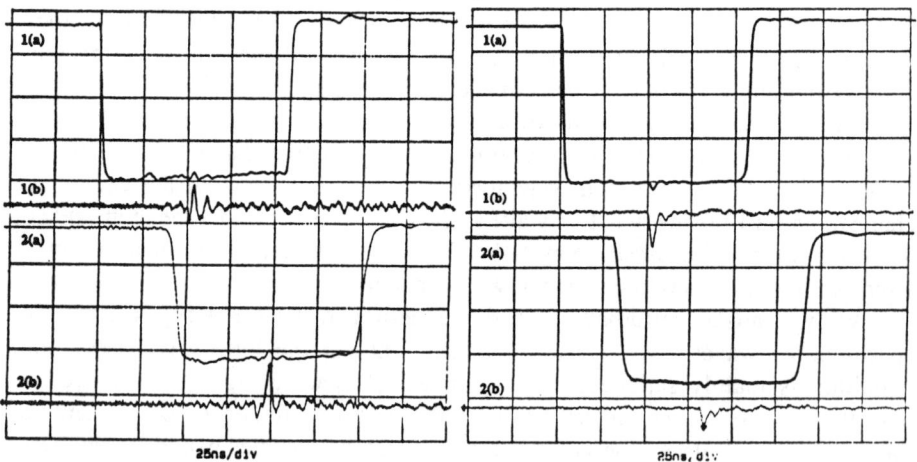

Fig. 11 Growth of a slow wave.   Fig. 12 Decay of a fast wave.

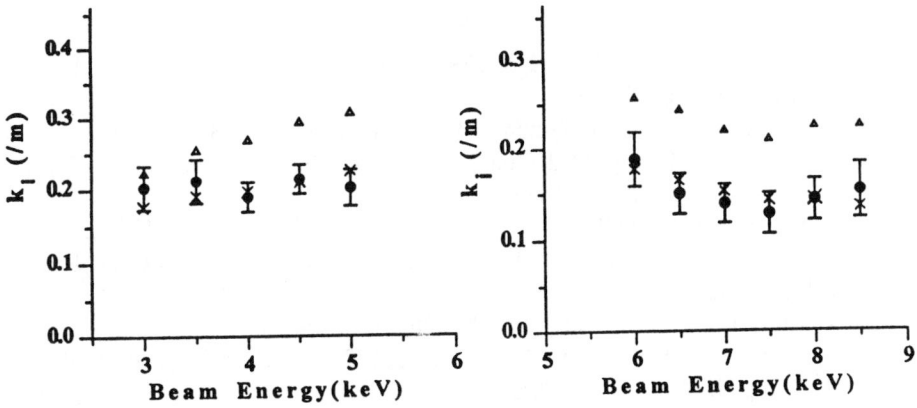

Fig. 13 Growth rate of slow waves.   Fig. 14 Decay rate of fast waves.

V. ON-GOING AND FUTURE WORK

An experimental program to systematically study the longitudinal instability in space-charge dominated electron beams has been conducted. The experiments on space-charge waves have generated single slow and fast waves, have measured the *g*-factor and determined its correct formula for space-charge dominated beams, and have shown the reflection of space-charge waves at bunched beam ends. The resistive-wall instability experiment has demonstrated the growth of slow waves and the decay of fast waves and has measured the growth (decay) rates of the instability for localized perturbations.

The ongoing and future work on the instability includes the measurement of the possible increase of longitudinal energy spread and transverse emittance due to the instability, the study of the effect of capacitive components on the instability, and the attempt of damping the instability by longitudinal energy spread or transverse emittance.

We also have an experimental program to study the longitudinal energy spread of space-charge dominated beams from a thermionic electron gun. The experiment has produced some interest results, which we can not interpret completely. The study will continue and will be reported in the future.

Acknowledgment:

The authors would like to acknowledge the contributions to the programs by Dr. D. X. Wang, now at CEBAF, and H. Suk, and the contributions to the related simulation work by Dr. I. Haber of NRL. This research is supported by the US Department of Energy.

REFERENCES

1. J. T. Seeman, AIP Conf. Proc. No. **253**, 129 (1991).
2. R. A. Jameson, ibid, p. 139.
3. R. O. Bangerter, IL Nuovo Cimento, **106 A** (11), 1445 (1993).
4. M. Reiser, C. R. Chang, D. Kehne, K. Low, T. Shea, H. Rudd, and I. Haber, Phys. Rev. Lett., **61**(26), 2933 (1988).
5. D. Kehne, M. Reiser, and H. Rudd, AIP Conf. Proc. No. **253**, 47 (1991).
6. J. G. Wang, D. X. Wang, and M. Reiser, Nucl. Instr. & Meth. in Phys. Res. A **316**, 112 (1992).
7. D. Neuffer, IEEE Trans. Nucl. Sci. **NS-26**, 3031 (1979).
8. J. G. Wang, D. X. Wang, and M. Reiser, Appl. Phys. Lett., **62**(6), 645 (1993).
9. D. X. Wang, J. G. Wang, D. Kehne, and M. Reiser, Appl. Phys. Lett., **62**(25), 3232 (1993).
10. A. Faltens, E. P. Lee, and S. S. Rosenblum, J. Appl. Phys. 61, 5219 (1987).
11. D. X. Wang, J. G. Wang, D. Kehne, M. Reiser, and I. Haber, Il Nuovo Cimento, **106 A**(11), 1739 (1993).
12. D. X. Wang, J. G. Wang, and M. Reiser, Phys. Rev. Lett., **73**(1), 66 (1994).
13. J. G. Wang, D. X. Wang, and M. Reiser, Phys. Rev. Lett., **71**(12), 1836 (1993).
14. V. K. Neil and A. M. Sessler, Rev. Sci. Instr. **36**(4), 429 (1965).
15. L. Smith, HIFAR-Note-98, 1986 (unpublished).
16. J. G. Wang, H. Suk, D. X. Wang, and M. Reiser, Phys. Rev. Lett., **72**(13), 2029 (1994).
17. E. Lee, in Proc. of the 1981 Linac Conf., Santa Fe, NM, 1981, p.263.
18. P. J. Channell, A. M. Sessler, and J. S. Wurtele, Appl. Phys. Lett. **39**(4), 359 (1981).
19. I. Hofmann, Z. Naturforsch, **37**a, 939 (1982).
20. D. A. Callahan, A. B. Langdon, A. Friedman, and I. Haber, in the Proc. of the 1993 PAC, 730 (1993).
21. J. G. Wang, D. X. Wang, H, Suk, and M. Reiser, Phys. Rev. Lett., **74**(16), 3153 (1995).
22. C. K. Birdsall, G. R. Brewer, and A. V. Haeff, Proc. IRE **41**, 865 (1953).
23. H. Suk, J. G. Wang, and M. Reiser, in the proc. of the 1995 PAC, Dallas, Texas, May 1-5, 1995.
24. J. G. Wang and M. Reiser, Phys. Fluids B **7** (7), 2286 (1993).
25. J. G. Wang, H. Suk, and M. Reiser, to appear in the J. of Fusion Eng. and Design.

Maryland Transport Experiment Comparison to Theory and Simulation

I. Haber,[*] D. A. Callahan,[**] A. Friedman,[**] D. P. Grote,[**]
D. Kehne,[+1] A. B. Langdon,[**] M. Reiser,[+] H. Rudd,[++2]
H. Suk,[+] D. X. Wang,[+1] and J. G. Wang[+]

[*]*Plasma Physics Division, Naval Research Laboratory, Washington DC 20375*
[**]*Lawrence Livermore National Laboratory, Livermore CA 94550*
[+]*Institute for Plasma Research, University of Maryland, College Park MD 20742*
[++]*Berkeley Research Associates, Springfield VA 22150*

Abstract. The Maryland Transport Experiment employs a beam of low energy electrons to study, on a scaled basis, the nonlinear physics of space-charge-dominated transport which is important to a range of accelerator systems. Several series of experiments are discussed, which have investigated significant aspects of both the transverse and longitudinal physics, along with the generally high level of agreement which has been attained in comparisons to theory and simulation.

INTRODUCTION

Over the past several years, the University of Maryland Transport Experiment has been used, in concert with theory and simulation, to explore the fundamental physics of a space-charge-dominated beam. As will be discussed below, because of the scalability of the Maxwell-Vlasov system of equations which describe the dynamics of a particle beam, the results which have been achieved are applicable to a wide range of accelerator systems, particularly those where required luminosities are sufficiently high that the beam is space-charge dominated in at least a part of the accelerator system.

An important characteristic of the experimental program has been the close coordination between experimental measurement and simultaneous investigations using theory and simulation. This has permitted the continuous exploitation and refinement of each of the methods to foster understanding of the detailed nonlinear dynamics of the beam. For example, the experiment is often limited in the detail and accuracy of possible measurements, but can be amenable to parameter searches. Simulations, on the other hand, generally require explicit inclusion of "relevant" physics not known a priori, and numerical techniques are not usually suitable for parameter searches, but offer the possibility of very detailed diagnostics, including arbitrary and sometimes "non-physical" modification of the model for diagnostic purposes. One of the primary

---
[1] Present address CEBAF, Newport News, VA 23606
[2] Present address Science Programs, Computer Sciences Corporation, Lanham, MD 20706

© 1996 American Institute of Physics

conclusions of the series of comparisons between experiment and simulation which have been performed over an extended period, is that the best results have resulted from a continuous process whereby the experiment, as well as the simulation techniques and the theoretical models, were continuously refined. This refinement has been used to increase the degree of agreement between the methods, as well as the level of understanding of the underlying physics. This process of continuous refinement will be illustrated in some of the examples presented below.

In the parameter regime that characterizes many intense beams, the beam emittances are sufficiently low that the space charge forces dominate the beam dynamics. If the longitudinal velocity spread is sufficiently small that the beam can be adequately described non-relativistically in the reference frame moving with the beam, then the full nonlinear time-dependent evolution of the beam will generally depend on only a small number of free parameters. For example, an important parameter which determines the degree to which the transverse space charge affects the beam dynamics, is the ratio of the beam plasma frequency to the betatron oscillation frequency characteristic of the focusing system. The longitudinal dynamics, on the other hand, will depend on λr_e, the line density times the classical electron (or ion) radius. That is, the longitudinal scaling is primarily with the beam line density or current, as opposed to the beam volume density in the transverse direction. These two dimensionless parameters are related by the beam radius, and therefore the geometry of the transport system.

Because of the practical limits on transport system design, most accelerators must operate in a similar range of scaled transverse parameters. For example, The requirement for stable transport of a space-charge-dominated beam[1] limits the maximum phase advance per cell which can be employed in the transport channel, while the need for efficient use of the focusing elements limits the minimum phase advance per cell. Similarly, transport system geometry limits the range of longitudinal scaled parameters because the nonlinearities which accompany large aperture transport limit the beam radius compared with magnet length, and therefore limit the transportable current. As a result of these scaling limits, it is found[2] that, for almost all space-charge-dominated beam systems, the space-charge waves which characterize the longitudinal dynamics, will travel on the order a beam radius per magnet period. Though a detailed discussion of the scaling of space-charge-dominated beam dynamics is beyond the scope of this report, the above points are presented to illustrate why many apparently dissimilar accelerator systems tend to operate in a very similar parameter regime from the standpoint of the space charge dominated physics.

The Maryland Transport Experiment employs a beam of low energy electrons as an inexpensive testbed which exploits the scalability of the space-charge-dominated physics. Several sets of measurements on this apparatus, which will be discussed below, have been employed to explore a significant range of the nonlinear phenomena which can occur in this regime.

TRANSVERSE DYNAMICS

The first series of experiments performed on the Maryland Transport Experiment were concerned with the fundamental transverse dynamics of a space-charge-dominated bunch. For example, one of the earliest questions was simply to determine whether the emittance of the beam after propagation down the transport line, consisting of 36 interrupted solenoid focusing magnets with 0.136 m length period (and two more used for matching), would be close to the intrinsic emittance that resulted from the temperature of the thermionic cathode.

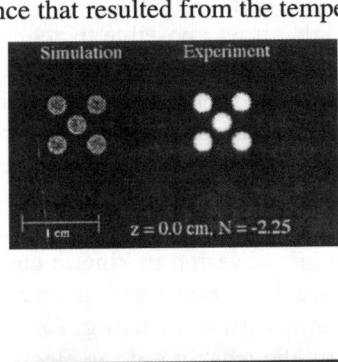
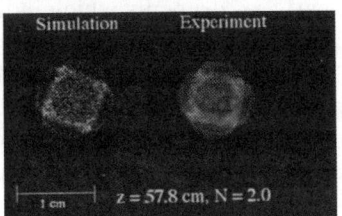
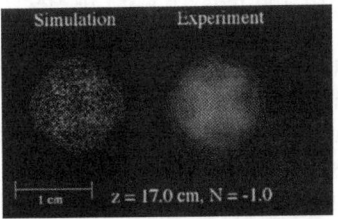
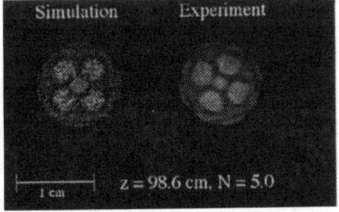
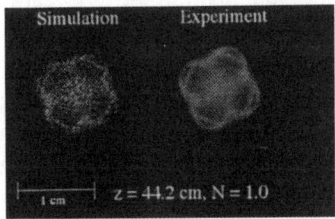
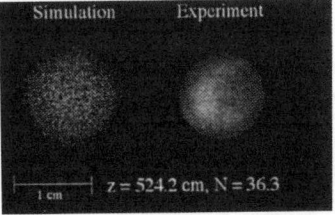

Figure 1. Simulation plots and fluorescent screen pictures of the beam profile at six different positions along the transport channel for the rms matched beam.

Initial measurements of the downstream beam emittances found values which were anomalously high. That is, not only were they somewhat larger than the intrinsic emittance, but were also large when compared to simulations which included a detailed model of the lenses, incorporating the nonlinearities in the focusing fields which had been carefully measured.[3] Simulations which introduced a misalignment comparable to what was expected to result from the tolerances in the experimental structure reproduced an emittance growth which was comparable to what was measured.[4] The gun structure was therefore more carefully aligned, and the downstream emittances then agreed closely to the prediction by the simulations that propagation would result in only a small degradation from the inherent gun emittance. In addition, the gun was mounted so that a small aiming error could be introduced to demonstrate that the previously measured emittance growth did in fact result from misalignment.

The early experiments which indicated that it was possible to achieve near-intrinsic downstream emittances, in agreement with the code predictions, were taken as an indication of the absence of any anomalous mechanisms for emittance growth not contained in the numerical model, and encouraged the assumption that the numerical model was capable of predicting the beam dynamics. Experiments were then conducted to test the theoretical prediction[5,6] that an initially nonuniform transverse current density would relax to a uniform distribution, and the excess potential energy would rapidly be converted to kinetic energy and cause the beam emittance to grow. The interest in merging several beams for use in a Heavy Ion Fusion facility provided motivation for using, for an initial nonuniformity with which to test the theory, the masking of the electron beam into five smaller beamlets. The potential energy in this configuration could be calculated analytically so that a theoretical prediction of the amount of emittance growth could be compared to simulation and experiment.

In addition to the downstream emittance measurement, a fluorescent screen which could be moved along the beam axis, was used to examine the evolution of the beam profile as the beam propagated down the transport system. These fluorescent screen images showed a complex intermingling of the beam images and the surprising reemergence of an image of the original five beam configuration some distance downstream. To examine this experiment numerically, the diagnostics in the simulation code were modified to simulate the fluorescent screen output by binning the simulation particle positions to obtain a local density and then attributing a gray scale to this density.

Figure 1 shows a comparison between the experimental and numerical beam configurations. No free parameters were used to obtain the agreement shown, except that the values of the magnetic fields were renormalized slightly so that the integrated axial field matched the experiment, in the sense that the beam rotation in the simulation matched the experimentally observed rotation. This was done because the analytic form which was used in the simulation to model the spatial variation in the magnetic field, was chosen to minimize the error in modelling the measured field, but was found to have an integrated axial field which did not match the actual lenses. Figure 2 is a plot of the evolution of the rms emittance as the simulated beam propagates down the transport system. Note that, as expected,[7] the rms emittance increases rapidly to its final value. The simulated final value was found to agree with both theory and measurement.

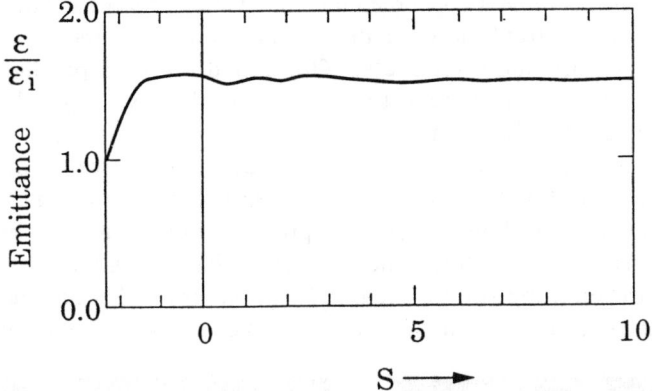

Figure 2. Variation in the rms emittance with propagation distance S, measured in lens periods, for the matched five beamlet configuration. The curve starts at the mask location with the location of S = 0 corresponding to the start of the periodic transport section. The region for S < 0 corresponds to the matching lenses between the mask and the transport line.

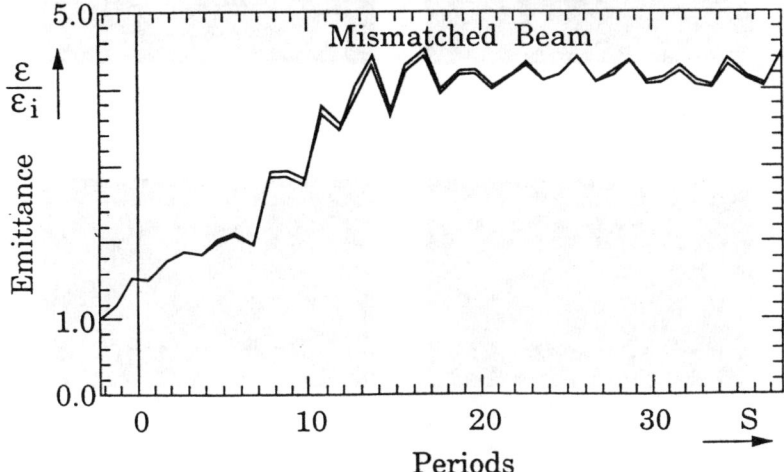

Figure 3. Variation in the rms emittance with propagation distance S, measured in lens periods, for the mismatched five beamlet configuration. The rapid growth in rms emittance that results from the homogenization of the beam cross section is followed by a slower growth resulting from the conversion of the mismatch energy into emittance. Note that only one point per period is plotted in this figure, and these are connected by straight lines.

A detailed description of the comparisons between the experimental and numerical results has been presented elsewhere,[8-10] and is outside the scope of this presentation. It is worth mentioning that the agreement between the fluorescent screen pictures and the numerically obtained density plots was actually better than can be presented here, because of the loss of some quality in the reproduction of the pictures. However, one interesting discrepancy between simulation and measurement concerned the observation that the beam is initially hollow, so that the central beamlet as the beam emerges form the mask has 20% less current than the outer beamlets. This could not be reproduced in the simulations, in the sense that starting the simulations with a hollow beam noticeably worsened agreement with the experimental data. This is taken as evidence that initial correlations in the beam distribution were not properly included in the initial distribution function used in the simulations. A proper description of the beam should, therefore, probably include a "first-principles" numerical description which traces the beam evolution through the gun structure starting from the emitting surface, in a manner similar to what has successfully been used in the ion source simulations and described elsewhere in this volume.[11]

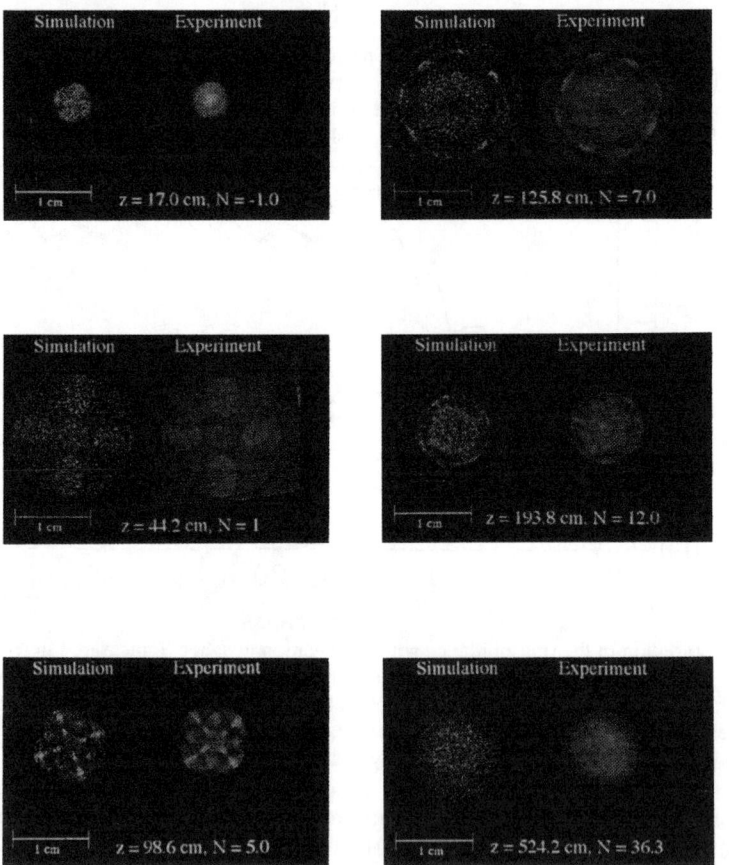

Figure 4. Simulation plots and fluorescent screen pictures of the beam profile at six different positions along the transport channel for the mismatched beam.

The development of a theoretical picture which predicts the evolution of emittance in an initially mismatched beam by assuming that the potential energy in the mismatch would be converted to beam kinetic energy,[12] was motivation for adjusting the matching lenses so that the five beamlet configuration was now mismatched to the downstream transmission line.[13] Simulations of this configuration verified that the initial mismatch energy would be converted to rms emittance on the scale of the experimental apparatus. This behavior can be seen from the simulated behavior of an initially mismatched five beamlet beam plotted in Fig. 3, which shows the evolution of the rms emittance. On a short time scale, just as in the matched case, the energy in the inhomogeneity causes a rapid rise in the rms emittance as it relaxes toward a uniform cross section. However, the emittance continues to grow, conforming to the prediction of the theory that the mismatch energy will be converted into kinetic energy.

An initial puzzle in the experiment was that the growth in rms emittance, beyond the component due to redistribution of the beam current cross section, which was predicted by the theory and seen in the simulation, was not observed, experimentally. Yet, comparison of the beam fluorescent screen images to those generated by the computer simulations, and shown in Fig. 4, were in excellent agreement. Upon further examination, this was found to occur because the primary emittance growth due to mismatch resulted from the acceleration of a relatively small number of beam particles into large amplitude betatron oscillations, so as to form a halo outside the main body of the beam distribution. Because of the low signal level compared to the background noise in the beam halo, the halo particles were not included in the value of the rms emittance calculated from the experimental data.

An important observation that emerged from this series of experiments is the contrast between the matched and mismatched behavior. Despite the highly nonlinear fields which are present during the beam relaxation from the five beamlet configuration, evidence of halo formation was not found unless the beam was mismatched. This can be seen by contrasting the pictures in Figs. 1 and 4, and can also be inferred from the agreement between measured and simulated rms emittances in the matched case, in contrast with the lack of agreement in the mismatched case where the portion of the distribution which contains only a small number of particles is excluded. It should also be noted that the halo particles are not easily visible in the experimental pictures presented here because of some loss of quality in the reproduction, but could be clearly observed on the original screen.

LONGITUDINAL DYNAMICS

In view of the successful use of the Maryland experimental apparatus to examine the transverse physics of a space-charge-dominated beam, the equipment was modified to conduct a series of experiments on the longitudinal physics. The axisymmetric solenoidal focusing system in the Maryland apparatus also is advantageous. Axisymmetry significantly facilitates simulation of the experiment by allowing the use, in the present case, of the r,z capability in the WARP[14] code. Also, as discussed below, it is also sometimes possible to accurately describe the beam dynamics in the experiment using a suitable one-dimensional model.

One of the primary modifications to the experimental apparatus which was made to facilitate the study of longitudinal bunch dynamics is the use of a gridded gun to modulate the beam current. In particular, this has allowed examination of flat and parabolic bunch profiles, as well as the controlled initiation of separate forward and backward travelling waves by initiating a a small perturbation to the beam. However, as will be discussed below, the lack of direct time-resolved measurement of either the transverse dynamics or the initial beam velocity distribution has complicated comparisons to the simulations.

In general, because the grid is grounded, the imposition of a grid to cathode voltage to perturb the beam, affects both the beam current and the beam energy. In view of the difficulty in accurately predicting the gun response to an initial perturbation, it is therefore difficult to infer the initial conditions from a directly measurable quantity. The simulations have therefore been used to infer the nature of an initial distribution which can explain the subsequent beam behavior. In some cases, particularly when examining the gross bunch dynamics, where the longitudinal behavior was found to be relatively insensitive to the details of the initial distribution this process has been largely successful. In other cases, such as the examination of the detailed evolution of an initial perturbation near the bunch end, comparison to simulation requires an extensive parameter search and this is an ongoing process.

Gross Bunch Dynamics

Because the bunch in the Maryland apparatus is highly space-charge dominated, the contribution to longitudinal beam expansion which results from the thermal spread in longitudinal velocities can be neglected when compared to the space-charge contribution. The gross longitudinal bunch dynamics therefore are largely independent of the details of the initial longitudinal velocity distribution. Since knowledge of the details of the initial longitudinal distribution is difficult to obtain experimentally, this independence provides a convenient parameter regime where experiment simulation and theory can be compared. Measurements which have been conducted of the gross longitudinal dynamics of the bunch have, in fact, been in generally good agreement with theory and simulation. Because of the inherently nonlinear nature of the longitudinal dynamics of a long bunch under most conditions, and the difficulty of performing analytical calculations in this nonlinear regime, the opportunities for comparisons to theory are limited. With a view towards maximizing the comparisons to theory, the experiments have therefore emphasized two cases. The first was a bunch with a line density that varies parabolically in z, the distance from the center of the bunch, so that, to the extent that the self-electric field can be described by the derivative of the line density, the self forces are linear and the beam expansion is self-similar and can be described by a longitudinal envelope equation.[15] The nature of the actual self-force, and how it compares to this assumption of proportionality to the derivative of the line density, will be discussed below. The second case is an initially flat-topped bunch with the current rapidly going to zero at the bunch ends and which can be described using the method of characteristics.[16]

Parabolic Bunch

An example of the data from the series of experiments examining the free expansion of a bunch with an initially parabolic variation in the line density is shown in Fig. 5, which shows the evolution of the current waveform as the bunch propagates down the transport channel. Comparisons of the expansion dynamics to the envelope solutions, and to one dimensional SHIFT-Z and two dimensional r,z WARP simulations, have been reported elsewhere,[17-20] and have generally exhibited good agreement. In addition, the process of comparing simulations to experiment has spurred some results of general applicability to the dynamics of a long bunch which will be mentioned here.

Because of the difficulty in obtaining time resolved measurements of the transverse beam characteristics it is difficult to obtain data on the degree that the beam is transversely matched at different positions along the bunch. A series of simulations was, therefore, performed to examine the sensitivity of the longitudinal dynamics to the degree of initial match. These simulations revealed, and this was verified experimentally by varying the matching lens settings, that the longitudinal beam expansion is surprisingly insensitive to the initial beam match. For example, if the beam in an r,z simulation is mismatched by initializing the bunch with a radius which is half its matched value, the rms bunch length after 6 meters of propagation, expands to 1.763 times its initial value compared to 1.757 times its initial value in the matched case, a difference of 0.3%. Similar insensitivity to initial match was also observed by simulating a beam which was matched only at the beam center but increasingly mismatched with distance away from the center. A diagnostic in the simulation which measured the ratio of the expansion rate of the mismatched bunch to the matched bunch showed that the observed insensitivity in longitudinal expansion results from the difference in time scales between the longitudinal and transverse motion. Though the two directions are coupled in the sense that the longitudinal self-force depends on the instantaneous variation in the beam radius, this effect averages out over the many transverse oscillations which occur before the beam has changed significantly in the longitudinal direction.

It should be mentioned however, that some details of the expansion of the beam do depend on the beam transverse match. A substantial mismatch can transversely accelerate a small fraction of the beam population into a halo outside the main body of the distribution function,[21,22] and this is, in fact, observed in the simulations where the beam is mismatched. Since the halo particles are radially outside the main body of the beam, they experience a reduced longitudinal force and are "left behind" as the beam expands. This effect can be observed in the longitudinal phase space plots of the mismatched expanding bunch.

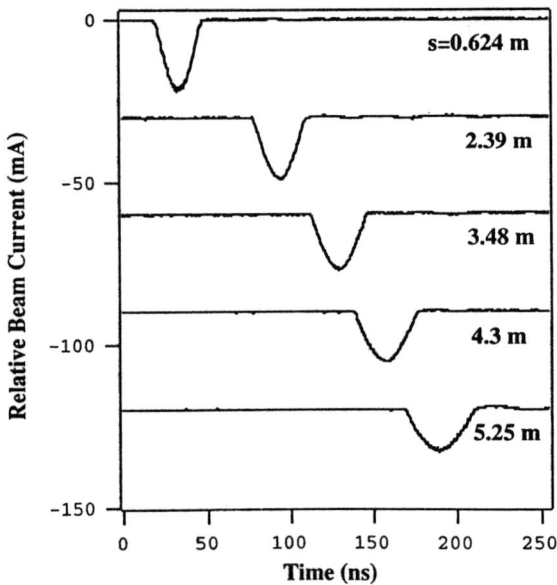

Figure 5. Current profiles of a longitudinally expanding, 2.5 KeV, 28 mA, parabolic bunch, measured at five different locations down the transport channel.

The insensitivity of the gross characteristics of the longitudinal expansion to transverse mismatch suggests that, at least over a fairly large parameter range, such gross longitudinal dynamics in a space- charge-dominated bunch can be adequately modelled by a one-dimensional description. This conjecture was reinforced by experimental observations[23] which verified that the propagation of waves on a space-charge-dominated bunch behaves as predicted by a one-dimensional description. This description, which assumes that the variation in the longitudinal self-electric field occurs on a scale long compared with the beam diameter, and the field can be described by the relation,

$$E_z = \frac{1}{4\pi\epsilon_0} g \frac{\partial \lambda}{\partial z} ,\qquad (1)$$

where $g(z) = 2 \ln(b/a)$, and ϵ_0 is the permitivity of free space. Note that the factor g in this expression is different from the expression $g = 1 + 2 \ln(b/a)$ which is obtained when the beam is assumed emittance dominated or embedded in a strong magnetic field[24] so that the particles can not move transversely. The line integral argument presented to derive this relationship[23] does not assume a small perturbation. Equation (1) can, therefore, be used to describe the gross bunch dynamics, in addition to providing an accurate prediction of wave phenomena. Furthermore, an implication of this method for calculating the self-field is that the longitudinal self-electric field within a slowly varying space-charge-dominated bunch is independent of the transverse position within the beam.

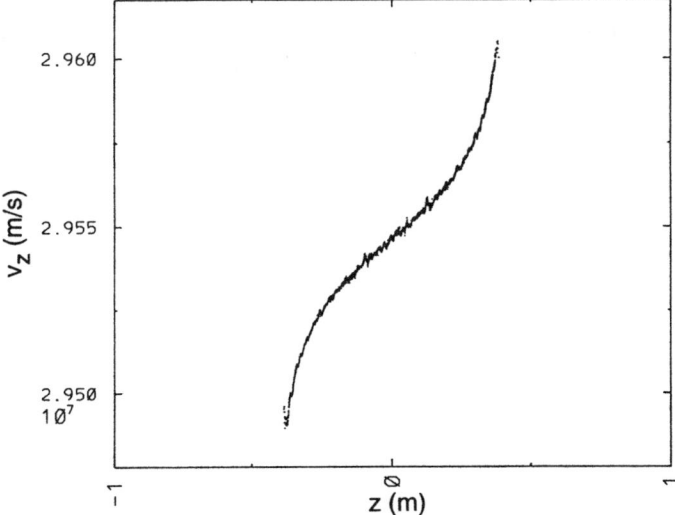

Figure 6. Plot of the "near-initial" z-v_z phase space from an r,z WARP simulation of a parabolic bunch after the beam has propagated only 60 mm. Because the beam density has not had time to evolve significantly from its initial state, the particle velocities are a good measure of the initial self-electric fields. The small scale fluctuations appear to result from imperfections in the initial load, in addition to grid-scale errors in the field calculation.

The degree of equivalence between the one-dimensional model and a full r,z description can be seen by comparing the longitudinal phase space plots in Figs. 6 and 7. Figure 6 is a "near initial" plot of the particle phase space taken from an r,z

WARP simulation of a parabolic bunch, and showing the beam after 60 mm of propagation. Such a phase space plot is a good "diagnostic" of the beam self field since the longitudinal density distribution has not yet had enough time to substantially evolve. The particle velocities in the plot are therefore a measure of the strength of the initial self-fields. The bunch is initially cold longitudinally, so that if all the particles at a particular z were to experience the same longitudinal force, the particles would fall on a curved line with the slight "s" shape. The deviation from the straight line which would result from a derivative force law, arises from the variation in the g factor, as defined in Eq. (1). The g factor changes as the beam radius varies with current in order to remain in transverse equilibrium with the uniform solenoidal focusing. The short scale deviations from the curved line appear to be due to numerical imperfections in loading the particles in r,z relative to the numerical grid which is employed in the PIC method in addition to errors in calculating the electric field. Figure 7, on the other hand, is a simulation of the same bunch using the SHIFT-Z longitudinal simulation code. The calculation of the self-field in this code is done using the relationship in Eq. (1) where the factor g is varied along the bunch assuming that the local beam radius is matched to the constant focusing force, assuming a constant volume density in the bunch. Note that there are no free parameters in the comparison of the two models, and the agreement between the two methods has been found to be quite good over a range of parameters.

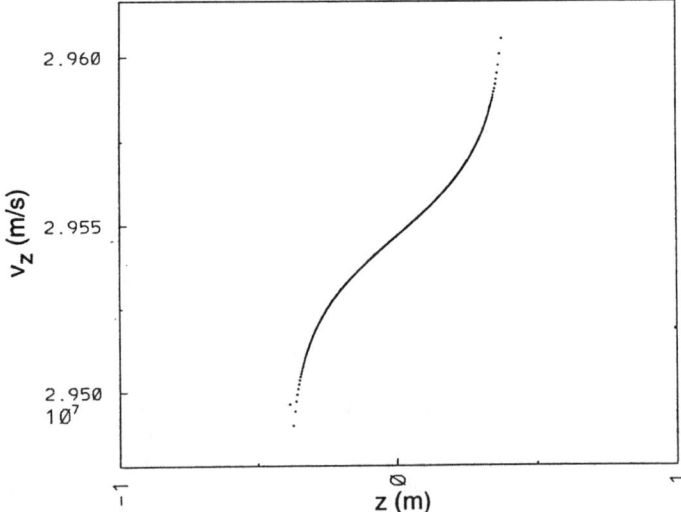

Figure 7. Plot of the z-v_z phase space for the same conditions as Fig. 6, but from a SHIFT-Z one-dimensional simulation where the local electric field is calculated using a derivative force law corrected for the variation in g along the bunch.

An interesting characteristic of the dynamics of the expansion of a parabolic bunch which was observed, is the linear shape to which the phase space evolves as the beam expands. This can be seen in Fig. 8, which is [B phase space plot of the same initially cold parabolic bunch shown in Fig. 6, but after the beam has propagated 6 m. The expanding bunch, despite the nonlinearities in the self force that result from the variation of g along the bunch, evolves to a self-consistent distribution which is characterized by linear self-forces and therefore a linear phase space.

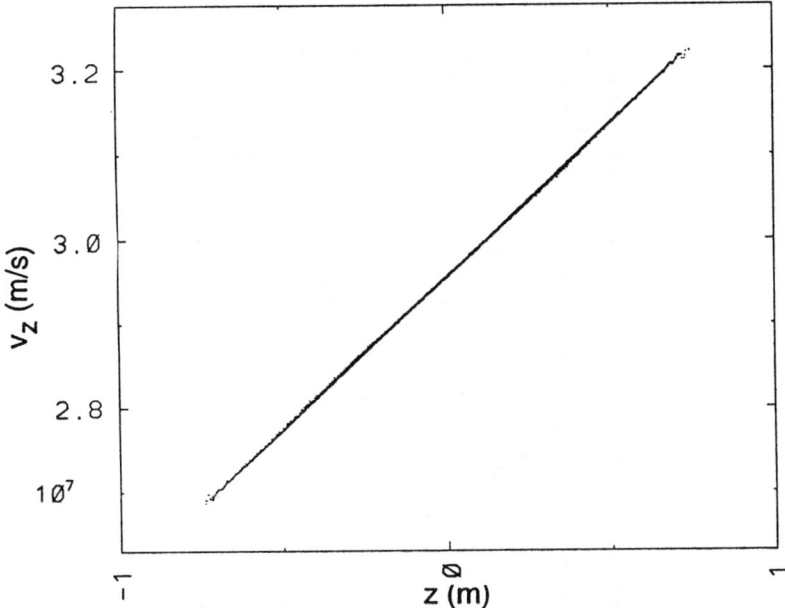

Figure 8. Plot of the z-v_z phase space of the beam, as shown in Fig. 6, after propagating 6 m. The beam profile has self-consistently evolved to a current density variation where the self-electric fields have become linear.

Flat Top Bunch

In addition to the investigation of the gross bunch dynamics of an initially parabolic bunch, experiments have also been conducted to examine the behavior of a flat-top current waveform with a rapid fall off in the current at the bunch ends. Such a waveform has the virtue that many features of the dynamics can be predicted using the method of characteristics.[16] Details of the comparison between experiment, and theory and simulation, have been presented elsewhere[20,25] and are beyond the scope of this report. However, a typical example which is illustrative of the general nature of the comparisons that have been performed is presented in Fig. 9, which is a plot of the current distribution, at several positions down the transport channel, of an initially flat-top bunch, allowed to expand in the longitudinal direction. The predictions of theory and simulation are overlayed for comparison.

Only a small sample of the extensive comparisons between experiment, theory and simulation have been shown here. In the flat top bunch, as in the previously discussed work on the initial parabolic bunch, the level of agreement which has been obtained has been sufficient to lend support to future use of both the theoretical and numerical tools for predicting the longitudinal dynamics of a space-charge-dominated bunch.

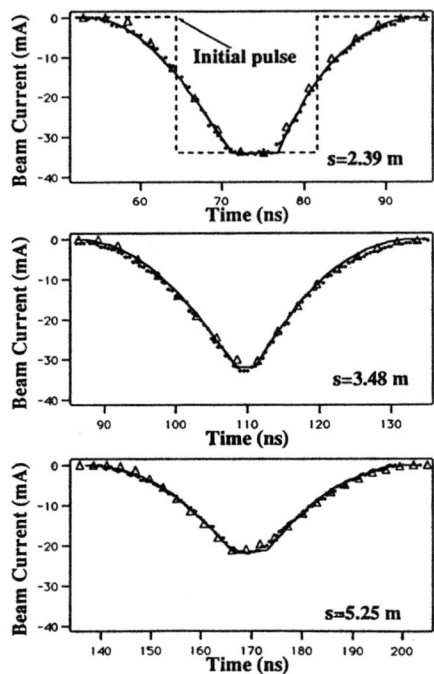

Figure 9. Measured current profiles at different locations in the transport channel of a longitudinally expanding, 2.5 KeV, 34 mA, rectangular bunch. Also plotted for comparison are predictions of the method of characteristics, as well as one-dimensional simulation.

Longitudinal Wave Phenomena

The longitudinal dynamics of a space-charge-dominated beam which is long compared with the beam radius, has important characteristics which are different from the transverse behavior. One of those differences, which can easily be derived[2] from the expression for the electric field in Eq. (1), is the propagation of space charge waves along the bunch. The excitation and propagation of these waves on a space-charge-dominated bunch can have significant implications to the beam dynamics. For example, details of the wave physics can be important to understanding the stability of the bunch in the presence of a longitudinal wall impedance, since the characteristics of space charge wave propagation can significantly alter bunch stability criteria.[26,27]

In the many cases where it is important to limit the longitudinal energy spread, it is also important to understand the dynamics of longitudinal waves, since these waves can be launched by any imperfection in the longitudinal waveform used to accelerate and contain the bunch. Furthermore, since space-charge dominance in the longitudinal direction implies that the longitudinal thermal velocity is less than the wave propagation velocity in the beam frame, these waves are not generally Landau damped and, once excited, can persist for a long time.

Several series of experiments were therefore conducted on the Maryland apparatus to study the characteristics of wave propagation on the flat top bunch. However, because investigation of the linear physics of wave propagation involves launching a small perturbation on the bunch, signal-to-noise ratios are generally less favorable than those obtained in studying the gross bunch dynamics. Obtaining data with a sufficient signal level to distinguish the desired effects from fluctuations in the beam characteristics as well as spurious noise can sometimes be difficult. These signal-to-noise considerations have been particularly significant in limiting the explicit measurement of the velocity distribution associated with the space charge waves.

In a bunch which is dominated by space charge, the gross bunch dynamics will usually be insensitive to the details of the initial beam velocity distribution. However, wave eigenfunctions are characterized by a perturbation in both the line density and wave velocity, so that ratio of the excitation of the slow and fast waves on the bunch depends on the velocity component of the initial excitation. Because of the difficulty in accurately measuring the initial velocity component, details of the initial conditions must often be inferred, either from a downstream measurement of the velocity distribution, or alternatively from the characteristics of the actual evolution of the slow and fast waves as they separate after the bunch has propagated some distance. Though ambiguity in the measured initial conditions has, in some cases, complicated comparison between simulation and experiment, and more detailed agreement is being sought in the ongoing program, substantial progress has been made in using the experiment in concert with theory to study the stability and dynamics of small perturbations on the bunch.

One experiment of a fundamental nature was the controlled launching of space-charge waves on a flat top bunch.[28] This experiment not only demonstrated that fast and slow waves could be launched, in verification of the theory, but also that the wave propagation velocity behaved in agreement[23] with the longitudinal force law given in Eq. (1). The modification of longitudinal dynamics that occurs as a result of the transverse space charge dominance was therefore demonstrated experimentally.

A subsidiary experiment, conducted because of the importance of bunch end reflection to beam stability,[26] involved the launching of the perturbation near the bunch end to observe the reflection of the wave off the bunch end.[29] This experiment, however, has not yet been verified by simulation because of the ambiguity, mentioned above, as to the detailed initial conditions appropriate to reproduction of the experimental data. Because of the large parameter space which must be accessed, this work is still in progress.

In the space-charge-dominated regime, the response of the beam to an external impedance can be strongly influenced by the excitation of waves on the bunch. A series of experiments, which are still underway, has been concerned with examining the characteristics of these unstable waves in a resistive beam pipe which impedes the flow of the beam return current. Figure 10 illustrates the degree to which agreement has been obtained between the measured decay of the forward-travelling fast wave, and the theoretically calculated decay rate, corrected for the dispersion in wave velocities and growth rates as the perturbation length becomes comparable to the beam pipe diameter.[30] Similar agreement has also been obtained comparing theory and experiment for the growth of the backward-travelling slow wave. In addition, further evidence of agreement between theory and experiment has been obtained by comparing the dependence of growth and decay rates on parameters such as beam current and pulse length.

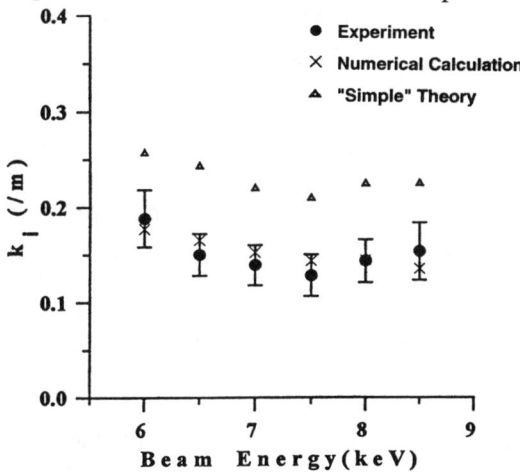

Figure 10. Comparison of the experimentally measured decay rate of a forward traveling perturbation, after propagation through a section of resistive beam pipe, compared to simple long-wavelength theory and a more complete numerical calculation which includes dispersion.

However, as was the case in some of the experiments described above, the lack of detailed knowledge of the initial conditions makes detailed comparisons with first-principles simulation difficult. Because of the short pipe length, the fast and slow waves do not have time to separate during the time they traverse the 1 m length of resistive beam pipe. Sensitivity studies using the r,z WARP code have, in fact, shown that the ratio of the fast and slow waves initially excited can strongly affect the growth or decay of a perturbation for the parameters present in the experiment. Further work is therefore needed to obtain simulations which agree with the experiment on some of the finer details of the data which has been obtained, such as the modification to the beam pulse shape by the resistive pipe.

ONGOING AND FUTURE WORK

The program of experimentation, as well as the comparisons of experimental results to theory and simulations is an ongoing one. Much of the close agreement between experiment and both theory and simulation has, in fact, resulted from the ongoing nature of this process, which has allowed the continued refinement of both experimental and simulation methods, as well as the development of the appropriate theory. Therefore, this process of refinement is being continued with a view to obtaining a continually better picture of the fundamental physics of a space-charge-dominated beam.

Some refinements in the experimental apparatus are in progress or planned which will permit removal of some of the ambiguities which have complicated a definitive comparison to theory. For example, plans to lengthen the region of resistive pipe will permit greater growth and decay of the slow and fast waves respectively, and therefore simplify quantitative measurement of the appropriate rates. This will permit, for example, a better estimate of the reactive component of the impedance seen by the beam. The longer resistive region will also allow the two waves to begin to separate so as to help remove any ambiguity as to the initial conditions. Similarly, simulations which more closely model the range of initial conditions,. and also a range of complex wall impedances, will also aid in refining the quest for accurate agreement.

Another set of experiments currently underway[31], which have not yet been mentioned here, have concentrated on accurate measurements of the longitudinal velocity distribution. These measurements should permit experimental investigation of any possible influence on the evolution of the longitudinal distribution of the collective electrostatic isotropization instability.[32-34]

Finally a major and exciting increase in experimental capabilities should result from construction of a circular transport system, which is currently in process. This will permit examination of the beam evolution over much longer timescales than now possible. The longer timescales will permit new classes of experiments, such as determination of the conditions required for the establishment of steady state longitudinal containment of a long bunch. In addition, it will become possible to examine the longitudinal impedance instability in a more tractable regime where the waves grow slowly during propagation. It should also generally be possible, because of the larger parameter range available on an experiment with a longer beam lifetime, to relax some of the ambiguities in measurement which limit the accuracy of comparisons with theory and simulation.

ACKNOWLEDGMENT

The authors acknowledge support for the research reported here by the United States Department of Energy at the Naval Research Laboratory under contracts DE-AI02-93ER40700 and DE-AI02-94ER54232, at the Lawrence Livermore National Laboratory under contract W-7405-ENG-48 and at the University of Maryland under contract DE-FG02-94ER40855.

REFERENCES

1. I. Hofmann, L. J. Laslett, L. Smith and I. Haber, "Stability of the Kapchinskij-Vladimirskij (K-V) Distribution in Long Periodic Transport Systems", *Particle Accelerators*, **13**, 145 (1983).
2. J. Bisognano, I. Haber, L. Smith, A. Sternlieb, "Non-linear and Dispersive Effects in the Propagation and Growth of Longitudinal Waves on a Coasting Beam," *IEEE Tran. Nucl. Sci.*, **NS28**, 2513 (1981).
3. P. Loschialpo, W. Namkung, M. Reiser, J. D. Lawson, "Effects of Space Charge and Lens Aberrations in the Focusing of an Electron Beam by a Solenoid Lens", *J. Appl. Phys.* **57**, 10 (Jan. 1985).
4. H. Rudd, I. Haber, C. R. Chang, D. Kehne, K. Low, M. Reiser, T. Shea, "Comparison of Simulation, Theory and Measurement for the Maryland Transport Experiment," *Nucl. Instr. and Methods in Phys. Res.* **A278** 198 (1989).
5. J. Struckmeier, J. Klabunde and M. Reiser, "On the Stability and Emittance Growth of Different Phase-Space Distributions in a Long Magnetic Quadrupole Channel", *Particle Accelerators*, **15**, 1344 (1984).
6. T. P. Wangler, K. R. Crandall, R. S. Mills, and M. Reiser, "Relation between Field Energy and RMS Emittance in Intense Particle Beams," *IEEE Trans. Nucl. Sci.* **NS32**, 2196 (1985).
7. O. A. Anderson, "Some Mechanisms and Time Scales for Emittance Growth," *Heavy Ion Inertial Fusion*, AIP Conf. Proc. **152**, ed. Martin Reiser, Terry Godlove & Roger Bangerter, p. 253, (AIP, New York, 1986).
8. M. Reiser, C. R. Chang, D. Kehne, K. Low, T. Shea, H. Rudd, and I. Haber, "Emittance Growth and Image Formation in a Nonuniform Space-Charge-Dominated Electron Beam," *Phys Rev. Lett.*, **61**, 2933 (Dec. 26, 1988).
9. I. Haber, D. Kehne, M. Reiser, and H. Rudd, "Experimental, Theoretical and Numerical Investigation of the Homogenization of Density Nonuniformities in the Periodic Transport of a Space-Charge Dominated Beam," *Phys. Rev.* **A15**, 44, 5194 (Oct. 15, 1991).
10. Martin Reiser, *Theory and Design of Charged Particle Beams* (Wiley, New York, 1994).
11. S. Yu, "Comparison of Experiment to Theory - LBL Experiment," in proceedings of this conference.
12. M. Reiser, "Free Energy and Emittance Growth in Nonstationary Charged Particle Beams," *J. Appl. Phys.* **70** 1919 (1991).
13. D. Kehne, M. Reiser and H. Rudd, "Experimental Studies of Emittance Growth in a Nonuniform, Mismatched, and Misaligned Space-Charge Dominated Beam in a Solenoidal Channel," *High-Brightness Beams for Advanced Accelerator Applications*, A.I.P Conf. Proc. 47, ed. W. W. Destler and S. K. Guharay, (A.I.P. New York, 1992)
14. Debra A. Callahan, A. Bruce Langdon, Alex Friedman and Irving Haber, "Longitudinal Beam Dynamics for Heavy Ion Fusion," *Proc. 1993 Particle Accel. Conf.*, 730 (IEEE, 1993).
15. David Neuffer, "Longitudinal Motion in High Current Ion Beams - A Self-Consistent Phase Space Distribution with an Envelope Equation," *IEEE Tran. Nucl. Sci.* **NS26** 3, 3031 (June, 1979).

16. D. D.-M. Ho, S. T. Brandon and E. P. Lee, "Longitudinal Compression of Heavy-Ion Beams," *Particle Accelerators,* **35** 15 (1991).
17. J. G. Wang, D. X. Wang, and M. Reiser, "Longitudinal Expansion of Space-Charge Dominated Drifting Beam Bunches with a Parabolic Line Charge Density Distribution," *Appl. Phys. Lett.,* **62**, 6, 645 (Feb. 8, 1993).
18. I. Haber, D. A. Callahan, A. B. Langdon, M. Reiser, D. X. Wang, and J. G. Wang, "Computer Simulation of the Maryland Transport Experiment," *Proc. of the 1993 Particle Accel. Conf.,* 3660, IEEE Catalog No. 93CH327779-7 (IEEE, NJ 1993).
19. D. X. Wang, J. G. Wang, D. Kehne, M. Reiser, I. Haber, "Experimental Studies of Longitudinal Beam Dynamics of Space-Charge Dominated Beams," *Proc. of the 1993 Particle Accel. Conf.,* 3627, IEEE Catalog No. 93CH327779-7 (IEEE, NJ 1993).
20. D. X. Wang, "Experimental Studies of Longitudinal Dynamics of Space-Charge Dominated Electron Beams," to be published in *Proc. of the 1995 Particle Accel. Conf.*.
21. J. S. O'Connell, T. P. Wangler, R. S. Mills, K. R. Crandall, "Beam Halo Formation from Space-Charge Dominated Beams in in Uniform Focusing Channels," *Proc. of the 1993 Particle Accel. Conf.,* 3657 IEEE Catalog No. 93CH327779-7 (IEEE, NJ 1993).
22. Robert A. Jameson, "Design for Low Beam Loss in Accelerators for Intense Neutron Source Applications," *Proc. of the 1993 Particle Accel. Conf.,* 3926, IEEE Catalog No. 93CH327779-7 (IEEE, NJ 1993).
23. J. G. Wang, H. Suk, D. X. Wang, and M. Reiser, "Determination of the Geometry Factor for Longitudinal Perturbations in Space-Charge Dominated Beams," *Phys. Rev. Lett.,* **72**, 13, 2029 (1994).
24. V. K. Neil and A. M. Sessler, "Longitudinal Resistive Instabilities of Intense Coasting Beams in Particle Accelerators', *Rev. Sci. Instrum.* **36** 429 (1965).
25. D. X. Wang, J. G. Wang, M. Reiser, I. Haber, "Longitudinal Expansion and Compression of Electron Bunches with Rectangular Line Charge Profiles," *Il Nuovo Cimento,*" **106 A**, 1739 (Nov. 1993).
26. J. Bisognano, I. Haber and L. Smith, "Threshold Behavior for Longitudinal Stability of Induction Linac Bunches", *IEEE Trans. Nucl. Sci.* **NS30**, 2501 (Aug. 1983).
27. J. G. Wang and M. Reiser, "Longitudinal Instability of Space-Charge Dominated Beams in Transport Channels with Complex Wall Impedances," *Phys. Fluids B*, **5** 2286 (1993).
28. J. G. Wang, D. X. Wang, and M. Reiser, "Generation of Space-Charge Waves Due to Localized Perturbations in a Space-Charge Dominated Beam," *Phys. Rev. Lett.,* **71** 12 1836 (1993).
29. J. G. Wang, D. X. Wang, H. Suk, and M. Reiser, "Reflection and Transmission of Space-Charge Waves at Bunched Beam Ends," *Phys. Rev. Lett.,* **74,** 16, 3153 (1995).
30. J. G. Wang, H. Suk, and M. Reiser, "Experiments on Space-Charge Waves and Longitudinal Instabilites in Space-Charge Dominated Beams," to be published in *Journal of Fusion Engineering Design*.
31. N. Brown, M. Reiser, D. Kehne, D. X. Wang, J. G. Wang, "Longitudinal Kinetic Energy Spread from Focusing in Charged Particle Beam," *Proc. of the 1993 Particle Accel. Conf.,* 62, IEEE Catalog No. 93CH327779-7 (IEEE, NJ 1993).
32. "Transverse-Longitudinal Energy Equilibration in a Long Uniform Beam," I. Haber, D. A. Callahan, A. Friedman, and A. B. Langdon, to be published in the *Proc. of the 1995 Particle Accel. Conf.*
33. I. Haber, D. A. Callahan, A. Friedman, D. P. Grote, and A. B. Langdon, "Transverse-Longitudinal Temperature Equilibration in a Long Uniform Beam," to be published in *Journal of Fusion Engineering Design*.
34. I. Haber, D. A. Callahan, C. M. Celata, W. M. Fawley, A. Friedman, D. P. Grote, and A. B. Langdon, "PIC Simulation of Short Scale-Length Phenomena," in proceedings of this conference.

Critical design issues of LEBTs for high-brightness ion beam transport and focusing

S. K. Guharay

Institute for Plasma Research
University of Maryland
College Park, Maryland 20742

I. Introduction

In the Workshop on Future Hadron Facilities in the U.S.[1], the injector working group studied the linac section of high energy colliders. It was concluded that further improvements, especially in the low energy section, are required to satisfy the stringent emittance budget of next-generation high luminosity colliders. Similar conclusions had emerged earlier in the context of developing Superconducting Super Collider (SSC).[2] The relevance of this problem extends to many modern applications of advanced accelerators, namely, radioactive waste transmutation, spallation neutron source, fusion materials irradiation tests, heavy ion fusion, etc. Although the state-of-the art of the individual components of a low energy section, namely, ion source, low energy beam transport (LEBT) and radio-frequency quadrupole accelerator (RFQ), has advanced significantly, an efficient handshake among the different units has not been demonstrated experimentally, especially in the context of transporting intense, high-brightness beams. The matching problem in the ion source-LEBT-RFQ section warrants detailed critical investigation. The crux of the problem can be translated to two basic questions:
(1) What are the input conditions for LEBT?
This is primarily determined by the beam from an ion source, namely, beam current, beam voltage, beam emittance, geometry and divergence.
H^- beams are commonly used in modern accelerators[3,5] Intense, space-charge dominated H^- beams are highly diverging (envelope divergence $\geq$ 50 mrad near the extraction region). Therefore, the LEBT has to deal with large, diverging beams.
(2) What are the matching conditions at the RFQ input (namely, the Twiss parameters)?
This defines the desired output parameters of the LEBT. A tight acceptance ellipse, especially in the case of high-frequency RFQs, demands small, highly converging beams. Typical parameters correspond to a beam radius of about 1.2 to 1.5 mm and beam convergence at maximum envelope radius of about -70 mrad.

The above constraints, along with the requirements of emittance preservation in high-brightness accelerators as well as good differential pumping between the ion source and the RFQ, set the basic design problems for LEBTs.

© 1996 American Institute of Physics

In this article, some general discussions of different LEBT schemes are given, and the critical beam dynamics issues are highlighted with an illustrative example.

II. General discussions on LEBT

The underlying principles of lens systems for focusing of charged particle beams are dealt in several text books.[6,7] The problem of beam transport was highlighted in an earlier article.[8]

A LEBT system, in general, can be developed using the following two basic schemes:
(a) magnetic lenses (assisted by gas neutralization), namely, solenoid or permanent magnetic quadrupole (PMQ) lenses,
(b) electrostatic lenses, namely, einzel lens, electrostatic quadrupole lens (ESQ), helical electrostatic quadrupole lens (HESQ), and combination of ESQ and einzel lenses.
Important characteristics of these focusing schemes are summarized below.

The strong points of the magnetic focusing scheme are: (i) lot of operational experience[9,10] exists especially with high-perveance beams, and (ii) it is attractive for handling beams at high energy. The main weak point of the magnetic focusing scheme is that strong magnetic fields are required in the case of low-velocity beams. This demand is often overcome by adding gas neutralization in a low-energy magnetic beam transport line. However, the physics problem becomes complex in this situation because many issues, namely, beam pulse length, neutralization time, beam loss during space-charge neutralization, beam-plasma instabilities, emittance blow-up due to space-charge decompensation near RFQs, etc., need to be carefully handled.

The major attractive features of the electric focusing scheme are: (i) it is very effective at low energy, (ii) it offers clean differential pumping between an ion source and an RFQ due to operation at very low base pressures, and (iii) guidance in experiments can be obtained through reliable simulations. The electric focusing scheme suffers primarily from spherical and chromatic aberrations. However, as the lens parameters can be defined very accurately, reliable simulation codes can be developed to design the lens system and evaluate its optimum parameters for minimizing the deleterious effects. The electric focusing scheme has been used very successfully in heavy-ion fusion research at LBL[11]; a long transport channel with electrostatic quadrupole lenses (ESQ) was developed for transport and focusing. Preliminary experiments using einzel lenses[12] and helical ESQ lenses[13] were conducted at the SSC Laboratory. Interesting experimental results with helical ESQ lenses have been reported in the context of transporting heavy ions.[14] A combination of ESQ lenses with a "ring" lens was used effectively in LBL[15]; the experiment was conducted with low-perveance beams.

At Maryland, the design problems of LEBTs for high-brightness beams were studied in detail. The aim has been to focus 30 mA, 35 kV H⁻ beams from an ion source in to an RFQ maintaining the emittance growth within about 50%; SSC RFQ[16] (427.6 MHZ) was considered. It was determined that a LEBT consisting of six ESQ lenses and a short einzel lens section offers an attractive solution.[17-19] This composite electrostatic LEBT scheme has merits to satisfy many desired conditions of a low-emittance injector for advanced accelerators. Detailed simulation results on beam dynamics were reported earlier.[17-19] Some essential illustrations are given below to highlight the guiding principles.

III. LEBT for a Low-Emittance Injector

H⁻ beams corresponding to three different types of high-brightness ion sources, namely, (i) Penning-Dudnikov type source at Los Alamos, (ii) volume ionization sources at Brookhaven and SSCL, and (iii) magnetron-type ion source at SSCL were considered[17-19]. The transport of 30 mA, 35 kV H⁻ beams from the SSCL volume source is discussed below.

Figure 1 shows schematics of the compact LEBT system developed at Maryland. A combination of six ESQ lenses and a short einzel lens has been cascaded to cover a length of about 30 cm. The LEBT system is made compact enough to insure minimum drift space between the lenses. The ESQ lens parameters were determined according to the following hierarchy ("simulation tree") of numerical schemes which were developed by and large at Maryland.

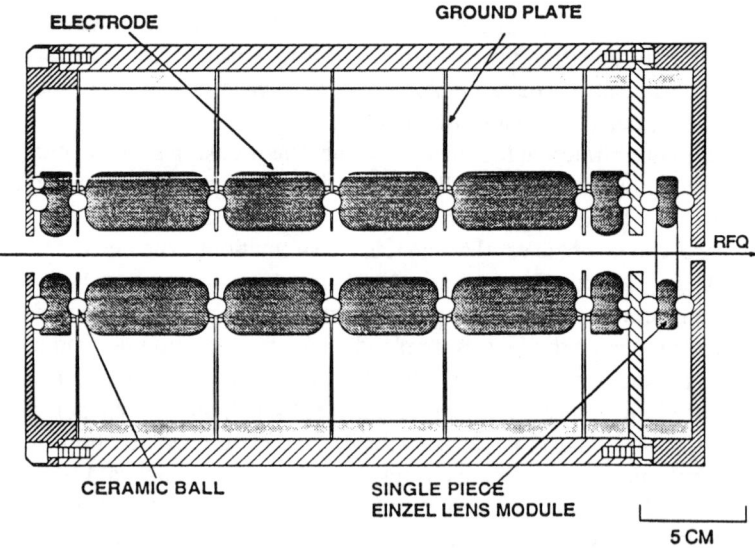

Fig. 1 Schematic diagram of an electrostatic LEBT system consisting of six ESQ lenses and an einzel lens.

SIMULATION TREE

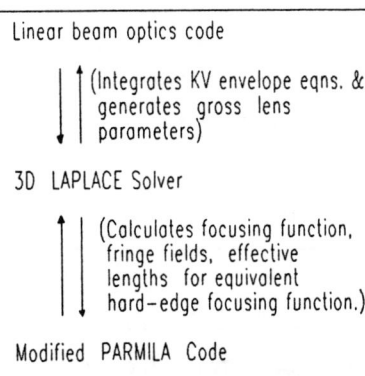

Linear beam optics code
(Integrates KV envelope eqns. & generates gross lens parameters)

3D LAPLACE Solver
(Calculates focusing function, fringe fields, effective lengths for equivalent hard-edge focusing function.)

Modified PARMILA Code
(Evaluates particle distribution in phase space and emittance values.)

This scheme is reiterated until a satisfactory solution is obtained.

A recent effort on computer-aided LEBT design has been made[20], and the initial results are encouraging. The ESQ lens parameters are given in table 1.

TABLE 1. ESQ lens parameters.

Lens No.	Spacing (mm)	Aperture radius (mm)	Electrode radius (mm)	Length (mm)
1 & 6	2.0	15.0	17.2	25.0
2 & 5	2.0	22.0	25.2	59.0
3 & 4	2.0	22.0	25.2	47.0

The electrode radius is 1.1468 times the aperture radius to minimize nonlinear fields. The spacing indicates the gap between an ESQ lens and its neighboring ground plate.

The parameters of the einzel lens in Table 2 have been determined using the SNOW-2D code.

TABLE 2. Einzel lens parameters

Aperture radius of the center electrode (mm)	Spacing (mm)	Thickness of the center electrode (mm)
17.0	7.0	10.0

The spacing indicates the gap between the center electrode (at high voltage) and its neighboring ground electrodes.

The K-V envelope solution through the ESQ lenses is shown in Fig. 2 for a 30 mA, 35 kV H⁻ beam; the beam parameters of the SSCL volume source are used here. A hard-edge type focusing function for the ESQ LEBT is assumed. Note that the beam at the input to the ESQ lens is large and highly diverging (beam radius $X_i \sim 5.6$ mm and divergence of the beam envelope $X_i' \sim 56$ mrad). The ESQ lens system transforms this beam into a converging beam.

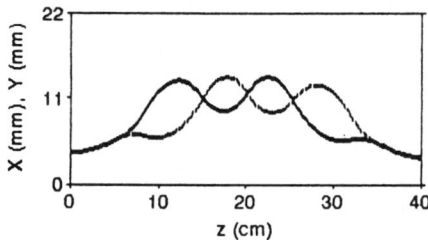

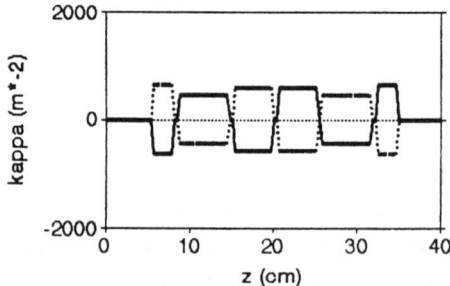

Fig. 2 K-V envelope solution (top) and hard-edge focusing function (bottom).

The final focusing to satisfy the matching conditions for RFQ ($\alpha = 1.26$, $\beta = 1.86$ cm/rad, $\tilde{\epsilon}_n = 0.2$ mm mrad) is accomplished by the einzel lens. The emittance through the ESQ lens system is evaluated using the particle simulation code, PARMILA. Figure 3 (a) shows evolution of the rms normalized emittance $\tilde{\epsilon}_n$, and the maximum beam excursions in the two orthogonal directions (X and Y) are shown in Fig. 3(b). Here, each set of two data points (open and closed

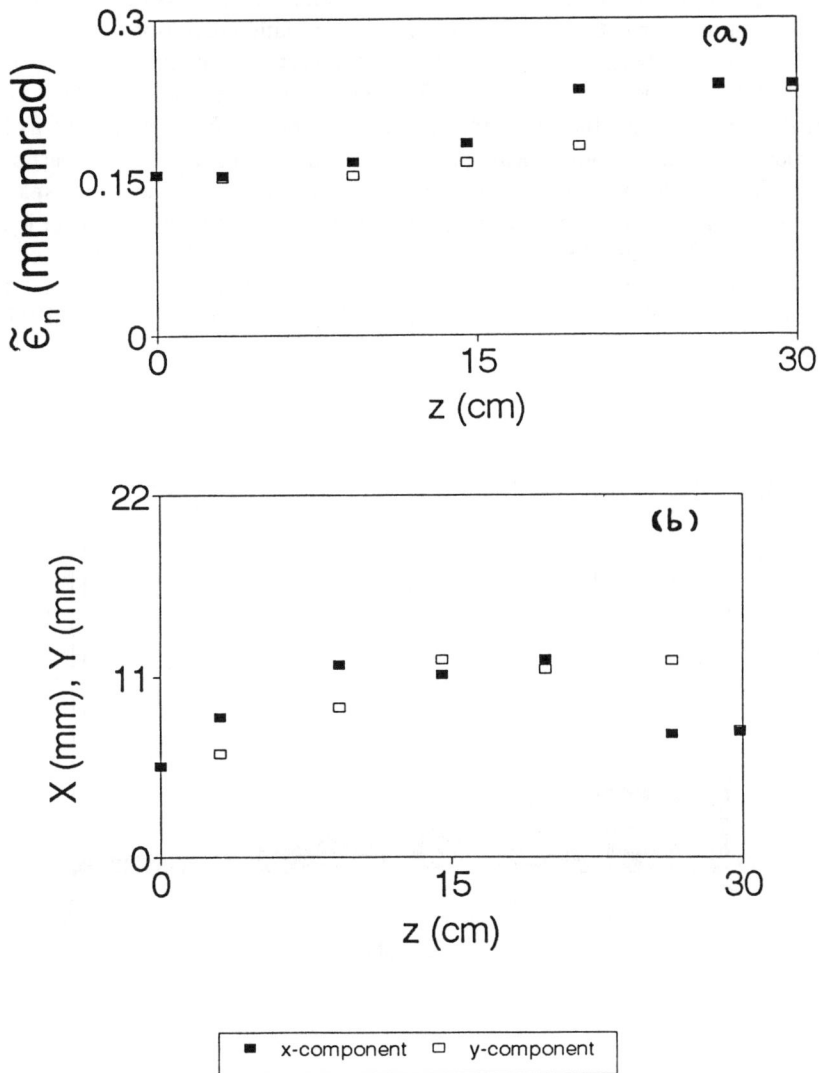

Fig. 3 Evolution of (a) rms normalized emittance, and (b) beam excursion through the ESQ lens system.

rectangles) corresponds to the values of the parameters in the two orthogonal directions. The data at z = 0 correspond to the input values; afterwards, each data set represents the values at the end of the subsequent lenses. Note that the emittance grows as the beam excursion through the lenses increases. This suggests that the external focusing force on the beam should be applied adiabatically to avoid any sudden increase of the amplitude of the beam envelope, and thereby the emittance growth can be controlled. In the above example, the emittance growth in the ESQ LEBT was estimated to be a factor of about 1.5. It has been analyzed that the emittance growth is essentially due to chromatic aberrations.

Next, the beam through the einzel lens section is examined using the SNOW-2D code. The output from the ESQ lens is coupled to the einzel lens. An equivalent ellipse describing the output beam distribution from the ESQ lens is constructed to define the input beam parameters for the einzel lens. The particle trajectories through the einzel lens are shown in Fig. 4, and the phase-space distributions of the particles are shown in Fig. 5. It is noted that the einzel lens does not introduce any distortions to the beam distribution. Hence, the overall emittance growth of the full LEBT section is estimated to be about a factor of 1.5. This result satisfied the emittance budget allowed in the SSCL ion source-LEBT-RFQ section.

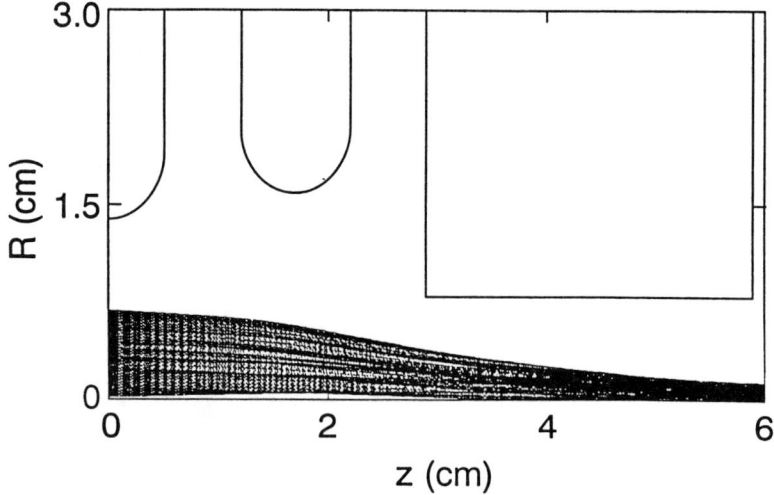

Fig. 4 Particle trajectories through the einzel lens. The center electrode is at high voltage.

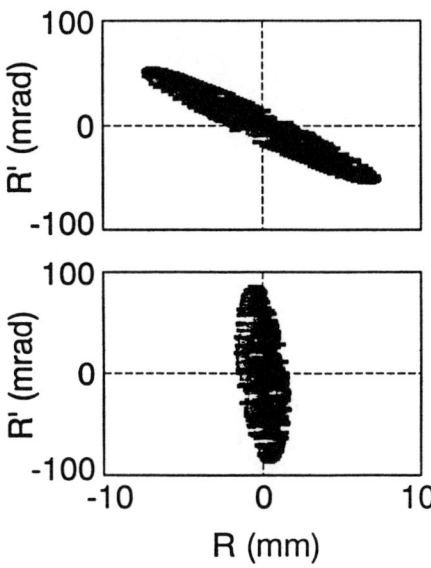

Fig. 5 Particle distributions through the einzel lens from SNOW-2D: input (top), output (bottom).

IV. Conclusions

In this article, the basic design principles for an efficient LEBT have been discussed. A combination of ESQ lenses and a short einzel lens is shown to satisfy a very stringent matching condition of an RFQ. Electrostatic LEBT schemes are expected to work well, especially for short-pulsed (< 50 µs) beams, with beam currents of ~ 30 mA at a beam voltage of ~ 35 kV. Higher beam currents can be also transported efficiently using electrostatic quadrupole lenses if the beam voltage is increased and the perveance remains close to typically the 30 mA, 35 kV case. Some notable features of the composite LEBT scheme are:

(1) A good buffer space between an ion source and an RFQ can be provided for differential pumping.

(2) A combination of first-order (ESQ lenses) and second-order (einzel lens) focusing elements offers flexibilities to satisfy extreme matching requirements without any significant emittance dilution.

(3) The lenses can be tuned easily by adjusting low-voltage power supplies of ESQ lenses.

(4) The beam dynamics can be simulated reliably. This provides a good guideline to recognize the sensitivity of the control knobs.

Finally, a strong thread between simulation and experiments must be established to enhance our understanding of the physics problem and strengthen the confidence level of simulations. Detailed experimental studies on different LEBT schemes are highly needed, especially in the context of intense, high-brightness beams.

Acknowledgments:

The author thankfully acknowledges many valuable discussions with Professor M. Reiser. Thanks are also due to Drs. K. Saadatmand, C. R. Chang, and C. Allen.

References:

1. Proc. Workshop on Future Hadron Facilities in the US, Bloomington, IN, July 6-10, 1994, Fermilab rept.# FERMILAB-TM-1907.
2. "Site-Specific Conceptual Design", SSCL report# SSCL-SR-1056, July 1990.
3. C. W. Schmidt and C. D. Curtis, Brookhaven National Lab. Report# BNL 50727, p.123.
4. J. G. Alessi, D. McCafferty, and K. Prelec, Proc. LINAC Conf. , Ottawa, Canada, Aug. 24-28, 1992, p.335.
5. H. Haseroth, et al., Proc. LINAC Conf. , Ottawa, Canada, Aug. 24-28, 1992, p. 58.
6. M. Reiser, Theory and Design of Charged Particle Beams, Wiley, N.Y., 1994.
7. J. D. Lawson, The Physics of Charged-Particle Beams, Clarendon Press, Oxford, 1988.
8. M. Reiser, Nucl. Instrum. and Meth. In Phys Res. B **56/57**, 1050 (1991).
9. J. G. Alessi, J. M. Brennan and A. Kponov, Rev. Sci. Instrum. **61**, 625 (1990); J. G. Alessi, J. M. Brennan, J. Brodowski, H. N. Brown, R. Gough, A. Kponov, V. LoDestro, P. Montemurro, K. Prelec, J. Staples, and R. Witkover, Proc. LINAC Conf., Newport News, VA, Oct. 3-7, 1988, p. 196.
10. R. Stevens, AIP Conf. Proc. No. 287, 1994, p. 646.
11. T. J. Fessenden, AIP Conf. Proc. No. 253, 1992, p. 160.
12. K. Saadatmand, J. E. Hebert and N. C. Okay, Rev. Sci. Instrum. **65**, 1173 (1994).
13. D. Raparia, Proc. LINAC Conf., 1990, p. 405; D. Raparia et al., Rev. Sci. Instrum. **65**, 1457 (1994).
14. Y. Mori, A. Takagi, T. Okuyama, M. Kinsho, H. Yamamoto, T. Ishida and Y. Sato, Proc. LINAC Conf., Ottawa, Canada, Aug. 24-28, 1992, p. 642.
15. O. A. Anderson, L. Soroka, J. W. Kwan and R.P. Wells, Proc. 2nd Europ. Particle Accelerator Conf., Nice, France, June 12-16, 1990, LBL Report-28962.
16. T. S. Bhatia, et al., Proc. 1991 Particle Accelerator Conf., San Francisco, CA, May 6-9, 1991, p. 1884.
17. S. K. Guharay, C. K. Allen, M. Reiser, K. Saadatmand and C. R. Chang, Rev. Sci. Instrum. **65**, 1774 (1994).

18. S. K. Guharay, C. K. Allen and M. Reiser, Nucl. Instrum. And Meth. In Phys. Res. A **339**, 429 (1994).
19. S. K. Guharay and M. Reiser, Nucl. Instrum. and Meth. In Phys. Res. A **363**, 135 (1995).
20. C. K. Allen and M. Reiser, Institute for Plasma Research, Univ. Maryland, CPBG Technical Rept. #95-011, August 22, 1995.

Space-charge dominated heavy ion beams in Electrostatic Quadrupole (ESQ) Accelerators

Simon Yu, Shmuel Eylon, Enrique Henestroza
Lawrence Berkeley National Laboratory
Berkeley, California

Dave Grote
Lawrence Livermore National Laboratory
Livermore, California

The Electrostatic Quadrupole (ESQ) Accelerator concept was first introduced and studied by Abramyan (Ref. 1). Subsequently, in the quest for high current negative ion injectors for magnetic fusion applications, the Neutral Beam Program at Lawrence Berkeley National Laboratory (LBNL) studied the ESQ for a number of years (Ref. 2). More recently, the Heavy Ion Fusion Program in the United States, in pursuit of a fusion driver-scale high current injector, constructed a pulsed 2MV, 800mA singly charged potassium ion machine (Ref. 3). This machine has been in operation for two years now. This paper summarizes the beam dynamics aspect of the ESQ accelerators that we have learned with the 2MV injector and related experiments, as well as associated theory and simulations.

The Heavy Ion Program, in addition to requiring a high current and high energy injector, also needs a beam of very low emittance. The design goal of normalized edge emittance of less than 1Π mm-mr allows at most a factor of 3 increase from the intrinsic beam emittance due to the finite temperature (960° C) at the source.

The combined requirement of high current and low emittance implies a highly space-charge-dominated beam. Typically, the beam envelope is determined totally by the balance of external (electrostatic) focusing and space charge effects, with the contribution from the finite emittance term being a minor perturbation ($\varepsilon^2/a^2 \sim 0.01\ k^2 a^2$). A key design consideration is the preservation of the low emittance throughout the injector.

The ESQ injector is a device in which a number of electrostatic quadrupoles are arranged in a configuration to provide simultaneous accelerations and strong focusing to a traversing ion beam (See Figure 1). The ESQ section is generally preceded by a diode in which the ion beam is extracted. The ESQ configuration has inherent advantages from the point of view of high voltage engineering due to the large transverse fields that sweep out unwanted electrons which would

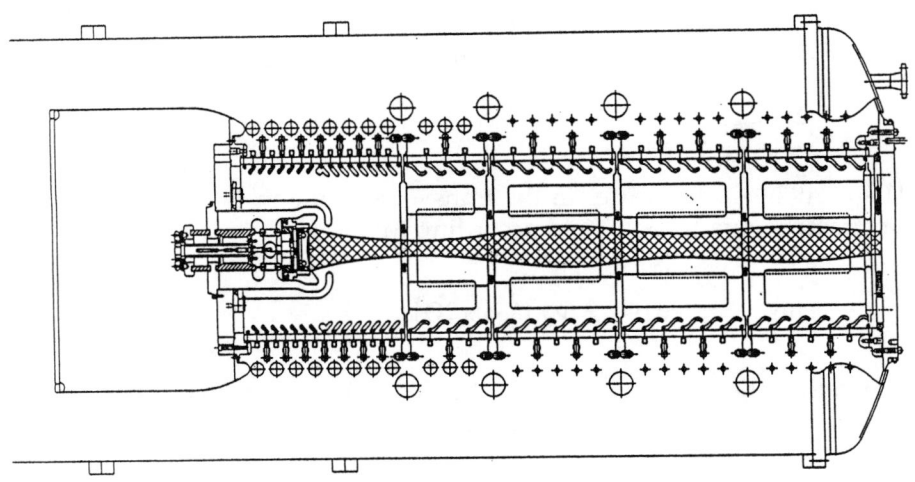

Fig.1 The ESQ-injector layout schematics.

otherwise initiate high voltage breakdown. On the other hand the ESQ geometry is manifestly 3-D, and is more complicated than the more conventional Pierce columns with axisymmetric structures. The interdigital structure leads to higher order multipoles in the focusing fields, and these nonlinear effects must be carefully avoided to minimize deleterious distortions of phase-space. In contrast to a purely focusing channel, the beam energy in a ESQ accelerator is monotonically increasing, and cancellations of multipoles in adjacent focusing and defocusing elements, which act to reduce the nonlinear effects in truly periodic systems, are not operative in ESQ accelerators.

In addition to field nonlinearities, there is an inherent kinematic effect associated with accelerators with low beam energy and strong electrostatic focusing which we call the "energy effect". The physical mechanism of this 3rd order nonlinearity is that particles in a fixed axial position along the accelerator do not all have the same energies, but may have significantly different potential (and therefore kinetic) energies depending on the relative proximity to the positive or negative electrodes in the transverse plane. The spread in energies leads in turn to variable betatron periods for different particles resulting in distortion of particle phase-space. This effect can be analyzed quantitatively, starting with the particle equation of motion

$$\frac{d^2x}{dz^2} = \frac{eE_x}{mv^2}$$

The particle energy varies in the transverse plane in the case of a strong electrostatic quadrupole field according to

$$\frac{1}{2}mv^2 \sim e\phi = e\left(\Phi_o + \Phi_q\, r^2 \cos 2\Theta\right)$$

Expanding the particle energy in the denominator, we have to leading order

$$\frac{d^2x}{dz^2} = \frac{eEx}{2T}\left(1 - \frac{e\Phi_q}{T} r^2 \cos 2\Theta\right)$$

where $T = e\Phi_o$ is the average particle energy on axis. Hence, even if the field is perfectly linear, the particle orbit has a leading third order aberration

$$\frac{d^2x}{dz^2} = -k\, x\left(1 - \frac{e\Phi_q}{T} r^2 \cos 2\Theta\right)$$

Note that the coefficient of the nonlinear term is proportional to Φ_q/T. Hence the effect is strongest for strong quadrupole focusing and/or low particle energy.

The interplay between the energy effect, the nonlinear components of the 3-D external field, all in the presence of strong electrostatic self-forces, is clearly a rather complicated beam dynamics problem. Thus, much of our earliest effort was invested in the development of the LLNL 3-D PIC code WARP3D (Ref. 4) into a design tool, and the final detailed design of ESQ configuration (diode energy, quadrupole strengths, geometries, etc.) was based on WARP3D predictions of minimal emittance growth (Figure 2).

Before we embarked on the construction of the full machine, we felt that it was necessary to experimentally verify the code predictions, and a "1/4-scale" experiment was conducted on the Single Beam Test Experiment (SBTE) facility at LBNL in the spring of 1993 (Ref. 5). The experiment was conducted with a 100 keV 10mA K+ beam with a

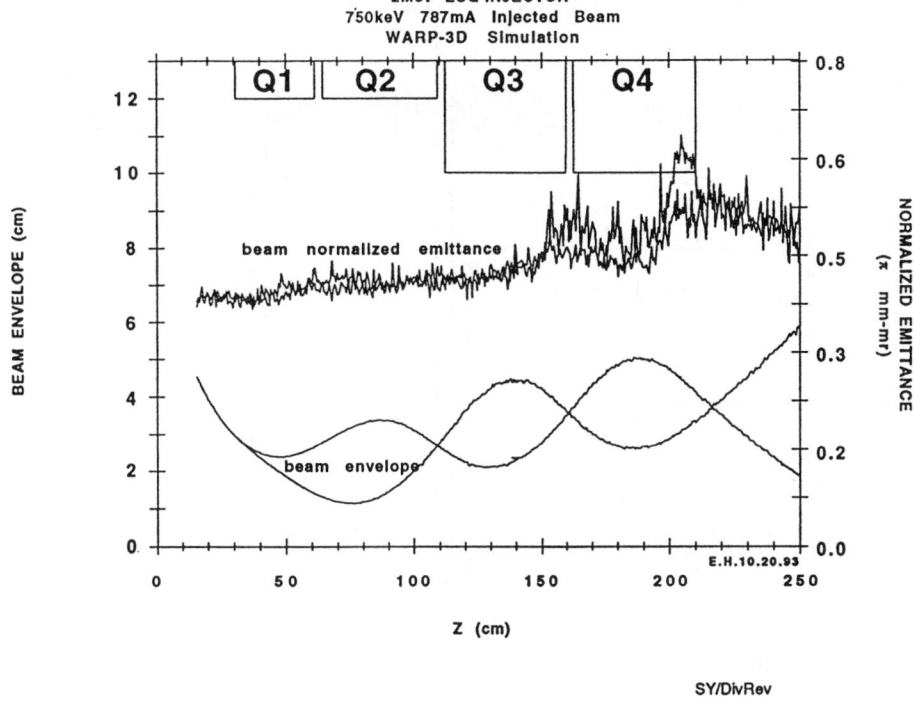

Fig.2 WARP3D simulation results of the beam envelope and normalized emittance along the ESQ section (z direction).

normalized emittance of $\varepsilon_n = 0.06 \, \Pi$ mm-mr. A special 4-quad ESQ accelerating unit was constructed (See Figure 3) and placed after the source extraction grid at the SBTE, and phase space distortions were measured over a large parameter space of diode and quad voltages. Theory predicts the invariance of beam envelope and phase-space distortions if the ratio of quad to beam voltage Φ_q/T and beam current to energy relationship of $I/T^{3/2}$ are kept constant. Hence, we are able to measure phase-spaces in a parameter regime directly relevant to the 2MV injector in the 100 keV experiment by appropriate scalings of quad voltages and current. The resulting measurements agree very well with the code predictions over the entire range of parameter space in the experiments (See Figure 4). We felt that our confidence in the code predictions for the full machine was well founded.

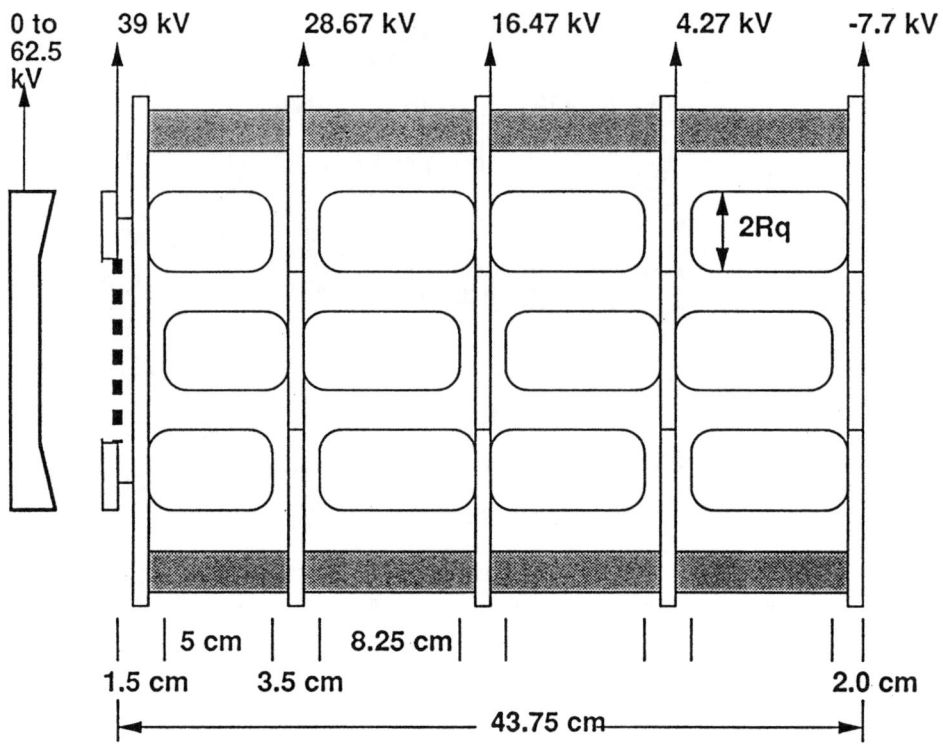

Fig.3 Scaled ESQ experiment layout schematics.

Particle phase space at the exit of a 4-quad ESQ structure:

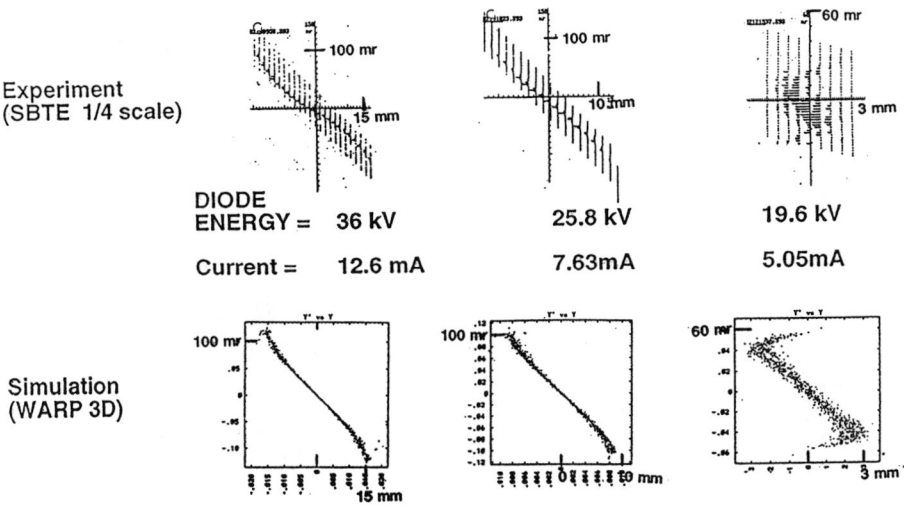

Fig.4 Scaled ESQ output beam phase space profiles, experimental and 3D simulation results.

On October 27, 1993, slightly less than a year after commencing the ESQ construction project, the injector was completed. On the first day of operation, October 28, 1993, we exceeded the design goals in beam energy and current (2.15 MV and 800 mA), and the ESQ column has proven to be robust against high voltage breakdowns as expected. Subsequently, the emittance was measured and the normalized edge emittance was indeed less than 1∏ mm-mr over a range of parameters (See Figure 5).

These successes not withstanding, the measurements for the 2MV injector, unlike the SBTE experiments, showed a major discrepancy when detailed comparisons with code predictions were made. While the trend of phase-space evolution with increasing current agrees with code predictions there seemed to be a systematic error in the absolute value of current and/or voltages (See Figure 6). Detailed checks were performed on all current and quad voltage measurements as well as the geometries of the mechanical components (quads and source), and no significant deviations from physics design specifications were uncovered.

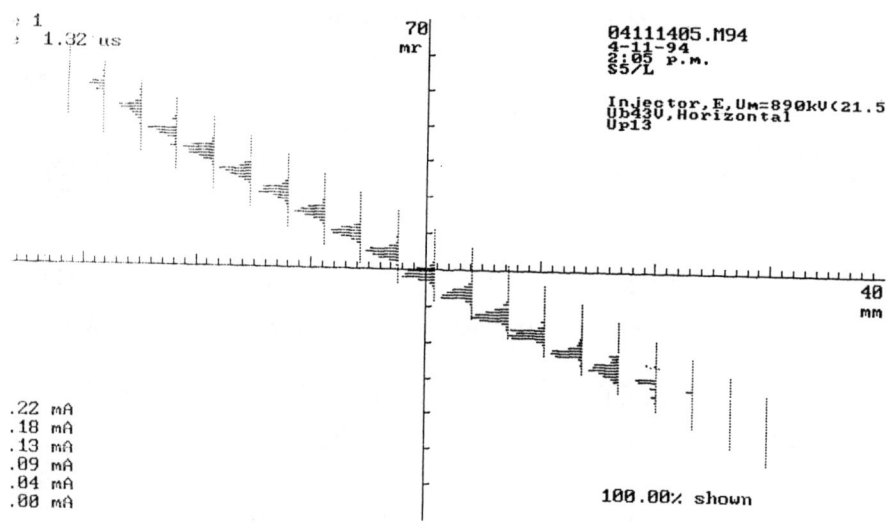

Fig.5 The ESQ-injector output phase space profile with measured normalized edge emittance of 0.65 π mm mr.

Up to this point in time, the simulations were conducted in two steps. First, the gun code EGUN was used to design the diode from the source to the diode exit. Secondly, an axial location within the diode is identified where the potential line lies on a vertical plane perpendicular to the axis. This line is then used to generate the entrance plane for particle injection in WARP3D. This two-step process is necessitated by the fact that the WARP3D code did not have the necessary particle-extraction algorithms implemented, while the EGUN code is axisymmetric, and incapable of simulating the 3-D ESQ structures. However, over the region of overlap of the 2 codes, we checked to see that the predicted beam envelope parameters (beam radius and angle of convergence) were consistent. Many WARP3D checks revealed a high level of sensitivity of final phase space at ESQ exit to the entrance conditions. Nevertheless, within the context of this 2-step modelling, we were not able to get agreement between ab initio code calculations and the experiments.

The solution came when source extraction algorithms were implemented in WARP3D, and the resulting predictions, to our own amazement, agree well with the experiments over a broad range of

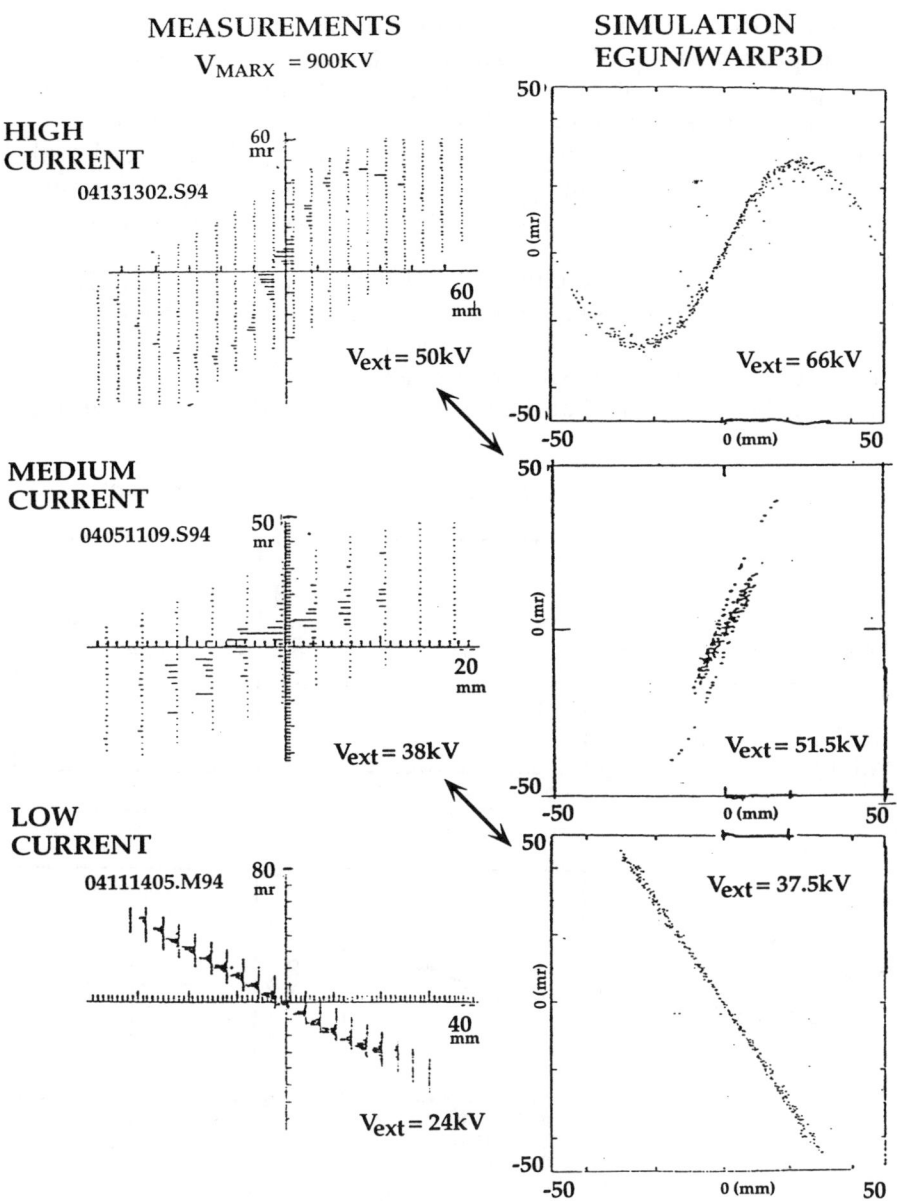

Fig.6 ESQ-injector output beam phase space profiles, experimental and early version of 3D simulation results compared.

parameters (See Figure 7). Subsequent code sensitivity studies reveal that the phase-space profile at exit depends not only on the initial envelope parameters at entrance, but also on the detailed particle and field distributions at entrance. This level of subtle sensitivity was hidden in the SBTE experiment where the presence of a physical grid removes much of the computational ambiguities. The 2MV injector was designed with no physical extraction grid because of component reliability considerations.

At this point in time, the injector performance had met all design goals, and the disagreement with code predictions have been resolved. Nonetheless, detailed profile density measurement at injector exit reveals a new source of potential troubles. A slit scan across the beam reveals that the beam at injector exit is hollow with a

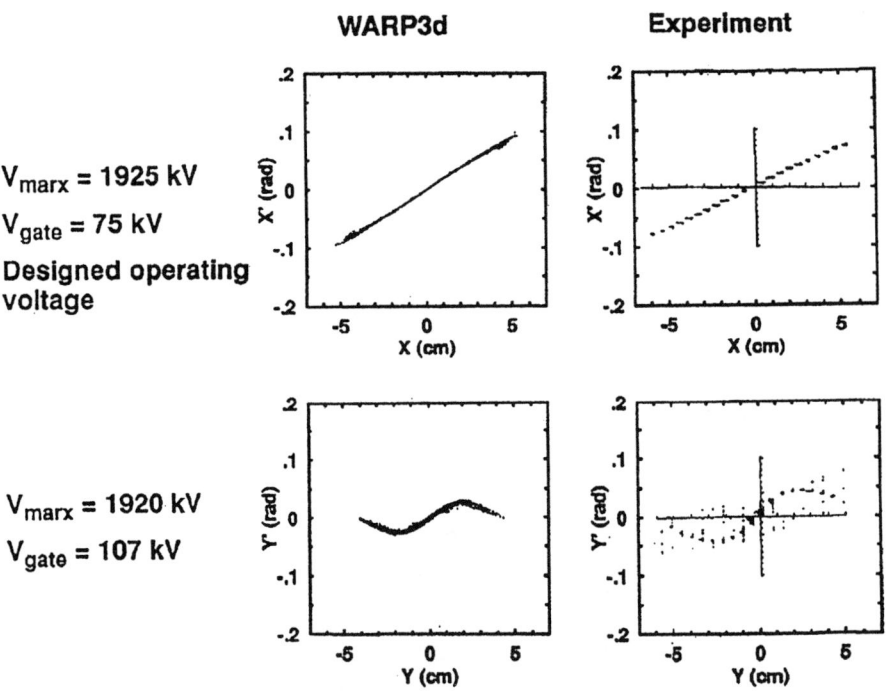

Fig.7 Two examples of ESQ-injector output beam phase space measurements, in agreement with simulations using the improved version of WARP3D which includes source extraction modeling.

very steep "wall" on the beam edge. The density profile over the beam body shows features of rippling, and haloes are detected on the beam edge (See Figure 8). Such nonlinear density profiles are sure to induce

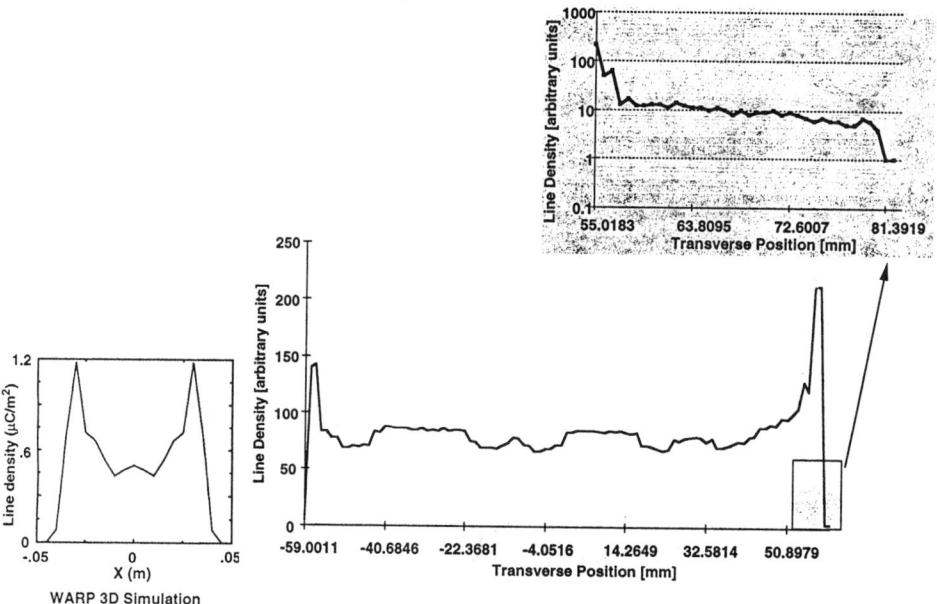

Fig.8 Injector output beam density profiles measurements and WARP3D simulations showing hollowing, rippling, and halos.

nonlinear fields, and even if the emittance is acceptable at injector exit, further transport down the beam line is bound to increase emittance.

This effect is under study on a 6-quad "matching section" recently installed at the end of the injector. The density profiles have been measured at various points in the matching section beamline, and rapid evolutions of the profile are indicated. (See Figure 9) Detailed emittance measurements are in progress.

There are some indications of hollowing effects from WARP3D, although the code gridding to date is not adequately fine to resolve sharp density structures. Work is also in progress from the theoretical end to resolve the origin of the observed density fluctuations. Whether these are space-charge waves or results of imperfect sources, or other effects, is a chapter to be closed in the future.

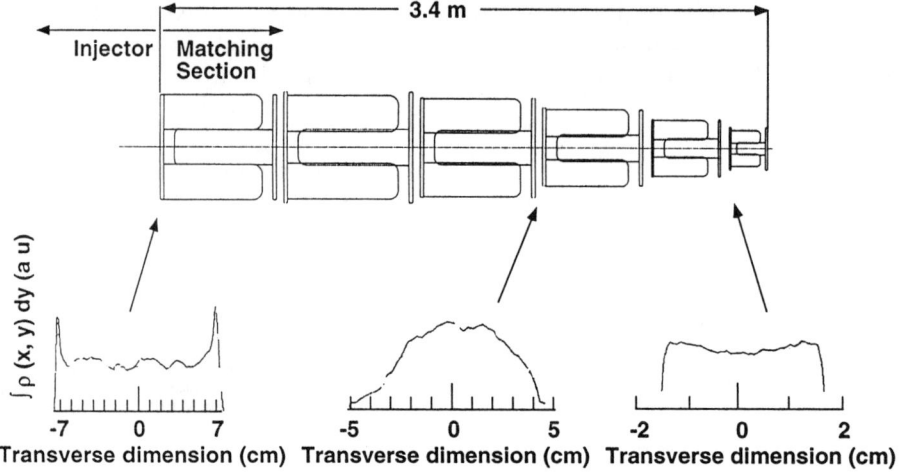

Fig.9 Beam density profile evolution measured along the matching section.

REFERENCES

1. E.A. Abramyan and V.A. Gapunou <u>Atomnaya Energiya</u> 20 385 (1966).
2. O.A. Anderson et al., Proceedings of the 1989 Particle Accelerator Conference, March 20-23, 1989.
3. S. Yu et al., Proceedings of the 1995 Particle Accelerator Conference, Dallas, Texas, 1995.
4. A. Friedman et al., <u>Phys. Fluids</u> <u>B4</u>, 2203 (1992).
5. E. Henestroza et al., Proceedings of the 1993 Particle Accelerator Conference, Washington, D.C. (1993).

PERFORMANCE and MEASUREMENTS OF THE AGS and BOOSTER BEAMS*

W.T. Weng

AGS Department, Brookhaven National Laboratory
Upton, New York 11973-5000

Abstract. In May 1995, the AGS reached its upgrade intensity goal of 6×10^{13} ppp, the highest world intensity record for a proton synchrotron on a single pulse basis. At the same time, the Booster reached 2.2×10^{13} ppp surpassing the design goal of 1.5×10^{13} ppp due to the introduction of second harmonic cavity during injection. The critical accelerator manipulations, such as resonance stopband corrections, second harmonics cavity, direct rf feedback, gamma-transition jump, longitudinal phase space dilution, and transverse instability damping, will be described as well as some beam measurements. Possible future intensity and brightness upgrades will also be reported.

INTRODUCTION

The achievable intensity in a proton synchrotron is limited by many accelerator physics reasons. Most notable and widely discussed is the incoherent space charge tune shift during injection (1,2). Due to the strong energy dependence of the amount of tune shift caused by the space charge force, raising injection energy can alleviate this limiting effect. Empirically, it has been shown both at the AGS Booster and the CERN PS Booster(3,4) that resonance stopband correction is effective in limiting the growth of betatron oscillations and hence retaining more particles inside the synchrotron under large space charge tune shifts. However, there are many possible mechanisms also playing a very important role in limiting the achievable intensity of a proton synchrotron. First, is the available horizontal aperture to accommodate the beam for its maximum momentum spread, usually occurring during injection or gamma-transition. Secondly, the beam loading effect from the beam on the accelerating cavity. If the current is such that the resultant accelerating bucket is insufficient to contain the beam, feedforward or feedback corrections have to be applied to raise the intensity. Thirdly, the transverse coupled bunch instability due to the resistive wall effect tends to occur on the order of a few 10^{12} ppp for proton synchrotrons in the few GeV range. Once an effective damping system is in place, this limitation can be easily eliminated. Fourthly, the loss introduced by the transition energy crossing, which can cause particle loss by a large momentum spread, large

---

*Work performed under the auspices of the U.S. Department of Energy.

dispersion function or exciting coherent instabilities. The fifth category includes the single bunch instabilities, such as microwave, head-tail, or mode-coupling instabilities.

There is no telling, a priori, that which of the above-mentioned limiting factors will come first to limit the achievable intensity of a given accelerator. Only careful calculations and machine studies can reveal the relative importance of each mechanism. In the following, we will use the AGS as a prototypical example to show specific effects of some of the factors mentioned above and methods introduced to combat them.

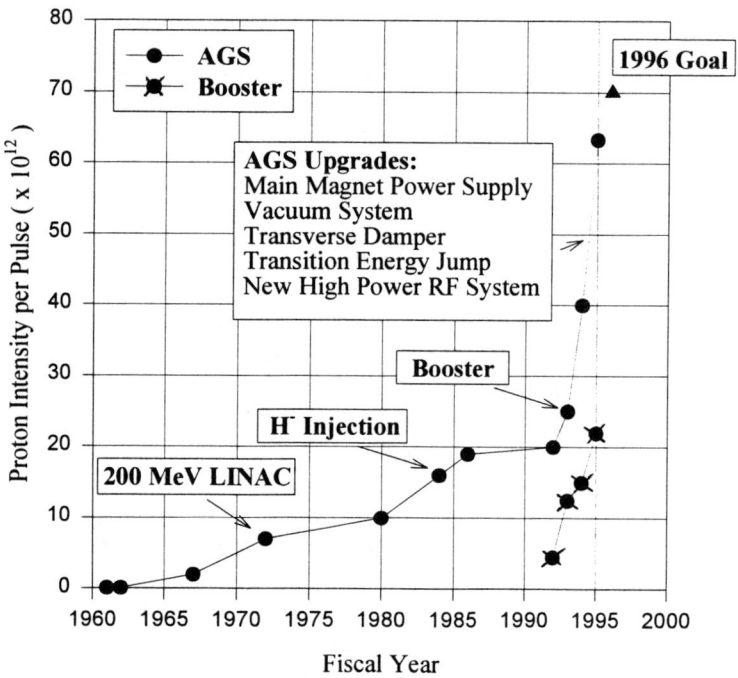

Figure 1.　　The evolution of the AGS proton intensity.

Shown in Figure 1 is the intensity evolution of the AGS since its completion in 1961. The accelerator was originally proposed with an intensity of a few 10^9 ppp in mind and it eventually reached 6.3×10^{13} ppp in 1995. Major improvements in the intensity record are summarized briefly in the following chronology.

Year	Parameters or Improvements	Intensity
1970	50 MeV Linac injector Space charge limited at Injection ($\Delta v_y \approx 0.3$)	3×10^{12}
1976	200 MeV Linac injector Resonance stopband correctors Transverse damper	10^{13}
1990	H^- injection Rf feedforward compensation $\Delta v_y \approx 0.7$	1.6×10^{13}
1995	1.5 GeV Booster injection Resonance stopband correctors Direct rf feedback γ-transition jump $\Delta v_y \approx 0.35$	6.3×10^{13}

It is clear that raising the injection energy is the most effective way to increase the achievable intensity for a space-charge limited low energy proton synchrotron. It is also clear that many accelerator physics manipulations have to come into play to keep all those particles inside the synchrotron. In the following we will describe some of those processes introduced in the AGS and Booster. A brief description of how the Booster and AGS are linked together in both proton and heavy ion operations, can be found in Reference 5. To assist the readers and facilitate the discussion, some of the relevant accelerator parameters are summarized below.

	Booster	AGS
Circumference, m	201.78	807.12
Injection energy, GeV	0.2	1.5
Extraction energy, GeV	1.5	28.0
v_x / v_y	4.82/4.83	8.7/8.8
x_p, m	2.9	2.2
γ_{tr}	4.5	8.5
Harmonic number, h	3(2)	12(8)
RF voltage, kV	90	400
Intensity, 10^{13} ppp	2.2	6.3
Estimated tune shift, Δv_y	0.35	0.25

CRITICAL PHYSICS PROCESSES

A. Resonance Stopband Corrections

The Booster working points are chosen to be about $v_x = 4.85$ and $v_y = 4.90$ at injection and the estimated space charge tune shift at full intensity is about $\Delta v_x = 0.25$ and $\Delta v_y = 0.35$. At high intensity, the tune of some of the particles can cross $2v_x = 9$, $2v_y = 9$, $v_x - v_y = 0$, $v_x + v_y = 9$, $3v_x = 14$, $3v_y = 14$, $v_x + 2v_y = 14$, $2v_x + v_y = 14$ lines. Examples of particle losses due to some resonance lines and the survival of beam after correction are shown in Figure 2 (3).

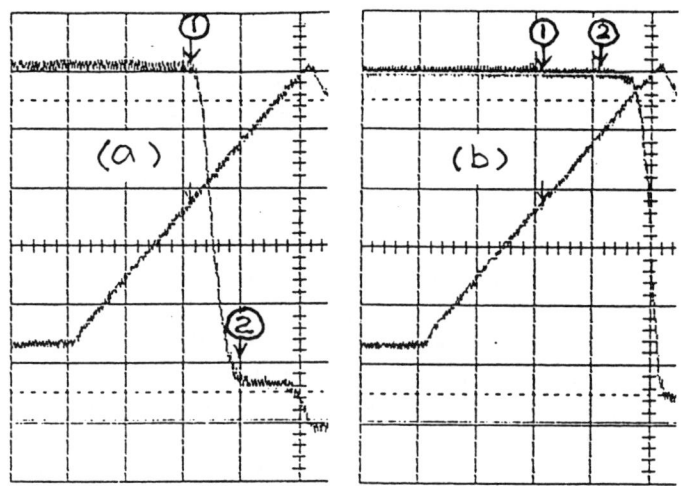

Figure 2. Stopband correction: (a) without correction and (b) with correction, (1) $2v_x = 0$ and (2) $v_x + v_y = 9$.

The left trace shows the beam intensity decrease when it encounters the first resonance of $2v_x = 9$ and further decreases when it encounters the second resonance $v_x + v_y = 9$. The right trace shows that the total beam intensity is almost constant in crossing those two resonances after the correction system is turned on. The same process is repeated for all the resonances listed above. Such a correction study is carried out at flattop by varying the tune. During acceleration, with or without correction, this can make a 5-10% difference for weak resonances and a 30-50% difference for strong resonances. Shown in Figure 3 is the comparison with and without stopband correction of the full beam intensity over the acceleration cycle. During 1995 running, it was discovered that octupole correction was needed for the AGS during injection. The effect of the octupole correction is shown in Figure 4.

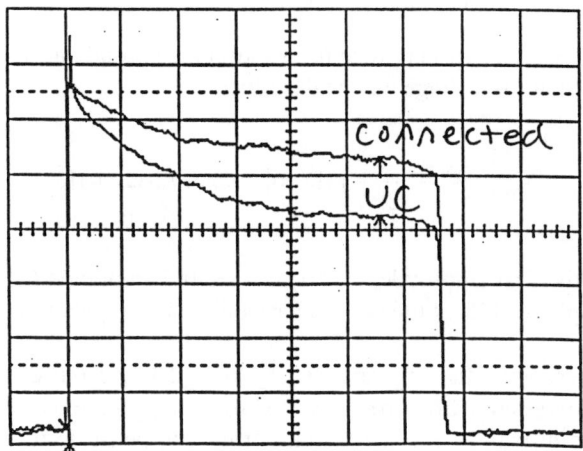

Figure 3. Booster intensity in one cycle with or without stopband corrections.

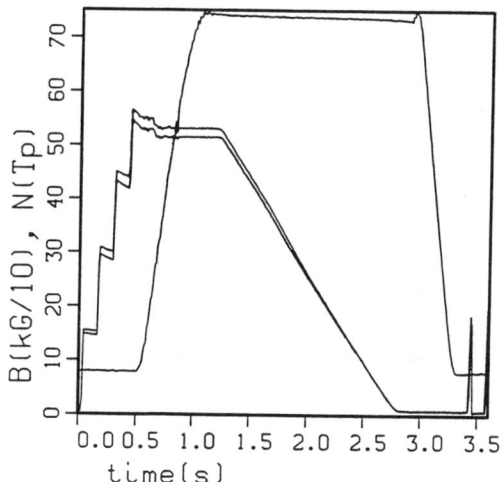

Figure 4. AGS intensity with and without octupole corrections.

B. Second Harmonic Cavity

If the accelerating rf system has only one and the same frequency, the resultant rf bucket usually assumes a parabolic shape with the maximum around the synchronous phase angle. The final charge distribution tends to have a sharp maximum in the middle and hence a larger space charge force.

Along with the main rf system, an additional second harmonic cavity can also be added, so that the voltage can be described as

$$V = \frac{e}{2\pi h} V_o [Sin\phi + rSin2(\phi - \delta)],$$

where V_o is the first amplitude, $r = r(t)$ is a second amplitude (as a fraction of V_o), $\delta = \delta(t)$ is a phase shift of the second harmonic with respect to the first harmonic.

The reason for the large space charge tune shift at capture is due to the charge inhomogeneity resulting from the single rf system. By judiciously choosing ϕ_s, r, and δ the charge distribution in the vicinity of ϕs can be made to be uniform, hence, reducing the bunching factor and space charge force. Thus, the relative deviation in the capture efficiency for the double system compared to the single system will be within the strip between the two curves $\delta\ell$ and δA in Figure 5. In other words, the capture efficiency of the double rf system should be better than that of the single rf system by about 20-30% (6).

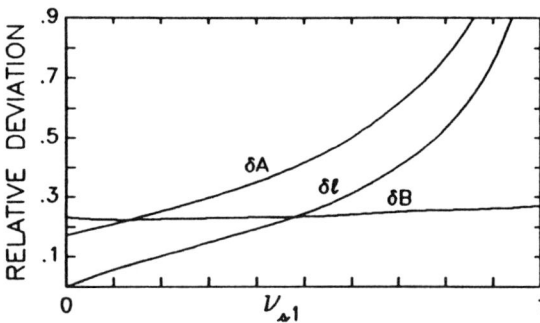

Figure 5. Relative deviations in bucket length, area and bunching factor.

The actual double voltage shape and resultant charge distribution in the AGS Booster is shown in Figure 6 (7), where the lower trace represents the rf waveform and the upper trace represents the beam current. It is clear that with a double rf system, the resultant current distribution is smoother in the middle. Such a second harmonic cavity raised the final intensity in the Booster from 1.5×10^{13} ppp in 1994 to 2.2×10^{13} ppp in 1995.

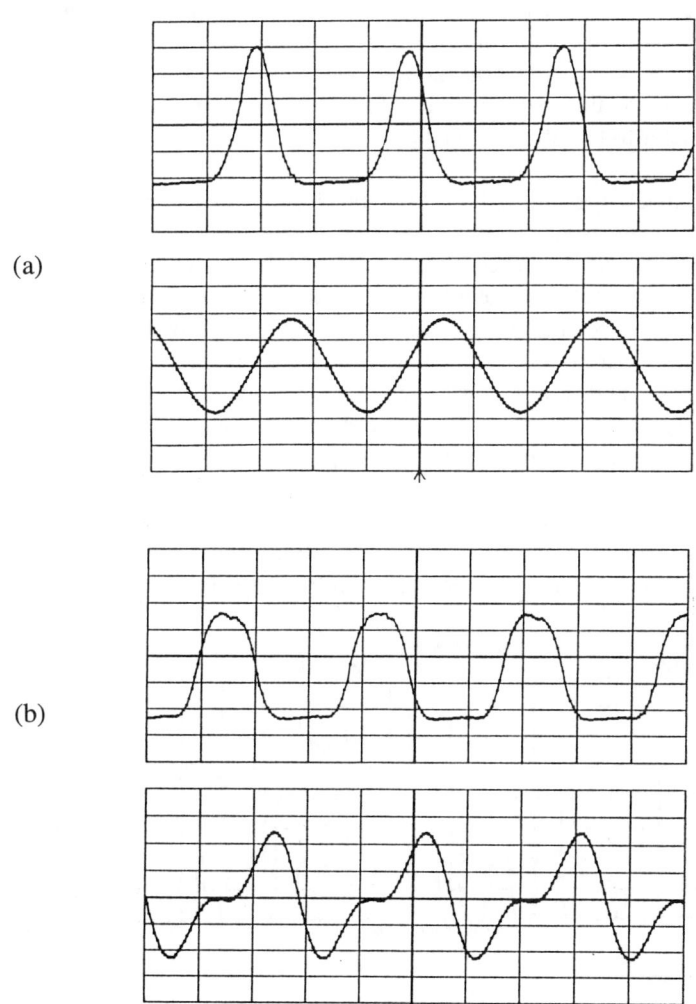

Figure 6. Rf voltage and beam current: (a) single rf system, (b) double rf system.

C. **Direct RF Feedback**

At injection into the AGS, the cavities are operated at 1.5 kV/gap, which requires 0.5 A of current from the power amplifier, (I_o). At 6×10^{13} ppp the rf beam current, (I_B), is 6.0 A, implying a beam loading parameter, I_B/I_o, of 12. It has been shown (8, 9) that when the beam loading parameter becomes greater than 2, the beam control loops, tuning, AVC, and phase, are cross-coupled and become unstable. RF feedback is needed to reduce the effective beam loading parameter. Feedback reduces the perturbations of the gap voltage by the value of the loop gain, and the beam current, seen from the control loops, is effectively reduced. Loop gains of 17 dB and greater (depending on the operating point of the tetrode) are used to reduce the beam loading parameter to less than 1.7 (10).

RF feedback does not reduce the impedance of the cavity but it does reduce the impedance that the beam "sees". It also reduces the beam intensity that the beam loops "see". To measure the effectiveness of the feedback the cavity was stimulated with beam of wide spectral content by using a single bunch, kept short by the other cavities. The short bunch makes spectral lines of almost constant amplitude over the first 17 revolution harmonics. The cavity is tuned to the eighth harmonic with the power amplifier off and shows a characteristic resonance response with a Q of 50 in Figure 7a. In Figure 7b, the power amplifier and the rf feedback have reduced the Q to 5, which is consistent with the feedback loop gain. These results agree well with low-level network analyzer measurements, but are more rigorous in that they stimulate the cavity directly at the gap with beam and are done at high level. At 25 MHZ, the beam signal is down by 20 dB, but is still useful in showing that the higher order modes on the busbars have been effectively damped (10).

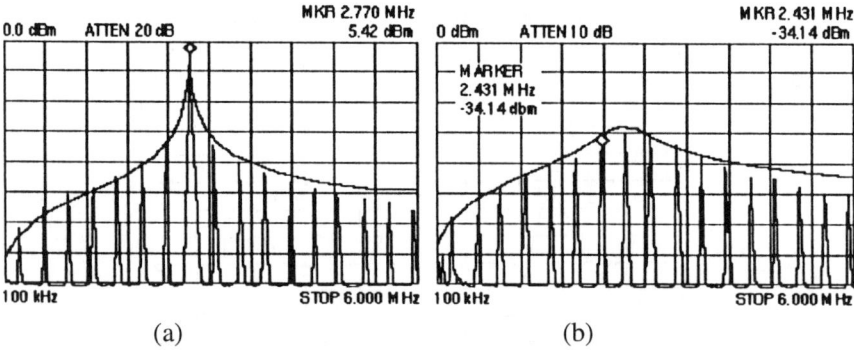

Figure 7.   Cavity impedance measurements: (a) cavity voltage with amplifier off and (b) cavity voltage with amplifier and rf feedback on.

Without such a direct feedback system, the beam loading effect will limit the achievable AGS intensity to about 2.5×10^{13} ppp.

D. <u>Longitudinal Phase Space Dilution</u>

Right after injection, the proton beam suffers longitudinal coupled bunch instabilities when the intensity exceeds 2×10^{13} in the AGS. The growth rate and mode of excitation depend strongly on the intensity and the initial injection errors from the Booster. The way we elect to combat the instabilities is to use the existing VHF cavity, operating at 93 to 100 MHZ, to create a controlled blowup of the longitudinal emittance of the beam. Once the space charge density is reduced, the beam stays stable during the 600 msec period of injection. Since those bunches injected earlier experience a longer time of dilution, the emittance will be larger than later bunches. The mountain range display of bunches inside the AGS from injection time to top energy is shown in Figure 8.

Another time the bunch dilution system is used is right after transition. At this time, the bunch length is the shortest, and the space charge effect is most severe. A strong transverse single bunch instability develops right after transition. Again, the VHF cavity is activated to blow up the longitudinal emittance to reduce the space charge effect and hence alleviate the violent beam blowup as shown in Figure 8.

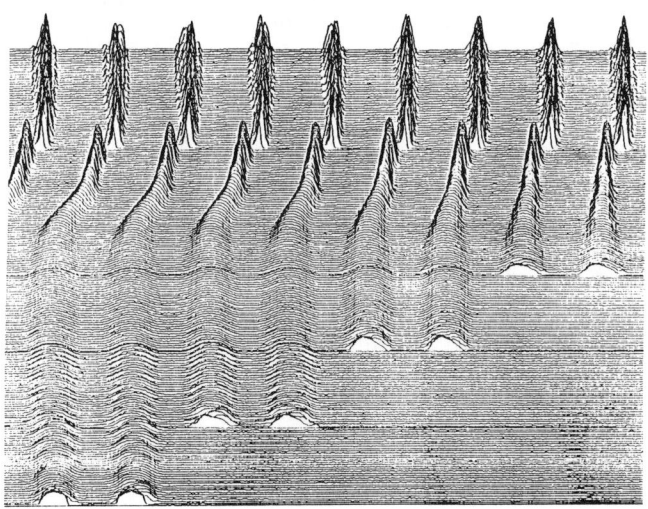

Figure 8.   AGS bunch dilution from injection to beyond transition.

E. Gamma-Transition Jump

At an intensity of approximately 1.5×10^{13} protons per pulse, AGS beam losses at transition are less than 5%. However, as improvement plans are implemented and the intensity is increased to 6×10^{13} protons per pulse, new mechanisms will become important and the losses will increase. A gamma-transition jump system has been built, minimizing these losses by speeding up passage through transition. It follows the work of Werner Hardt (11) at CERN.

Part of the work involves minimizing losses at transition caused by the negative mass instability. Pulsed quadrupole doublets are used to speed up passage through transition. Existing magnets separated by 3/2 betatron wavelength appear adequate for the purpose. Computer modeling has been carried out to determine the rise time and strength of the quadrupoles.

Hardt's idea, which has been implemented at the CERN PS, was based on the observation that quadrupole pairs separated by ½ betatron wavelength and configured as doublets can alter γ_t of a synchrotron without affecting its tune. By arranging to cross transition while γ_t is rapidly decreasing, the bunch area blowup caused by the negative mass instability can be substantially reduced. The criterion for no blowup due to negative mass instability is shown in Figure 9 (12). Here the attainable AGS intensity is plotted as a function of bunch area for several crossing speed enhancement factors, f'.

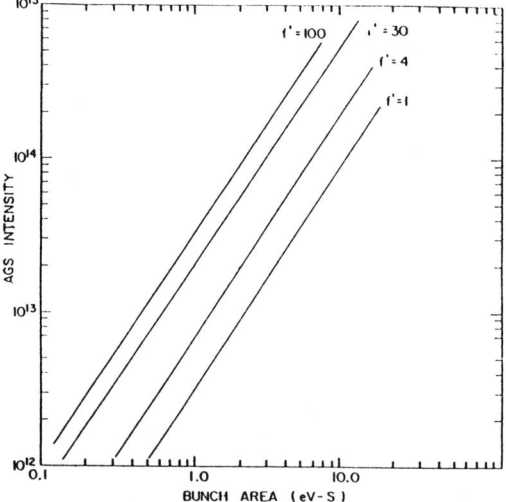

Figure 9. AGS intensity for lossless transition as a function of bunch area and crossing speed enhancement factor, f'.

As can be seen from Figure 9, to pass through transition at 6×10^{13} ppp without appreciable loss requires $f' \approx 30$ with bunch area about 2.0 eV-sec, which is conveniently satisfied by the VHF cavity after injection. The effect on AGS bunches with or without γ_{tr}-jump can be seen in Figure 10 (13).

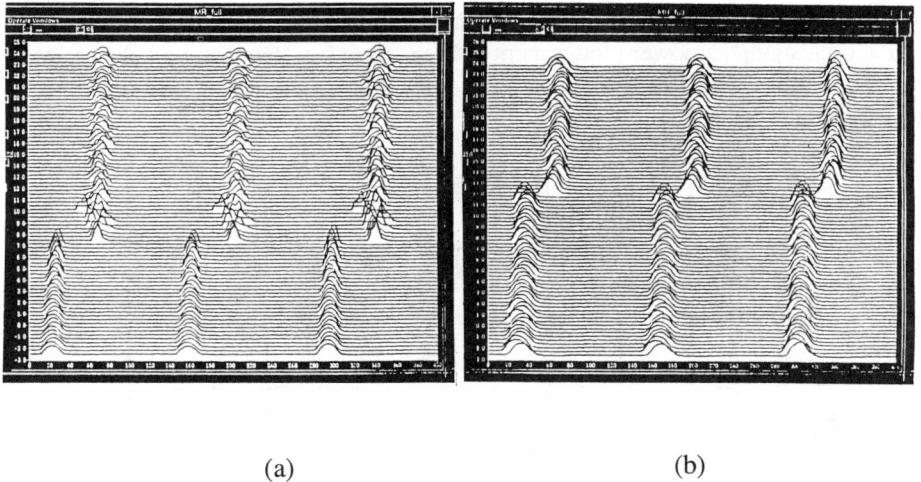

(a) (b)

Figure 10. Bunch shape before and after phase transition: (a) without gamma-transition jump and (b) with gamma-transition jump.

Currently, the AGS still suffers beam losses at transition crossing. This is partially due to the fact that the present gamma-transition jump system does not preserve the dispersion function. In fact, it even increases the dispersion function from 2 m to about 6 m. During transition crossing, the momentum spread is much larger and hence the horizontal beam size due to the dispersion function. An improved system with acceptable dispersion function increase is under investigation for future operations. Another reason for the beam loss during transition in the past is the shrinkage of the rf bucket due to severe beam loading, which has been improved with direct rf feedback.

F. Transverse Coupled-Bunch Instability and Its Damping

It has been estimated that the threshold for transverse coupled bunch instability excited by the resistive wall is at about 4-5 10^{12} ppp. A damper system has been constructed to damp such an instability when it occurs. In Figure 11, the upper trace is the position signal of the vertical orbit and the lower trace is the current of the beam. When the instability occurs, about 60% of the beam is lost. The suppression of coherent motion had been tried successfully by the transverse damper system. The actual threshold of vertical instability has been found to be about 7×10^{12} ppp, when $\upsilon_x = 4.94$ and $\xi_x = -0.25$, which can be avoided by adjusting the tune and chromaticity of the machine. Active damping is necessary when the beam intensity is larger than 10^{13} ppp (14). By supplying a constant amplitude of damping, instead of proportional to the oscillation, the power requirement of the damping system can be reduced by a factor of four. The effectiveness of the constant amplitude method has also been tested in the Tevatron.

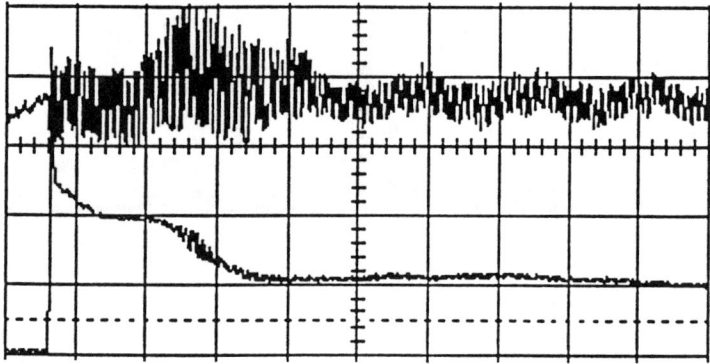

Figure 11.   Signal of transverse instability and intensity.

FUTURE IMPROVEMENTS

A. Stopband Corrections

During 1995 operations, it was found that octupole corrections were needed during the injection process in the AGS. At that time, only two octupoles were available. A more complete correction system with 8 octupoles will be installed for future operations. Out of the 16 sextupoles used for stopband corrections in the past, some were found to develop shorts in the coil winding. These will be repaired and put back in service.

B. New Gamma-Transition Jump Configuration

As mentioned above, the present gamma-transition jump system induces a dispersion increase which in turn limits the horizontal aperture available. A new configuration without undue dispersion increase will be implemented for further intensity increases.

C. Barrier Cavity

During the 1994 and 1995 proton runs, a slow loss observed on the AGS injection porch severely limited the number of protons that could in principle be accelerated. This loss was drastically reduced when the rf was turned off and, to a lesser extent, when the bunch length was increased. A barrier cavity system in the AGS might increase the intensity by avoiding bunched beam operation during injection. This method is discussed in more detail in this Workshop by M. Blaskiewicz (15).

D. AGS Accumulator Ring

To further increase the intensity in the AGS to beyond 10^{14} ppp, an accumulator ring inside the AGS tunnel is needed. Currently, the AGS accepts only 4 pulses from the Booster, which can be operated at 7.5 Hz. If an accumulator can be employed to accept beam continuously without interruption during 1.2-2.5 sec of the AGS cycle times, 2-3 times more beam can be accumulated and delivered into the AGS ring every cycle.

E. Emittance Preservation and Particle Loss Control

So far we have only focused on the attainment of high intensity in the AGS with little regard to the preservation of emittance or particle losses. A better understanding of the critical physics processes can lead to improvement of those two problems. Particle loss is important for current high intensity synchrotrons, such as: AGS, ISIS, and PSR. These accelerators run at a few percent loss during one cycle, which is barely tolerable from the personnel safety and component protection point of view. In all discussions of the next generation pulsed neutron source, a few parts in 10^4 efficiency is required to keep the total particle losses comparable to today's practice.

Preservation of emittance is important for injection into colliders, such as: RHIC, LHC, and the Tevatron. More theoretical exploration, computer simulation, and machine studies are required to improve the understanding of the fundamental physics processes involving space charge effects and achieve design goals for these colliders.

ACKNOWLEDGMENTS

The work discussed in this report has been performed collectively by the accelerator staff in the AGS Department over the past several years.

REFERENCES

1. L.J. Laslett, On Intensity Limitations Imposed by Transverse Space-charge Effects in Circular Particle Accelerators, Proc. 1963 Summer Study on Storage Rings, Accelerators and Experimentation at Super-high Energies, BNL Report 7534, 324-67.

2. W.T. Weng, Space Charge Effects--Tune Shifts and Resonances, in Physics of Particle Accelerators, Ed. M. Mont, AIP Conf. Proc. 154 (AIP, New York, 1987).

3. C. Gardner, et al., Observation of Correction of Resonance Stopbands in the AGS Booster, Proc. 1993 PAC, p. 3633.

4. J. Gareyte, et al., Beam Dynamics Experiments in the PS Booster, Proc. 1975 PAC, p. 1855.

5. W.T. Weng, Operation of the Brookhaven AGS with the Booster, Proc. 1993 PAC, p. 3726.

6. J.M. Kats and W.T. Weng, Effects of the Second Harmonic Cavity on RF Capture and Transition Crossing, Proc. XVth Int. Conf. on High Energy Accelerators, Hamburg, Germany, July 20-24, 1992, p. 1052.

7. J.M. Brennan, private communication, BNL, 1995.

8. F. Pedersen, Beam Loading Effects in the CERN PS Booster, IEEE Trans. Nucl. Sci., NS-22, 1975, p. 1906.

9. D. Boussard, Design of a Ring RF System, Proc. CERN School on RF Engineering, p. 474, Oxford, England, April 1991.

10. J.M. Brennan, The Upgraded RF system for the AGS and High Intensity Proton Beams, Proc. 1995 PAC.

11. W. Hardt, Gamma Transition-jump Scheme of the CPS, Proc. Ninth Intl. Conf. on High Energy Part. Accel., p. 434, Stanford, 1974.

12. P. Yamin, et al., A Fast Transition Jump Scheme at the Brookhaven AGS, Proc. 1987 IEEE PAC, p. 1216.

13. W.K. van Asselt, et al., The Transition Jump System for the AGS, Proc. 1995 PAC.

14. D. Russo, M. Brennan, M. Meth, T. Roser, Results from the AGS Booster Transverse Damper, Proc. 1993 PAC, p. 2286.

15. M. Blaskiewicz, Barrier Cavity Longitudinal Dynamics, these proceedings.

Space Charge Induced Non Structure Resonances

S. Machida and Y. Shoji*
National Laboratory for High Energy Physics (KEK)
1-1 Oho, Tsukuba-shi, Ibaraki-ken, 305 JAPAN[†]

December 15, 1995

Abstract

In this paper, we emphasize the space charge force as a resonance source. Like an ordinary resonance excited by lattice imperfections, space charge induced resonances are categorized into structure and non structure resonances. Space charge induced non structure resonances are excited when a beam envelope is perturbed by quadrupole errors in a lattice.

1 Introduction

In a proton synchrotron, repulsive force among particles, namely space charge force, limits the total number of particles in the machine. It is said that imperfections of lattice magnets excite a betatron resonance and a particle which has a large tune shift crosses the lower order resonances and eventually gets lost. Usually, a particle cannot survive for long time at resonances of and lower than fourth order, so that the allowed tune spread is 0.25. Therefore, the total number of particles which gives the tune shift of 0.25 is about the maximum limit.

In those modelling of space charge effects, resonance sources are imperfections of lattice magnets. In other words, if one could correct the lattice imperfections by some correction magnets up to some orders, the allowed tune shift can be enlarged and the maximum possible number of particles in the machine increases.

Parzen [1] and Machida [2], however, recently pointed out that even without any imperfections of lattice magnets, the space charge force itself could be a source of resonances. Since the space charge force is modulated according to the beam envelope, it has strong harmonic components which are superperiod times integer. Even

---

*Present address: Himeji Institute of Technology, Shosya, Himeji-shi, 671-22 JAPAN
[†]E-mail: machida@kekvax.kek.jp and shojiy@kekvax.kek.jp

© 1996 American Institute of Physics

if one could correct the lattice imperfections completely, a particle would be trapped in that space charge induced resonances and it sets a limit to the number of particles. Notice that space charge force can excite only even order resonances because of the symmetry of the force. Also, their simulation results show that fourth order effects are most significant.

These resonances should be called, more precisely, space charge induced *structure* resonances. By that effects, a particle can be trapped in the resonances which are excited by the same beam. The only way to reduce the effects is to have higher superperiodicity, which enlarge the tune space area free from the resonances.

Although the main harmonic components of the space charge force comes from the lattice superperiodicity, other harmonics can be also excited if the beam envelope is perturbed by quadrupole errors. In an actual machine, there is more or less such an envelope perturbation and the space charge force could excite *non structure* resonances, too. We will show the evidence of the space charge induced *non structure* resonances, especially fourth order, experimentally and by simulation.

In the following, we will first show some experimental evidence of that space charge induced non structure resonances in the KEK PS, more specifically one dimensional fourth order resonance. Secondly, we will estimate the strength of the space charge induced resonances with a realistic lattice perturbation. Finally, some results of simulation study will be described.

2 Experimental Evidence

In the KEK PS, it has been known for long time that the beam survival ratio; the beam intensity at 350 ms after injection to that right after injection, becomes worse at and above the vertical tune of 7.25. The survival ratio becomes even worse when the beam intensity is increased. First, it was thought that the lattice has octupole errors somewhere and the stop band of $\nu_y = 7.25$ is strong. The correction of the resonance was tried by several octupole magnets and the resonance strength was estimated. It turns out that the lattice imperfections were not the source of the bad survival.

To see the resonance structure in more detail, the vertical tune was scanned with the fixed horizontal tune of $\nu_x = 7.12$ and saw the survival ratio with the number of particles per bunch as a parameter. Figure 1 shows the results. The horizontal axis is the coherent tune measured at each beam intensity. When the number of particles is 1.0×10^{11}, the survival ratio becomes gradually worse towards half integer resonance at $\nu_y = 7.50$, but no fine structure is observed. When the number of particles is 3.2×10^{11}, a dip is observed just above $\nu_y = 7.25$. Above that tune, the survival ratio becomes gradually worse as similar to the lower intensity case. When the number of particles becomes more, the dip near $\nu_y = 7.25$ is shifted to higher tune and becomes deeper. At the same time, the dip shape is destroyed on the higher tune side. The survival ratio is bad all the way from the dip to the half integer.

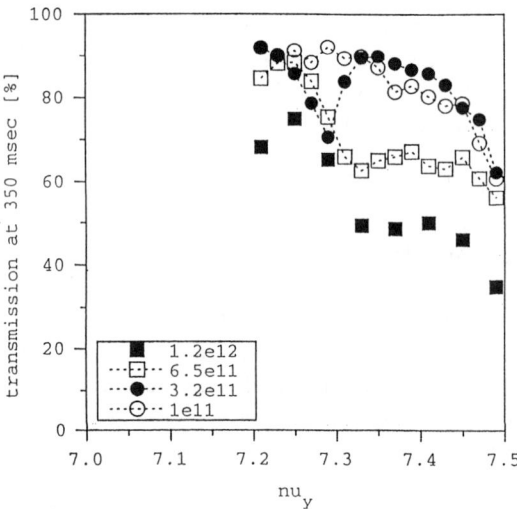

Figure 1: We observed the beam survival ratio at 350 ms from the injection as a function of the vertical tune. Four different beam intensity (number of particles in a bunch) was taken. A dip near $\nu_y = 7.25$ emerges when the beam intensity is increased.

From those results, we thought that there is a fourth order resonance at $\nu_y = 7.25$, whose strength depends on the beam intensity. At that point, we began to investigate the space charge induced non structure resonance.

3 Quick Look at Resonance Width

We first look at the resonance width of the space charge induced resonances as follows. The Hamiltonian including magnet imperfections and space charge force is written as

$$
\begin{aligned}
H &= \frac{1}{2}p_x^2 + \frac{1}{2}p_y^2 + \frac{1}{B\rho}[\frac{1}{2!}\frac{\partial B_y}{\partial x}(x^2 - y^2) \\
&+ \frac{1}{3!}\frac{\partial^2 B_y}{\partial x^2}(x^3 - 3xy^2) \\
&+ \frac{1}{4!}\frac{\partial^3 B_y}{\partial x^3}(x^4 - 6x^2y^2 + y^4) + ...] \\
&+ \frac{r_p \lambda}{\beta^2 \gamma^3 B}[-\frac{1}{\sigma_x(\sigma_x + \sigma_y)}x^2 - \frac{1}{\sigma_y(\sigma_x + \sigma_y)}y^2 \\
&+ \frac{2\sigma_x + \sigma_y}{12\sigma_x^3(\sigma_x + \sigma_y)^2}x^4 + \frac{1}{2\sigma_x\sigma_y(\sigma_x + \sigma_y)^2}x^2 y^2
\end{aligned}
$$

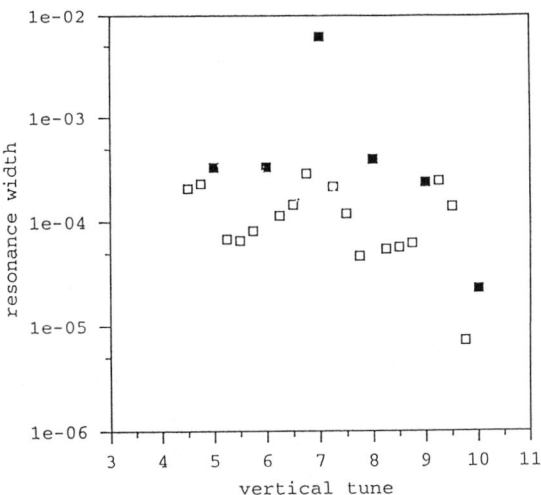

Figure 2: The width of the space charge induced resonances defined as the formula below. Since the KEK PS has fourfold symmetry, every integer tune is a fourth order structure resonance. A quadrupole imperfection is included in the lattice so that the beam envelope is perturbed. That is why non structure resonances, as well as structure resonances, are excited by the space charge force.

$$+\frac{2\sigma_y + \sigma_x}{12\sigma_y^3(\sigma_x + \sigma_y)^2}y^4 + ...].$$

We assumed that the beam distributions are Gaussian in the horizontal and vertical planes and a coasting beam in the longitudinal. The σ_x and σ_y are the horizontal and vertical rms beam size, respectively, and λ is a line charge density. The first square bracket shows the magnetic force terms in lattice elements. They are essentially zero if the lattice is constructed in a perfect manner except a quadrupole term which provides focusing force and a sextupole one which is introduced intentionally to correct chromaticity. The second square bracket shows the space charge terms of both linear and nonlinear contributions. Those exist even if the lattice is perfectly constructed.

Once we know the Hamiltonian, the fourth order resonance width can be estimated as

$$\Delta e_s = \frac{2J_y}{4\pi}\int \beta_y^2 \frac{1}{\beta^2\gamma^3 B}\frac{2\sigma_y + \sigma_x}{12\sigma_y^3(\sigma_x + \sigma_y)^2}e^{-i\cdot 4\phi_y}ds,$$

where $2J_y$ is the Courant-Synder invariant, β_y is the vertical amplitude function and ϕ_y is the vertical phase of betatron oscillations.

Figure 2 shows the width of the space charge induced resonances, both structure and non structure resonances in the KEK PS. Since the KEK PS has fourfold symmetry, at every integer tune, a fourth order resonance becomes a structure resonance. The modelled lattice has a localized quadrupole imperfection such that the quadrupole resonance width at $\nu_y = 7.5$ is about 0.04. That is the major source for the perturbation of the beam envelope. The rms normalized emittance and the number of particles per bunch are assume 6×10^{-6} π·m·rad for horizontal, 3×10^{-6} π·m·rad for vertical, and 1.0×10^{12} (the total number of particles is 9 times larger), respectively. Besides the space charge force, there is no fourth order resonance source in the lattice.

As expected, structure resonances at every integer has stronger width compared with nearby non structure resonances. At $\nu_y = 7$, it has the maximum strength where the phase advance of the FODO unit cell is 90 degree so that the periodicity of the driving force (space charge force) modulated by the beam envelope matches the fourth order phase. We call it "super structure resonance". In addition, the space charge induced non structure resonances between every structure resonances are clearly shown and can be as strong as other structure resonances. In particular, the ones near "super structure resonance" are strong and become weaker as the distance from "super structure resonance" is increased.

4 Simulation Results

The estimate of the space charge induced resonance width supports our conjecture that there is such a resonance in the KEK PS. To get more confidence, macro particle simulation has been performed to reproduce the experimental result.

We take two kinds of modelling for space charge force [3]. One way, which is the simplest one, is to assume that beams have always Gaussian distributions in the six dimensional phase space and no emittance growth occurs. That is similar to the weak-strong approximation in a study of beam-beam effects. The space charge force is included as an external force. In this way, a single particle simulation can be performed.

The other way is more or less self consistent. Although the Gaussian distributions are assumed, the transverse emittance is updated every turn using a bunch of macro particles, say 1000 particles. In this way, we can include the time evolution of emittance, and therefore space charge force can be time dependent.

Figure 3 shows the simulation results which should corresponds to the beam study results mentioned above. The later modelling of space charge force is used. In that figure, the vertical axis is the emittance at 5 ms after the injection. When a resonance is excited and some particles are trapped, the emittance growth occurs. It appears as a bump, instead of a dip in the beam study. The emittance growth near integer $\nu_y = 7$ is manifest, which is excited by "super structure resonance".

Since simulated time evolution is not long enough; 5 ms compared with 350 ms of the the beam study, the emittance growth around $\nu_y = 7.25$ is not clear, but small

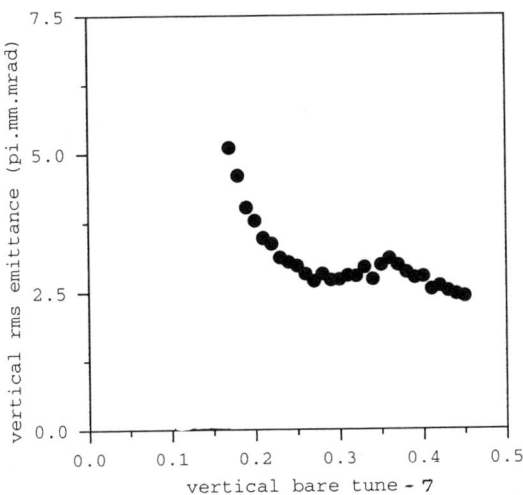

Figure 3: A simulation result corresponding to Figure 1. The vertical axis is the emittance at 5 ms after injection. A dip in Figure 1 appears as a bump near $\nu_y = 7.35$.

bump just above $\nu_y = 7.35$ is visible. Note that the horizontal axis is the bare tune, not the coherent tune, so that the resonance at $\nu_y = 7.25$ becomes visible when the bare tune is $\nu_y = 7.35$ or above. The maximum incoherent tune shift is about -0.4 with the intensity assumed.

In fact, the resonance structure becomes more clear when we look at the Poincare map of a two sigma amplitude particle as shown in Figure 4, where we use the former modelling of space charge force. The bare tune is set at $\nu_y = 7.36$ and the loaded tune is shifted to $\nu_y = 7.25$, where the space charge induced non structure resonance is excited. That makes small bump in Figure 3.

According to the estimate of the resonance width above, the space charge non structure resonance becomes weaker when we choose the tune far from "super structure resonance". We did the same exercise of Figure 3 except that the integer part is chosen to be 5. We expected the space charge non structure resonance at $\nu_y = 5.25$ should be weaker than that at $\nu_y = 7.25$. Actually, Figure 5 shows that no bump around $\nu_y = 5.35$. Also, as shown in Figure 6, the Poincare map of a two sigma amplitude particle does not show any resonance structure either. We are trying to see the difference of $\nu_y = 7.25$ and $\nu_y = 5.25$ experimentally.

5 Summary

We observed the fourth order resonance ($\nu_y = 7.25$) with a beam intensity dependence in the KEK PS. One possible explanation is that the resonance is excited by

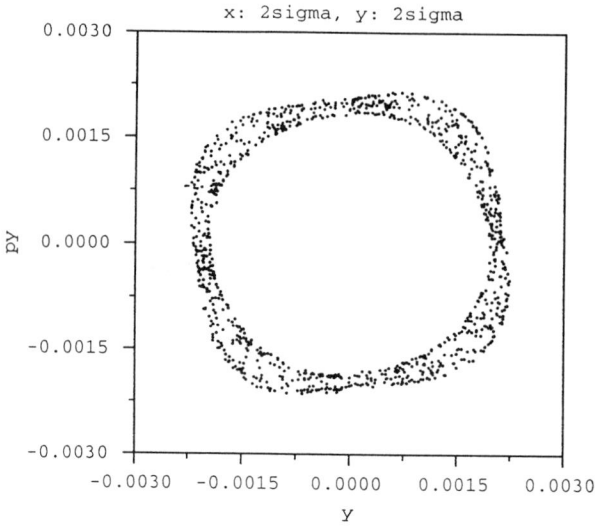

Figure 4: Poincare map of a two sigma amplitude particle at $\nu_y = 7.36$ in Figure 3. Fourth order resonance structure is shown.

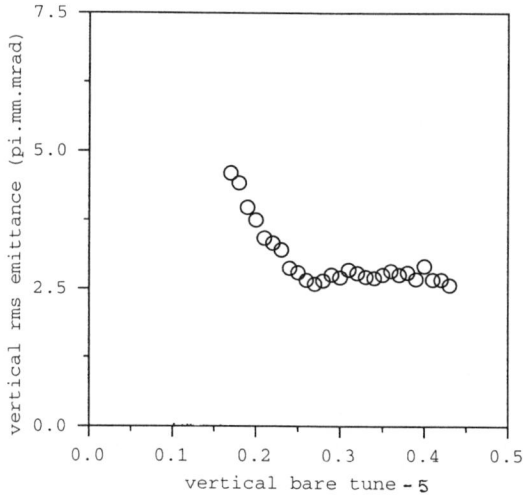

Figure 5: A simulation result corresponding to Figure 1 except that the integer part of the vertical tune is 5. The vertical axis is the emittance at 5 ms after injection. No bump appears near $\nu_y = 5.35$.

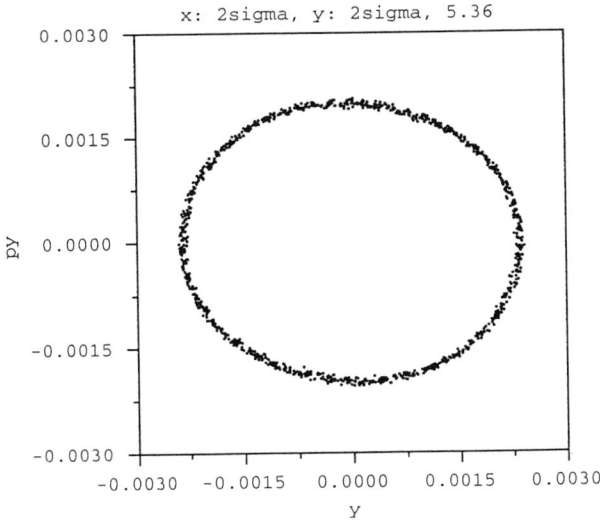

Figure 6: Poincare map of a two sigma amplitude particle at $\nu_y = 5.36$ in Figure 5. No resonance structure is shown.

the space charge force. To confirm that, simple estimate of the resonance width and macro particle simulation have been done. It was shown that the non structure resonances can be excited by space charge force together with the beam envelope perturbations by quadrupole errors.

As the space charge induced *structure* resonance has its maximum strength when the unit FODO cell has 90 degree phase advance ($\nu_y = 7$ in the KEK PS), the space charge induced *non structure* resonance near that phase advance becomes stronger than others. Its strength decreases as its phase advance is apart from 90 degree. In the KEK PS, for example, the space charge induced non structure resonance with the integer part of tune 5 is supposed to be weaker than that of integer part 7.

Acknowledgements

The experiment was done by the KEK PS intensity upgrade task force. Among them, Drs. M. Shirakata, T. Toyama, M. Yoshii, K. Takayama, J. Kishiro, and H. Sato took a major part to take and analyse the data. The authors would like to appreciate their efforts and discussions. The authors would like to thank Dr. I. Yamane for his encouragement during this study.

References

[1] G. Parzen, Nuclear Instruments and Methods in Physics Research A281 (1989) p.413.

[2] S. Machida, Nuclear Instruments and Methods in Physics Research A309 (1991) p.43.

[3] S. Machida, "The Simpsons Program", AIP Conference Proceedings 297, p.459, 1993.

Halo Formation and Chaos in Space-Charge-Dominated Beams

C. Chen
Plasma Fusion Center
Massachusetts Institute of Technology
Cambridge, Massachusetts 02139

ABSTRACT

The mechanisms of halo formation and chaotic particle motion are explored for space-charge-dominated beams propagating through a linear, periodic focusing channel provided by an applied periodic solenoidal magnetic field. For root-mean-squared (rms) matched beams with nonuniform density profiles, it is shown in a test-particle model that nonlinearities in the self field forces not only can result in chaotic particle motion but also can cause a halo to develop. The size of the halo is found to be bounded by a Kolmogorov-Arnold-Moser (KAM) surface. For rms mismatched beams, it is confirmed in the computer simulations that, for beams mismatched into the periodic focusing channel, the beam envelope exhibits nonlinear resonances and chaotic behavior, as predicted by the previous analysis of the beam envelope equation. As a result of emittance growth, transient effects are observed in the rms beam dynamics, and halos can develop. The halo characteristics for a periodic focusing channel are found to be qualitatively the same as those for a uniform focusing channel. A threshold condition is obtained numerically for halo formation for mismatched beams in the uniform focusing approximation. These results indicate that both precise rms beam matching and precise density profile control are required to prevent space-charge-dominated beams from developing halos.

I. INTRODUCTION

Halo formation is an important issue in the design of advanced high-current, high-power particle accelerators in wide applications such as high-energy physics research, tritium production, and heavy ion fusion as well as in the development of advanced radiation sources and high-power, high-resolution radar. The consequences of halo production range from emittance growth in the beam, to the buildup of radioactivity in the accelerator, to the meltdown of system components, to mention a few examples. It has been recognized recently [1-6] that halos around space-charge-dominated beams are attributed to chaotic particle orbits induced by *intrinsic* nonlinearities in the beam dynamics. Such chaotic particle orbits not only are sensitive to initial conditions, but also occupy a larger region in phase space than regular particle orbits, leading to halo development as well as the growth of the total beam emittance.

In this paper, we explore the mechanisms of halo formation and chaotic behavior in continuous, space-charge-dominated beams propagating through a linear, periodic focusing channel provided by a periodic solenoidal magnetic field. A space-charge-dominated beam satisfies the condition

$$\frac{SK}{4\sigma_0 \epsilon} > 1,$$

where K is the normalized beam perveance, ϵ is the unnormalized rms emittance of the beam, S is the periodicity of the focusing channel, and σ_0 is the vacuum phase advance over one period. For an electron beam,

$$\frac{SK}{4\sigma_0 \epsilon} = 2.9 \times 10^{-5} \frac{1}{\sigma_0} \left(\frac{S}{\epsilon_n}\right) \frac{I_b}{\gamma_b^2 \beta_b^2},$$

where I_b is the beam current in Amperes, $\epsilon_n = \gamma_b \beta_b \epsilon$ is the normalized rms emittance, and $\gamma_b = (1 - \beta_b^2)^{-1/2}$ is the (average) relativistic mass factor of the electrons. For an ion beam,

$$\frac{SK}{4\sigma_0 \epsilon} = 1.6 \times 10^{-8} \frac{1}{\sigma_0 A} \left(\frac{q}{e}\right) \left(\frac{S}{\epsilon_n}\right) \frac{I_b}{\gamma_b^2 \beta_b^2},$$

where q/e and A are the ion charge state and atomic mass, respectively, I_b is the beam current in Amperes, $\epsilon_n = \gamma_b \beta_b \epsilon$ is the normalized rms emittance, and $\gamma_b = (1 - \beta_b^2)^{-1/2}$ is the (average) relativistic mass factor of the ions.

After presenting a test-particle model and a self-consistent model in Sec. II, we use the test-particle model to study the dynamics of space-charge-dominated beams which is rms matched into the periodic focusing channel but has a nonuniform density profile transverse to the direction of propagation (Sec. III). It is shown that self-field nonlinearities due to the density nonuniformity induces chaotic particle motion which causes some of the particles to escape from the beam interior to form a halo. Because rms beam matching, which does not guarantee necessarily exact beam matching except for the ideal Kapchinskij-Vladimirskij (KV) distribution, is widely utilized in accelerator design, this halo mechanism is of particular importance.

The self-consistent two-dimensional macroparticle model is used to investigate the evolution of both rms beam quantities and the particle phase-space distribution in the regime where a mismatch between the beam and the focusing channel occurs (Sec. IV). As predicted by the previous envelope analysis [7-9], nonlinear resonant and chaotic phenomena in the envelope evolution are confirmed in the computer simulations, supporting the expectation that such nonlinear phenomena should be experimentally observable. Halo formation is investigated. While the analytical model [1, 10] for envelope-mismatched beams without emittance growth does not provide an escape mechanism for core particles to move into the halo, the emittance growth and transient effects in the self-consistent model provide escape mechanisms. The size of the halo is estimated. The halo characteristics for periodic focusing configurations are found to be qualitatively the same as those for uniform focusing configurations. A threshold condition for halo formation is obtained numerically.

These results together with qualitatively the same results obtained for the alternating-gradient quadrupole focusing configuration [2-4] indicate that both precise rms beam matching and precise density profile control are required to prevent space-charge-dominated beams from developing halos.

II. MATHEMATICAL MODELS

We consider a thin, continuous, intense charged-particle beam propagating with average axial velocity $\beta_b c \vec{e}_z$ through an axisymmetric, linear focusing channel provided by the applied, periodic solenoidal magnetic field

$$\vec{B}_0(x,y,s) = B_z(s)\vec{e}_z - \frac{1}{2}B_z'(s)(x\vec{e}_x + y\vec{e}_y) \tag{1}$$

and

$$\vec{B}_0(x,y,s+S) = \vec{B}_0(x,y,s), \tag{2}$$

as illustrated schematically in Fig. 1. Here, $s = z$ is the axial coordinate, S is the fundamental periodicity length of the focusing field, the 'prime' denotes derivative with respect to s, and c is the speed of light in *vacuo*. To study the beam dynamics, we present a test-particle model and a self-consistent model as follows.

A. Test-Particle Model

In the present test-particle model, it is assumed that the beam is rms matched into the periodic focusing channel but has a *nonuniform* density profile transverse to the direction of beam propagation. It is also assumed that the beam density profile is given *a priori* by the parabolic form

$$n_b(r,s) = \begin{cases} \hat{n}_b + \delta\hat{n}_b - 2\delta\hat{n}_b r^2/r_b^2, & \text{for } r < r_b(s) \\ 0, & \text{otherwise.} \end{cases} \tag{3}$$

In Eq. (3), $r = (x^2 + y^2)^{1/2}$ is the radial coordinate, $r_b(s) = r_b(s+S)$ is the radius of the rms matched beam, $N = \int n_b(x,y,s)dxdy = \pi\hat{n}_b(s)r_b^2(s) = $ const. is the number of particles per unit axial length of the beam, and $\delta\hat{n}_b(s) = \delta N/\pi r_b^2(s)$ is a measure of the nonuniformity of the beam density and is allowed to be in the range $0 \le \delta\hat{n}_b \le \hat{n}_b$.

Following Sacherer [11], it is readily shown that the rms-matched beam radius $r_b(s) = r_b(s+S)$ satisfies the envelope equation

$$\frac{d^2 r_b}{ds^2} + \kappa_z(s)r_b - \frac{gK}{r_b} - \frac{(4g\epsilon)^2}{r_b^3} = 0, \tag{4}$$

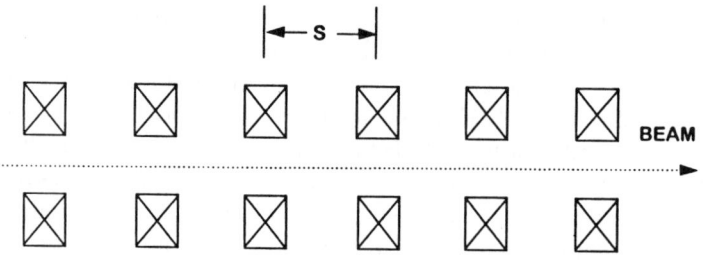

Figure 1: Schematic of intense charged-particle beam propagation through a periodic solenoidal magnetic field.

where

$$\kappa_z(s) = \left[\frac{qB_z(s)}{2\gamma_b\beta_b mc^2}\right]^2 = \kappa_z(s+S) \tag{5}$$

is the focusing parameter squared, $K = 2q^2 N/\gamma_b^3\beta_b^2 mc^2$ is the normalized beam perveance, m and q are the rest mass and charge of the particle, respectively, $\beta_b c$ is the average axial beam velocity, $\gamma_b = (1-\beta_b^2)^{-1/2}$ is the relativistic mass factor, $g = (1 - \delta N/3N)^{-1}$ is the density shape factor, and ϵ is the unnormalized rms emittance. The emittance ϵ is taken to be constant because the present test-particle model is aimed at the *onset* of halo formation, where the rms properties of the beam are not expected to vary appreciably.

It should be recalled that for a round beam in the solenoidal focusing channel, the relation $\epsilon = \epsilon_{\tilde{x}} = \epsilon_{\tilde{y}}$ holds, where

$$\epsilon_{\tilde{x}} = (\langle \tilde{x}^2\rangle\langle \tilde{x}'^2\rangle - \langle \tilde{x}\tilde{x}'\rangle^2)^{1/2} \tag{6}$$

and

$$\epsilon_{\tilde{y}} = (\langle \tilde{y}^2\rangle\langle \tilde{y}'^2\rangle - \langle \tilde{y}\tilde{y}'\rangle^2)^{1/2} \tag{7}$$

are the unnormalized rms emittances [12] in the $\tilde{x}$- and $\tilde{y}$-directions in the Larmor frame [13] of reference, respectively. In Eqs. (6) and (7), $\langle\cdots\rangle$ denotes the ensemble average over the beam particle distribution, and the coordinate in the Larmor frame of reference, $(\tilde{x},\tilde{y})$, is related to the coordinate in the laboratory frame of reference, (x,y), by the relations

$$\tilde{x}(s) = x(s)\cos[\phi(s)] - y(s)\sin[\phi(s)], \tag{8}$$

$$\tilde{y}(s) = x(s)\sin[\phi(s)] + y(s)\cos[\phi(s)], \tag{9}$$

where $\phi(s) = \int_{s_0}^{s} ds[\kappa_z(s)]^{1/2}$.

In the paraxial approximation, the equilibrium Poisson equation can be expressed as

$$\frac{1}{r}\frac{\partial}{\partial r}r\frac{\partial}{\partial r}\Phi^{(s)}(r,s) = -4\pi q n_b(r,s), \tag{10}$$

where $\partial^2/\partial s^2 \cong 0$. From Eq. (10), the scalar potential for the self-electric field associated with the beam space-charge is found to be

$$\Phi^{(s)}(r,s) = \begin{cases} -q(N+\delta N)r^2/r_b^2(s) + q(\delta N/2)r^4/r_b^4, & \text{for } r < r_b(s) \\ -qN - q\delta N/2 - 2qN\ln[r/r_b(s)], & \text{otherwise.} \end{cases} \tag{11}$$

The vector potential for the self-magnetic field associated with the beam current is defined by

$$\vec{A}^{(s)}(r,s) = \beta_b \Phi^{(s)}(r,s)\vec{e}_z. \tag{12}$$

Therefore, the self-electric and self-magnetic fields are defined, respectively, by $\vec{E}^{(s)} = -\nabla_\perp \Phi^{(s)}$ and $\vec{B}^{(s)} = \nabla_\perp \times \vec{A}^{(s)}$, where $\nabla_\perp = \vec{e}_x \partial/\partial x + \vec{e}_y \partial/\partial y$.

In the Larmor frame of reference, it can be shown that the transverse equations of motion for an individual test particle in the combined periodic solenoidal and self fields $\vec{E}^{(s)}$ and $\vec{B}_0 + \vec{B}^{(s)}$ are

$$\frac{d^2\tilde{x}}{ds^2} + \kappa_z(s)\tilde{x} + \frac{q}{\gamma_b^3\beta_b^2 mc^2}\frac{\partial}{\partial \tilde{x}}\Phi^{(s)}(\tilde{x},\tilde{y},s) = 0, \tag{13}$$

$$\frac{d^2\tilde{y}}{ds^2} + \kappa_z(s)\tilde{y} + \frac{q}{\gamma_b^3\beta_b^2 mc^2}\frac{\partial}{\partial \tilde{y}}\Phi^{(s)}(\tilde{x},\tilde{y},s) = 0, \tag{14}$$

where $\Phi^{(s)}(\tilde{x},\tilde{y},s) = \Phi^{(s)}(r,s)$ is defined in Eq. (11), and $r = (\tilde{x}^2 + \tilde{y}^2)^{1/2}$. It is readily shown that the canonical angular momentum is conserved, i.e.,

$$P_\theta = \gamma_b m \beta_b c(\tilde{x}\tilde{y}' - \tilde{y}\tilde{x}') = \text{const.}, \tag{15}$$

because of the axisymmetry.

It is important to specify initial conditions for test particles that are consistent with the density profile assumed in Eq. (3). This is accomplished by the particular choice of initial distribution function at $s = s_0$,

$$f(\tilde{x},\tilde{y},\tilde{x}',\tilde{y}',s_0) = \frac{N - \delta N}{16\pi^2 g^2 \epsilon^2}\delta(W-1) + \frac{\delta N}{8\pi^2 g^2 \epsilon^2}H(W). \tag{16}$$

In Eq. (16), $\delta(x)$ is the Dirac δ-function, $H(x)$ is the function defined by $H(x) = +1$ for $0 \le x \le 1$, and $H(x) = 0$, otherwise, and W is the variable defined by

$$W = \frac{\tilde{x}^2}{r_b^2} + \frac{\tilde{y}^2}{r_b^2} + \frac{1}{16g^2\epsilon^2}\left[(r_b\tilde{x}' - \tilde{x}r_b')^2 + (r_b\tilde{y}' - \tilde{y}r_b')^2\right]. \tag{17}$$

Here, r_b and r'_b denote the "initial" values at $s = s_0$. It is readily verified that $n_b(x,y,s_0) = \int f d\tilde{x}' d\tilde{y}'$ indeed yields the parabolic density profile in Eq. (3), and that $4\pi g\epsilon$ and $4\pi g\epsilon$ are the maximum initial areas occupied by the beam particles in the phase planes $(\tilde{x}, \tilde{x}')$ and $(\tilde{y}, \tilde{y}')$, respectively.

The dynamical equations (13) and (14) together with Eqs. (4), (11), and (16) completely describe the model and will be used in Sec. III to explore the mechanism of halo production in intense charged-particle beams that are rms matched into *periodic* solenoidal focusing channels with charge density nonuniformities.

B. Self-Consistent Model

In the present self-consistent two-dimensional macroparticle model, the beam is represented by N_P macroparticles. The beam density is approximated by

$$n_b(x,y,s) = \frac{N}{N_P} \sum_{i=1}^{N_P} \delta[x - x_i(s)]\delta[y - y_i(s)], \qquad (18)$$

where (x_i, y_i) is the transverse position of the i-th macroparticle, and $\delta(x)$ is the Dirac delta function. For such a beam of N_P macroparticles moving in the combined periodic solenoidal and self-consistent self fields $\vec{E}^{(s)}$ and $\vec{B}_0 + \vec{B}^{(s)}$, the transverse equations of motion for the i-th macroparticle can be expressed in the Larmor frame of reference as [6]

$$\frac{d^2 \tilde{x}_i}{ds^2} + \kappa_z(s)\tilde{x}_i + \frac{q}{\gamma_b^3 \beta_b^2 mc^2} \frac{\partial}{\partial \tilde{x}_i} \Phi^{(s)}(\tilde{x}_i, \tilde{y}_i) = 0, \qquad (19)$$

$$\frac{d^2 y_i}{ds^2} + \kappa_z(s)\tilde{y}_i + \frac{q}{\gamma_b^3 \beta_b^2 mc^2} \frac{\partial}{\partial \tilde{y}_i} \Phi^{(s)}(\tilde{x}_i, \tilde{y}_i) = 0, \qquad (20)$$

where $i = 1, 2, \cdots, N_P$, and

$$\Phi^{(s)}(\tilde{x}_i, \tilde{y}_i) = \frac{qN}{4\pi N_P} \sum_{j=1(j\neq i)}^{N_P} \ln[(\tilde{x}_i - \tilde{x}_j)^2 + (\tilde{y}_i - \tilde{y}_j)^2] \qquad (21)$$

is the self-consistent scalar potential experienced by the i-th macroparticle and is obtained self-consistently from Eq. (18) and the Poisson equation.

As described by Eqs. (19)-(21), the self-consistent two-dimensional macroparticle model for intense charged-particle beams involves $2N_P$ second-order ordinary differential equations which can be integrated numerically with a computer code. The present macroparticle (direct interaction) model, which is equivalent to particle-in-cell (PIC) models, is more straightforward but requires more computations than corresponding PIC models.

It is inevitable that round-off errors and discrete particle effects generate noise in computer simulations of charged-particle beams, regardless of whether the present macroparticle model or a PIC model is used. Therefore, it is important to validate simulation results, which is done in part by simulating beam propagation with the matched KV equilibrium distribution. The KV equilibrium [14-16] is the only known Vlasov equilibrium for periodically focused intense charged-particle beams and has been discussed extensively [8] for the periodic solenoidal magnetic field configuration. In the configuration space, a stable KV beam equilibrium corresponds to a solid beam with a uniform transverse density profile and a radius which varies periodically in the direction of propagation with the same periodicity as the focusing lattice.

It is demonstrated in our benchmark simulations that, with as many as 10^3 macroparticles in the present self-consistent model, the properties of stable KV beam equilibria are preserved over propagating distances at least on the order of 100 focusing periods. In particular, the computer simulations show that the particles are well confined within the periodically varying outermost beam envelope, and that the beam emittance remains constant within expected relative fluctuations of order of $N_P^{-1/2}$.

III. CHAOS AND HALOS IN RMS MATCHED BEAMS

In the section, use is made of the test-particle model in Sec. III.A to show that chaotic particle motion induced by the nonlinearities in the self-field forces is responsible for particle escape from the beam interior to form a halo around an intense charge-particle beam which is rms matched into a *periodic* focusing channel but has a *nonuniform* density profile transverse to the direction of beam propagation.

For a uniform intense beam with $\delta N = 0$ (i.e., with $g = 1$), the self-field forces in Eqs. (13) and (14) have linear dependences on x and y within the beam envelope [i.e., $r \leq r_b(s)$]. In fact, for $W \leq 1$, Eqs. (13) and (14) are decoupled (linear) Hill's equations, which are integrable. Any particle with initial conditions such that $W(s = s_0) \leq 1$ is well behaved and well confined within the beam envelope, provided the latter is stable. Therefore, in the context of the test-particle model, no halo can develop around such a uniform beam which is a KV beam equilibrium. It should be pointed out that for $W > 1$, particle orbits can become chaotic because Eqs. (13) and (14) describe a Hamiltonian system with one and one-half degrees of freedom and are no longer integrable.

For a *nonuniform* intense beam with $\delta N > 0$ (i.e., with $g > 1$), however, Eqs. (13) and (14) are nonintegrable due to the nonlinearities in the self-field forces, regardless of the value of W. In this case, particles with initial conditions $W(s = s_0) \leq 1$ can become chaotic. Some of such chaotic particles escape from the beam interior to form a tenuous halo around the intense beam.

To illustrate the phase space structure for a nonuniform beam, Fig. 2 shows

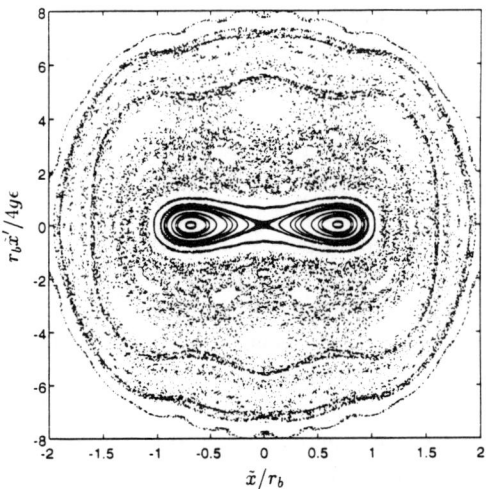

Figure 2: Poincaré surface-of-section plot as obtained from Eq. (13) for beam propagation through a step-function lattice. Here, the choice of system parameters corresponds to $\sigma_0 = 89°$, $\eta = 0.2$, $KS/\epsilon = 10.0$, and $\delta N/N = 0.2$.

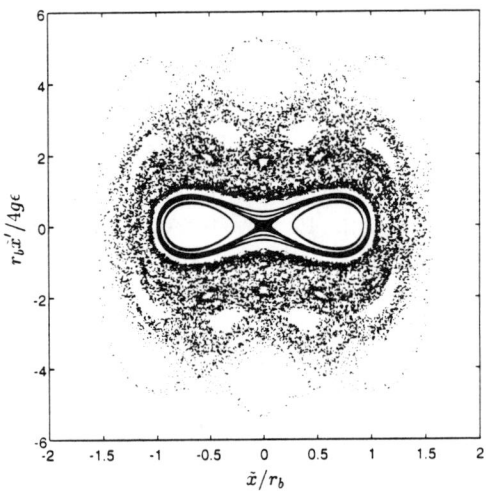

Figure 3: Poincaré surface-of-section plot of 21 test particles as obtained from Eq. (13) for the same choice of system parameters as in Fig. 2. Here, the particles are initially distributed evenly on the ellipse $W_{\tilde{x}}(0) = \tilde{x}^2(0)/r_b^2(0) + r_b^2(0)\tilde{x}'^2(0)/16g^2\epsilon^2 = 1$.

the Poincaré surface-of-section plot for the phase plane $(\tilde{x}, \tilde{x}')$, as obtained from Eq. (13) for beam propagation through a step-function lattice. In Fig. 2, 41 particles are followed for 1500 lattice periods starting from $s = 0$, and the points represent the phase-space trajectories of the particles at $s = 0, S, 2S \cdots$ The initial conditions are: $\tilde{x}(0)/r_b(0) = 0.1n$ ($n = 0, \pm 1, \pm 2, \cdots, \pm 20$), $\tilde{x}'(0) = 0$, $\tilde{y}(0) = 0$, and $\tilde{y}'(0) = 0$. The initial condition $\tilde{y}(0) = 0 = \tilde{y}'(0)$ assures $\tilde{y}(s) = 0 = \tilde{y}'(s)$. The choice of system parameters corresponds to: $\sigma_0 = 89°$, lattice filling factor $\eta = 0.2$, $KS/\epsilon = 10.0$, and $\delta N/N = 0.2$ (i.e., $g \cong 1.07$). A symmetric lattice with $\kappa_z(s) = \kappa_z(-s)$ is chosen here such that $r_b'(0) = 0$ [7].

There is an (unstable) X-point at the origin $(\tilde{x}, \tilde{x}') = (0, 0)$, and two stable points at $(\tilde{x}, \tilde{x}') = (\pm 0.7 r_b, 0)$ in Fig. 2. Despite existence of the X-point, particles near the origin are well confined and have regular orbits. However, because of the X-point, the chaotic region has engulfed part of the regular region near the origin and becomes accessible to particles that are initially located on the ellipse $W_x(0) \lesssim 1$ and that would otherwise be well confined inside a uniform density beam. It is the loss of the regular phase space region with $W_x(0) \lesssim 1$ that causes a nonuniform density beam to develop a halo.

Figure 3 shows the Poincaré surface-of-section plot for 21 test particles with initial conditions such that

$$W_{\tilde{x}}(0) = \frac{\tilde{x}^2(0)}{r_b^2(0)} + \frac{r_b^2(0)\tilde{x}'^2(0)}{16g^2\epsilon^2} = 1 \quad (22)$$

with $r_b'(0) = 0$. The test-particle trajectories are followed by numerically integrating Eq. (13) for propagation over 4000 lattice periods ($s = 4000S$). There are two classes of particles in Fig. 3. Particles with large initial displacement (i.e., in the range $0.5 \leq |\tilde{x}(0)/r_b(0)| \leq 1$) are located inside the regular region and their orbits are regular. However, particles with small initial displacement (i.e., with $|\tilde{x}(0)/r_b(0)| < 0.5$) enter the chaotic region, becoming halo particles. Since the chaotic region in phase space is connected between $|\tilde{x}/r_b| < 1$ and $|\tilde{x}/r_b| > 1$, the chaotic behavior in the particle motion leads to the *rapid* escape of particles from the beam interior. The chaotic particle orbits are bounded by a KAM surface [17] at a large radius outside the beam (i.e., at $|\tilde{x}/r_b| \cong 1.5$), as indicated by the dense layer at $|\tilde{x}/r_b| \cong 1.5$ shown in Fig. 2. Although the chaotic test particles escape from the beam interior rapidly on the order of a few lattice periods, it takes the order of 1000 lattice periods for them to diffuse up to the KAM surface at $|\tilde{x}/r_b| \cong 1.5$, forming a well developed halo.

IV. CHAOS AND HALOS IN RMS MISMATCHED BEAMS

The beam self fields induce a rich variety of nonlinear resonances and chaotic behavior in the envelope oscillations of mismatched, space-charge-dominated beams propagating through a periodic solenoidal focusing channel.

This was first predicted based on an envelope analysis [7, 8] in which the beam emittance was assumed to be constant and the effect of emittance growth was ignored. In this section, we review results of the numerical verification of the prediction and the simulation study of the evolution of the particle phase-space distribution [6], using the self-consistent model described in Sec. II.B.

In the remainder of this section, we introduce the dimensionless variables and parameters defined by

$$\frac{s}{S} \to s, \quad \frac{x}{\sqrt{4\epsilon_0 S}} \to x, \quad \frac{y}{\sqrt{4\epsilon_0 S}} \to y, \quad S^2 \kappa_z \to \kappa_z, \quad \frac{SK}{4\epsilon_0} \to K, \quad (23)$$

where $\epsilon_0 = \epsilon_{\tilde{x}}(0) = \epsilon_{\tilde{y}}(0)$ is the initial rms KV beam emittance which is assumed to be the same for the $\tilde{x}$- and $\tilde{y}$-directions. Unless specified otherwise, the above dimensionless variables and parameters will be used hereafter.

To make direct comparisons with the earlier envelope analysis [7, 8], we consider here a specific periodic focusing channel described by

$$\kappa_z(s) = [a_0 + a_1 \cos(2\pi s)]^2 . \quad (24)$$

The vacuum phase advance over one period of such a focusing lattice is given approximately by

$$\sigma_0 = \left[\int_0^1 \kappa_z(s) ds\right]^{1/2} = \left(a_0^2 + \frac{a_1^2}{2}\right)^{1/2} . \quad (25)$$

Furthermore, we define the effective (total) beam radius as

$$r_b = \sqrt{2\langle r^2 \rangle} = \sqrt{2\langle x^2 + y^2 \rangle} , \quad (26)$$

which is $\sqrt{2}$ times the rms beam radius $\langle r^2 \rangle^{1/2}$. For the special case of the matched KV equilibrium distribution [8], the effective beam radius is equal to the outermost beam radius, because the beam is round with uniform density.

A. RMS Beam Dynamics

Figures 4 and 5 show, respectively, the evolution of the emittance and effective radius, [computed as rms quantities using Eqs. (6), (7) and (26)], of a mismatched, space-charge-dominated beam propagating through the focusing channel, as obtained from the simulation for the following choice of system parameters: $K = 3$, $a_0 = a_1 = 0.648$ ($\sigma_0 = 45.5°$), and $N_P = 1024$. The beam is loaded initially according to a KV distribution but the beam radius is mismatched outward initially by 75% from the equilibrium beam radius, i.e., $\delta r_b/\bar{r}_b(0) = 0.75$, where $\delta r_b = r_b(0) - \bar{r}_b(0)$ is the initial beam radius mismatch and $\bar{r}_b(0)$ is the initial beam radius of the corresponding matched KV beam equilibrium.

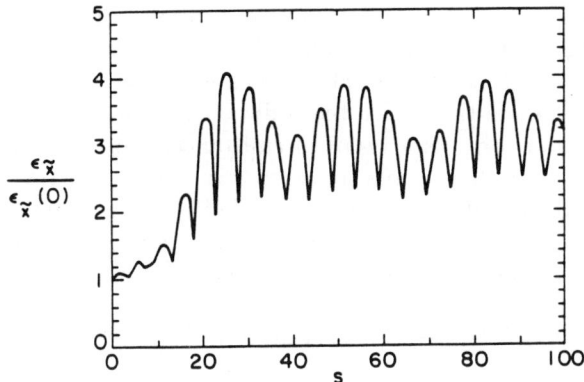

Figure 4: Evolution of the relative rms emittance of a mismatched, space-charge-dominated beam in the focusing channel, as obtained from the simulation for the following choice of parameters: $K = 3$, $a_0 = a_1 = 0.648$ ($\sigma_0 = 45.5°$), $\delta r_b/\bar{r}_b(0) = 0.75$, and $N_P = 1024$.

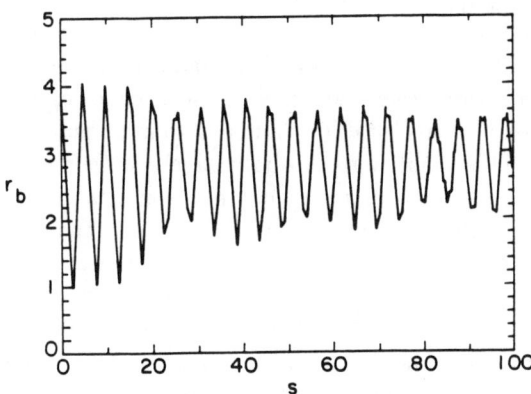

Figure 5: Evolution of the effective beam radius for a mismatched, space-charge-dominated beam in the focusing channel, as obtained from the simulation for the same choice of system parameters shown in Fig. 4.

It is evident in Fig. 4 that the emittance $\epsilon_{\tilde{x}}$ varies significantly as the beam propagates through the focusing channel. When the emittance reaches its maximum at $s \cong 23$, it has increased by as much as a factor of three from its initial value. The emittance for the $\tilde{y}$-direction evolves in a similar way as that for the $\tilde{x}$-direction. As a result of emittance growth, transient effects are observed in the envelope evolution shown in Fig. 5, particularly in the early stage of development from $s = 0$ to 25. Both the emittance and the effective beam radius oscillate back and forth once approximately every five lattice periods. The oscillation period is given approximately by [7]

$$\lambda = 2\pi [4\sigma_0^2 + K^2 - K(4\sigma_0^2 + K^2)^{1/2}]^{-1/2}. \qquad (27)$$

Because the beam emittance increases on average by a factor of two from its initial value, the effective value of K is $K = 1.0$ for the case shown in Fig. 5. Substituting $\sigma_0 = 45.5°$ and $K = 1.0$ into Eq. (27) yields the oscillation period $\lambda = 4.9$ (i.e., 4.9 lattice periods), which is in good agreement with the simulation result.

From the data shown in Fig. 5, the effective beam radius is differentiated with respect to s, and the Poincaré surface-of-section plot is generated to better visualize the resonant behavior in the envelope oscillations. The result is the separatrix of the fifth-order nonlinear resonance shown in Fig. 6, where the effective beam radius r_b and its derivative $r_b' = dr_b/ds$ are plotted in the plane (r_b, r_b') at $s = 26, 27, \cdots, 75$. For a clear view of the nonlinear resonance structure, the 50 points in Fig. 6 are connected by five contours, each of which traces 10 points that are separated longitudinally by about five lattice periods with random fluctuations seen typically inside a chaotic, slightly broadened separatrix. A chain of five stable islands is found inside the five contours shown in Fig. 6. This result agrees qualitatively with the earlier prediction based on the envelope analysis, as one compares present Fig. 6 with Fig. 2(b) in [7]. Both analyses show weakly chaotic behavior in the envelope evolution. (Note that the overall structure of the nonlinear resonance depends crucially on σ_0 and K [7] but does not change qualitatively from a sinusoidal to step-function focusing lattice.) Although not shown in present Fig. 6, a fourth-order nonlinear resonance is also found for larger initial envelope mismatches [e.g., $\delta r_b/\bar{r}_b(0) = 1.6$], as predicted by the earlier envelope analysis. Of course, the main advantage in the present analysis is that the beam emittance is allowed to evolve self-consistently. More importantly, the simulation results presented in Figs. 4-6 show that, after emittance growth and transient effects, the nonlinear resonances and chaotic behavior in the envelope evolution should be experimentally observable for mismatched, space-charge-dominated beams propagating through a periodic focusing channel.

Although more pronounced chaotic behavior was predicted in the envelope oscillations for $\sigma_0 > 90°$ [7-9], a direct confirmation of such chaotic envelope

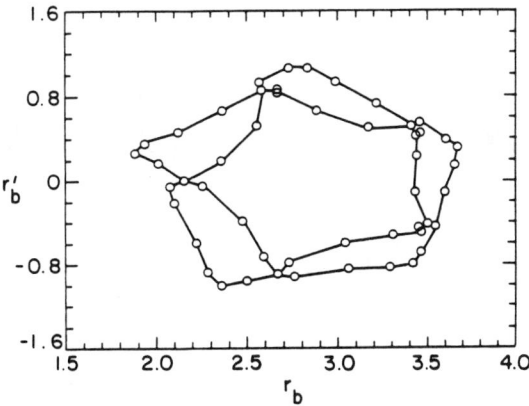

Figure 6: Poincare surface-of-section plot generated from the data in Fig. 5 showing the separatrix of the fifth-order nonlinear resonance for the envelope evolution from $s = 26$ to 75.

oscillations remains challenging. This is because emittance growth is found to be so pronounced in this regime that initially space-charge-dominated beams tend to evolve rapidly into emittance-dominated beams.

B. Halo Formation

The rms properties of mismatched, space-charge-dominated beams vary smoothly and exhibit nonlinear resonances and weakly chaotic behavior, as discussed in Sec. IV.A. Once mismatch causes a beam to form a dense core and a tenuous halo, however, the rms description of the beam becomes inadequate. Under such circumstances, we must also examine the self-consistent evolution of the beam particle distribution in the phase space (x, y, x', y'). From the point of view of accelerator design, of particular interest are the condition for halo formation and the size of a beam halo relative to the effective beam radius.

Shown in Fig. 7 are plots of the particle phase plane (x, y) at (a) $s = 38$, (b) $s = 39$, (c) $s = 40$, (d) $s = 41$, (e) $s = 42$, and (f) $s = 43$, for the same choice of system parameters shown in Figs. 4-6. Approximately, five percent of macroparticles are in the beam halo. The halo particles are chaotic. The halo radius (i.e., maximum radius achieved by halo particles) is approximately constant as the beam propagates through the focusing channel. The ratio of the halo radius to the maximum effective beam radius is found to be about 1.6, as seen from Figs. 5 and 7(d), whereas the ratio of the halo radius to the minimum effective beam radius is about 3.8, as seen from Figs. 5 and 7(a).

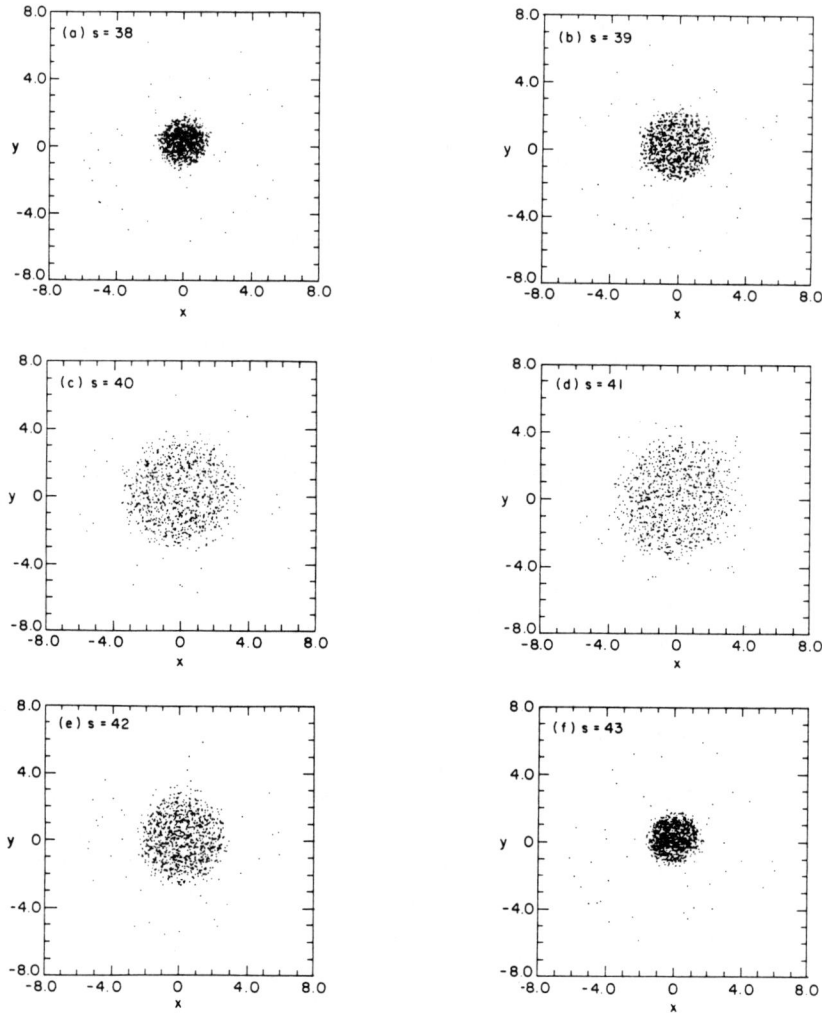

Figure 7: Plots of 1024 macroparticles in the phase plane (x, y) at (a) $s = 38$, (b) $s = 39$, (c) $s = 40$, (d) $s = 41$, (e) $s = 42$, and (f) $s = 43$, for the same choice of system parameters shown in Figs. 4-7.

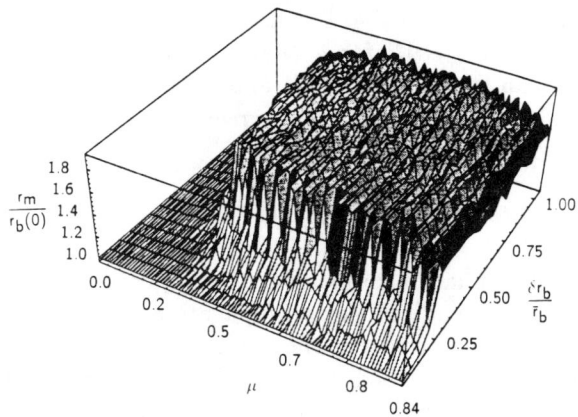

Figure 8: Relative maximum radius $r_m/r_b(0)$ achieved by the beam particles as a two-dimensional function of the relative mismatch amplitude $\delta r_b/\bar{r}_b$ and the space-charge parameter $\mu = 1 - (\sigma/\sigma_0)^2$, obtained from simulations for the case of a uniform focusing channel.

It is found that the characteristics of halos do not change qualitatively from a periodic to uniform focusing channel. This is perhaps because the focusing parameter of a periodic focusing channel, $\kappa_z(s)$, may be averaged over one focusing period to yield a uniform focusing channel with the effective (dimensional) focusing parameter squared $\kappa_{z0} = \int_0^S \kappa_z(s)ds = (\sigma_0/S)^2 = $ const.

Finally, we estimate the mismatch threshold for halo formation by means of computer simulations. For simplicity, this is done for the case of an initially KV distribution propagating through a uniform focusing channel with $\kappa_z(s) = $ const. In such a simulation, the maximum radius achieved by the beam particles is determined after the beam has propagated more than 50 periods of mismatched envelope oscillations. The simulations are performed over a wide range of K and $\delta r_b/\bar{r}_b$, where $\delta r_b = r_b(0) - \bar{r}_b \geq 0$, $r_b(0)$ is the initial beam radius, and $\bar{r}_b = $ const. is the radius for the matched beam in the uniform focusing channel. The results [6] are shown in Fig. 8, where the maximum radius achieved by the beam particles, r_m, is plotted relative to the initial beam radius $r_b(0)$ as a two-dimensional function of the relative mismatch amplitude $\delta r_b/\bar{r}_b$ and the space-charge parameter $\mu = 1 - (\sigma/\sigma_0)^2$. Here, σ_0 and $\sigma = (1/2)[(K^2 + 4\sigma_0^2)^{1/2} - K]$ are the vacuum and space-charge-depressed phase advances per unit axial length for the matched beam, respectively. Note that $\mu \to 1$ ($K/\sigma_0 \gg 1$) for space-charge-dominated beams, whereas $\mu \to 0$ ($K/\sigma_0 \to 0$) for emittance-dominated beams.

The onset of a plateau in Fig. 8 defines the threshold for halo formation, which occurs at $\delta r_b/\bar{r}_b \cong 0.2$ for $\mu > 0.4$ but becomes increasing large as $\mu \to 0$. The plateau shown in Fig. 8 is approximately flat and has a vertical height of $r_m/r_b(0) \cong 1.6$, which is in agreement with the results shown in Figs. 5 and 7(d).

V. CONCLUSION

The mechanisms of halo formation have been identified with chaotic behavior in space-charge-dominated beams propagating through a linear, periodic focusing channel provided by an applied periodic solenoidal magnetic field.

First of all, for rms matched beams with nonuniform density profiles, it was shown in a test-particle model that nonlinearities in the self-field forces not only can result in chaotic particle motion but also can cause a halo to develop. The size of the halo was found to be bounded by a KAM surface. The numerical results showed that the chaotic particles escape from the beam interior rapidly on the order of a few lattice periods but take the order of 1000 lattice periods to diffuse up to the KAM surface to form a well-developed halo. This halo mechanism does not occur for rms matched beams in uniform focusing channels. Self-consistent computer simulation studies of this halo mechanism are in progress.

Secondly, a self-consistent two-dimensional macroparticle model was presented for studies of the dynamics of intense charged-particle beams. It was confirmed in the computer simulations that, for beams mismatched into the periodic focusing channel, the beam envelope exhibits nonlinear resonances and chaotic behavior, as predicted by the previous analysis of the beam envelope equation [7-9]. As a result of emittance growth, transient effects were observed in the rms beam dynamics. These results further support the expectation that the nonlinear resonances and chaotic behavior in the envelope evolution should be experimentally observable after the emittance growth and transient effects for mismatched, space-charge-dominated beams propagating through a periodic focusing channel. Also investigated were the self-consistent evolution of the particle distribution in the phase space and halo formation. Unlike the analytical model [1, 10] for envelope-mismatched beams without emittance growth which does not provide an escape mechanism for core particles to move into the halo, the emittance growth and transient effects in the self-consistent model provided escape mechanisms. The halo size was estimated. The halo characteristics for a periodic focusing channel were found to be qualitatively the same as those for a uniform focusing channel. A threshold condition was obtained numerically for halo formation for mismatched beams in the uniform focusing approximation.

Finally, based on the results in this paper and qualitatively the same results obtained for beam propagation through the alternating-gradient magnetic quadrupole focusing channel [2-4], we conclude that both precise rms beam matching and precise density profile control are required to prevent space-charge-dominated beams from developing halos.

ACKNOWLEDGMENTS

The author wishes to acknowledge numerical assistance provided by Yoel Fink, and that many of the ideas presented in this article resulted from stimulating discussions and vigorous collaboration with Ron Davidson, Qian Qian, and Bob Jameson. This work was supported in part by the Department of Energy, Office of High Energy and Nuclear Physics, Grant No. DE-FG02-95ER40919 and in part by the Air Force Office of Scientific Research, Grant No. F49620-94-1-0374.

References

[1] J.S. O'Connell, T.P. Wangler, R.S. Mills, and K.R. Crandall, Proc. 1993 Part. Accel. Conf. (IEEE Service Center, Piscataway, New Jersey, 1993), Vol. 5, p. 3657.

[2] Q. Qian, R.C. Davidson, and C. Chen, Phys. Plasmas **1**, 3104 (1994).

[3] Q. Qian, R.C. Davidson, and C. Chen, Phys. Rev. **E51**, 5216 (1995).

[4] Q. Qian, R.C. Davidson, and C. Chen, Phys. Plasmas **2**, 2674 (1995).

[5] J.M. Lagniel, Nucl. Instrum. Methods Phys. Res. **A345**, 1516 (1994).

[6] C. Chen and R.A. Jameson, Phys. Rev. **E52**, 3074 (1995).

[7] C. Chen and R.C. Davidson, Phys. Rev. Lett. **72**, 2195 (1994).

[8] C. Chen and R.C. Davidson, Phys. Rev. **E49**, 5679 (1994).

[9] S.Y. Lee and A. Riabko, Phys. Rev. **E51**, 1609 (1995).

[10] R.L. Gluckstern, Phys. Rev. Lett. **73**, 1247 (1994).

[11] F.J. Sacherer, IEEE Trans. Nucl. Sci. **NS-18**, 1105 (1971).

[12] P.M. Lapostolle, IEEE Trans. Nucl. Sci. **NS-18**, 1101 (1971).

[13] J.D. Lawson, *Physics of Charged-Particle Beams* (Oxford Science Publications, New York, 1988).

[14] I.M. Kapchinskij and V.V. Vladimirskij, Proc. Int. Conf. on High Energy Accelerators (CERN, Geneva, 1959), p. 274.

[15] I. Hofmann, L.J. Laslett, L. Smith, and I. Haber, Part. Accel. **13**, 145 (1983).

[16] R.C. Davidson, *Physics of Nonneutral Plasmas* (Addison-Wesley, Reading, MA, 1990), Chapter 10.

[17] A.J. Lichtenberg and M.A. Lieberman, *Regular and Chaotic Dynamics*, 2nd Edition (Springer-Verlag, New York, 1992).

Overlapping resonances and chaos in Halo formation[1]

A.Riabko, S.Y.Lee

Department of Physics, Indiana University, Bloomington, IN 47405

The "particle-core" model in the framework of Kapchinskij-Vladimirskij (KV) distribution function approach is used in order to study dynamics of a test particle in a space charge dominated beam propagating through time-dependent FOFO focusing channel. A simple analytical method is suggested in order to find a closed orbit solution for the envelope; the question about stability of small amplitude oscillations of weakly mismatched beam around the closed orbit solution is considered. The test particle motion is described by the methods of Hamiltonian dynamics; it is shown that a set of parametric resonances of the particle Hamiltonian can be generated by mismatched envelope oscillations in the time-dependent focusing channel. It is demonstrated that the onset of global chaos in the particle phase space exhibits a sharp transition when the amplitude of envelope oscillations is larger than some critical value. Implications caused by the time dependence of the linear focusing field are discussed. The relation between the critical envelope mismatch for the halo formation and the space charge perviance is obtained numerically and explained qualitatively for few values of the linear focusing field amplitude of the FOFO channel.

I. INTRODUCTION

One of the traditional ways to analyze the transport properties of space-charge dominated beams is based on KV distribution function approach [1]. This approach provides a self-consistent equilibrium distribution function, dynamics of which is determined by a combination of the external focusing force and self-induced space charge force. The motion of the test particle is considered in the framework of the "particle-core" model, which assumes that the envelope dynamics determines the test particle motion, but the envelope evolution in time is not affected by the test particle. The envelope oscillations of initally mismatched beam produce a set of parametric resonances in the particle phase space, which may overlap and create a large stochastic area in the particle phase space [2,3]. Results of PIC simulations [4,5] and "particle-core" model simulations [3,6,7] have indicated that the halo formation arises mainly

---

[1]Work supported in part by grants from NSF PHY-9221402 and DOE DE-FG02-93ER40801.

from resonance excitations. Taking into consideration the fact that every realistic particle distribution at finite temperature has a diffusive tail, we expect that the motion of the particles from the diffusive tail will be strongly affected by the system of overlapping resonances. The logical conclusion is that the thermal diffusion process produces the tail distribution and the overlapping resonances facilitate halo formation.

The paper, which extends our work to the periodic focusing system, is organized as follows. In Sec II, we review the properties of the envelope Hamiltonian, suggest a simple analytical approximation for the closed orbit solution and study the stability of small envelope oscillations in the case of weakly mismatched beam in FOFO focusing channel. In Sec III, we employ the action-angle variables approach for the analysis of test particle Hamiltonian. We will show that parametric resonances are excited by the envelope modulation of the weakly mismatched beam; these resonances can interact with the structure resonances due to time-dependence of the focusing channel. In Sec IV, we consider a possible scenario of transition to global chaos. Conclusion is given in Sec V.

II. ENVELOPE DYNAMICS

The envelope function of a beam in axially symmetric FOFO focusing channel is given by

$$R_b'' + k(z) R_b - \frac{K_b}{R_b} - \frac{\epsilon^2}{R_b^3} = 0, \tag{1}$$

where z is the longitudinal distance, ϵ is the emittance of the beam and $R(z)$ is the envelope radius. The linear focusing field $k(z)$ and the generalized perviance K_b are given by

$$k(z) = \alpha \sum_{n=-\infty}^{\infty} \delta(z - n), \quad K_b = \frac{2 N r_{cl}}{\beta^2 \gamma^3}, \tag{2}$$

where α is the strength of the focusing field, r_{cl} is the classical radius of the particle, β and γ are relativistic factors of the beam, and N is the number of particles per unit length. Using the dimensionless envelope coordinate and normalized perviance

$$R = \frac{R_b}{\sqrt{\epsilon}}, \quad K = \frac{K_b}{\epsilon}, \tag{3}$$

the envelope equation of motion can be derived from the envelope Hamiltonian [2]

$$H_e = \frac{P_R^2}{2} + \frac{k(z)}{2} R^2 - K \ln R + \frac{1}{2 R^2}. \tag{4}$$

Particle phase advance in zero space charge limit ($K \to 0$) is given by:

$$\mu = \cos^{-1}\left(1 - \frac{\alpha}{2}\right).\qquad(5)$$

We separate the original envelope Hamiltonian H_e into a sum of time-independent Hamiltonian H_0 and time-dependent perturbation H_1 where

$$H_0 = \frac{P_R^2}{2} + V_e(R) = \frac{P_R^2}{2} + \frac{\mu^2 R^2}{2} - K \ln R + \frac{1}{2R^2},\qquad(6)$$

$$H_1 = \frac{k(z) - \mu^2}{2} R^2.\qquad(7)$$

Hamiltonian H_0 describes the beam dynamics in the equivalent uniform focusing channel with linear focusing field μ^2. The original time-dependent linear focusing channel $k(z)$ and a uniform focusing channel with focusing strength μ^2 are considered to be equivalent because we obtain identical expressions for particle phase advance in zero space charge limit in both cases. This substitution of the original time-dependent focusing channel by an equivalent uniform focusing channel is called Floquet transformation and can be used in order to obtain a simple analytical approximation for the closed orbit solution $\tilde{R}(z)$.

H_0 can be linearized around a single minimum of the effective envelope potential $V_e(R)$

$$R_0 = \left(\frac{\sqrt{K^2 + 4\mu^2} + K}{\mu^2}\right)^{\frac{1}{2}},\qquad(8)$$

and the corresponding linear equation of motion is given by

$$R'' + \nu_e^2 (R - R_0) + \left(k(z) - \mu^2\right) R = 0,\qquad(9)$$

where

$$\nu_e = \left(4\mu^2 - \frac{2K}{R_0^2}\right)^{\frac{1}{2}}\qquad(10)$$

is a linear tune of time-independent Hamiltonian H_0. Solving Eq. (9) we obtain the following approximation for the closed orbit trajectory:

$$\tilde{R}(z) = \frac{\nu_e^2}{\nu_e^2 - \mu^2} R_0 - r_o \cos \nu_e \left(z - \frac{1}{2}\right),\qquad(11)$$

where $z \in [0..1]$ and r_o is chosen so that $\tilde{R}(z)$ satisfies the closed orbit trajectory condition

$$\tilde{R}'(0+) = -\frac{\alpha}{2} \tilde{R}(0),\qquad(12)$$

and given by

$$r_o = \frac{\alpha \nu_e^2 R_o}{2(\nu_e^2 - \mu^2)} \frac{1}{\nu_e \sin \frac{\nu_e}{2} + \frac{\alpha}{2} \cos \frac{\nu_e}{2}}. \tag{13}$$

If the beam is initially mismatched, small radial amplitude oscillations of the envelope radius $R(z)$ around the closed orbit trajectory $\tilde{R}(z)$ are excited:

$$R(z) = \tilde{R}(z) + \zeta(z), \tag{14}$$

where $|\zeta(z)| \ll \tilde{R}(z)$ and the linearized equation for $\zeta(z)$ is

$$\zeta'' + \left(k(z) + \frac{K}{\tilde{R}^2(z)} + \frac{3}{\tilde{R}^4(z)}\right) \zeta = 0. \tag{15}$$

If we employ "smooth" approximation (replacing $\tilde{R}(z)$ by $\overline{\tilde{R}(z)} = R_0$) we obtain

$$\zeta'' + \nu_e^2 \zeta = 0, \tag{16}$$

which is an adequate approximation in most cases.

A. Linear instability region and its analytical estimations

However, if the point (α, K) is close to the linear Mathieu type resonance ($\nu_e \approx \pi$), more adequate approximation is necessary in order to estimate the instability region (we have to take time dependence of $k(z)$ and $\tilde{R}(z)$ into consideration). A number of such approximate analytical estimations, all of which are based on the stability analysis of an "equivalent" Mathieu's equation, is presented below.

First we consider linearized equation

$$\zeta'' + \nu_e^2 \zeta + (k(z) - \mu^2)\zeta = 0. \tag{17}$$

Transition matrix T over the period of the lattice is given as

$$T = \begin{bmatrix} \cos(\sqrt{\nu_e^2 - \mu^2}) & \frac{\sin(\sqrt{\nu_e^2 - \mu^2})}{\sqrt{\nu_e^2 - \mu^2}} \\ -\sqrt{\nu_e^2 - \mu^2} \sin(\sqrt{\nu_e^2 - \mu^2}) & \cos(\sqrt{\nu_e^2 - \mu^2}) \end{bmatrix} \begin{bmatrix} 1 & 0 \\ -\alpha & 1 \end{bmatrix} \tag{18}$$

and condition on instability of linear motion:

$$\frac{1}{2} Tr(T) = \cos\sqrt{\nu_e^2 - \mu^2} - \frac{\alpha}{2\sqrt{\nu_e^2 - \mu^2}} \sin\sqrt{\nu_e^2 - \mu^2} < -1. \tag{19}$$

We can also approximate Eq. (17) by Mathieu's equation using the fact that

$$k(z) = \alpha + 2\alpha \cos(2\pi z) + ..., \tag{20}$$

so finally we obtain

$$\zeta'' + \left(\nu_e^2 + \alpha - \mu^2 + 2\alpha\,\cos(2\pi z)\right)\zeta = 0. \quad (21)$$

Here we can use well-known conditions on instability of the solution of Mathieu's equation [8]:

$$u'' + (a + 2q\,\cos(2t))\,u = 0 \quad (22)$$

$$B(q) < a < A(q) \quad (23)$$

$$A(q) = 1 + q - \frac{q^2}{8} - \frac{q^3}{64} - \frac{q^4}{1536} + \cdots \quad (24)$$

$$B(q) = 1 - q - \frac{q^2}{8} + \frac{q^3}{64} - \frac{q^4}{1536} + \cdots, \quad (25)$$

where

$$a = \frac{1}{\pi^2}\left(\nu_e^2 + \alpha - \mu^2\right) \quad (26)$$

$$q = \frac{\alpha}{\pi^2}. \quad (27)$$

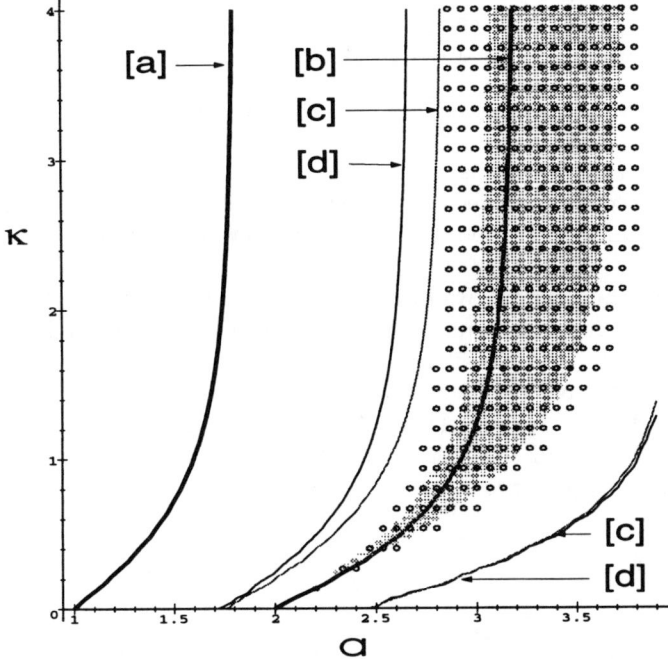

FIG. 1. a). 1 : 3 envelope resonance, b). 1 : 2 linear envelope resonance, c). corresponds to Eq. (19), d). corresponds to Eq. (21), shaded area correspond to Eq. (28), circles(o) mark numerically found instability region.

Another relatively simple approach, which demonstrates a slightly better agreement with results of numerical simulations, is based on approximation of the original Eq. (15) for ζ by a corresponding Mathieu's equation

$$\zeta'' + \tfrac{1}{\pi^2}\left(\alpha + g_0 + 2\left(\alpha + \tfrac{g_1}{2}\right)\cos(2z)\right)\zeta = 0\,, \qquad (28)$$

where g_0 and g_1 are obtained from the following Fourier expansion

$$\frac{3}{\tilde{R}^4(z)} + \frac{K}{\tilde{R}^2(z)} = g_0 + g_1\cos(2\pi z) + \dots\,. \qquad (29)$$

Figure 1 presents the instability region obtained from numerical simulations and compares its boundary with different analytical estimations based on Mathieu's equation approach. It should be noted that the last approach provides a correct limiting condition at $K \to 0$, where there is no Mathieu instability in zero space charge limit.

B. Envelope oscillations around the closed orbit trajectory

We assume in this section that the linear Mathieu's instability is not essential and the operational point (α, K) is far from any linear or non-linear structure resonance. In the case of weakly mismatched beam we have the following harmonic approximation for the envelope radius

$$R(z) = \tilde{R}(z) + M\,R_0\,\cos\nu_e\,z\,, \qquad (30)$$

where $M \ll 1$ is a mismatch parameter and $\tilde{R}(z)$ is the closed orbit trajectory. Thus, provided the matched orbit $\tilde{R}(z)$ does not introduce any harmful effects, the particle motion is modulated at frequency ν_e with smallness parameter M.

However, the original Hamiltonian H_e is not linear and therefore more adequate treatment is required in the case of finite-amplitude envelope oscillations around the closed orbit trajectory $\tilde{R}(z)$. Fortunately, the time-dependent perturbation H_1 is a linear Hamiltonian, therefore non-linearity of time-dependent H_e is equivalent to that of time-independent Hamiltonian H_0, which may be analyzed by the method of action-angle variables. The action of a given envelope torus is

$$J_e = \frac{1}{2\pi}\oint P_e dR_e\,, \qquad (31)$$

and H_0 can be approximated by

$$H_0 = \nu_e J_e + \frac{\alpha_e}{2} J_e^2 + \dots\,, \qquad (32)$$

where non-linear detuning parameter α_e can be derived using the methods of standard perturbation theory [3].

Thus, the tune of the envelope oscillations $Q_e(J_e) = \frac{\partial H_0(J_e)}{\partial J_e}$ is amplitude dependent and may be approximated by

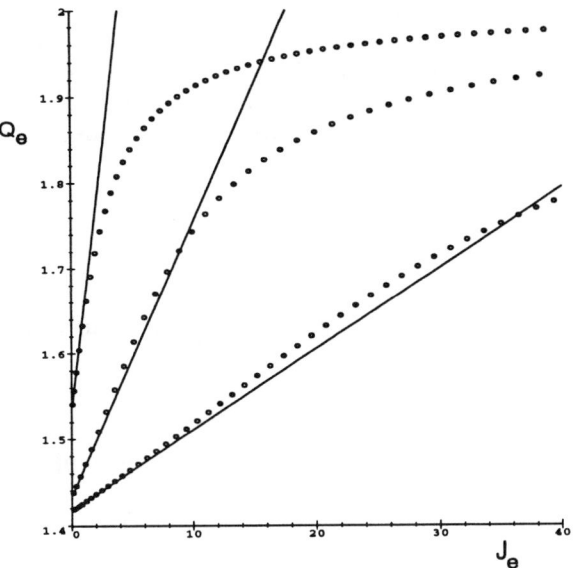

FIG. 2. Q_e vs. J_e and approximate formula $Q_e = \nu_e + \alpha_e J_e$ for $\frac{K}{2\mu} = 1, 3, 9$.

$$Q_e(J_e) \approx \nu_e + \alpha_e J_e \approx \nu_e + \frac{\alpha_e}{2}\left(\frac{{\zeta'}^2(0)}{\nu_e} + \nu_e \zeta(0)^2\right). \tag{33}$$

Figure 2 compares this approximate formula with results of numerical simulations for time-independent Hamiltonian H_0; it also illustrates the fact that for large envelope oscillations ($J_p \to \infty$) when the space-charge effects are not important and the envelope dynamics is linear, the envelope tune Q_p equals 2μ.

III. PARTICLE DYNAMICS

The equation for particle motion in KV beam with radius $R(z)$

$$\begin{cases} x'' + \left(k(z) - \frac{K}{R^2(z)}\right)x = 0, & |x| < R(z) \\ x'' + k(z)x - \frac{K}{x} = 0, & |x| > R(z) \end{cases} \tag{34}$$

can be derived from the Hamiltonian:

$$H_p = \frac{P_x^2}{2} + \frac{k(z)x^2}{2} - \frac{Kx^2}{2R^2(z)}\Theta(R(z) - |x|)$$
$$- K\left(\ln x + \frac{1}{2}\right)\Theta(|x| - R(z)), \tag{35}$$

where $\Theta(x)$ is the step function ($\Theta(\xi) = 1$ for $\xi \geq 0$ and 0 otherwise).

All the particles can be classified as "core" particles, which stay all the time inside the core, or as "external" particles which can be outside of the beam. The motion of the "core" particles is described by the linear equation of motion and their coordinates always preserve the Courant-Snyder invariant even in the case of mismatched beam:

$$\frac{1}{2}\left(\left(\frac{x}{R(s)}\right)^2 + \left(R'(s)x - R(s)x'\right)^2\right) = J_x. \tag{36}$$

The K-V particle distribution is defined by the following condition:

$$\rho(J_x, J_y) = \rho_0 \delta(J_x + J_y - \frac{1}{2}). \tag{37}$$

We replace the original equations of motion by the approximate ones:

$$\begin{cases} x'' + \left(\mu^2 - \frac{K}{R_0^2}\right)x + (k(z) - \mu^2)x - \left(\frac{K}{R^2(z)} - \frac{K}{R_0^2}\right)x = 0, & |x| < R_0 \\ x'' + \mu^2 x - \frac{K}{x} + (k(z) - \mu^2)x = 0, & |x| > R_0 \end{cases} \tag{38}$$

The corresponding Hamiltonian can be represented as a sum of time-independent particle Hamiltonian H_0 and time-dependent perturbations H_1 and H_2 where

$$H_0 = \frac{P_x^2}{2} + \frac{\mu^2 x^2}{2} - \frac{Kx^2}{2R_0^2}\Theta(R_0 - |x|) - K \ln x \, \Theta(|x| - R_0), \tag{39}$$

$$H_1 = (k(z) - \mu^2)\frac{x^2}{2} - \left(\frac{K}{\tilde{R}^2(z)} - \frac{K}{R_0^2}\right)\frac{x^2}{2}\Theta(R_0 - |x|), \tag{40}$$

$$H_2 = \frac{K\zeta(z)}{R_0^3}\frac{x^2}{2}\Theta(R_0 - |x|). \tag{41}$$

Here H_2 is generated by the envelope mismatch and proportional to $\zeta(z) = M R_0 \cos(\nu_e z)$ - small amplitude envelope oscillations around the closed orbit trajectory $\tilde{R}(z)$. The relative importance of this term depends on the mismatch parameter $M \ll 1$. On the other hand, H_1, which is produced by the time dependence of the linear focusing channel $k(z)$ and, consequently, time dependence of the closed orbit solution $\tilde{R}(z)$, does not have a corresponding smallness parameter, so its relative importance can not be controlled by any external parameter and is determined only by how well the closed orbit solution follows the external focusing field. It should be noted, however, that such a separation is meaningful only for sufficiently small envelope mismatch parameters $M \approx 0.2 \div 0.3$.

A. Action-angle variable, the system of resonances and resonance strength

The primary Hamiltonian H_0 is time-independent and can be analyzed in terms of action-angle variables. The action for a torus of the unperturbed Hamiltonian is given by

$$J_p = \frac{1}{2\pi} \oint p_x dx . \tag{42}$$

Using the generating function

$$F_2(x, J_p) = \int_{x_0}^{x} p_x dx . \tag{43}$$

the conjugate angle variable is given by $\psi_p = \frac{\partial F_2(x, J_p)}{\partial J_p}$. The particle tune is amplitude-dependent and is given by

$$Q_p(J_p) = \frac{\partial H_0(J_p)}{\partial J_p} . \tag{44}$$

$Q_p(J_p)$ inside of the envelope is constant:

$$Q_p(J_p \leq \frac{1}{2}) = \nu_p = \left(\mu^2 - \frac{K}{2 R_o^2}\right)^{\frac{1}{2}} . \tag{45}$$

In the limit of large amplitude particle oscillations the space charge effects are not important, the particle motion is harmonic and the particle tune is again constant

$$\lim_{J_p \to \infty} Q_p(J_p) = \mu . \tag{46}$$

In order to understand the effect of time-dependent perturbations H_1 and H_2 we expand these terms in the action-angle variables (J_p, ψ_p) of the unperturbed Hamiltonian H_0. Assuming a weakly mismatched beam, we obtain the following approximation for H_p

$$H_p = H_o(J_p) + \sum_{n,m=-\infty}^{\infty} g_{n,m}(J_p) e^{(2\pi n z - m \psi_p)}$$

$$+ M\kappa \cos(\nu_e z) \sum_{n=-\infty}^{\infty} G_n(J_p) e^{in\psi_p} , \tag{47}$$

where the resonance strengths $G_n(J_p)$ and $g_{n,m}(J_p)$ are given by

$$G_n(J_p) = \frac{1}{2\pi} \int_{-\pi}^{\pi} \left(\frac{x^2}{R_o^2} - 1\right) \Theta(R_0 - |x|) e^{-in\psi_p} d\psi_p , \tag{48}$$

$$g_{n,m}(J_p) = \alpha_n T_m^*(J_p) + \beta_n K_m^*(J_p), \tag{49}$$

where α_n, $T_m(J_p)$, β_n and $K_m(J_p)$ are Fourier coefficients in the following expansions

$$k(z) - \mu^2 = \sum_{n=-\infty}^{\infty} \alpha_n e^{2i\pi n z}, \tag{50}$$

$$\frac{K}{\tilde{R}^2(z)} - \frac{K}{R_0^2} = \sum_{n=-\infty}^{\infty} \beta_n e^{2i\pi n z}, \tag{51}$$

$$\frac{x^2}{2} = \sum_{m=-\infty}^{\infty} T_m(J_p) e^{im\psi_p}, \tag{52}$$

$$\frac{x^2}{2} \Theta(R_0 - |x|) = \sum_{m=-\infty}^{\infty} K_m(J_p) e^{im\psi_p}. \tag{53}$$

The Hamiltonian of Eq. (47) shows the resonance structure explicitly. A coherent perturbation to a torus of Hamiltonian H_0 is present when the resonance condition $n\psi_p \approx m\nu_e$ for particle-envelope resonances and $m\psi_p \approx 2\pi n$ for particle-lattice resonances. In order to demonstrate the effect of a single particle-lattice primary resonance, we perform another canonical transformation to the resonance rotating frame using the generating function

$$F_2 = \left(\psi_p - \frac{m}{n} \nu_e z \right) J_p. \tag{54}$$

New canonically conjugate variables are given by

$$I_p = J_p, \quad \phi_p = \psi - \frac{m}{n} \nu_e z. \tag{55}$$

Neglecting all time-dependent terms, we obtain the time averaged Hamiltonian in the resonance rotating frame as

$$\langle \bar{H}_p \rangle = H_o(I_p) - \frac{m}{n} \nu_e I_p + M \kappa G_n(I_p) \cos n \phi_p. \tag{56}$$

The fixed points of this time averaged Hamiltonian $\langle \bar{H}_p \rangle$ can be found from

$$nQ_p(I_{FP}) - m\nu_e \pm M \kappa G'_n(I_{FP}) = 0, \tag{57}$$
$$\sin n \phi_p = 0, \tag{58}$$

where the prime corresponds to the derivative with respect to I_p. thus, the resonance island structure with n stable and n unstable fixed points is formed by the time-dependent perturbation, qualitatively changing the topology of the particle phase space. The particle-lattice resonances due to H_1 can be treated the same way using the generating function

$$F_2 = \left(\psi_p - \frac{2\pi m}{n} z\right) J_p, \tag{59}$$

we obtain the corresponding time-averaged Hamiltonian near $m:n$ resonance as

$$\langle \bar{H}_p \rangle = H_o(I_p) - \frac{2\pi m}{n} I_p + g_{m,n}(I_p) \cos n\phi_p. \tag{60}$$

The locations of particle – lattice resonances due to H_1 are given by:

$$2\pi n - m Q_p(I_{FP}) \pm g'_{n,m}(I_{FP}) = 0, \tag{61}$$
$$\sin n\phi_p = 0. \tag{62}$$

For a weakly mismatched beam ($M \ll 1$) the particle-envelope resonance condition (57) becomes

$$\frac{Q_p}{\nu_e} = \frac{m}{n}. \tag{63}$$

Assuming that $g'_{n,m}(J_{FP})$ in (61) is a small function of particle action, we obtain the approximate condition for the particle-lattice resonances:

$$\frac{Q_p}{2\pi} = \frac{m}{n}. \tag{64}$$

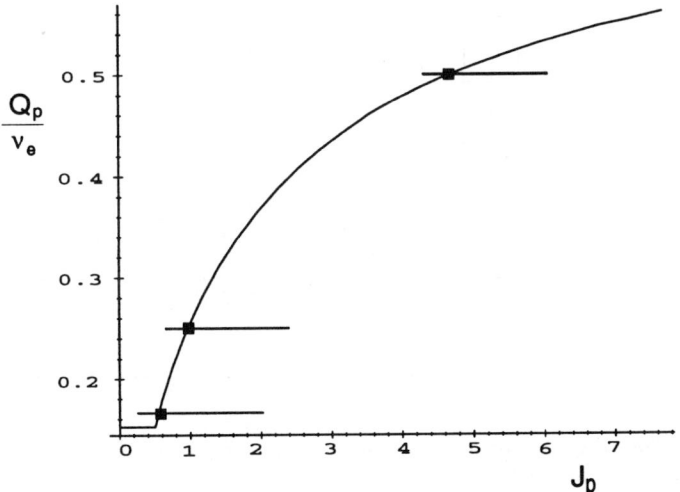

FIG. 3. $\frac{Q_p}{\nu_e}$ vs J_p for $\kappa = 2.19$. Intersections of horizontal lines with $\frac{Q_p}{\nu_e}$ mark locations of parametric resonances.

Figure 3 shows the ratio $\frac{Q_p(J_p)}{\nu_e}$ as function of particle action J_p for $\kappa = \frac{K}{2\mu} = 2.19$. The horizontal lines mark the locations of primary parametric resonances. Since the ratios $\frac{\nu_p}{\nu_e}$ depends only on κ, the threshold on existence of a particular primary envelop-particle resonance is determined by a single parameter of the system. Figure 4 presents the range of possible ratios $\frac{Q_p}{\nu_e}$ in dependence on κ; $\frac{\nu_p}{\nu_e}$ is the lower solid curve and $\frac{\mu}{\nu_e}$ is the upper solid curve. Horizontal lines correspond to the primary particle-envelope resonances (2 : 1, 4 : 1 and 6 : 1). It is important to note that 2 : 1 resonance exists for all values of space charge parameter. Other primary particle-envelope resonances appear as κ increases. Threshold of existence of n-th order particle-envelope resonances for weakly mismatched beam ($M \ll 1$):

$$\frac{\nu_p}{\nu_e} = \frac{1}{n}. \tag{65}$$

Therefore as the space charge parameter K increases more high order resonances appear (Fig. 4). In general n-th order primary resonance exists only if

$$\kappa > \frac{n^2 - 4}{\sqrt{8(n^2 - 2)}}. \tag{66}$$

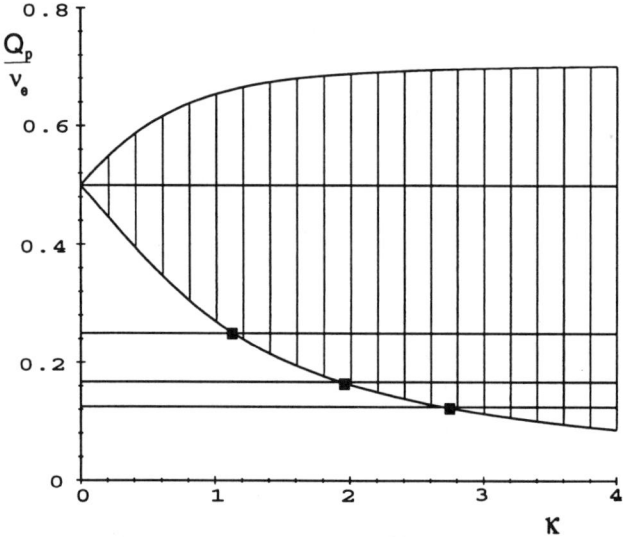

FIG. 4. The range of ratios of the particle tune to the envelope tune as function of effective space charge strength κ.

In general true linear particle tune depends also on envelope mismatch parameter M and can be estimated as follows in the case of a uniform focusing channel. The equation of motion for the core particles is linear and periodic with the period of envelope oscillations (Hill's type equation). So, because

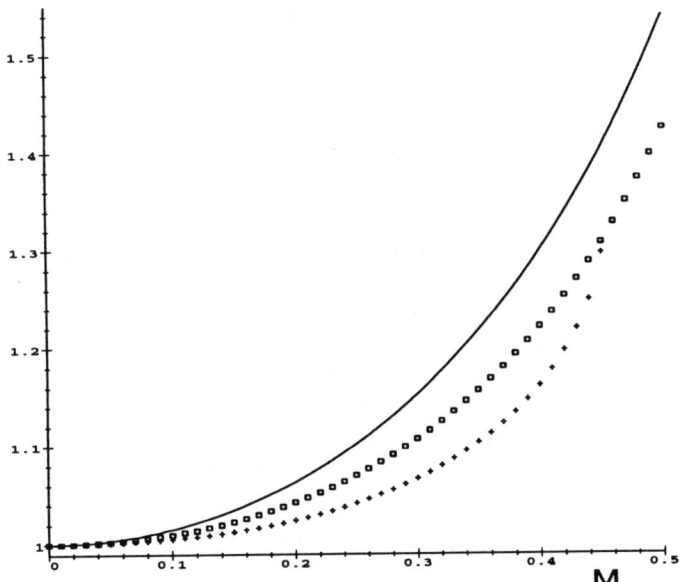

FIG. 5. Eq. (68) and numerically obtained $\frac{\tilde{\nu}_p}{\nu_p}\frac{\nu_e}{Q_e(M)}$ as function of mismatch parameter M for $\kappa = 1$ (crosses) and $\kappa = 4$ (boxes).

the envelope radius $R(z)$ is equivalent to $\sqrt{\beta(z)}$, we conclude that we can estimate particle phase advance as

$$\tilde{\nu}_p = \frac{1}{2\pi} \int \frac{dz}{R^2(z)}, \qquad (67)$$

where $R(z) \approx R_0 (1 - M \cos \nu_e z)$. This integral can be evaluated analytically and gives the following result for the linear tune of the "core" particles as function of mismatch parameter M

$$\tilde{\nu}_p(M) = \frac{\nu_p}{(1-M^2)^{\frac{3}{2}}}. \qquad (68)$$

It also should be pointed out that envelope tune Q_e is amplitude dependent (see Eq. (33)) and depends on envelope mismatch parameter M as

$$Q_e(M) = \nu_e \left(1 + \frac{\alpha_e R_0^2}{2} M^2\right), \qquad (69)$$

so the resonance condition can approximately be expressed by the following formula

$$\frac{\tilde{\nu}_p(M)}{Q_e(M)} = \frac{\nu_p}{(1-M^2)^{\frac{3}{2}}} \frac{1}{\nu_e \left(1 + \frac{\alpha_e R_0^2}{2} M^2\right)} \approx \frac{\nu_p}{\nu_e} \left(1 + \left(\frac{3}{2} - \frac{\alpha_e R_0^2}{2}\right) M^2\right) \qquad (70)$$

Figure 5 presents the results of numerical simulations for the uniform focusing channel for $\kappa = 1$ and $\kappa = 4$ compared with Eq. (68).

IV. OVERLAPPING RESONANCES AND TRANSITION TO GLOBAL CHAOS

All the particles initially located inside the envelope execute linear motion. Therefore the Hamiltonian tori for those particles can be distorted but not destroyed by the beam mismatch. For the particles initially positioned outside the envelope the Hamiltonian flow has a different structure. The dominant period 2 particle-envelope resonance exists for all values of space charge parameter κ. Depending on the values of κ and particle phase advance in the zero space charge limit μ, there exists a set of higher order particle-envelope and particle-lattice resonances, which may overlap and create a large stochastic area in the phase space, destroying corresponding Hamiltonian tori of the unperturbed Hamiltonian. Briefly, the possible scenario for Halo formation can be described as follows. Every realistic particle distribution at finite temperature has a diffusive tail. The motion of the particles from the diffusive tail will be strongly affected by the system of overlapping resonances. Thus, we may suggest that the diffusion process produces the tail distribution and the overlapping resonances facilitate halo formation. With this qualitative picture in mind, we investigate the dependence of the critical mismatch parameter M_c, which is defined as the envelope mismatch for the onset of global chaos, on the parameters of the system.

A. Transition to global chaos in the uniform focusing channel

The main purpose of this work is to understand the effects of time-dependence of the linear focusing field on the mechanism of Halo formation. The case of a smooth focusing channel was studied in details elsewhere [3]. However, for the sake of clarity, we reproduce some of the results of [3]. The border of the "core" region is given by the envelope ellipse

$$\frac{\hat{x}^2}{R^2(0)} + R^2(0)\, \hat{p}_x^2 = 1, \qquad (71)$$

where $R(0)$ is the beam radius (provided $R'(0) = 0$) and $\hat{x}$ and $\hat{p}_x$ are particle initial phase space coordinates. Though the energy is not conserved in this case because of the time dependence of the Hamiltonian, we may consider consider only the Poincare surfaces of section at the minimum radius location. The maximum energy at $x = 0$ is defined as the Poincare energy for the envelope particle given by

$$E_P = \frac{\nu_p}{2(1-M)^2}. \qquad (72)$$

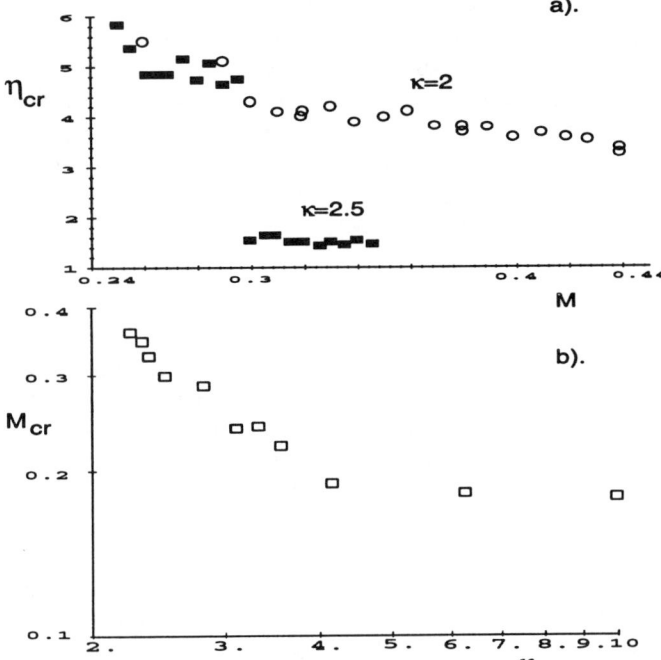

FIG. 6. a). η_{cr} as function of M for $\kappa = 2$ (boxes) and 2.5 (circles). b). M_{cr} vs space charge parameter κ obtained from numerical simulations.

The Poincare energy for the particle initially located outside the envelope is larger than E_P and given by $E = \eta E_P$ with $\eta > 1$. For a given mismatch parameter M there is a critical number η_c such that all the particles with $\eta > \eta_c$ will be trapped by halo and orbit around period 2 resonance. Though some particle with $\eta < \eta_c$ can be affected by a local chaos, they are separated from period 2 resonance by a stable Hamiltonian torus and can not participate in halo formation.

Figure 6-a shows η_c as a function of mismatch parameter M for two values of space charge parameter κ. A "smooth" dependence for $\kappa = 2$ should be compared with sharp drop of η_c for $\kappa = 2.5$ at $M \approx 0.29$. This fact can be understood as follows. As it was mentioned above, there are very few resonances for a small κ. The width of period 2 resonance and width of the chaotic layer around the separatrix of this resonance are smooth functions of mismatch parameter M (energy width is proportional to $\sqrt{M}$ [9] and M respectively). Therefore η_c will decrease smoothly as the mismatch parameter M increases. Some stepwise decrease is expected from the overlap of the period 2 resonance with high-order secondary resonances. On the other hand, in the case of larger κ more low-order primary resonances are present. Because in general the resonance strength decreases as the order of the resonance increases, low-order resonances are wider and can affect the particle motion stronger. These primary resonances interact and create a system of higher order resonances, which overlap and create a region of chaotic motion in the vicinity of the stable core. As this local chaos overlaps with chaotic region formed around the separatrix of the period 2 resonance, the transition to global chaos occurs and sudden jump in a functional dependence of $\eta_c(M)$ is observed. This sharp transition provides a way to determine a critical mismatch parameter M_c for particular value of κ.

B. Time-dependent focusing channel

The scenario of halo formation described above is still valid even in the case of time-dependent focusing channel provided the particle phase advance in the zero space charge limit is small. In order to justify this statement we have to point out that, in the first approximation, the only effect of time dependence of the linear focusing field, compared with the case of the uniform focusing channel, is the introduction of additional time-dependent perturbation H_2. This term reveals itself in the resonance structure, which it creates. However, for small μ only very high-order (and consequently, weak) primary particle-lattice resonances are present; these resonances can not modify the structure of the particle phase space significantly. This conclusion is illustrated by Figure 7-a, which represents the functional dependence $M_c(\kappa)$ for $\mu = 45°$; this is almost identical to the results for the uniform channel. Also, as space charge parameter κ increases, the linear particle tune is suppressed, thus making particle-lattice resonances even less important.

However, for the stronger linear focusing field and correspondingly larger μ the conditions for relatively low order primary envelope-lattice and particle-lattice resonances can be satisfied. The presence of such resonances can substantially modify the general structure of the particle phase space and greatly enhance the mechanisms of halo formation. The tolerance on beam mismatch may be substantially reduced compared to that for the equivalent uniform focusing channel.

1. Enhancement of halo formation due to envelope-lattice resonances

Generally it is required that the envelope oscillations be stable (e.g. far from the resonance $n\nu_e = 2\pi m$) what can be ensured by the appropriate choice of parameters of the system (α, K). However, as it was demonstrated in [2], the tune of envelope oscillations $Q_p(J_p)$ is amplitude dependent and increases with mismatch parameter M; therefore even if the resonance condition is not satisfied for the linear envelope tune ν_p, it may be satisfied for some orbit that corresponds to some finite value of the mismatch parameter M. In this case the envelope-lattice resonance structure is formed and the envelope motion can not be described as small oscillations around the closed orbit even if the beam is initially weakly mismatched: the resonance islands have a finite width what may result in relatively large deviation of the beam radius from the closed orbit solution. The motion of the test particle is also strongly modulated; this leads to qualitative changes in dependence $M_c(\kappa)$ compared to the case of the uniform focusing channel.

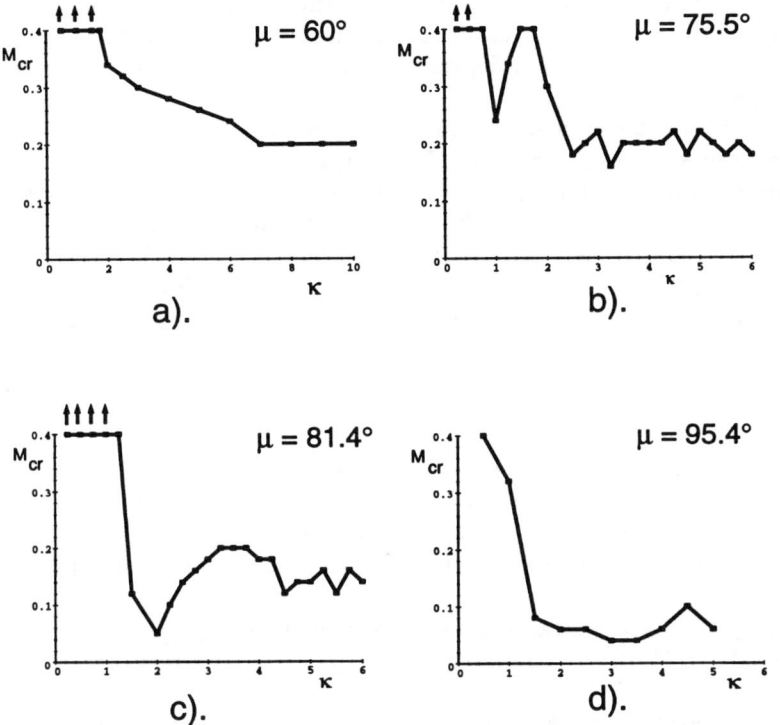

FIG. 7. $M_{cr}(\kappa)$ for a). $\alpha = 1$ b). $\alpha = 1.5$ c). $\alpha = 1.7$ d). $\alpha = 2.2$. Arrows directed upwards indicate that the critical mismatch amplitude $M_{cr}(\kappa)$ is actually larger than shown.

Examples of the effect produced by the presence of the envelope-lattice resonance can be found on the Figure 7-b,c. For example, a sharp drop in mismatch tolerance is seen on [b] at $\kappa \approx 1$ which can not be explained by particle-envelope resonances alone. In the vicinity of this point ($\alpha = 1.5, \kappa \approx 1$) the particle-lattice resonance island traps the envelope orbits with initial mismatch $M \approx 0.2$. The same effect can be seen of Figure 7-b with the only difference that for stronger focusing field the islands are created at smaller mismatch amplitudes and higher values of space charge parameter κ. Finally we conclude that we have to avoid not only the envelope-lattice resonance condition for the linear tune, but also for the beams with finite values of mismatch parameter.

2. Enhancement of halo formation due to particle-lattice resonances

If linear focusing field is strong enough (so, the condition $nQ_p = 2\pi m$ is valid for low-level primary resonances), a system of particle-lattice resonances is created and produces a substantial effect on the motion of a test particle. This system of so called structure resonances exists even in the case of a perfectly matched beam. If the beam is mismatched, the presence of structure resonances can ease the overlap of particle-envelope resonances what results in large changes of $M_{cr}(\kappa)$ dependence compared to the case of the uniform focusing channel.

Example of the effect produced by the presence of the particle-lattice resonance is presented on the Figure 7 [d]. In this case $\alpha = 2.2$, so 1:2 and higher order structure resonances can be observed. The overlap of these resonances with ones due to envelope modulation makes transition to global chaos much easier and reduces the critical mismatch parameter to ≈ 0.05 for large κ (compare to ≈ 0.20 for the uniform channel). This effect is important only for the focusing channel with zero space charge phase advance greater than 90°.

V. CONCLUSION

In conclusion, the beam transport problem in the time-dependent focusing channel is studied in the framework of KV distribution and "particle-core" model. Some analytical results concerning the closed orbit solution for the envelope equation are presented and the question about stability of small envelope oscillations of weakly mismatched beam is discussed. Analytical estimations of linear instability region are obtained by means of deriving an "equivalent" Mathieu's equation and compared with results of numerical simulations.

The major part of the paper is devoted to the study of the dynamics of the test particle, motion of which is determined by a combination of external linear focusing force and self-electromagnetic force, produced by the spatial space charge. The particle motion is described by the methods of Hamiltonian dynamics; it is shown that a set of parametric resonances of the particle Hamiltonian can be generated by mismatched envelope oscillations in the time-dependent focusing channel. These resonances can be classified as particle-envelope resonances, due to existence of resonance condition between particle motion and modulation, produced by the envelope, or as structure particle-lattice resonances, due to resonance interaction between the particle and the linear focusing channel. 2 : 1 particle-envelope resonance is particularly important because it occurs at all values of space charge perviance parameter and gives rise to halo. The other resonances may overlap, if the mismatch parameter is greater than some critical value, and create a large stochastic region in the particle phase space, providing a possible mechanism for halo formation. It was found from the numerical simulations and qualitatively understood using the methods of Hamiltonian dynamics, that the presence of structure resonances in time-dependent system can greatly reduce the value of critical mismatch parameter compared with a case of uniform focusing channel with the same value of space-charge parameter. Though the structure resonances alone are not able to produce halo, they can ease the process of overlapping of particle-envelope resonances. The picture can further be complicated by the presence of envelope-lattice resonance islands which may exist for finite values of mismatch parameter M even if the closed orbit trajectory is stable. Finally, the relation between critical mismatch amplitude M_{cr} and parameters of the system (α, K) is obtained numerically for different cases.

Though this paper discusses only a very special case of time-dependent focusing channel, it is reasonable to expect that qualitative conclusions about implications, introduced by time-dependency of the linear focusing field, are relevant even for more realistic models like beam transport in FODO channel.

REFERENCES

1. I.M. Kapchinskij and V.V. Vladimirskij, *Proceedings of the International Conf. on High Energy Accelerators*, p. 274 (CERN, Geneva, 1959).
2. S.Y.Lee, A.Riabko, Phys.Rev. **E51**, 1609 (1995).
3. A.Riabko, *et al.*, Phys. Rev. **E51**, 3529 (1995).
4. R.A. Jameson, in *Proc. of the 1993 Part. Accel. Conf.* p.3926 (IEEE, Piscataway, 1993).
5. J.S. O'Connell, T.P. Wangler, R.S. Mills and K.R. Crandall, in *Proceedings of the Particle Accelerator Conf.* edited by J. Bisognano, p.3657 (IEEE, Piscataway, NJ, 1993).
6. J.M. Lagniel, Nucl. Inst. and Meth. in Phys. Res. **A345**, 46 (1994), ibid. 405 (1994).
7. I. Hofmann, L.J. Laslett, L. Smith, and I. Haber, Particle Accelerators, **13**, 145 (1983); J. Struckmeier and M. Reiser, Particle Accelerators, **14**, 227 (1983).
8. M. Abramovich and I.A. Stegun, Handbook on Mathematical Functions (National Bureau of Standards, Washington, DC, 1975).
9. A.Riabko, *et al.*, PAC 95 Conference Proceedings (1995).

Nonlinear resonances, chaos and halo formation in space-charge dominated beams

J-M Lagniel

Commissariat à l'Energie Atomique - Direction des Sciences de la Matière
Groupe d'Etudes et de Conception d'Accélérateurs
(CEA-DSM-GECA)

CEA Saclay - Laboratoire National Saturne
91191 Gif-sur-Yvette Cedex - FRANCE

- Abstract -

A significant breakthrough in the understanding of the mechanisms leading to charge redistribution and halo formation in space-charge dominated beams has been achieved in the past few years. The first aim of this paper is to summarize the main results which have been obtained in recent studies. The nonlinear resonances excited by space charge forces are analysed, then it is shown that they can induce chaotic trajectories, therefore halo formation. The second objective is to present some ideas in order to try to promote new successful studies, as well as to point out the ambiguous character of some theories used in this field without any serious justification.

- I - Introduction.

In numerous nonlinear dynamical systems studied in various disciplines (fluid dynamics, celestial mechanics, chemistry, biology, ecology, economy...), chaotic (stochastic) motions are generated by the dynamics itself whereas no random force is present. The chaotic behaviour of particle trajectories has been studied for a long time in the accelerator field to understand the beam-beam effects in colliders and to determine the dynamic apperture of storage rings (see ref.[1] for example). During the 1972 linac conference [2], A. Sessler writted :

> "one can conjecture that the phenomenon studied in [Numerical calculations of the effects of space charge..., R. Chasman] is the result of nonlinear space charge forces causing particles to be above the stochasticity limit. It would be most illuminating to undertake analyses, analogous to those in [Research concerning the theory of non-linear resonance and stochasticity, B.V. Chirikov] and [Non-linear space charge effects, E. Keil, Arnol'd diffusion lifetime in storage rings, E. Keil]".

Nevertheless, nobody followed this idea at that time (it could be interesting to analyse the reason why), and it was only 21 years later that the stochastic (chaotic) behaviour of the particle trajectories induced by space charge force has been demonstrated for the first time [3].

The first aim of this paper is to summarize the main results which have been obtained from recent studies on unbunched beams evolving in continuous focusing and FODO channels. The nonlinear resonances excited by space charge force are analysed in section II, then it is shown how these nonlinear resonances can induce chaotic trajectories in section III. Charge redistribution and halo formation are analysed in section IV. Numerical simulation results showing the particle diffusion (due to the presence of nonlinear resonances and enhanced in the chaotic areas) are presented. On the basis of these studies, some ideas are also proposed in order to try to promote new succesful studies, as well as to point out the ambiguous character of ideas sometimes applied to this field without any real justification.

- II - Nonlinear resonances induced by space charge.

For azimuthally symetric beams evolving in a <u>continuous focusing channel</u>, the dimensionless equations of motion for both beam core envelope and individual (test) particles depend only on the space-charge tune depression $\eta = \sigma_t / \sigma_{0t}$ where σ_t and σ_{0t} are the phase advances *per unit length* respectively with and without space charge [4]. When the beam is mismatched, resonances between the particle motion and the space-charge force (a "nonlinear oscillating perturbing force") can be characterized by the tune $v = \sigma_{particle} / \sigma_{core}$. The core envelope oscillation being characterized by $\sigma_{core} = \sqrt{2(\sigma_{0t}^2 + \sigma_t^2)}$, the tunes are given by :

$v_i = \sigma_t / \sigma_{core} = \eta / \sqrt{2(1+\eta^2)}$ for particles injected inside the core for uniform distributions, or for particles traveling near the axis of nonuniform beams if σ_t is the phase advance on the axis,

$v_\infty = \sigma_{0t} / \sigma_{core} = 1 / \sqrt{2(1+\eta^2)}$ for particles injected further and further from the core because the influence of the space-charge force becomes smaller and smaller [5].

For beams with uniform or monotonically-decreasing distributions, the tune of the test particles is then such that $v_i \le v < v_\infty$. Figure 1 (from Ref.[5]) gives the tune spread of the particles versus the space-charge tune depression η, then the range of nonlinear resonances which are excited by the beam core oscillation for a given η value. This analysis already presented in Ref.[5] and rediscovered in Ref.[6] shows that the strong $v = 1/2$ resonance is always present. For beams evolving in a continuous focusing channel, the excitation of these resonances is a function of the mismatching level.

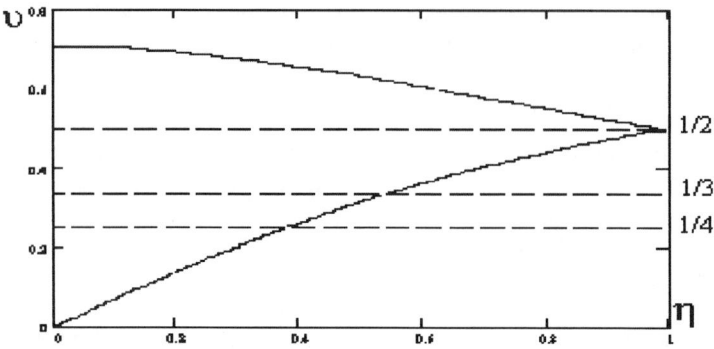

Fig. 1 : Tune spead of the particles versus space-charge tune depression

Figure 2 (from Ref.[5]) shows calculations of phase advances ($\sigma = 2\pi\nu$) for $\eta = 0.1$ and $\eta = 0.5$. To establish this figure, the evolution of particles injected with increasing amplitudes and x'= 0 is computed over one beam-core oscillation, then σ is calculated. The results are given for both uniform (KV) and Gaussian distributions with the same particle density on the axis. Obviously, the nonlinear resonances can be excited into the beam core for nonuniform densities.

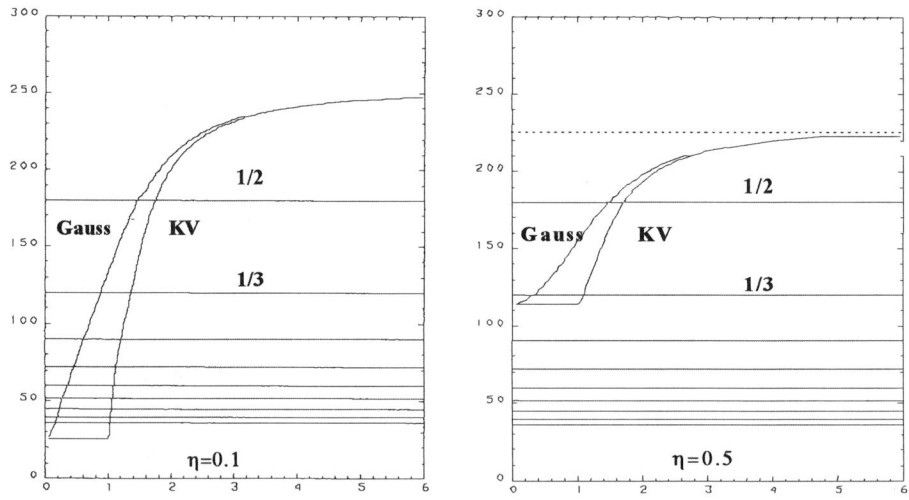

Fig.2 : Phase advances : $\sigma = 2\pi\nu$ (°) vs the particle amplitude
for $\eta = 0.1$ (left) and $\eta = 0.5$ (right)
for KV and Gaussian distributions with Ro = Rrms = 1.
The nonlinear resonances 1/2, 1/3 ... are also ploted.

For a <u>matched beam in a FODO channel</u> [7-8-9], again both order and number of resonances are determined by the choice of the phase advances *per focusing period* σ_{0t} and σ_t. For uniform or monotonically-decreasing distribution functions for which the phase advance on the axis is σ_t, the tune of the particles which travel near the axis is close to $v = \sigma_t/2\pi$. When the transverse energy of the test particles is increased, the space-charge effect becomes more and more negligeable, the tune of the particles which travel far from the axis is then close to $v_\infty = \sigma_{0t}/2\pi$. The parametric resonances which are *excited by the FODO channel* when the beam is matched are then in the range :

$$\sigma_t/2\pi \leq v < \sigma_{0t}/2\pi$$

For <u>mismatched beams in a FODO channel</u>, the main nonlinear resonances excited by the beam modulation due to the quadrupolar focusing are still present and, in addition, new resonances are excited by the two eigen frequencies of the mismatched envelope oscillation [9].

For weak mismatches, the envelope equations in smooth approximation can be linearized and, following I. Hofmann [10], the two eigen modes (even and odd) can be calculated. They are given by :

$$\sigma_e = \sqrt{2(\sigma_{0t}^2 + \sigma_t^2)} \text{ and } \sigma_o = \sqrt{\sigma_{0t}^2 + 3\sigma_t^2}$$

The rms envelope equations have been numerically integrated (without smooth approximation) for $\sigma_{0t} = 62°, \sigma_t \sim 20°$, and a weak initial mismatch (10% in the (x,x') plane). The Fourier spectrum of the envelopes (figure 3) shows that, even for this weak mismatch, the amplitudes of the odd and even modes (the two peaks at low frequency) are already large compared to the one of the main mode (f=1). These eigen frequencies are very close to those calculated using the theoretical formulas derived using the smooth approximation. It must be pointed out that for large mismatches, the envelope oscillations become rapidly chaotic [11], a more complicated frequency spectrum can be observed.

Above the main resonances excited by the natural beam modulation due to the quadrupolar focusing system ($\sigma_t/2\pi \leq v < \sigma_{0t}/2\pi$), new resonances are then excited by the two eigen frequencies of the envelope oscillations when the beam is mismatched. These additional resonances are in the range

$$\sigma_t/\sigma_{e,o} \leq v_{e,o} < \sigma_{0t}/\sigma_{e,o}$$

then : $\quad \eta/\sqrt{2(\eta^2+1)} \leq v_e < 1/\sqrt{2(\eta^2+1)} \quad$ for the even mode

and $\quad \eta/\sqrt{3\eta^2+1} \leq v_o < 1/\sqrt{3\eta^2+1} \quad$ for the odd mode [9].

Figure 4 shows the range of resonances which can be excited by the two modes as a function of the tune depression η. As in the case of a continuous focusing channel, the strong $v = 1/2$ resonance is always present.

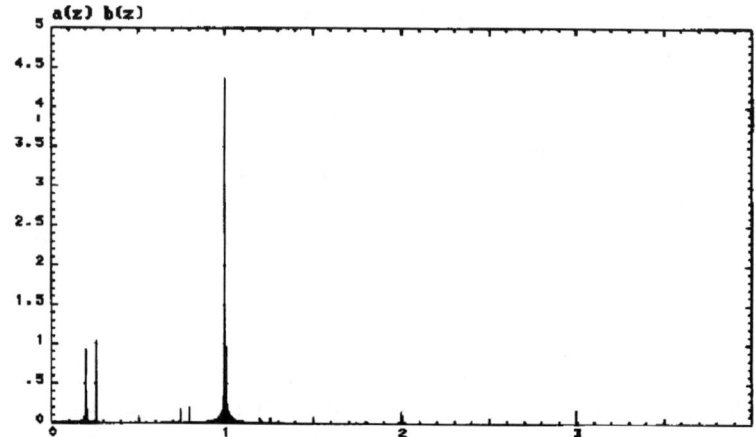

Fig.3 : Fourier spectrum of a mismatched beam envelope.
The frequency of the envelope oscillation induced by the quadrupoles is f = 1, the two low frequency peaks are the two eigen frequencies of the mismatched envelope oscillation.

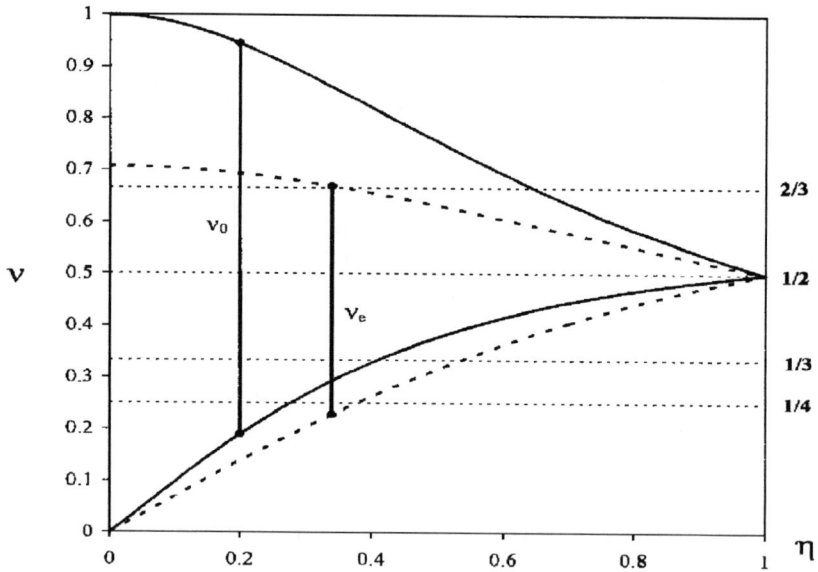

Fig.4 : Range of resonances excited by a mismatch versus η.
Each pair of curves defines the range of the nonlinear resonances ($v = 2/3$, 1/2, 1/3...) which are excited by the odd (v_o) and even (v_e) modes. For $\eta \leq 0.4$, the strong resonances $v = 2/3$, 1/2, 1/3 and 1/4 are excited.

- III - Chaos induced by the nonlinear resonances -

The very first observation of chaotic particle trajectories has been presented in Ref.[3] for space-charge dominated beams evolving in a <u>continuous focusing channel</u>. These analyses [3] have been confirmed in Ref.[4] for different values of the tune depression and for a nonuniform distribution. This significant breakthrough in the understanding of the mechamisms leading to charge redistribution and halo formation in high-intensity beams has been achieved thanks to the Particle-Core Model (PCM) [12]. As usually done to study *complex systems*, the PCM is a simplified model which keeps the dominant properties of the real physical system and which allows an analysis of the basic phenomena.

In Ref.[3], it has been demonstrated that the *resonance overlap mechanism* can lead to the formation of a halo area where the particle trajectories are stochastic. This chaotic behaviour has been clearly observed using the Poincaré surface of section technique (figure 5). Sensitive dependance on initial conditions (figure 6) and intermittencies which characterize chaotic systems have also been shown. These results have been confirmed by analytical studies [13] as well as by numerous simulations done using multiparticle PIC codes [14].

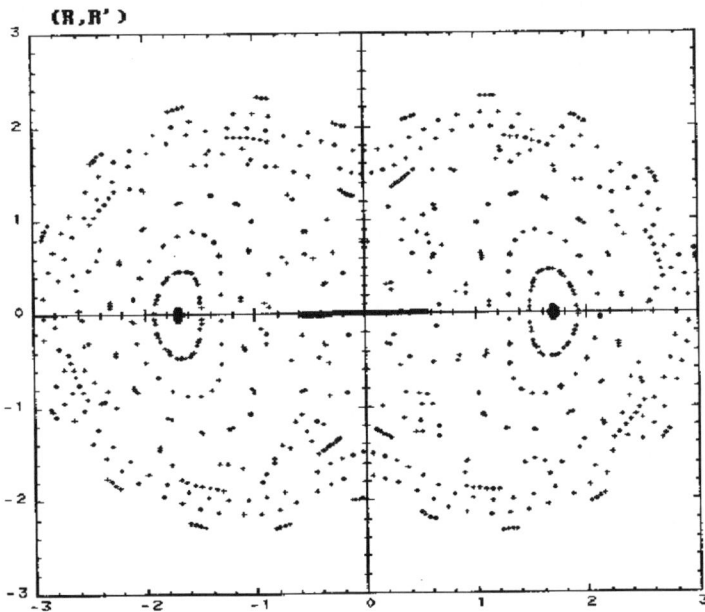

Fig. 5 : Poincaré surface of section for $\eta = 0$ (from Ref.[3])
The two $\nu = 1/2$ islands are localized around $R = \pm 1.7$ and $R' = 0$.
The outer curves are "KAM surfaces" which limit the chaotic area.

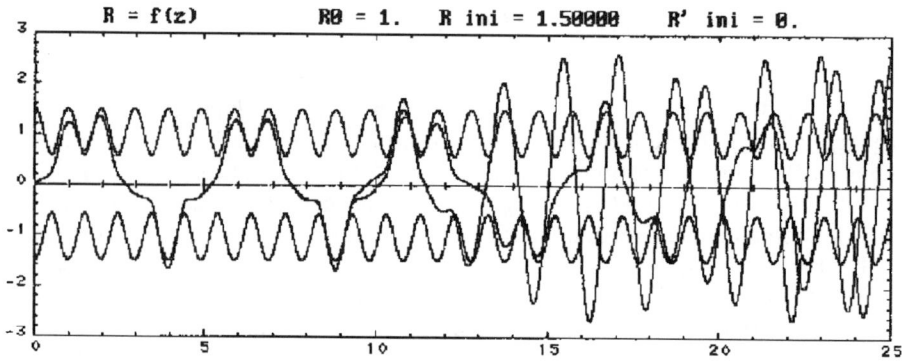

Fig. 6 : Sensitive dependence on initial conditions.
The trajectories of two particles injected at x = 0, x' = 0.10000 and 0.10001 (initial dx'/x' = 10^{-4}) are completely different after 10 beam core oscillations.

The PCM has also been used to study the behaviour of <u>matched beams in a FODO channel</u> [7-8-9]. The Poincaré surface of section technique is used again to analyse the phase space topology. Figure 7 clearly shows the position of some resonances in the (x,x') phase plane for uncoupled (y = y' = 0) particles.

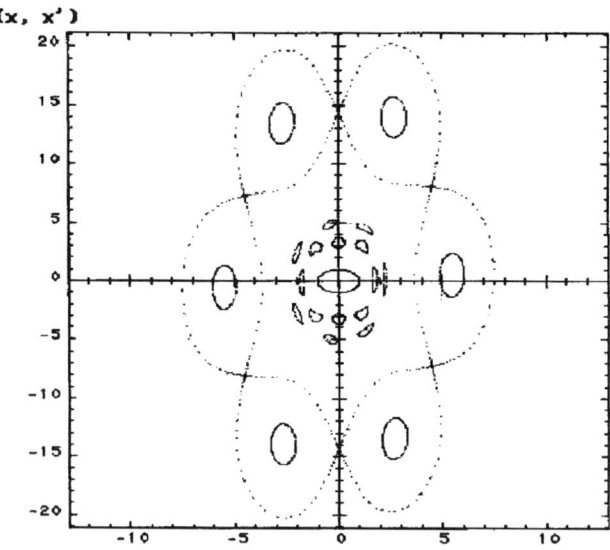

Fig. 7 : Poincaré surface of section (x,x') for $\sigma_{0t} = 62°$ and $\sigma_t = 20°$

212

As pointed out in section II, the $v = 1/6$ resonance is excited far from the beam core because it corresponds to a phase advance $\sigma = 60°$ close to $\sigma_{0t} = 62°$. The size of the chaotic areas is limited when the strong nonlinear resonances $v = 1/4$ and $v = 1/5$ are avoided, that is when $\sigma_{0t} < 72°$. For the parameters choosen to draw figure 7, thin chaotic zones (not shown in this figure) are present only on the periphery of the uniform-density beam core.

To take into account the coupling force induced by space charge leads to the analysis of a nonautonomous system with N = 2.5 degrees of freedom : (x,x')+(y,y')+z. In this case, resonances form a dense "Arnol'd web" into which particles can diffuse (Arnol'd diffusion) as demonstrated in Ref.[8].

For <u>mismatched beams in a FODO channel</u>, additional resonances are excited by the even and odd modes of the envelope oscillations (see section II and figure 4). Figure 8 illustrates the fact that "The more there are oscillators, the more they couple between themselves, and the more one can anticipate chaos to be observed" [15]. Drawn with $\sigma_{0t} = 62°$ and $\sigma_t = 20°$ (same parameters as figure 7), figure 8 shows the wide chaotic area formed around the beam core for a weak mismatch (+10% in x, -10% in y) which excites only the odd mode. The $v_0 = 1/2$ resonance is clearly shown by this Poincaré section [9]. At a larger amplitude, the main resonance $v = 1/6$ (figure 7) is not affected by the weak mismatch.

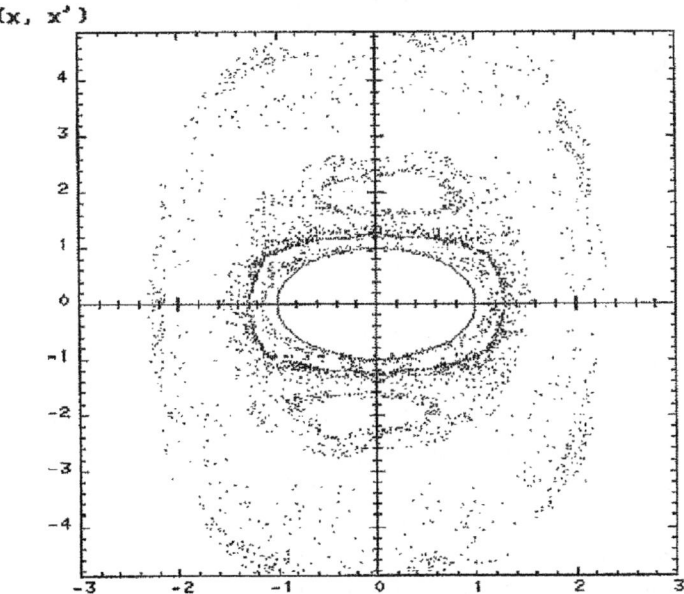

Fig. 8 : Poincaré section for an uncoupled particule, $\sigma_{0t} = 62°$ and $\sigma_t = 20°$ mismatch = +10% in x, -10% in y (odd mode only)

- IV - Halo formation induced by nonlinear resonances and chaos.

It is clear that nonlinear resonances can scatter particles around the beam core, this halo formation is highly enhanced when chaotic areas are formed :

Nonlinear resonances   =>   Chaos   =>   Halo formation

The links between these phenomena are now known : the resonance overlap mechanism explains the formation of stochastic areas, and diffusion in these stochastic areas (Arnol'd web for N > 2) increases particle redistribution and halo formation.

This knowledge has been used to find a way to avoid halo formation. The basic idea is simple : as halo comes mainly from particle diffusions in chaotic areas, halo formation will be avoided if the resonances do not overlap. This has been demonstrated in Ref.[8] where defocusing octupoles were used to cancel emittance growth and halo formation in a FODO channel tuned with $\sigma_{0t} = 100°$. Other types of nonlinear correctors can also be used, modified octupoles have been proposed in Ref.[8], effects of duodecapoles are demonstrated in Ref.[16].

For realistic accelerator parameters ($\sigma_{0t} < 90°$ and $\sigma_t > 0°$ on the axis), we know that the stochastic web is thin, then the diffusion rates could be very low. Nevertheless, particle losses must be extremely low in the new generation of high-current accelerators. Arnol'd diffusion rates are low at large amplitude but relative losses lower than 10^{-7}/m must be achieved [17]. We must keep in mind that, for example, if one particle diffuses from an inner stochastic area to an outer one after 1000 or 10000 FODO periods, there is a high probability that one particle diffuses in few periods if the inner stochastic area is filled with 1000 or 10000 particles (ergodic hypothesis).

Anticipating *a priori* that such a type of diffusion is too slow to induce halo formation is not a scientific attitude. So, we have done a large number of simulations using the PCM to answer the question : can we observe diffusions along the Arnol'd web which drive particles from the beam core vicinity towards a resonance located far from it? To do this, up to 80000 test particles per simulation have been injected in the (x,y) plane (x' = y' = 0) and followed along 200 and 600 FODO periods. For each particule, the maximum radius (Rmax) reached during both 200 and 600 periods was stored in order to calculate $\Delta = $ (Rmax - Ro) / Ro which is a measure of the diffusion for a particle injected with an initial radius Ro. These studies have been done for different mismatching values and two tunes : $\sigma_{0t} = 62°$ and $\sigma_t = 20°$ for which the $\nu = 1/6$ resonance is located far from the beam core (see figure 7), $\sigma_{0t} = 75°$ and $\sigma_t = 20°$ for which, this time, the $\nu = 1/5$ ($\sigma = 72°$) resonance is located far from the beam core.

Figure 9 gives a 3D representation of Δ versus the initial positions of the test particules in the ranges 0 < x < 6 and 0 < y < 6 with x' = y' = 0 and dimensions normalized with respect to the beam core radius. The large stochastic area located near the beam core and the $\nu = 1/6$ resonance area can be clearly localized but no diffusion from one area to the other can be observed.

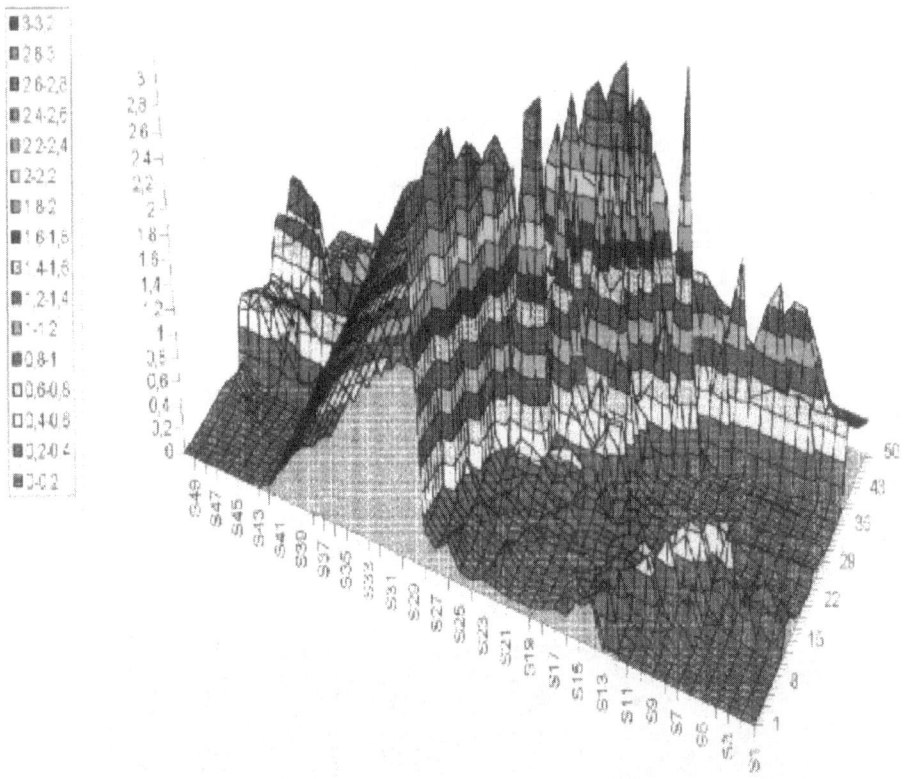

Fig. 9 : Example of simulation result for a mismatched beam in FODO channel Δ (vertical axis), computed over 200 FODO periods, is ploted versus the test particle initial position ($0 < x < 6$ and $0 < y < 6$ with Rcore ~ 1) $\sigma_{ot} = 62°$ and $\sigma_t = 20°$, the mismatch is +10% in x and y

- IV - Conclusion -

Table 1 summarizes a large number of simulation results obtained using the PCM for weakly or strongly mismatched beams. The values of Δ which are reported in this table concern only the particles injected in the stochastic area surrounding the beam core ($1 < Ro < 2$, see figures 8 and 9). From this initial positions, Δ must reach values around 4 or 5 if the test particules diffuse towards the $\nu = 1/6$ or $\nu = 1/5$ resonances.

Δ for mismatch X-Y	$\sigma_{0t} = 62°, \sigma_t = 20°$		$\sigma_{0t} = 75°, \sigma_t = 20°$	
	# = 200	# = 600	# = 200	# = 600
1.1-1.1	0.8	0.8		
1.1-0.9	0.8	1.2		
1.3-0.7	1.0	1.4		
1.0-1.1	0.8	1.0	1.2	1.4
1.0-1.2	1.0	1.2	1.6	1.6
1.0-1.3	1.0	1.2	1.4	1.8
1.0-1.4	1.0	1.4	1.8	2.0
1.0-1.5	1.2	1.4 (zoom =1.8)	1.6	2.2

Table 1 : Values of Δ versus the mismatching factor for two tunes
is the number or FODO periods for the calculation of Δ

The main result of these studies is that we never observed particles jumping from the beam core vicinity to a strong resonance (1/6 or 1/5) located far from it. The maximum amplitude reached by particles injected close to the uniform beam core (1 < Ro < 2) is limited to approximatively 3 times the beam core radius. Nevertheless, a zoom showing the values of Δ in a narrow area located on the beam core border (figure 10) recalls us that the system we are studying is chaotic, therefore very sensitive to the initial conditions.

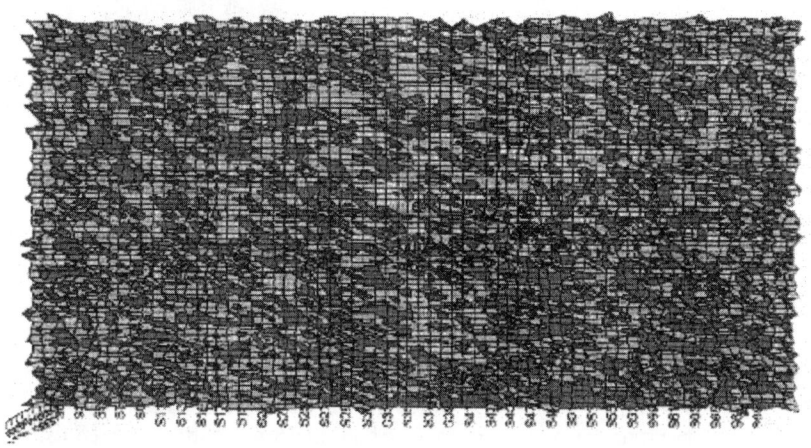

Fig. 10 : Zoom on initial positions 0.06 < x < 0.22 and 1.05 < y < 1.12
for $\sigma_{0t} = 62°$, $\sigma_t = 20°$, x mismatch = 1 (0%) and y mismatch = 1.5 (50%).
In this chaotic area, the Δ value (0 < Δ < 1.8)
is very sensitive to the initial conditions.

Then, even with the PCM, our "toy model" which is used to simplify the studies, it is necessary to inject a very large number of test particles to check that a very low fraction ($\sim 10^{-7}$/m) of the beam does not diffuse at large amplitudes. An accurate code able to handle a very large number of particles ($>10^6$) is developed by R. Ryne at LANL. From the studies presented here, it seems that it is the only way to obtain the results needed for the next generation of high-current linacs.

Another remark can be made. For *collisionless* beams evolving in FODO channels (without acceleration), we are dealing with *conservative Hamiltonian systems*. With the nonlinear space-charge forces, these systems are *nonintegrable*. We have shown some characters which are common to this class of systems :

-1- The global properties of the system are determined by the location and the size of the nonlinear resonances and by the *resonance overlap* mechanism which leads to the formation of stochastic (chaotic) areas.

-2- Particle diffusion and halo formation are *enhanced* in the chaotic areas (Arnol'd web for $N > 2$).

-3- Beams which are not a priori unstables (beam current around 100 mA, $\sigma_i > 0$ on the axis) are in a *regime of weak chaos*. The system contains a mixture of *quasiperiodic (KAM) and chaotic orbits*, and a mixture of *stable and unstable periodic orbits* (fixed points).

-4- The flow is *ergodic only on a subspace* of the phase space (the chaotic regions) but isolated adiabatic islands exist in these regions. Fortunately, *the system is not in a regime of strong chaos*.

At this point, following J.D. Lawson [18], "it may be asked whether the concepts of statistical mechanics and thermodynamics can yield new insights or put on a more firm footing recent work on emittance growth" (and halo formation). This question addresses in fact one of the most challenging problems studied (since 150 years !) by theoretical physics : the bridge between dynamics and thermodynamics. This bridge is under construction on the basis of the chaos theory which shows that the structure of the equations of motion with "deterministic randomness" on the microscopic level emerges as irreversibility on the macroscopic level. In other words, Boltzmann irreversibility is a consequence of molecular chaos "superimposed" on the equations of dynamics; in a purely dynamical system, both randomness and irreversibility can be consequences of the equations of motion. But, the deterministic description in terms of Liouville equation can be transformed into a Markov chain *only* if the system is strongly chaotic. Then, it seems that the Fokker-Planck equation cannot be used to study the global evolution of beams which are not unstables. These arguments also push us to think that, as in the past, to regard the beam as a 'drifting gas' described in terms of thermodynamic variables will not lead to any new insights of practical value.

It must be added that particles evolving in an accelerator are usually accelerated. Then, the system becomes a *nonconservative nonintegrable system*. Its behaviour is very different from the one described here (to be published).

Bibliography

[1] L.R. Evans and J. Gareyte, "Beam-beam effects", 1985 CERN Accelerator School, CERN 87-03, 1987.

[2] A. M. Sessler, "Collective phenomena in accelerators", LINAC72 Conf. Proc., October 1972, p.291.

[3] J-M. Lagniel, "On halo formation from space-charge dominated beams", LNS/SM/93-35, August 1993 and NIM-A Vol.345 (1994) No.1, p.46.

[4] T.P. Wangler, "Dynamics of beam halo in mismatched high-current charged-particle beams", LA-UR-94-1135, March 1994.

[5] J-M. Lagniel and A-C. Piquemal, "On the dynamics of space-charge dominated beams", LINAC94 Conf. Proc., August 1994.

[6] see the works presented during this workshop by S.Y. Lee and A. Riabko.

[7] J-M. Lagniel, "Chaotic behaviour and halo formation from 2D space-charge dominated beams", LNS/SM/93-42, December 1993 and NIM-A Vol.345 (1994) No.3, p.405.

[8] J-M. Lagniel, "Chaotic behaviour induced by space charge", EPAC94 Conf. Proc., June 1994, p.1177.

[9] J-M. Lagniel and D. Libault, "Chaos, a source of charge redistribution and halo formation in space-charge dominated beams, PAC95, May 1995.

[10] I. Hofmann, "Transport and focusing of high-intensity unneutralized beams", in Applied charged particle optics (C), A. Septier, Academic press, 1983]

[11] C. Chen and R.C. Davidson, "Non-linear resonances and chaotic behavior in a periodically focused intense charged particle beam", Physical Review Letters, Vol.72 (1994), No.4, p.2195.

[12] J.S. O'Connell, T.P. Wangler, R.S. Mills and K.R. Krandall, "Beam halo formation from space-charge dominated beams in uniform focusing channels", PAC93 Proc, p.3657.

[13] R.L. Gluckstern, "Analytic model for halo formation in high current ion linacs", Physical Review Letters, Vol.73, No.9, August 1994.

[14] see T.P. Wangler, R. Ryne, A. Piquemal, C. Chen ..., this workshop.

[15] D. Ruelle, "Hasard et chaos", Editions Odile Jacob, 1991.

[16] see the works presented during this workshop by Y. Batygin

[17] J.R. Delayen et al., "Design considerations for high-current superconducting ion linacs", PAC93 Proc. p.1715 and earlier references therein.

[18] J.D. Lawson, "The emittance concept", High-brightness beams for advanced accelerator applications, AIP Conf. Proc. No.253, 1991.

Lowest-Order Phase Space Structure of a Simplified Beam Halo Hamiltonian*

David L. Bruhwiler
Advanced Technology and Development Center, Northrop Grumman Corporation,
Princeton NJ 08540-6620, USA

ABSTRACT

Hamiltonian perturbation theory is applied to the particle-core model for zero-angular-momentum test-particles in the limit of small mismatch and moderate space charge. A first-order treatment captures the lowest-order averaged dynamics arising from the dominant 2:1 parametric resonance, neglecting any chaotic effects that might arise from the overlap of higher-order resonances. The analysis shows that test-particles from a matched Kapchinskij-Vladimirskij (KV) distribution are driven into the halo by the oscillations of the mismatched core KV distribution, if the mismatch factor exceeds a critical value which depends on the space charge parameter μ. This dynamical effect persists, although the time scale grows without bound, even in the limit $\mu \to 0$. A symplectic test-particle code and self-consistent particle simulations both show good agreement with the analysis.

I. INTRODUCTION

The particle-core model (PCM) of Ref. [1] is a simplified model for studying the dynamics of an unbunched, mismatched beam in a linear, continuous focusing channel. Rather than treating the full problem, the model postulates a stable, oscillating beam core consisting of a KV distribution [2]. Test-particles feel the space charge forces generated by the core, which are linear but oscillating inside the core and nonlinear but constant in time outside of the core. The oscillating core pumps energy into some of these test-particles, driving them out to twice the maximum core amplitude -- such particles are said to constitute the halo. We define the mismatch factor $m \equiv R_m/a - 1$, where a is the matched core radius and R_m is the maximum radius of the mismatched core. The PCM has been carefully studied previously, both theoretically and numerically (see e.g. Ref. [3] and references therein).

Our analysis begins with the Hamiltonian that describes the test-particle motion, treating it perturbatively under the assumption that the beam mismatch is small and the nonlinear effects are weak [4]. The first approximation implies $m \ll 1$, while the second approximation limits our analysis to test-particles near the beam core. Although we are unable to follow particles out into the halo, the range of validity extends far enough beyond the core for us to establish the conditions under which particles near the core will be swept out into the halo.

Our approach is analogous to that of Gluckstern [5]. We average away all of the fast oscillations in the problem to find a Hamiltonian which governs the lowest-order dynamics of the system. Like Gluckstern, we find that this dynamics is dominated by a strong 2:1 resonance between the oscillations of the beam core and the oscillations of test-particles outside of the core. Our analysis is complementary to most previous work on the PCM [3], which addresses chaotic motion arising due to the overlap of higher-order resonances in the space charge dominated limit.

---

*Supported by the IFMIF program of DOE through Oak Ridge National Laboratory, subcontract 15X-SN293C, and by Northrop Grumman Corporation.

We consider the dynamics of test-particles with zero angular momentum p_ϕ to make the analysis tractable. Particles with small p_ϕ pass close to the center of the beam and feel most strongly the perturbing effect of the oscillating core (as described e.g. in Ref. [6]) and are a good diagnostic for the onset of halo formation.

Our averaged Hamiltonian is obtained from the initial Hamiltonian via a series of canonical transformations. By transforming the initial phase space coordinates of zero-angular-momentum test-particles from a matched KV distribution to the phase space of this averaged Hamiltonian, we are able to show that there is a critical value of the mismatch parameter m_c above which some of the particles have initial phase space coordinates that lie outside of the inner separatrix. Such particles are swept slowly out into the halo. Contrary to the physical picture developed by previous workers, in which beam halo occurs only for beams with strong space charge, this dynamical behavior persists even as $\mu \to 0$.

The test-particles we consider in finding m_c are not part of the beam core, because they are drawn from a matched KV distribution, while the core consists of a mismatched KV distribution. Test-particles drawn from the core distribution remain inside the core. However, the "matched" test-particles have the same transverse energy as core particles, and they have initial oscillation amplitudes that are smaller than the maximum core oscillation amplitude -- the important difference is that some of them are initially oscillating out of phase with the core. The motivation for treating such particles is that in practical situations, a continuous ion beam is usually generated with a pre-halo arising from aberrations in the extraction optics of the ion source. This pre-halo could then evolve into a true halo when the beam is injected into an accelerating structure.

A symplectic test-particle integrator is used to test the analytical prediction for $m_c(\mu)$ and to verify the persistence of halo growth in the limit of very small μ. The Topkark code [7] is used to show that the predicted dynamics is readily observed in self-consistent particle simulations. Also, the halo particles in the Topkark simulations have finite angular momentum, indicating that our results are not strictly confined to the limit $p_\phi=0$.

The notation used is as follows. The zero-current phase advance is denoted by k_0, and the full unnormalized emittance by ε. The generalized perveance is $\kappa = 2I/I_0(\beta\gamma)^3$, where I is the beam current and $I_0=4\pi\varepsilon_0 mc^3/q$ the characteristic current. The matched beam radius for a KV beam in a linear, continuous focusing channel, denoted by a, is defined by $(k_0 a)^2 \equiv \kappa/2 + [(\kappa/2)^2 + (k_0\varepsilon)^2]^{1/2}$. The space charge parameter is $\mu=\kappa/(k_0 a)^2$ ($0\leq\mu\leq1$). The full current phase advance is denoted by k, and the tune depression is $k/k_0=(1-\mu)^{1/2}$. See also Chap. 4 of Ref. [8].

II. INITIAL HAMILTONIAN

We begin our analysis with the Hamiltonian describing transverse particle dynamics in the Larmor precessing frame [9]:

$$H_i(p_r, p_\phi, r, z) \approx \frac{1}{2}\left(p_r^2 + k_0^2 r^2 + p_\phi^2/r^2\right) - \frac{1}{2}\kappa\left[r^2/R^2(z)\right]\Theta[R(z)-r] \\ - \frac{1}{2}\kappa\left[1 + 2\ln(r/R(z))\right]\Theta[r-R(z)], \quad (1)$$

where p_r and p_ϕ are the radial and angular momentum, Θ is the Heavyside function, and R(z) is the envelope radius of an idealized, mismatched KV beam which is presumed to be driving the single particle dynamics. Because the Hamiltonian is independent of angle ϕ, p_ϕ is a constant of the motion.

If the KV core is slightly mismatched, then R(z) oscillates about a mean value (see Ref. [10] for a detailed treatment), with a lowest order behavior given by $R(z) \approx a[1-m\cos(k_e z)]$, where $k_e^2 = 2(k_0^2 + k^2)$. We insert this approximation in Eq. (1) and expand to first order in m. This approximation limits us to small values of m. It also limits us to small values of μ, because the core oscillations become more complicated in the space charge dominated limit (μ close to unity).

For our analysis, we consider only particles with $p_\phi = 0$. Thus, we can restrict our attention to the x-plane, letting $p_r \to p_x$ and $r \to |x|$ such that Eq. (1) becomes:

$$H_i(x, p_x, z) \approx \frac{1}{2}(p_x^2 + k^2 x^2) - m\kappa(x^2/a^2)\cos(k_e z)\Theta(a - x^2) \quad (2)$$
$$- \frac{1}{2}\kappa[1 - (x^2/a^2) + \ln(x^2/a^2)]\Theta(x^2 - a).$$

The far right term of Eq. (2) vanishes for $|x|<a$, but this nonlinear term becomes large as $|x|/a$ becomes large. Because we treat this term as a perturbation below, we require $|x|/a<2$ for our analysis to remain valid. For this reason, we further simplify our analysis below by expanding the logarithm in the small parameter $|x|/a - 1$.

We also normalize this initial Hamiltonian by the factor $E_{KV} = k^2 a^2/2$, which is the energy of all particles in a matched KV distribution. We normalize x to the matched beam radius a, and to maintain the Hamiltonian form we must subsequently normalize p_x to $k^2 a/2$. This scaling yields the following:

$$K(q, p, z) \approx \frac{1}{4}k^2 p^2 + q^2 - 2m\frac{\kappa}{k^2 a^2}q^2\cos(k_e z)\Theta(1 - q^2) \quad (3a)$$
$$+ \frac{\kappa}{k^2 a^2}\left(\frac{8}{3} - 6|q| + 4q^2 - \frac{2}{3}|q|^3\right)\Theta(q^2 - 1) ;$$
$$= K_0(q, p) + K_m(q, z) + K_1(q) , \quad (3b)$$

which is the starting point of our analysis. In the unperturbed limit, K_m and K_1 vanish, and the motion is linear: q oscillates sinusoidally between $\pm q_{max} = \pm h_0^{1/2}$, where h_0 is the numerical value of K_0. For a KV beam, $h_0 = q_{max} = 1$.

III. CLASSICAL AND SECULAR PERTURBATION THEORY

In this section, we apply first-order classical and secular perturbation theory to the Hamiltonian of Eq. (3). In doing so, we treat the z-dependent part K_m and the nonlinear part K_1 as small perturbations to the zeroth-order part K_0.

First, we transform to action-angle variables [11] of the zeroth-order Hamiltonian $K_0(q,p)$. We define the action $J(h_0) \equiv (1/2\pi)\oint dq P(h_0, q)$, and the function P as the inverse of the equation $K_0[q, P(h_0, q)] \equiv h_0$. Integration yields $h_0 = kJ$. Neglecting K_m and K_1, we find to lowest order $q = (kJ)^{1/2}\cos\theta$ and $p = -2(J/k)^{1/2}\sin\theta$, where we have chosen the convention that $\theta = 0, \pi$ corresponds to $q = (kJ)^{1/2} \equiv q_{max}$ and $p = 0$. Transforming the Hamiltonian to these action-angle variables yields:

$$K(\theta,J,z) \approx kJ - 2m\frac{\mu}{1-\mu}kJ\cos^2\theta \cos(k_e z) \Theta(1-kJ\cos^2\theta)$$

(4a)

$$+\frac{\mu}{1-\mu}\left(\frac{8}{3}-6\sqrt{kJ}|\cos\theta|+4kJ\cos^2\theta-\frac{2}{3}(kJ)^{3/2}|\cos^3\theta|\right)\Theta(kJ\cos^2(\theta)-1);$$

$$= K_0(J) + K_m(\theta,J,z) + K_1(\theta,J), \quad (4b)$$

where we have used the same notation for the functions here as in Eq. (3), because K_0, K_m and K_1 have the same value, although they have new functional forms.

We now apply classical perturbation theory to $K_1(\theta,J)$ in Eq. (4). This term must be Fourier expanded in θ, being careful to account for the effect of the Heavyside function. The first Fourier component is

$$K_{10}(J) = \frac{8}{3\pi}\frac{\mu}{1-\mu}\left[(4+3kJ)\arccos\left(\frac{1}{\sqrt{kJ}}\right)-\frac{1}{3}(19+2kJ)\sqrt{kJ-1}\right]\Theta(kJ-1). \quad (5)$$

Note that the Heavyside function in Eq. (6) is now independent of θ.

The θ-dependent components in the Fourier expansion of $K_1(\theta,J)$ are removed with a canonical transformation from $(\theta,J) \to (\bar{\theta},\bar{J})$, using a mixed-variable generating function of the second kind [11]:

$$F_2(\theta,\bar{J}) = \bar{J}\theta - \sum_{n=2,4,6...}\frac{1}{nk}K_{1n}(\bar{J})\sin(n\theta) ; \quad (6a)$$

$$J = \frac{\partial}{\partial\theta}F_2(\theta,\bar{J}) = \bar{J} - \sum_{n=2,4,6...}\frac{1}{k}K_{1n}(\bar{J})\cos(n\theta) ; \quad (6b)$$

$$\bar{\theta} = \frac{\partial}{\partial \bar{J}}F_2(\theta,\bar{J}) = \theta - \sum_{n=2,4,6...}\frac{1}{nk}\frac{\partial}{\partial \bar{J}}K_{1n}(\bar{J})\sin(n\theta) . \quad (6c)$$

The far right terms of Eq.'s (6) are assumed to be small or first-order quantities, as are K_1 and K_m. When writing the transformed Hamiltonian below, we neglect all second-order and higher terms. The transformed Hamiltonian is:

$$K(\bar{\theta},\bar{J},z) \approx -2m\frac{\mu}{1-\mu}k\bar{J}\cos^2(\bar{\theta})\cos(k_e z) \Theta(1-k\bar{J}\cos^2(\bar{\theta}))$$

(7a)

$$+k\bar{J}+\frac{4}{3\pi}\frac{\mu}{1-\mu}\left[(4+3k\bar{J})\arccos\left(\frac{1}{\sqrt{k\bar{J}}}\right)-\frac{1}{3}(19+2k\bar{J})\sqrt{k\bar{J}-1}\right]\Theta(k\bar{J}-1)$$

$$= K_m(\bar{\theta},\bar{J},z) + K_0(\bar{J}) . \quad (7b)$$

We have in effect averaged away the sinusoidal oscillations of K_1, leaving only the constant term of the Fourier expansion, which governs the averaged dynamics.

Next we treat **K_m** perturbatively. As above, we first Fourier expand it:

$$\mathbf{K}_m(\theta, \mathbf{J}, z) = -\frac{1}{2} m \frac{\mu}{1-\mu} \, k\mathbf{J} \left\{ f_0(\mathbf{J}) \cos(k_e z) \right.$$
$$\left. + \sum_{n=1}^{\infty} f_{2n}(\mathbf{J}) \left[\cos(2n\theta + k_e z) + \cos(2n\theta - k_e z) \right] \right\} \quad (8)$$

where the first few nonzero Fourier components are given by:

$$f_0(\mathbf{J}) = 2 - \frac{4}{\pi} \left[arcos\left(\frac{1}{\sqrt{k\mathbf{J}}}\right) - \frac{1}{k\mathbf{J}} \sqrt{k\mathbf{J}-1} \right] \Theta(k\mathbf{J}-1) \; ; \quad (9a)$$

$$f_2(\mathbf{J}) = 1 - \frac{2}{\pi} \frac{1}{k\mathbf{J}} \left(1 + \frac{2}{k\mathbf{J}}\right) \sqrt{k\mathbf{J}-1} \; \Theta(k\mathbf{J}-1) \; ; \quad (9b)$$

$$f_4(\mathbf{J}) = -\frac{8}{3\pi} \left[3 \; arcos\left(\frac{1}{\sqrt{k\mathbf{J}}}\right) + \frac{1}{(k\mathbf{J})^2}\left(\frac{4}{k\mathbf{J}}-1\right)\sqrt{k\mathbf{J}-1} \right] \Theta(k\mathbf{J}-1) \; . \quad (9c)$$

A straightforward attempt to treat $\mathbf{K}_m$ with classical perturbation theory results in a problem with small denominators. Chapter 2 of Ref. [11] discusses this issue in detail. The problematic term is $f_2(\mathbf{J})cos(2\theta-k_e z)$, which has a coefficient containing the denominator $2\omega_0(\mathbf{J})-k_e$, where

$$\omega_0(\mathbf{J}) = \frac{\partial}{\partial \mathbf{J}} \mathbf{K}_0(\mathbf{J}) \; . \quad (10)$$

This denominator vanishes for values of $\mathbf{J}$ that are relevant to the halo dynamics. At phase space locations where this occurs, our perturbation expansion would be invalid and our analysis would break down. The physical reason for this mathematical difficulty is that there is a strong 2:1 resonance between the oscillating KV core, which has wavenumber k_e, and the test-particle oscillating with wavenumber $\omega_0(\mathbf{J})$. This is the same parametric resonance identified by Gluckstern in Ref. [5].

To avoid the problem of small denominators and to capture the dynamics of the parametric resonance, we must first apply secular perturbation theory, which means simply that we must transform into a frame rotating with this resonance before applying classical perturbation theory. This is accomplished with the following generating function of the second kind:

$$F_2(\mathcal{J}, \theta, z) = \mathcal{J}(2\theta - k_e z) \; ; \quad (11a)$$

$$\phi = \frac{\partial}{\partial \mathcal{J}} F_2(\mathcal{J}, \theta, z) = 2\theta - k_e z \; ; \quad (11b)$$

$$\mathbf{J} = \frac{\partial}{\partial \theta} F_2(\mathcal{J}, \theta, z) = 2\mathcal{J} \; . \quad (11c)$$

The resulting Hamiltonian is

$$\mathcal{H}(\mathcal{J}, \phi, z) = \mathbf{K}[\theta(\phi, z), \mathbf{J}(\mathcal{J}), z] + \frac{\partial}{\partial z} F_2[\mathcal{J}, \theta(\phi, z), z]$$

$$\approx -(k_e - 2k)\mathcal{J} + \frac{4}{3\pi}\frac{\mu}{1-\mu}\left[(4+6k\mathcal{J})\arccos\left(\frac{1}{\sqrt{2k\mathcal{J}}}\right) - \frac{(19+4k\mathcal{J})}{3}\sqrt{2k\mathcal{J}-1}\right]\Theta(2k\mathcal{J}-1)$$

$$-\frac{m\mu}{1-\mu}k\mathcal{J}\left\{f_0(2\mathcal{J})\cos(k_e z) + \sum_{n=1}^{\infty}f_{2n}(2\mathcal{J})[\cos(n\phi+(n+1)k_e z) + \cos(n\phi+(n-1)k_e z)]\right\}$$
(12a)

$$= \mathcal{H}_0(\mathcal{J}) + \mathcal{H}_m(\phi,\mathcal{J},z) .$$
(12b)

The resonant term is the far right term inside the summation with n=1; it is the only component of $\mathcal{H}_m$ independent of z. In applying classical perturbation theory, we will average over all of the fast oscillating terms that go like $cos(nk_e z)$, while the remaining $cos(\phi)$ term governs the dynamics of the slow averaged motion.

We now make use of the generating function $F_2(\mathcal{J},\phi,z) = \mathcal{J}\phi + F_m(\mathcal{J},\phi,z)$, where $\Omega_0(\mathcal{J})\partial F_m/\partial\phi + \partial F_m/\partial z = \mathcal{H}_m(\phi,\mathcal{J},z) + m(\mu/(1-\mu))k\mathcal{J} f_2(2\mathcal{J})cos(\phi)$ and we define Ω_0 to be:

$$\Omega_0(\mathcal{J}) = \frac{\partial}{\partial \mathcal{J}}\mathcal{H}_0(\mathcal{J}) .$$
(13)

To first order, this yields the Hamiltonian

$$\mathcal{H}(\phi,\mathcal{J}) = -(k_e - 2k)\mathcal{J}$$
$$+ \frac{4}{3\pi}\frac{\mu}{1-\mu}\left[(4+6k\mathcal{J})\arccos\left(\frac{1}{\sqrt{2k\mathcal{J}}}\right) - \frac{1}{3}(19+4k\mathcal{J})\sqrt{2k\mathcal{J}-1}\right]\Theta(2k\mathcal{J}-1) \quad (14)$$
$$- m\frac{\mu}{1-\mu}k\mathcal{J}\cos(\phi)\left[1 - \frac{1}{\pi k\mathcal{J}}\left(1+\frac{1}{k\mathcal{J}}\right)\sqrt{2k\mathcal{J}-1}\ \Theta(2k\mathcal{J}-1)\right] ,$$

which is independent of z and, hence, is a constant of the motion.

Finally, we make a scaling transformation to simplify the notation: let $w=2k\mathcal{J}$, $H=(2k/k_0)\mathcal{H}$, $\psi=\phi$, and nondimensionalize the longitudinal position to the zero-current phase advance by letting $\tau=k_0 z$. The final Hamiltonian is:

$$H(\psi,w) = -2(\sqrt{1-\mu/2} - \sqrt{1-\mu})w$$
$$+ \frac{8}{3\pi}\frac{\mu}{\sqrt{1-\mu}}\left[(4+3w)\arccos\left(\frac{1}{\sqrt{w}}\right) - \frac{1}{3}(19+2w)\sqrt{w-1}\right]\Theta(w-1) \quad (15)$$
$$- m\frac{\mu}{\sqrt{1-\mu}}w\cos(\psi)\left[1 - \frac{\sqrt{w-1}}{2\pi w}\left(1+\frac{1}{w}\right)\Theta(w-1)\right] ,$$

with the following equations of motion:

$$\frac{dw}{d\tau} = -\frac{\partial H(\psi,w)}{\partial \psi} = -m\frac{\mu}{\sqrt{1-\mu}}w\sin(\psi)\left[1-\frac{\sqrt{w-1}}{2\pi w}\left(1+\frac{1}{w}\right)\Theta(w-1)\right] ; \quad (16a)$$

$$\frac{d\psi}{d\tau} = \frac{\partial H(\psi,w)}{\partial w} = -2\left(\sqrt{1-\mu/2} - \sqrt{1-\mu}\right)$$

$$- m\frac{\mu}{\sqrt{1-\mu}}\cos(\psi)\left[1 - \frac{w - \frac{1}{2} - \frac{3}{2w} + \frac{2}{w^2}}{2\pi\sqrt{w-1}}\Theta(w-1)\right] \quad (16b)$$

$$+ \frac{8}{3\pi}\frac{\mu}{\sqrt{1-\mu}}\left[3\arccos\left(\frac{1}{\sqrt{w}}\right) - \frac{w+1-\frac{2}{w}}{\sqrt{w-1}}\right]\Theta(w-1).$$

Eq. (15) makes it clear that there are only two dimensionless parameters in this problem: the mismatch factor *m* and the space charge parameter μ.

Motion in the w-ψ phase plane describes the slow, averaged evolution of the amplitude w and phase ψ of the oscillations of a test-particle. The Hamiltonian H does not depend on the independent variable τ, so H is a constant of the motion. This means phase space trajectories simply follow contours of H in the phase plane.

IV. ANALYSIS OF THE TRANSFORMED HAMILTONIAN

The phase space topology of the Hamiltonian in Eq. (15) is determined by the locations and types of any fixed points in the plane. There is a stable fixed point at w=0, which corresponds to a test-particle moving straight down the beam axis with no transverse motion. In this limit, ψ is not well defined. To find the other fixed points, we must find all values of (w,ψ) such that dw/dτ=dψ/dτ=0. This leads us to two more fixed points, a stable one at (w=w_s,ψ=0) and an unstable saddle point, or x-point, at (w=w_x,ψ=π). We find thus that the phase space contains a separatrix with one lobe (the inner separatrix) inside of another lobe (the outer separatrix), just like the topology shown in Fig.'s 2 and 4 of Ref. [5].

Unfortunately, w_s and w_x must be obtained numerically by finding the roots of Eq. (16b) for ψ=0,π respectively. This has been done for many values of μ and *m*. In each case, we found the value of the Hamiltonian at the x-point, h_x=H(π,w_x). By plotting this contour in a polar projection of the w-ψ phase plane, we can see the inner separatrix and the beginning of the outer separatrix. Such contour plots are shown in Fig. 1 below for μ=0.2 and *m*=0.1 and 0.2. We don't extend these plots out to values of w large enough to see the full outer separatrix, because then K_1 of Eq. (3) is no longer small compared to K_0, and our perturbative treatment of the nonlinearities breaks down. However, our perturbative analysis is valid out to the inner separatrix.

The thicker curves in Fig. 1 show the initial phase space positions of zero-angular-momentum test-particles from a matched KV distribution in this transformed phase space. For a KV distribution, the unperturbed Hamiltonian of Eq. (3) has the value h_0=1. This corresponds to the transformed Hamiltonian of Eq. (15) having the value h_{kv}=-2[(1-μ/2)$^{1/2}$-(1-μ)$^{1/2}$). Neglecting nonlinearities, such particles would have initial phase space positions lying on the curve:

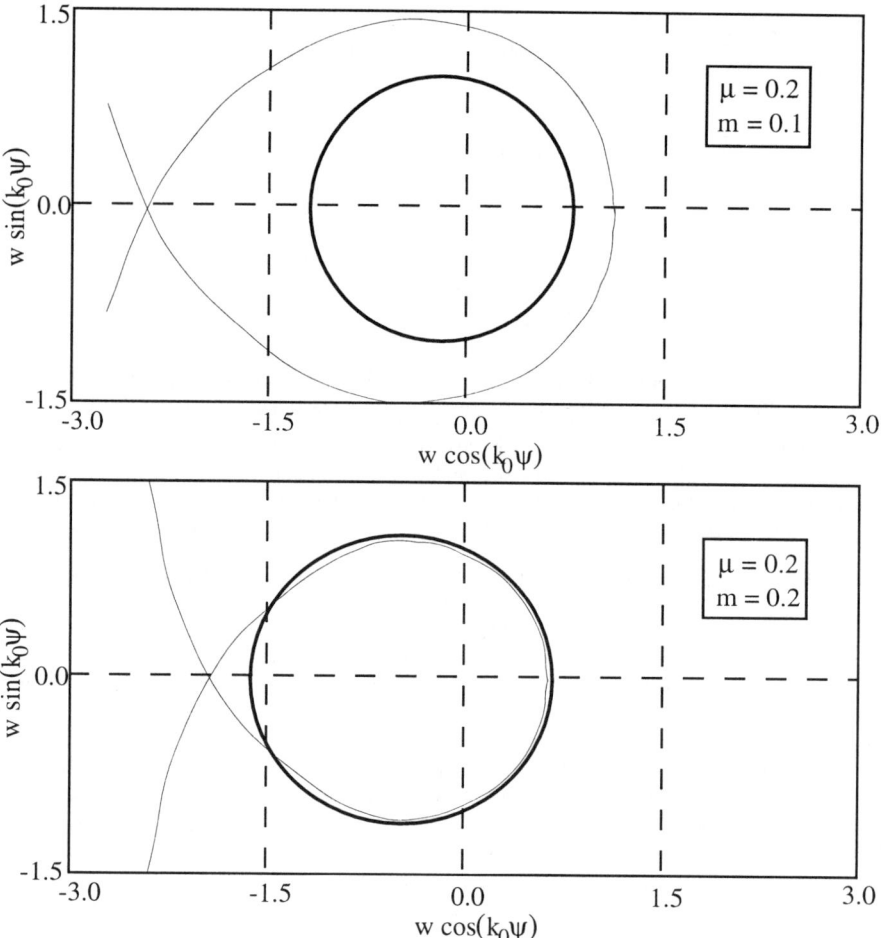

Fig. 1. Phase space separatrix arising from mismatched KV core, shown in polar coordinates with initial conditions of a matched KV beam: μ=0.2; m=0.1, 0.2. For m=0.2, some initial conditions lie outside of the inner separatrix.

$$w(\psi) = \left[1 + \frac{m}{\Delta} \cos(\psi)\right]^{-1} ; \quad \Delta \equiv \frac{2}{\mu} \sqrt{1-\mu} \left(\sqrt{1-\mu/2} - \sqrt{1-\mu}\right), \quad (17)$$

where Δ is the same as that defined in Eq. (2.10) of Ref. [5].

Figure 1 shows how, for μ=0.2, the ring of initial conditions for a KV distribution overlaps the inner separatrix as *m* is increased. Thus, for sufficiently large values of *m*, we find that the amplitude oscillation for test-particles in an initially-matched KV distribution, which have the appropriate slow phase ψ, will slowly grow as their location in the w-ψ phase plane sweeps along the outside of the inner separatrix, past the x-point and along the inside of the outer separatrix into the halo. This process may take a long time, because motion near the x-point is very slow.

Our analysis provides a simple numerical method for obtaining $m_c(\mu)$, the value of m above which certain zero-angular-momentum core particles will be swept out into the halo. The curve in Fig. 2 indicates where the ring of initial conditions defined in Eq. (17) first overlaps the inner separatrix. This separatrix structure persists in the $\mu \to 0$ limit, because in carrying out secular perturbation theory above we subtracted a factor from the lowest-order Hamiltonian such that all three terms in our final Hamiltonian of Eq. (15) vanish linearly with μ as $\mu \to 0$. Thus, the perturbative terms do not become negligibly small compared to the lowest-order term, contrary to what one might have expected.

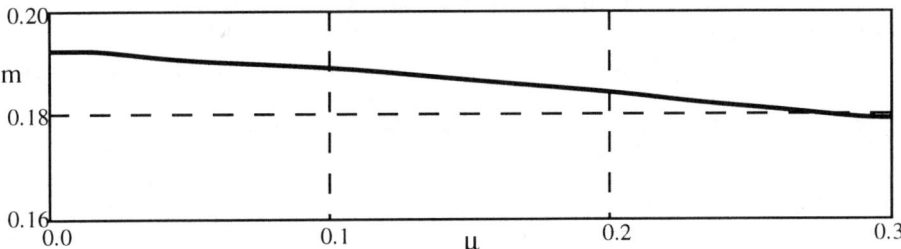

Fig. 2. Halo stability boundary in the μ-m plane for zero-angular-momentum test-particles from a matched KV distribution.

Although particles can be swept out into the halo for arbitrarily small values of μ, the time scale for this to occur grows arbitrarily long. The Hamiltonian equations are readily solved in the linear, unperturbed limit, for which only the first term on the right hand side of Eq. (15) is kept. Doing so shows the motion is oscillatory with a period scaling as μ^{-1} [4]. Although the motion of particles swept out along the separatrix is more complicated than that of core particles, the μ^{-1} scaling of the time scale still holds. We numerically confirm this assertion below.

V. NUMERICAL CONFIRMATION OF THE ANALYSIS

Here we present simulations with a symplectic test-particle code that verify our analysis, particularly in the limit of small μ. Below, we also present the results of self-consistent particle simulations with the Topkark code, again confirming our analysis and further showing that the halo dynamics we have described is not limited to particles with zero angular momentum.

There is some danger in making a simple approximation for the dynamics of the core oscillations, so we treat the equations for the full PCM, and because 10^4 core oscillations are required for a test-particle to enter the halo when $\mu \sim 0.001$, we use a 4th-order symplectic integration scheme to obtain good accuracy and eliminate spurious damping or driving terms. The Hamiltonian describing the PCM is

$$H(Q,P,q,p) = \frac{1}{2}\left[P^2 + Q^2 - \mu \ln(Q^2) + \frac{(\mu-1)}{Q^2}\right] + \frac{1}{2}\left[p^2 + q^2 - \mu F(q;Q)\right], \quad (18a)$$

where

$$F(q;Q) = q^2/Q^2 \text{ for } q^2 < Q^2; \quad F(q;Q) = 1 + \ln(q^2/Q^2) \text{ for } q^2 > Q^2. \quad (18b)$$

The independent variable is $\tau=k_0 z$; Q is the core radius, normalized to *a;* P=dQ/dτ; q=x/a is the normalized position of the test-particle; p=dq/dτ. The core dynamics (Q,P) feeds into the test-particle dynamics (q,p) via the function F(q;*Q*); however, the functional dependence of F on Q does not affect the core dynamics. We have attempted to indicate this distinction by italicizing Q when it is used as an argument for the function F. We have again assumed the limit of zero angular momentum.

This Hamiltonian can be split into four simple Hamiltonians, for which the motion can be readily solved analytically. Thus, it is ideal for use with the 2nd and 4th-order symplectic integration schemes developed by Forest, Ruth and Yoshida [12]. Also, several test-particle trajectories can be integrated simultaneously, because the test-particle motion does not couple into the core dynamics.

We implemented a 4th-order integrator and conducted simulations for many values of *m* and μ (shown in Fig. 3), typically using two test-particles, each consistent with zero angular momentum and a matched KV distribution: one with initial conditions q=1, p=0 (corresponding to $\psi=\pi$) and one with q=0, p=$(1-\mu)^{1/2}$ (corresponding to $\psi=0$). The core is started with its maximum radius, Q=1+m, and P=0. The first test-particle ($\psi=\pi$, dashed line in Fig. 3) remains inside the core, feels only linear forces, and undergoes stable amplitude oscillations governed by the Mathieu equation. The second test-particle ($\psi=0$, dot-dashed line in Fig. 3) first reaches its maximum amplitude (q=1) when the core first reaches its minimum amplitude (Q=1-m): at this point, it is outside the core and thus subject to nonlinear space charge forces.

Figure 3a shows a distinct change in the dynamics of the second test-particle (for $\mu=0.001$) when *m* is increased from 0.25 to 0.26, because this small change in *m* moves the initial phase space location from just inside the inner separatrix to just outside the inner separatrix. Thus, we find $m_c \approx 0.25$; whereas, we see from Fig. 2 that the analysis predicts $m_c \approx 0.19$ for $\mu=0$. The relative error is ~25%, which is reasonable for a first-order perturbation theory with a small parameter *m*~0.25. However, Fig.'s 3b and 3c show m_c increasing slightly with μ, while Fig. 2 shows m_c decreasing slightly. This discrepancy arises because the space charge nonlinearities increase and the core oscillations become more complicated as μ grows, so that our perturbation theory begins to break down.

These numerical simulations do confirm very clearly, both qualitatively and quantitatively, the dynamical picture described above. The particle with initial conditions outside the inner separatrix starts with an initial oscillation amplitude of unity, then this amplitude slowly grows to become as large as 2.5, which is roughly twice the maximum radius of the oscillating core. Furthermore, Fig.'s 3 show that the phase space topology is essentially unchanged as $\mu \to 0$, but that the time scale for the test-particle to reach its furthest extent in the halo scales as μ^{-1} (200 core oscillations for $\mu=0.1$, but 2×10^4 core oscillations for $\mu=0.001$). The slight deviations from this scaling result from differences in how closely each test-particle approaches the x-point.

As a test, the PCM equations were integrated with a 4th-order Runge-Kutta integrator. It was found that the maximum core oscillation amplitude (solid line in Fig. 3) would droop significantly after 10^4 oscillations, due to spurious damping.

In addition to test-particle runs, we conducted self-consistent particle simulations with the Topkark code [7] for $\mu=0.1$ and various values of *m*. The parameters correspond to a continuous 20 MeV, 102 mA deuterium beam, with a full un-

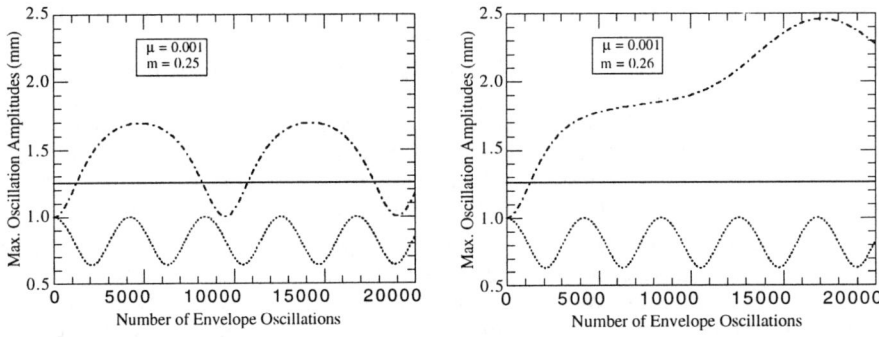

Fig. 3a. Results of symplectic PCM simulation for $\mu=0.001$ and m=0.25, 0.26. Solid line shows maximum amplitude of core oscillation; dashed line shows oscillation amplitude of a test-particle remaining inside the core; dot-dashed line shows amplitude of a test-particle that leaves the core. For m=0.25 the second particle is just inside the inner separatrix; for m=0.26 this particle is just outside the inner separatrix and moves out to the halo.

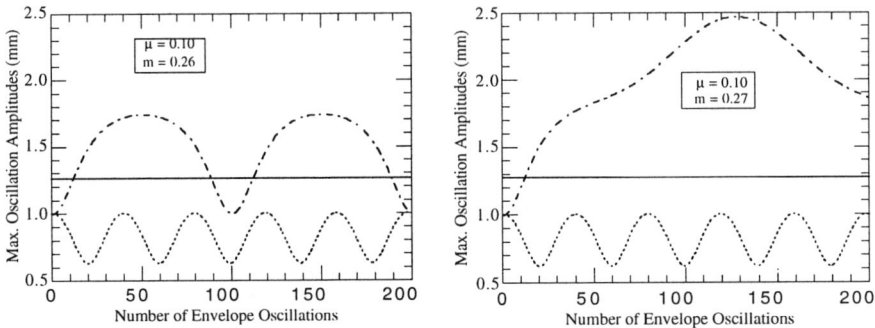

Fig. 3b. Same as Fig. 3a, but for $\mu=0.1$ and m=0.26, 0.27.

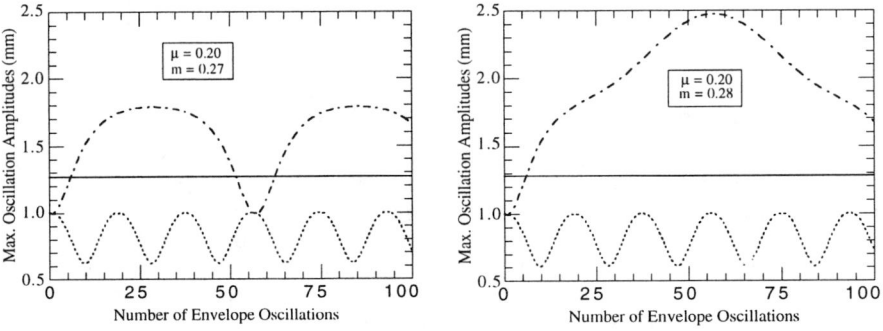

Fig. 3c. Same as Fig. 3a, but for $\mu=0.2$ and m=0.27, 0.28.

normalized emittance of $\varepsilon=3.05$ mm-mrad, in an infinitely long 5.90 T solenoid. The matched core radius is $a=1.0$ mm; the core oscillation wavelength is 1.0 m ($k_e=2\pi$ m^{-1}). The initial beam core is represented by 10^4 particles from a KV distribution (using algorithm of Ref. [13]), with a radius of $(1+m)$ mm and a Twiss α of zero. In addition, 100 particles were drawn from a matched KV distribution (i.e. initial radius of 1 mm). These 100 particles, which interacted fully in the self-consistent simulations, played the role of the test-particles in our analysis, except that they each had a finite angular momentum consistent with the KV distribution.

Because the simulation was axisymmetric, the space charge forces were modeled with Gauss' law, accounting appropriately for relativistic effects. The enclosed charge and current were calculated for each particle by simply counting the number of particles with smaller radius. This model would be noisy near the axis, where very few smaller-radius particles would be counted, so an explicitly linear model was used for particles with radii smaller than 1/16 of the RMS core radius.

Figure 4 shows the oscillations of the beam core radius (heavy line), which is defined to be $2^{1/2}$ times the RMS radius of the full particle distribution. The thin line shows the maximum radius of all the particles. When the core radius reaches its first minimum, one or more of the 100 out-of-phase particles (those with ψ initially close to zero) reach a maximum amplitude of 1 mm. As the core oscillates, it slowly pumps energy into some of these particles, and the maximum particle radius begins to grow well beyond the maximum core radius. Also, Fig. 4 shows these growing particle oscillations start to come into phase with the core oscillations, because the particles are sweeping around the inner separatrix towards the vicinity of the x-point, where $\psi \approx \pi$. Figure 4 shows the fast-time dynamics, which was averaged over in our analysis, and which we neglected to show in Fig. 3.

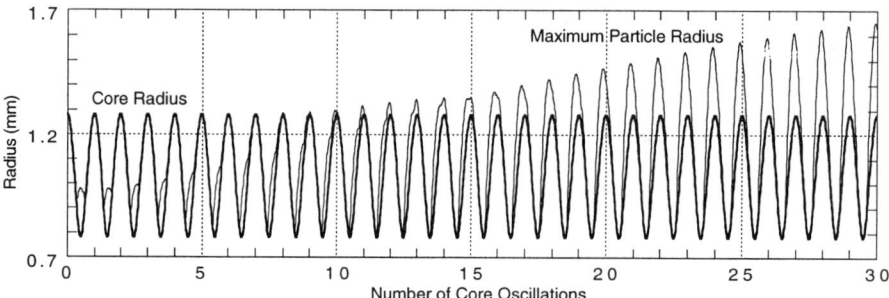

Fig. 4. Topkark simulation showing initial transfer of energy from mismatched beam core to single particle: $\mu=0.1$ and $m=0.28$.

In Fig. 5a, we show puncture plots of the initial particle positions and momenta for the simulation of Fig. 4. The 100 out-of-phase particles are not seen in the position plot, because they are from a matched distribution and have a smaller maximum radius than the core particles. However, because they have the same transverse energy as the core particles and a smaller maximum radius, they must have a larger maximum momentum. Thus, some of them appear as a diffuse cloud outside of the core distribution in the momentum plot.

Figure 5b shows puncture plots of the final positions and momenta, after 60 core oscillations, for a simulation with $m=0.27$. The particles are limited to radii of

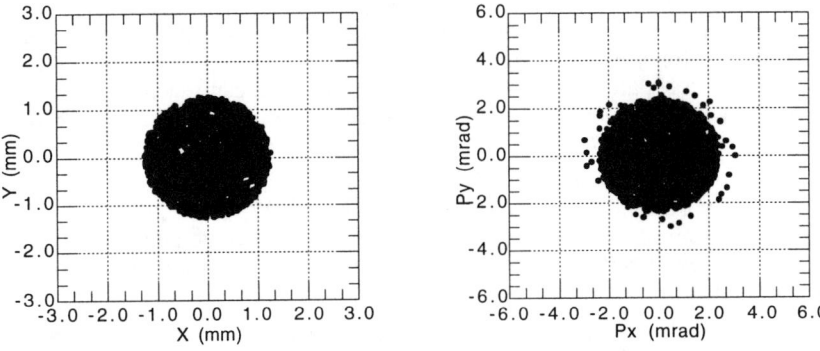

Fig. 5a. Initial conditions for simulation with $\mu=0.1$ and $m=0.28$.

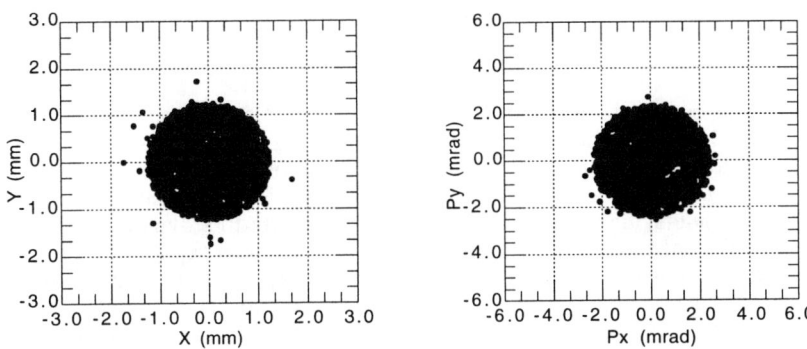

Fig. 5b. Final conditions, after 60 core oscillations, for simulation with $\mu=0.1$ and $m=0.27<m_c$. All particles are confined inside the inner separatrix.

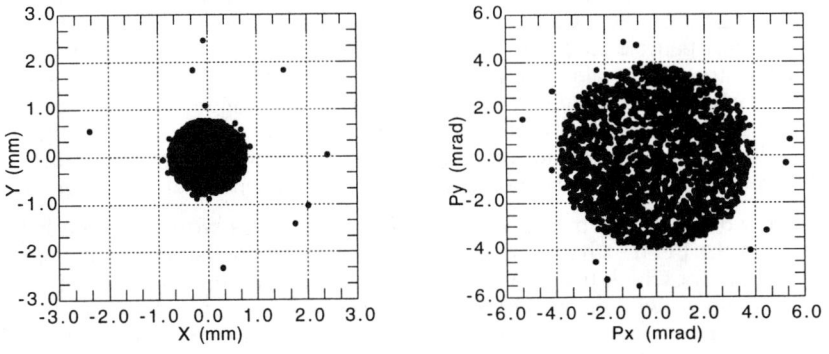

Fig. 5c. Final conditions, after 120.5 core oscillations, for simulation with $\mu=0.1$ and $m=0.28>m_c$. Some particles are outside of the inner separatrix.

≈1.8 mm or less, indicating they are all confined within the inner separatrix. This simulation corresponds roughly to the test-particle run presented on the left in Fig. 3b, which was for $m=0.26$. Figure 5c shows that with a slight increase in m to 0.28, some particles are outside the inner separatrix and can reach radii as large as ≈2.5 mm. This run is analogous to the test-particle run presented on the right in Fig. 3b, which was for $m=0.27$.

Thus, the self-consistent simulations confirm the results from the test-particle runs, with only a slight change in the value of m_c (≈0.27 vs. ≈0.26). More importantly, because all of the 100 out-of-phase particles had finite angular momentum, these results show that the dynamical picture described by our analysis is not strictly limited to particles with zero angular momentum.

Figure 5b shows the particles when the core is at its maximum radius, because particles inside the inner separatrix reach their maximum oscillation amplitude when they are near the x-point, for which $\psi=\pi$. Figure 5c shows the particles when the core is at its minimum radius, because halo particles moving along the outer separatrix reach their maximum oscillation amplitude when $\psi=0$.

VI. CONCLUSION

A perturbative Hamiltonian analysis of the particle-core model (PCM) in the limit of small mismatch and moderate space charge shows that zero-angular-momentum test-particles from a matched KV distribution are driven into the halo by oscillations of the mismatched core, if the mismatch factor m exceeds a threshold $m_c(\mu)$. The underlying phase space structure governing this dynamics is a separatrix arising from the 2:1 parametric resonance, with an inner separatrix nested inside of the outer separatrix. As m is increased, the inner separatrix squeezes down such that some of the test-particles have initial phase space coordinates lying just outside of it, thus allowing them to move slowly out along the outer separatrix into the halo. This phase space structure, and the resulting dynamics, persists even as the space charge parameter μ becomes arbitrarily small, although the time scale becomes arbitrarily long in this limit.

These analytical results have been confirmed numerically, both with a symplectic treatment of the PCM and with self-consistent particle simulations. The symplectic test-particle code verified that the dynamics persists in the limit of small μ, and that the time scale increases as μ^{-1}. The analysis predicts $m_c \approx 0.2$ for $0<\mu<0.3$, while the simulations show $m_c \approx 0.26$ in this range, which is reasonable quantitative agreement for a first-order treatment. The self-consistent simulations showed that the analysis is not limited to test-particles and, more importantly, that the dynamics of finite-angular-momentum particles is qualitatively the same.

Our work is complementary to much of the previous work on beam halo dynamics and the PCM, which has concentrated largely on the space charge dominated regime, for which μ approaches unity. The surprising result of our work is that beam halo growth remains a relevant issue even in the limit of low space charge and small beam mismatch. Although core particles do not themselves enter the halo in this regime, we show that test-particles with the same transverse energy as core particles, and with smaller initial maximum oscillation amplitudes then some core particles, can be swept out into the halo if they are oscillating out of phase with the core. This occurs because such particles exit the core briefly when it is at its minimum radius, and thus experience the nonlinear space charge forces.

The out-of-phase particles we have considered could arise in practice, for example, from transverse aberrations in the extraction optics of an ion source. As we have shown, particles from a pre-halo of this sort can be swept out into a true halo under conditions where core particle motion is completely stable. Thus it is reasonable to conclude that beam dynamics in the low-energy injector of high-power linacs where halo growth is a concern should be optimized to limit the development of a pre-halo.

ACKNOWLEDGMENTS

This research effort was formulated with guidance from R. Jameson and J. Cary, and I thank them both for numerous helpful discussions. M. Reusch helped develop the symplectic test-particle code and provided several insightful comments and criticisms regarding the analysis. Discussions with C. Chen, A. Riabko and R. Ryne at the ICFA workshop in Bloomington, Indiana were also helpful.

REFERENCES

[1] J. S. O'Connell, T. P. Wangler, R. S. Mills and K. R. Crandall, *Proc. 1993 Part. Accel. Conf.* (Washington DC, 1993), p. 3657.

[2] I. M. Kapchinskij and V. V. Vladimirskij, *Proc. International Conf. on High Energy Accel.* (CERN, Geneva, 1959), p. 274.

[3] R. A. Jameson, "Self-Consistent Beam Halo Studies & Halo Diagnostic Development in a Continuous Linear Focusing Channel," Los Alamos National Laboratory #LA-UR-94-3753 (1994).

[4] D. L. Bruhwiler, "IFMIF Beam Halo Study; Final Report," in "IFMIF Industrial Support Subcontract / Technical Status Report" (Sep. 1995).

[5] R. L. Gluckstern, Phys. Rev. Lett. **73** (1994), pp. 1247-1250.

[6] B. I. Bondarev, A. P. Durkin, B. P. Murin, "Halo Production in Charge-Dominated Beams / Single Particle Interactions," Moscow Radiotechnical Institute report to Los Alamos National Laboratory (September 15, 1993).

[7] D. L. Bruhwiler and M. F. Reusch, in Computational Accelerator Physics, AIP Conf. Proc. **297**, R. Ryne ed. (AIP, New York, 1993), p. 524.

[8] M. Reiser, *Theory and Design of Charged Particle Beams* (John Wiley and Sons, New York, 1994).

[9] C. Chen and R. C. Davidson, Phys. Rev. E **49** (1994), p. 5679.

[10] J.-M. Lagniel, Nucl. Instr. and Meth. in Phys. Res. A **345** (1994), p. 46.

[11] A. J. Lichtenberg and M. A. Lieberman, *Regular and Chaotic Dynamics*, 2nd edition (Springer-Verlag, New York, 1992).

[12] E. Forest and R. Ruth, Physica **D43**, 105 (1990); H. Yoshida, Phys. Lett. A **150**, 262 (1990).

[13] Y. Batygin, in Computational Accelerator Physics, AIP Conf. Proc. **297**, R. Ryne ed. (AIP, New York, 1993), p. 419.

The Envelope Dynamics of Intense Charge Particle beams in a FODO Channel

S.Y. Lee

Department of Physics, Indiana University, Bloomington, IN 47405

K.Y. Ng

Fermi National Acclerator Laboratory,[1] Batavia, IL 60510

Families of envelope parametric resonances generated by the periodic perturbation of FODO transport channel for space charge dominated beams are studied. Effects of envelope coupling between the horizontal and vertical envelope oscillations are analyzed.

I. INTRODUCTION

The envelope function of the Kapchinskij-Vladimirskij (KV) distribution, where charged particles are distributed uniformly in the emittance shell, satisfies a self-consistent evolution KV equation [1], which includes the space charge force. Furthermore, the rms values of *any* particle distribution have been shown to satisfy the KV equation [2]. In the past, numerical studies of the KV equation have shown that the envelope function of particle beams with a large space charge perveance can encounter nonlinear resonances due to the self space charge force. Since the tolerance for beam loss of high brightness beam is small, studies of nonlinear dynamics relevant to the KV envelope equation for an intense charged particle beam are important in order to preserve beams with high current and high brightness. For a mis-matched beam, the envelope function presumably obeys the KV equation with a periodic longitudinal modulation. Therefore, solutions of such a KV equation, in the presence of self-electromagnetic fields, can offer important insight toward understanding high brightness beam transport problems [3].

Although there are many numerical simulations in the past, analytic solution to understand the envelope dynamics is important. In Ref. [4,5], we have introduced the KV Hamiltonian to describe the KV envelope equation. There we have set up a procedure to evaluate the parametric resonance strength systematically. The *envelope Hamiltonian* can be decomposed into an energy independent autonomous part and a perturbation. The action-angle variables

---

[1] Operated by the Universities Research Association, under contracts with the U.S. Department of Energy.

for the unperturbed integrable Hamiltonian are introduced. The perturbation is expanded in action-angle variables, where parametric resonances appear naturally [6]. Since many low-energy beam transports use quadrupole focusing elements, it is logical to extend our envelope dynamics for intense charge particle beams to the quadrupole focusing systems. Here the horizontal and the vertical envelope functions are coupled through Coulomb force. Two dimensional nonlinear resonances may play important roles in the envelope dynamics.

This paper is organized as follows. Section II examines the envelope Hamiltonian. The envelope dynamics for the quadrupole focusing channel will be discussed in Sec. III. The conclusion is given in Sec. IV.

II. THE ENVELOPE HAMILTONIAN IN QUADRUPOLE FOCUSING CHANNEL

Since quadrupoles are widely used in most of the beam transport channel, it is useful to study the envelope dynamics in the quadrupole focusing channel. The horizontal and the vertical beam envelope functions R_{xb}, R_{zb} of the KV beam satisfy the envelope equations,

$$
\begin{cases}
R''_{xb} + k_x(s) R_{xb} - \dfrac{\epsilon_x^2}{R_{xb}^3} = \dfrac{2K_b}{R_{xb} + R_{zb}}, \\
R''_{zb} + k_z(s) R_{zb} - \dfrac{\epsilon_z^2}{R_{zb}^3} = \dfrac{2K_b}{R_{xb} + R_{zb}},
\end{cases}
\quad (1)
$$

where $k_x(s) = -k_z(s)$ for the transport channel without dipoles, $\epsilon_{x,z}$ are respectively the horizontal and vertical emittances, and K_b is the space charge perveance of the beam. For simplicity, we consider the case of round beam configuration with $\epsilon_x = \epsilon_z = \epsilon$ and equal focusing strengths for the horizontal and the vertical oscillations. Using the normalized coordinates,

$$
\theta = 2\pi \frac{s}{L}, \; k(s) = k_x L^2, \; K = \frac{K_b L}{\epsilon}, \; R_x = \frac{R_{xb}}{\sqrt{\epsilon L}}, \; R_z = \frac{R_{zb}}{\sqrt{\epsilon L}}, \quad (2)
$$

the Hamiltonian for the envelope equation of motion is given by

$$
H = \frac{1}{4\pi} \left[P_x^2 + P_z^2 + k(\theta)(R_x^2 - R_z^2) + \frac{1}{R_x^2} + \frac{1}{R_z^2} - 4K \ln(R_x + R_z) \right]. \quad (3)
$$

Again, we perform Floquet transformations to the horizontal and the vertical betatron oscillations to obtain the phase advance parameter μ_x and μ_z. For simplicity, we assume equal horizontal and vertical focusing strengths with $\mu = \mu_x = \mu_z$. The average envelope radii are given by

$$
R_{x0} = R_{z0} = R_0 = \left(\frac{\sqrt{1+\kappa^2} + \kappa}{\mu} \right)^{1/2}, \quad (4)
$$

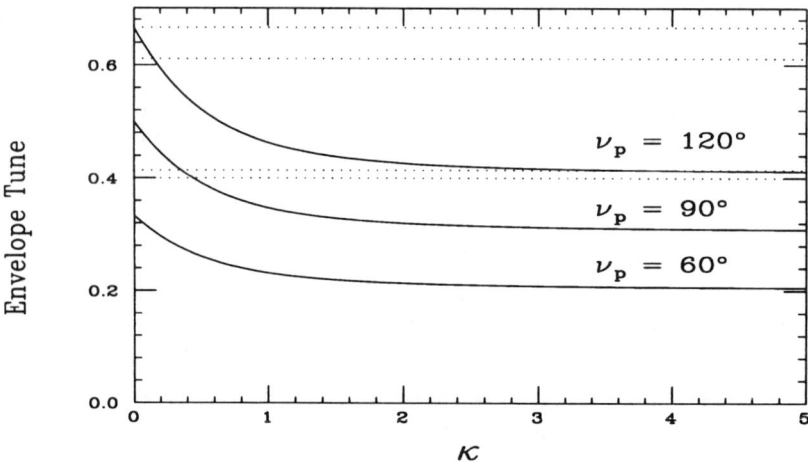

FIG. 1. The envelope tune as a function of the space charge parameter κ for the FODO transport channel with particle phase advance $\mu = 60°, 90°$ and $120°$ respectively.

where κ is the effective space charge parameter given by

$$\kappa = \frac{K}{2\mu}.$$

The envelope tunes near the matched beam radii are given by

$$\nu_x = \nu_z = \nu = \frac{2\mu}{2\pi}\left[1 - \frac{5}{4}\frac{\kappa}{\sqrt{1+\kappa^2}+\kappa}\right]^{1/2}. \quad (5)$$

which is slightly lower than the envelope tune of the paraxial system. Figure 1 shows $\nu_p = \nu_x = \nu_z$ as a function of the space charge parameter κ. The dotted line shown on the figure corresponds to the boundary of Mathieu instability to be discussed later.

Now, we expand the Hamiltonian into an unperturbed integrable Hamiltonian and a perturbation, $H = H_{0x} + H_{0z} + \Delta H$, with

$$H_{0x} = \frac{1}{4\pi}\left[P_x^2 + \mu^2 R_x^2 + \frac{1}{R_x^2} - 4K\ln(R_x + R_0)\right] \quad (6)$$

$$H_{0z} = \frac{1}{4\pi}\left[P_z^2 + \mu^2 R_z^2 + \frac{1}{R_z^2} - 4K\ln(R_0 + R_z)\right] \quad (7)$$

$$\Delta H = \frac{K}{\pi}\ln\frac{(R_x + R_0)(R_0 + R_z)}{R_x + R_z} + \frac{1}{4\pi}(k - \mu^2)R_x^2 + \frac{1}{4\pi}(-k - \mu^2)R_z^2. \quad (8)$$

Here we note that the space charge force has introduce a time independent perturbation to the envelope Hamiltonian. In particular, there is a linear coupling term, which can change the closed orbit solution of the system.

A. Linear envelope coupling due to the space charge force

We first consider the small orbit deviation from the matched envelope radii with

$$R_x = R_0 + X, \quad R_z = R_0 + Z. \tag{9}$$

The linearized time independent Hamiltonian is given by

$$H = \frac{1}{4\pi}(P_x^2 + P_z^2) + \pi\nu^2(X^2 + Z^2) + \frac{K}{4\pi R_0^2}XZ, \tag{10}$$

where ν is the small amplitude envelope tune of Eq. (5). Because of the linear coupling, the envelope motion can be divided into two normal modes similar to the linear betatron coupling. The tunes of these two modes are

$$\nu_\pm^2 = \nu^2 \pm \frac{K}{8\pi^2 R_0^2}. \tag{11}$$

The tunes of two eigen modes are shifted away from the degenerate horizontal and the vertical envelope tune. The + mode corresponds the breathing mode or the σ mode with an envelope tune identical to the paraxial system with angular momentum zero. The tune of − mode or the π-mode, where the horizontal and the vertical envelope radii are oscillating out of phase, is lower. In the case of equal horizontal and vertical envelope unperturbed tunes, the envelope functions are fully coupled.

To illustrate the effect of linear space charge coupling, we consider a uniform focusing approximation to the transport line described Eq. (10). For a beam with an initial mismatched horizontal envelope, the envelope oscillation will be coupled to the vertical envelope mismatch. For a beam with initial envelope mismatch given by

$$X(\theta=0) = X_0, \dot{X}(\theta=0) = 0, \ Z(\theta=0) = 0, \dot{Z}(\theta=0) = 0,$$

the envelope function becomes

$$X = \frac{1}{2}X_0(\cos\nu_+\theta + \cos\nu_-\theta); \quad Z = \frac{1}{2}X_0(\cos\nu_+\theta - \cos\nu_-\theta). \tag{12}$$

Thus an initial horizontal envelope mismatch can produce vertical envelope mismatch and vice versa. The linear envelope coupling resembles the linear betatron coupling.

To minimize the linear envelope coupling, we can use unequal focusing strength functions for horizontal and vertical oscillations. The linearized Hamiltonian is given by

$$H = \frac{1}{4\pi}(P_x^2 + P_z^2) + \pi(\nu_x^2 X^2 + \nu_z^2 Z^2) + 2\pi C X Z, \tag{13}$$

where $C = \frac{K}{2\pi^2(R_{x0}+R_{z0})^2}$. The envelope tunes of the normal modes become

$$\nu_\pm = \frac{1}{2}(\nu_x^2 + \nu_z^2) \pm \frac{1}{2}\sqrt{(\nu_x^2 - \nu_z^2)^2 + 4C^2}. \tag{14}$$

For an initially horizontally envelope mismatched beam, the envelope oscillations are given by

$$X = \frac{X_0}{\sqrt{(\nu_x^2 - \nu_z^2)^2 + 4C^2}} \left[(\nu_+^2 - \nu_z^2)\cos\nu_+\theta - (\nu_-^2 - \nu_z^2)\cos\nu_-\theta\right] \tag{15}$$

$$Z = \frac{CX_0}{\sqrt{(\nu_x^2 - \nu_z^2)^2 + 4C^2}} \left[\cos\nu_+\theta - \cos\nu_-\theta\right]. \tag{16}$$

In particular, if $|\nu_x^2 - \nu_z^2| \gg |C|$, then the horizontal mismatched envelope oscillations will not be strongly coupled to the vertical envelope oscillations. However, this may not have advantage in the transport of intense charge particle beams to be discussed later.

B. The envelope closed orbit

For the quadrupole focusing channel, the closed orbit can be obtained by expanding the Hamiltonian up to the linear order, i.e.

$$H_{\text{linear}} = \frac{1}{4\pi}(P_x^2 + P_z^2) + \pi\nu^2(X^2 + Z^2) + 2\pi CXZ$$
$$+ \frac{k - \mu^2}{4\pi}(R_0 + X)^2 + \frac{-k - \mu^2}{4\pi}(R_0 + Z)^2. \tag{17}$$

The envelope closed orbits, $R_x = R_0 + X, R_z = R_0 + Z$, are given by the solutions of the coupled linear equations

$$\begin{cases} \ddot{R}_x + \left[\nu^2 + \dfrac{k(\theta) - \mu^2}{4\pi^2}\right]R_x + CR_z = (\nu^2 + C)R_0, \\ \ddot{R}_z + \left[\nu^2 + \dfrac{-k(\theta) - \mu^2}{4\pi^2}\right]R_z + CR_x = (\nu^2 + C)R_0. \end{cases} \tag{18}$$

The closed orbit solution of the above equation is coupled through space charge force. The σ and π modes of the linear coupling discussed in last section will be changed by the alternative gradient focusing function. Here $(X \pm Z)$ are not eigen-function anymore. The actual linear coupling is changed by the dependence of the phase advance on the coordinate θ.

For a piecewise constant focusing function, the closed orbit can be solved by matrix multiplication method. In particular, the solutions of the inhomogeneous equations can be expressed in terms of $R_x = R_{x0} + X_{co}$ and $R_z = R_{z0} + Z_{co}$, where R_{x0} and R_{z0} are time independent solution, and X_{co} and Z_{co} satisfy the inhomogeneous equations

$$\begin{cases} \ddot{X}_{co} + \left[\nu^2 + \dfrac{k(\theta) - \mu^2}{4\pi^2}\right]X_{co} + CZ_{co} + \dfrac{k(\theta)}{4\pi^2}\hat{R} = 0, \\ \ddot{Z}_{co} + \left[\nu^2 + \dfrac{-k(\theta) - \mu^2}{4\pi^2}\right]Z_{co} + CX_{co} - \dfrac{k(\theta)}{4\pi^2}\hat{R} = 0, \end{cases} \tag{19}$$

with the periodic closed orbit conditions:

$$X_{co}(\theta) = X_{co}(\theta + 2\pi), \quad Z_{co}(\theta) = Z_{co}(\theta + 2\pi),$$

and

$$\hat{R} = \frac{\nu^2 + C}{\nu^2 + C - \mu^2/4\pi^2} R_0.$$

For piecewise constant focusing function $k(\theta)$, the closed orbit solutions can easily be obtained by matrix products. The envelope radii obtained from integrating the KV equation agree well with the solution of Eq. (19).

C. Envelope mismatch oscillations and Mathieu instability

Let x, z be envelope mismatch coordinates with

$$x = R_x - R_{x0} - X_{co}, \quad z = R_z - R_{z0} - Z_{co},$$

where X_{co} and Z_{co} are closed orbit solutions. The equations of motion for x and z, which are small amplitude solutions of the envelope function around the closed orbit, are given by

$$\begin{cases} \ddot{x} + \left[\nu^2 - \left(\frac{\mu}{2\pi}\right)^2 + \frac{k(\theta)}{4\pi^2}\right]x + \frac{K}{8\pi^2 R_0^2}z = 0, \\ \ddot{z} + \left[\nu^2 - \left(\frac{\mu}{2\pi}\right)^2 - \frac{k(\theta)}{4\pi^2}\right]z + \frac{K}{8\pi^2 R_0^2}x = 0. \end{cases} \quad (20)$$

To study the Mathieu instability, we expand the focusing function in Fourier harmonics

$$k(\theta) = \sum_{\ell=0}^{\infty} k_\ell \cos \ell\theta, \quad (21)$$

where $k_0 = 0$ for equal horizontal and vertical tunes. The lowest order perturbation to the envelope oscillation arises from the $\ell = 1$ term. Now we use the variable $\phi = \theta/2$, the linearized envelope equations become

$$\ddot{x} + (p + 2q \cos 2\phi)x + 4Cz = 0, \quad \ddot{z} + (p - 2q \cos 2\phi)z + 4Cx = 0. \quad (22)$$

where the dots correspond to derivatives with respective to the variable ϕ, and

$$p = 4\nu^2 - \left(\frac{\mu}{\pi}\right)^2, \quad q = +\frac{k_1}{2\pi^2}, \quad C = \frac{K}{8\pi^2 R_0^2}$$

The stability of the generalized Mathieu equation (GME) can be analyzed by harmonic linearization method. Let the solution of the GME be given by

$$x = a \cos \phi + b \sin \phi, \quad z = c \cos \phi + d \sin \phi,$$

where a, b, c and d are slowly varying with respect to the time ϕ. Neglecting the second order derivatives and assuming the solutions of a, b, c and d are in the form of $e^{s\phi}$, the eigenvalues are given by

$$s_{1,2} = \frac{1}{2}(\sqrt{q^2 - (p-1)^2} \pm i4C) \quad s_{3,4} = -\frac{1}{2}(\sqrt{q^2 - (p-1)^2} \pm i4C). \quad (23)$$

The corresponding eigenfunctions are

$$e_1 = \begin{pmatrix} q_+ \\ q_- \\ -iq_- \\ iq_+ \end{pmatrix} \quad e_2 = \begin{pmatrix} q_+ \\ q_- \\ iq_- \\ -iq_+ \end{pmatrix} \quad e_3 = \begin{pmatrix} -q_+ \\ q_- \\ -iq_- \\ -iq_+ \end{pmatrix} \quad e_4 = \begin{pmatrix} -q_+ \\ q_- \\ iq_- \\ iq_+ \end{pmatrix}, \quad (24)$$

where

$$q_\pm = \sqrt{q \pm (p-1)}.$$

Using these eigenfunctions, the evolution of the envelope oscillations can be obtained.

Now we like to discuss the threshold of instability. When the real part of the eigenvalues is positive, the system would encounter Mathieu instability. This corresponds the region of parametric space in (κ, μ) such that

$$q^2 > (p-1)^2.$$

Figure 2 shows center of the Mathieu instability marked 2:1 and the 3rd order instability marked 3:1. The dotted line corresponds to the lower boundary of the Mathieu instability.

It is worth mentioning that the region of Mathieu instability extends to $\kappa = 0$. In reality, at $\kappa = 0$, the system is linear and there is no instability. The approximation that we have used breaks down. The result can be resolved by expanding the KV equation:

$$\begin{cases} \ddot{R}_x + \frac{1}{4\pi^2}\left[k_x(s)R_x - \frac{1}{R_x^3} - \frac{2K}{R_x + R_z}\right] = 0, \\ \ddot{R}_z + \frac{1}{4\pi^2}\left[k_z(s)R_z - \frac{1}{R_z^3} - \frac{2K}{R_x + R_z}\right] = 0. \end{cases} \quad (25)$$

Small amplitude oscillations around the closed orbit is given by

$$\begin{cases} \ddot{x} + \frac{1}{4\pi^2}\left[k_x(s) + \frac{3}{R_{xco}^4} + \frac{2K}{(R_{xco} + R_{zco})^2}\right]x + \frac{2K}{4\pi^2(R_{xco} + R_{zco})^2}z = 0, \\ \ddot{z} + \frac{1}{4\pi^2}\left[k_z(s) + \frac{3}{R_{zco}^4} + \frac{2K}{(R_{xco} + R_{zco})^2}\right]z + \frac{2K}{4\pi^2(R_{xco} + R_{zco})^2}x = 0, \end{cases} \quad (26)$$

Since the driven term for the Mathieu instability in the above equation vanishes at the zero space charge limit. Thus the region of the Mathieu instability is overestimated in the small κ region [7].

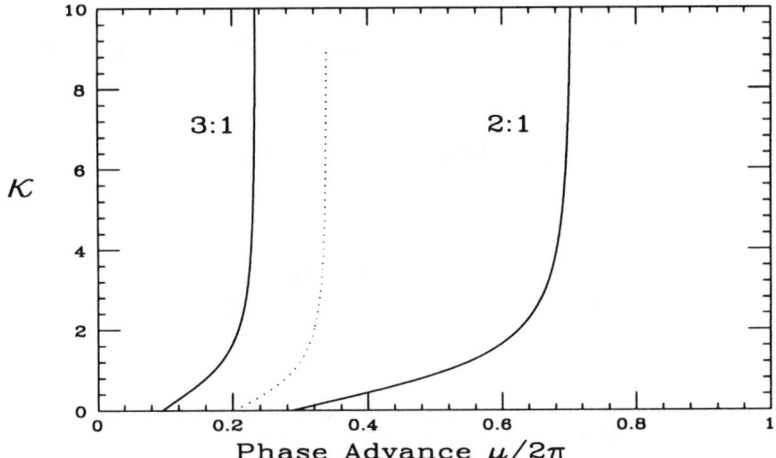

FIG. 2. The center (marked 2:1) and the boundary (dotted line) of the Mathieu instability are plotted in the parametric space κ vs μ. The 3rd order resonance (marked 3:1) is also plotted for reference.

We see from the eigen-frequencies in Eq. (23) that there are significantly only two modes, the σ and π. The other two modes correspond to the mirror image of them in the negative-frequency space. When the space-charge perveance is large enough, these two modes are separated and are stable. As the space-charge perveance decreases, these two modes merge together and become unstable.

III. THE NONLINEAR DETUNING OF THE ENVELOPE TUNE

Because of nonlinearity arising from the Coulomb force, the betatron tunes depend on the amplitudes of oscillations. To calculate the nonlinear detuning, we need to envelope detuning parameter we need to find the deviation of the closed orbit. Expanding the Hamiltonian up to the fourth order, we obtain nonlinear potential:

$$V_{nl} = -\left(\frac{1}{\pi R_0^3} + \frac{K}{24\pi R_0^3}\right)(x^3 + z^3) - \frac{K}{8\pi R_0^3}(x^2 z + xz^2)$$
$$+ \left(\frac{5}{4\pi R_0^6} + \frac{K}{64\pi R_0^4}\right)(x^4 + z^4) + \frac{K}{64\pi R_0^4}(4x^3 z + 6x^2 z^2 + 4xz^3) \quad (27)$$

Small amplitude oscillation around the closed orbit can be expanded in action angle variables:

$$x = \sqrt{\frac{J_x}{\pi \nu}} \cos \phi_x, \quad z = \sqrt{\frac{J_z}{\pi \nu}} \cos \phi_z, \quad (28)$$

Performing the canonical perturbation to cancel the leading order and averaging the nonlinear potential over the betatron phases, the nonlinear potential can be expressed as

$$\langle V_{nl} \rangle = \frac{1}{2}(\alpha_{xx} I_x^2 + \alpha_{zz} I_z^2) + \alpha_{xz} I_x I_z. \qquad (29)$$

where

$$\alpha_{xx} = \alpha_{zz} = \frac{\mu}{96\pi} \frac{\kappa[876 + 7\kappa(\kappa + \sqrt{1+\kappa^2})]}{[4 + 3\kappa(\kappa + \sqrt{1+\kappa^2})]^2},$$

$$\alpha_{xz} = -\frac{\mu}{48\pi} \frac{\kappa[108 - 7\kappa(\kappa + \sqrt{1+\kappa^2})]}{[4 + 3\kappa(\kappa + \sqrt{1+\kappa^2})]^2}. \qquad (30)$$

Along with the nonlinear detuning, the higher order term also generate nonlinear resonances. These nonlinear resonances are similar to those of nonlinear betatron resonances. The resonance condition is determined by the unperturbed betatron tunes.

IV. CONCLUSION

In conclusion, we have reformulated the KV envelope equation for the FODO transport channel in Hamiltonian dynamics and applied the parametric resonance analysis method to determine the dynamic behavior of the system. We have analyzed the generalized Mathieu instability in the two dimensional coupled system. We find that the threshold of the two dimensional coupled Mathieu instability is independent of the coupling. One can understand the Mathieu instability as mode coupling. In general, there are four mode frequencies in the coupled envelope equation; but two are just mirror image of the others in the negative-frequency space. The threshold of Mathieu instability corresponds to the condition that two of them (either positive or negative frequencies) become degenerate. We also calculate the nonlinear detuning parameters, which are important in stabilizing the Mathieu instability and higher order parametric resonances. Comprehension of the dependence of beam envelope properties on the machine and the beam parameters may lead to an understanding of the mechanism of halo formation, as well as finding methods of prevention.

ACKNOWLEDGMENTS

Work supported in part by grants from the NSF PHY-9221402 and the DOE DE-FG02-93ER40801.

REFERENCES

1. I.M. Kapchinskij and V.V. Vladimirskij, *Proceedings of the International Conf. on High Energy Accelerators*, p. 274 (CERN, Geneva, 1959).
2. P.M. Lapostolle, IEEE Trans. Nucl. Sci. **NS-18**, 1101 (1971); F.J. Sacherer, ibid. 1105 (1971); J.D. Lawson, P.M. Lapostolle, and R.L. Gluckstern, Part. Accel. **5**, 61 (1973); E.P. Lee and R.K. Cooper, ibid. **7**, 83 (1976).
3. J.D. Lawson, *Physics of charged particle beams*, 2nd ed. (Oxford Univ. Press, N.Y. 1988); R.C. Davidson, *Physics of nonneutral plasma*, (Addison-Wesley, reading, 1990); A.W. Chao, *Physics of collective instabilities in high energy accelerators*, (J. Wiley & Sons, Inc, New York, 1993).
4. S.Y. Lee and A. Riabko, Phys. Rev. E **51**, 1609 (1995).
5. A. Riabko *et al.*, Phys. Rev. E **51**, 3529 (1995).
6. S.Y. Lee *et al.*, Phys. Rev. E**49**, 5717 (1994).
7. A. Riabko and S.Y. Lee, in these proceedings (1995).

PIC Simulation of Short Scale-Length Phenomena

I. Haber,* D. A. Callahan,** C. M. Celata,+ W. M. Fawley,+
A. Friedman,** D. P. Grote,** and A. B. Langdon**

*Plasma Physics Division, Naval Research Laboratory, Washington, DC 20375
+Lawrence Berkeley National Laboratory, Berkeley, CA 94720
**Lawrence Livermore National Laboratory, Livermore, CA 94550

Abstract. The characteristics of a beam propagating in a channel of large aperture quadrupole magnets, and an anisotropy driven instability which transfers thermal energy from the transverse to longitudinal directions, are presented as two examples of phenomena requiring fine numerical resolution to be adequately described. In the first case, adequate resolution of the thin beam-edge sheath is found necessary to accurately simulate the particles whose orbits are driven unstable by the focusing nonlinearity. In the latter case, the tendency of the unstable mode to occur at longitudinal wavelengths comparable to beam diameter, requires adequate longitudinal resolution for the instability to be observed.

INTRODUCTION

Particle in cell (PIC) simulation techniques are a numerical method for advancing the equations of motion of a large number of particles in their self-consistent electromagnetic fields. The electromagnetic interaction of the ensemble of particles is long range, that is, each particle is influenced most strongly by the collection of particles in its vicinity rather than by the interaction with its nearest neighbors. The PIC methodology exploits this long-range nature of the forces by accumulating the charge densities and currents associated with an ensemble of particles onto a grid and using these grid-defined quantities to solve the partial differential equations which define the electromagnetic fields. The resulting grid-defined electromagnetic fields are then interpolated onto the particle positions and used to advance the orbits of the individual particles. A significant consequence of following this procedure is that the computational effort will generally scale linearly with the number of particles being integrated, rather than quadratically, as would result from the necessity to calculate the direct interaction between each particle and all the others.

PIC techniques have been extensively deployed for simulations of neutral plasmas,[1] and have also proved suitable for the examination of space-charge-dominated beams, which are often described as non-neutral plasmas. Their particular suitability for the examination of the beams in an accelerator, arises

© 1996 American Institute of Physics

from the fact that these beams are typically, as will be discussed below, only several Debye lengths in extent and have lifetimes of only several plasma periods.

PIC simulation of a particle beam employs a finite number of particles on a finite grid, to represent a physical system which has a much larger number of particles and whose behavior may involve scale lengths shorter than the grid spacing used in the numerical representation. Furthermore, both the number of simulation particles and the number of grid points used in describing the electromagnetic fields can be chosen separately. Inadequate numbers of either can therefore limit the accuracy of the numerical representation.

The particle distribution function in a PIC simulation has a nonphysical granularity which arises from the necessity to employ far fewer particles in the simulations than in an actual beam. This effect is usually identified as numerical collisionality because of the similarity to what would actually result from the Coulomb collisions in a plasma with a relatively small number of particles per Debye length.[1] In many instances where PIC methods are used to simulate space-charge-dominated beams, the number of particles which can be employed within available computational resources represents the practical limit on the accuracy of the simulations.[2,3] The emphasis below, however, is rather on the examination of short scale-length phenomena where accuracy of the simulation can be limited by the number of grid points, rather than the more usual limit imposed by the number of particles which can be employed.

Two examples are presented below which illustrate the importance of adequately resolving the self-consistent space charge forces that act on the beam particles. The first example discusses the propagation of a space-charge-dominated beam in a transport system of large-aperture quadrupole magnets and the limits to beam transport which result from the nonlinearities in such large-aperture focusing elements. The second example discusses the transfer of anisotropic thermal energy, from the transverse to the longitudinal directions, via a collective electrostatic instability driven by the anisotropy. This instability is found to occur at longitudinal wavelengths which are comparable to the beam radius and, therefore, will not be adequately represented without sufficient numerical resolution in the longitudinal direction. The two examples, therefore, represent physics whose accurate representation can be limited by the finiteness of the transverse and longitudinal gridding respectively. Both examples are examined in a frame moving with the beam, and the particle velocities in this frame are assumed to be sufficiently low that the self fields can be represented electrostatically.

LARGE APERTURE TRANSPORT

Currently envisioned applications of accelerator technology, such as the use of a beam of heavy ions to ignite a thermonuclear pellet, require the acceleration and transport of a high current beam, but without substantial growth in the beam emittance. In fact, for many high luminosity accelerator systems with even modest current requirements, the low energy transport of sufficient current without degradation in beam quality, can still be problematic.

If the beam intensity is sufficiently high, then the emittance term in the envelope description can be neglected relative to the space charge term. Furthermore, because of possible space-charge driven instabilities,[4] it is usually necessary to limit the phase advance per magnet period to somewhat less than 90 degrees. It is then possible to assume, for calculating the transportable current, that the envelope excursions in the alternating gradient period are small enough that the focusing force can be replaced by its value averaged over the focusing period. The transportable current can then be approximated by,[5]

$$I = \sigma_0^2 \left[\frac{a}{L} \right]^2 V^{3/2} \pi \epsilon_0 \left[\frac{2qe}{M_0 A} \right] , \qquad (1)$$

where σ_0 is the low current phase advance per lattice period of length 2L, a the mean beam edge radius, V the voltage to which the beam has been accelerated, ϵ_0 the permitivity of free space, qe the particle charge, and $M_0 A$ the mass.

Because, as stated above, the maximum σ_0 is limited by the requirement for beam stability, Eq. (1) illustrates the importance of maximizing the aperture ratio a/L to maximize the current transported. However, large aperture transport can be problematic from a beam dynamics perspective because the focusing nonlinearities that are present in large aperture lenses can cause emittance growth. Additionally, particles can leave the body of the beam distribution, eventually intercepting the beam pipe as their orbits are driven unstable by the nonlinear focusing forces.

Investigation of the dynamics of a space-charge-dominated beam in a large-aperture focusing system requires the choice of a suitable lens design to model the transport system. The model lens system should be simple enough to have general applicability but must also be realistic enough to be relevant to a real transport system. In particular, the focusing system should be realizable from a practical winding configuration. Also, the lenses in a large aperture transport system will generally be closely packed in order to maximize the focusing force, and the fringe fields of adjacent lenses then interact. It therefore becomes important to properly include in the lens description the interaction of neighboring lenses in order to properly model the lens nonlinearities. Such a model lens system, which is made up of a superposition of sinusoidal harmonics in z of the strength of the quadrupole gradient, and which also has some characteristics which are optimal for large aperture transport[6,7] was suggested by Laslett.[8] For simplicity, we will employ only the first harmonic in the expansion which can be used to generate this class of lenses. The magnetic fields can then be derived from the magnetic potential,

$$\psi = \psi_0 \cos(kz)\, I_2(kr)\, \sin(2\phi), \qquad (2)$$

where $k = 2\pi/L$, L is the magnet period, and the magnetic field can then be found as $\mathbf{B} = -\nabla\psi$.

In order to investigate propagation of a space-charge-dominated beam in such a large aperture quadrupole channel, a series of PIC simulations was performed, using the two-dimensional SHIFT-XY code.[2] The beam parameters used for these simulations consisted of 4 μC of 4 MeV, singly charged ions with 20 AMU mass. The transport system magnets had a peak field gradient 13.5 T/m and a full period of 1. The emittance of 6×10^{-4} m·rad corresponds to a matched major radius approximately 0.1 m and enough space charge to depress phase advance from 72° to 7.2° if the beam were a K-V distribution. However, the simulations described below were actually performed employing an rms equivalent semi-Gaussian, or thermal, initial distribution which is uniform in space but has a Gaussian velocity distribution with a temperature that is uniform across the beam.

The simulations were typically performed using 128K particles. The magnetic field at each particle location was calculated using a Taylor series expansion for the modified Bessel functions, so that the focusing forces do not depend on the choice of a numerical grid. A typical grid spacing of 2 mm was used, and the implications of this choice will be discussed below. For reference, it should be noted that the plasma frequency of the beam is such that the beam travels 3.55 m during a plasma period. The Debye length corresponding to the temperature in the initial thermal distribution is 1.7 mm, which is a measure of the thickness of the sheath which forms at the edge of the beam. Because the edge of the beam, and therefore the beam sheath, is also in the region which sees the most significant lens nonlinearities, the beam behavior can be expected to be strongly influenced by the details of this sheath region.

A detailed discussion of the numerous simulations which have been performed is outside the scope of this report. However, some of the features of the nonlinear transport are worth discussing here, especially those which are similar to what has been previously observed on a simpler nonlinear transport system[9] and, therefore, appear to have general applicability. As the current which is to be transported is increased, the matched radius of the beam also increases. Particles near the outer edge of the beam eventually experience sufficient nonlinearity in the external force that their orbits become unstable, and these particles increase in orbital radius until they intercept the beam pipe. This appears to be the most significant long term source of beam emittance. Two other sources of emittance growth are: the redistribution of an initially uniform space charge distribution so that the beam is in detailed local force balance with the external focusing, and the conversion of any of the potential energy in an initial mismatch[10] to kinetic energy by the lens nonlinearities.

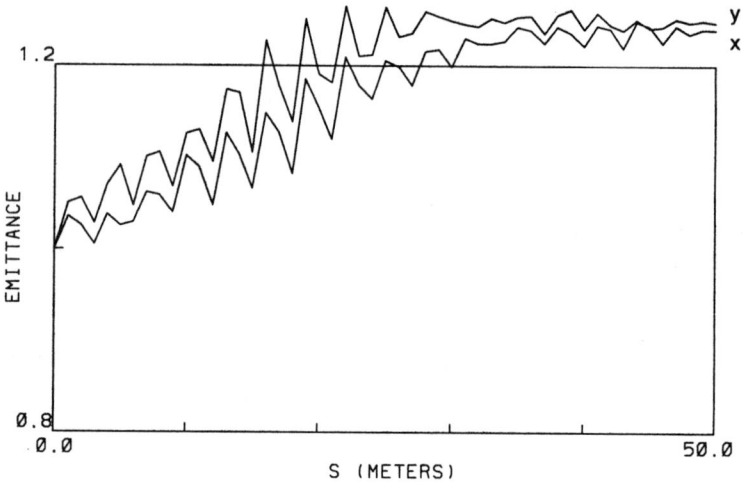

Figure 1. Plot of the evolution of the x and y rms emittances as a function of propagated distance, for a beam with a line-charge density of 4 μC/m propagating in a transport system with a 1 m full period. The initial beam radii of 102.4 mm and 56.3 mm correspond to a beam matched to the transport channel in the absence of the nonlinearities. Note that the emittance growth is capped by the removal of energetic particles from the system as they intercept the beam pipe.

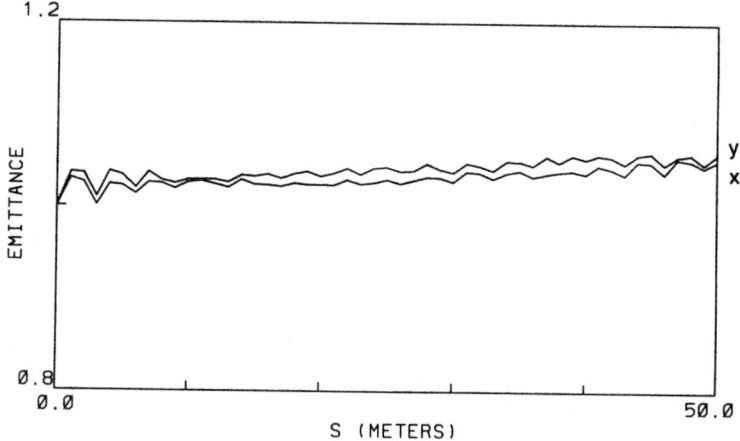

Figure 2. Plot of the x and y rms emittances for the same conditions as in Fig. 1, except that the initial radii of 99.3 mm and 59.1 mm are the values which minimize the change in the rms beam radius during propagation over the first period, so that the beam is rms matched to the lens system including the nonlinearities in focusing force.

Figure 1 shows the evolution of the rms emittance plotted as a function of the the distance propagated, for a beam whose radius is matched to a focusing system with the same linear parameters as those near the axis of the nonlinear system. The emittance growth can be seen to saturate as energetic particles are removed from the calculation of the rms emittance upon intercepting the wall of the transport system. This eventually balances the growth in emittance resulting from the increase in orbital radii of the remaining unstable particles.

As has been discussed previously,[9] the emittance growth tends to separate into different timescales which depend on the cause of the growth. The shortest timescale growth is generally the result of the transverse redistribution of the initially uniform beam density as it relaxes to a detailed force-balance with the external focusing. The longest timescale is usually due to the increase in orbital radii of the unstable particles, and intermediate is the conversion of the mismatch energy to emittance.

It should be noted that a substantial source of the emittance growth in this particular simulation results from the initial mismatch which occurs because the beam was matched to the linear part of the channel, even though, the average focusing force has been modified slightly by the focusing nonlinearity. The extent to which this effect is important can be seen by contrasting this case with the emittance growth in Fig. 2, which shows the evolution of a beam, which has been rematched to include the nonlinearities by adjusting the initial radius to minimize the change in the rms radii of the beam after its propagation through the first lens period. Note that the beam radius in Fig. 2 is approximately 0.2 of the lens half-period, so that with the beam rms matched in this way, transport over several lens periods can potentially be achieved with only modest growth in the rms emittance.

Another finding of the simulations that is worth mentioning, is that it is possible to further minimize the emittance growth of such a beam if the beam is accelerated during transport. Though a detailed discussion is outside the scope of the present paper, it should be noted that, since large aperture transport is most likely to be employed at relatively low beam energy, it is often possible to provide sufficient acceleration to have a substantially mitigating effect.

The onset of particle loss and the degradation of emittance occur quickly as the beam current, and therefore matched radius, are increased, causing the beam edge to intercept a region with sufficient nonlinearity to substantially affect the beam dynamics. It is, therefore, relatively easy to determine the threshold current at which beam degradation begins to occur. This has been verified by sensitivity tests which varied the numerical accuracy of the simulations, as well as by comparisons between different codes.[6] However, because the actual rates of emittance growth and particle loss depend on the interaction between the transport system nonlinearities and particles in the beam sheath region, determination of these rates requires adequate numerical resolution of the thin sheath region and can therefore be somewhat more sensitive to the choice of numerical parameters.

Determining the resolution which is adequate for describing the particle loss rate is not just a numerical issue, but is also related to how sensitive the actual orbital instability is to the shape of the sheath. That is, the necessity to employ sufficient numerical resolution of the sheath to determine beam behavior, implies that a self-consistent description of the shape of the beam sheath may be necessary to accurately predict the dynamics of orbital instability. It should be noted that the particle loss rate, rather than the rate of emittance growth, is usually a more appropriate diagnostic for determining beam behavior. This is because the amount of rms emittance growth which results from the instability-induced growth in the radii of individual particle orbits, strongly depends on the radius of beam pipe at which particles are removed from the system.

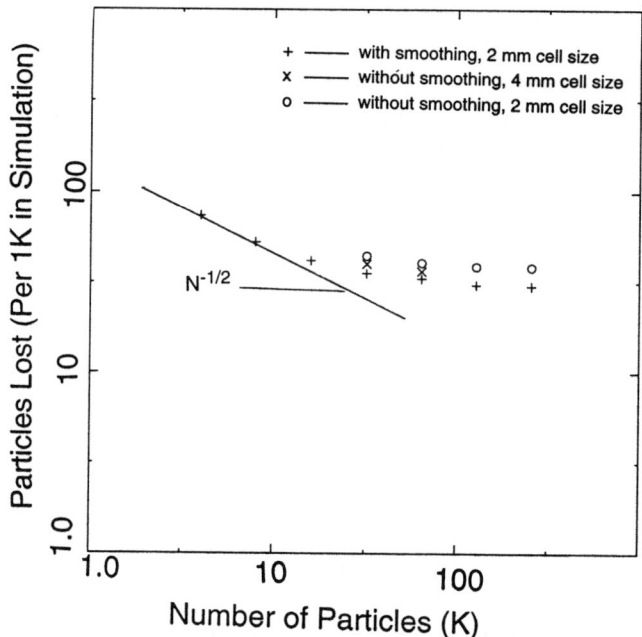

Figure 3. The logarithm of the number of particles lost, per 1k in the simulation, after 100 m of propagation, plotted against the logarithm of the number of particles employed in the simulation, at varying numerical resolutions. Numerical smoothing, can be seen to reduce the number of particles required to model the loss rate, however, it also appears to substantially reduce the effective resolution.

A test of the hypothesis that the detailed characteristics of the sheath may strongly influence the particle loss rate, was therefore performed by varying the numerical resolution used in the simulation. It was found, for the simulation parameters discussed above, that doubling the numerical cell size from 2 mm to 4 mm almost halved the rate at which particles, after an initial transient, intercepted a numerical pipe placed at 1.2 × the initial major radius of the beam. Going from 2 mm to 1 mm resolution also affected the particle loss rate, although the change was somewhat less than a factor of two, so that the effect appeared to saturate when a few cells per Debye length were employed. It was therefore concluded that a further doubling of the number of grid cells to 0.5 mm would probably result in little further change. However, this would result in too large a simulation to run easily with available computational resources.

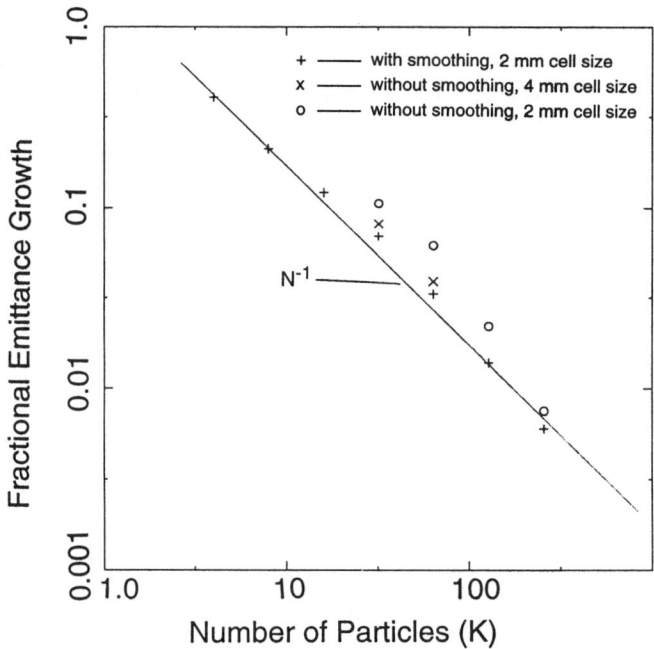

Figure 4. Logarithm of the residual growth in rms emittance (growth after the initial rapid transient) during 100 m of propagation, plotted against the number of particles in the simulation for various numerical grid sizes. Note that the lack of any evidence of leveling off as the number of particles is increased, is evidence that all of this residual growth is collisional.

Accordingly, a series of simulations was performed on a beam system where the beam emittance was doubled. Since the transport is space-charge dominated, and the matched radius is little changed by this increase in emittance, the thermal velocity, and therefore the Debye length, was also doubled to approximately 3.4 mm. The results of this series of simulations are summarized in the Fig. 3, which shows the fraction of particles which intercept the wall during 100 m of transport, as a function of the size of the grid cells and the number of particles. As would be expected, the fraction of particles lost does, at first, decrease as the number of particles is increased and the simulated distribution is rendered less granular. When a sufficient number of particles are employed, the fractional loss becomes independent of the number of particles. It is seen that this saturated loss rate changes in going from 4 mm to 2 mm.

A method for reducing the granularity of the distribution in a PIC simulation[1] is to smooth the electric fields by employing Gaussian shaped macroparticles. The results of some simulations which employed smoothing where the particle shapes are down by a factor of 1/e in half a cell are also shown on this plot. Though the smoothing does reduce the effective granularity of the distribution and therefore the numerical collision rate, it also reduces the numerical resolution, as measured by the saturated particle loss rate, when a sufficient number of particles are employed, by a little over a factor of two. It should be noted that simulations were also run with the grid spacing reduced to 1 mm, but these points were not plotted, because they fell on top of the points with 2 mm resolution. Because the 2 mm resolution appeared adequate to resolve the sheath, in the sense that a further increase did not seem to affect the particle loss rate, it was also concluded that the lower emittance simulations, described previously, were accurately described by 1 mm resolution.

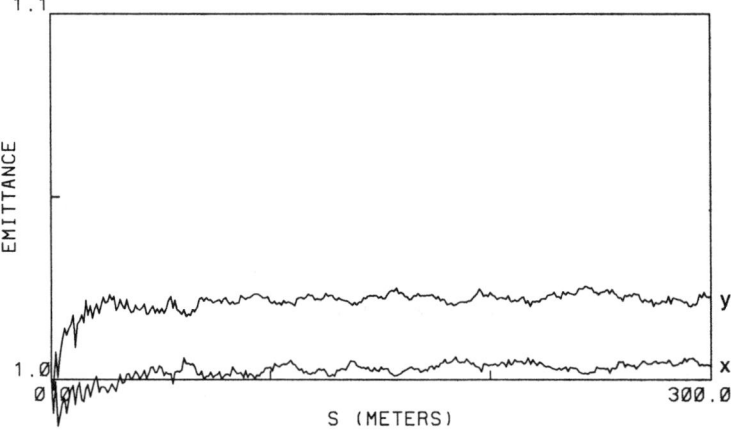

Figure 5. Evolution of the x and y rms emittances of a beam represented by 256K particles, 1 mm cell size, showing the rapid saturation of the emittance growth. Note that particles are removed as they intercept a beam pipe radius 1.2 × the initial major radius.

Figure 4 shows how the residual growth in the rms emittance, i.e., the growth after the initial transient during which the beam relaxes to local force balance with the nonlinear external forces, behaves as the numerical resolution and the number of particles are varied. In all cases shown this "residual" emittance growth continues to decrease as the number of particles is increased. This appears to indicate that the source of this residual growth is entirely numerical, and suggests that the beam system has, after an initial transient, entered a state where a small number of particles is being lost due to orbital instability, but the bulk of the distribution has reached a steady state where the increase in transverse kinetic energy as these particles go unstable, is matched by the energy removed from the beam as these particles intercept the beam pipe.

This apparently steady state behavior, after an initial transient, can be seen from the plot, in Fig. 5, of the rms emittance as a function of distance propagated. This steady state behavior can also be seen in the the plot, in Fig. 6, of particles lost as a function of propagation distance, which shows a constant loss rate after the initial transient. Note that Figs. 5 and 6 are plots from the simulation which is the "best" numerically, employing 256 K particles and a 1 mm grid spacing. However, this behavior is also seen over a range of numerical parameters.

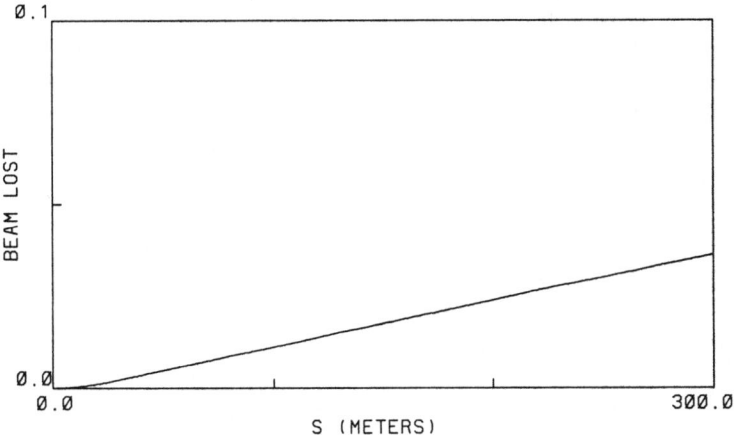

Figure 6. Fraction of particles lost after they intercept the simulation "wall" placed at 1.2× the inital major radius. Note that the loss rate is almost constant after the initial transient.

TEMPERATURE ISOTROPIZATION INSTABILITY

An example was shown above where the accuracy in simulation of the transverse physics can be limited by inadequacy in resolution of the detailed behavior of the beam sheath. An example is presented here to illustrate the importance of also using adequate longitudinal resolution in the simulation of multi-dimensional beam dynamics. The particular example presented here illustrates beam physics which is dominated by longitudinal modes with wavelengths on the order of the beam radius. These modes were found to be excited even on a bunch which is long compared with the beam radius, and with no external forces present which drive the beam at a frequency which would indicate the necessity to resolve these short longitudinal wavelengths. As will be illustrated below, these short wavelength oscillations in the beam represent a collective space-charge mode which oscillates at frequencies that can be comparable to the single particle betatron frequencies. These modes represent a class of electrostatic space-charge resonances which can occur at sufficiently high beam currents, but which are not usually considered in beam dynamics calculations. Furthermore, these modes will be absent from beam simulations unless the longitudinal phenomena are adequately resolved.

Instability Mechanism

In the absence of longitudinal focusing to constrain the beam length, addition of equal energy to each particle during acceleration decreases the longitudinal velocity spread in the beam frame, ($\delta v_z = \delta E/mv_0$), but not the transverse. This can cause the development of a substantial temperature anisotropy in the beam frame, particularly in the source region. Beam manipulations, which often increase the transverse emittance, but not the longitudinal, can also cause anisotropy. This anisotropy can cause a collective electrostatic mode, which resembles a short-scale perturbation of the beam envelope radius, to go unstable and couple energy to the longitudinal direction. This mechanism is similar to the Harris[11] instability which occurs in an unbounded plasma. In the current case, however, the cyclotron motion in the applied magnetic field which excites the Harris mode, is replaced by the betatron oscillations which characterize the transversely bounded beam, and the modal structure characteristic of the geometry substantially modifies the unstable dynamics. For simplicity this instability has been examined in axisymmetric geometry using the r,z capabilities of the WARP[12] simulation code. Because of the short longitudinal scale of the unstable modes, periodic boundary conditions in the longitudinal direction are employed to examine a small (several beam radii long) section of the beam.

The general characteristics of this instability have been previously reported[13,14] and will only be summarized here. In the parameter regime where there is a substantial anisotropy, and where the space charge depression in the phase advance is moderate, i.e., by a ratio of approximately two or three, the unstable mode appears to grow rapidly, with growth time comparable to the space-charge-depressed betatron oscillation time of the individual particles. It is worth noting that this instability does not seem to depend strongly on the details of the initial distribution function. So, for example, very similar behavior is seen when the initial transverse distribution is either a K-V distribution or a semi-Gaussian. However, for purposes of the present discussion, a particularly significant feature, which appears to be inherent in the unstable modes, is that their wavelength is comparable to beam radius, and these modes resemble a localized disturbance to the beam envelope. Because the resultant envelope-like oscillations are localized, they oscillate at a frequency which is different from the more rapid (in the beam frame) conventional long-wavelength envelope oscillations, and can therefore have oscillation frequencies which are comparable to the single particle depressed betatron frequency.

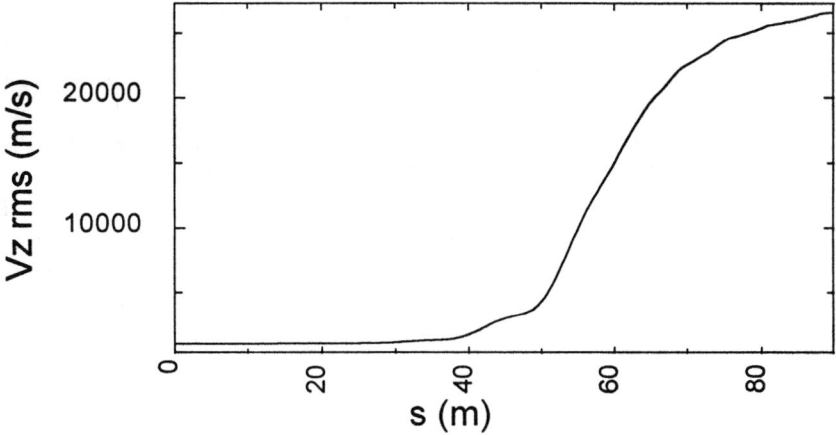

Figure 7. Longitudinal rms velocity in the beam frame plotted against the distance propagated by the beam, showing the longitudinal heating as energy is transferred to the longitudinal direction. Note the delayed onset of the heating as the unstable fields grow from initial small random fluctuations.

The simulation example presented here consists of a 4.8 A beam consisting of 10 MeV singly charged, 12 AMU ions. The 23 mm radius beam is matched to a uniform focusing system in a conducting pipe of 50 mm radius. In the example presented here the beam has an initially K-V distribution with an unnormalized emittance of 2.2×10^{-4} m·rad, sufficient to result in a space charge depression of the transverse betatron frequency by a factor of 2.35. The initial longitudinal thermal velocity is 10^3 m/s. The numerical grid used in this example employed 256 axial zones and 32 radial zones. Note that Gaussian smoothing, of the type discussed above, was employed to suppress a longitudinal grid instability sometimes seen in these simulations. This particular simulation employed 112K particles.

Figure 7 is a plot of the evolution of the longitudinal rms velocity of the initially anisotropic beam system, showing the transfer of energy to the colder direction by the electrostatic instability. The assertion that the transfer is via an instability is supported by the plot in Fig. 8 of the natural logarithm of the absolute value of the longitudinal electric field at a typical point in the beam. Note that the growth, as measured by the evolution of the peaks, is linear on this plot, indicating exponentiation in the electric field. From this plot, it is possible to infer that the real and imaginary parts of the unstable eigenfrequency are comparable and the oscillation period is comparable to the 16.9 m space-charge depressed oscillation period of the beam particles.

For the purposes of the discussion here, the nature of the unstable modes can be illustrated in Fig. 9, which is a $z-v_z$ phase space plot after the beam has propagated 35 m. This plot was taken from a simulation where the instability is artificially seeded by an initial local perturbation in the radius of the beam envelope at z=0. The initial perturbation is seen to grow at the original site, at the same time as it propagates outward, so that the instability appears to be primarily absolute rather than convective and the wavelength is comparable to the 23 mm beam radius. Further insight into the nature of the unstable mode structure can be seen by examining the r,z configuration space plot of the beam in Fig. 10. Note that this plot is after the beam has propagated 42.5 m, which is later in the beam evolution than the previous figure, so that the structure has had more time to become visible. In this plot, the variation in the beam radius is clearly exhibited.

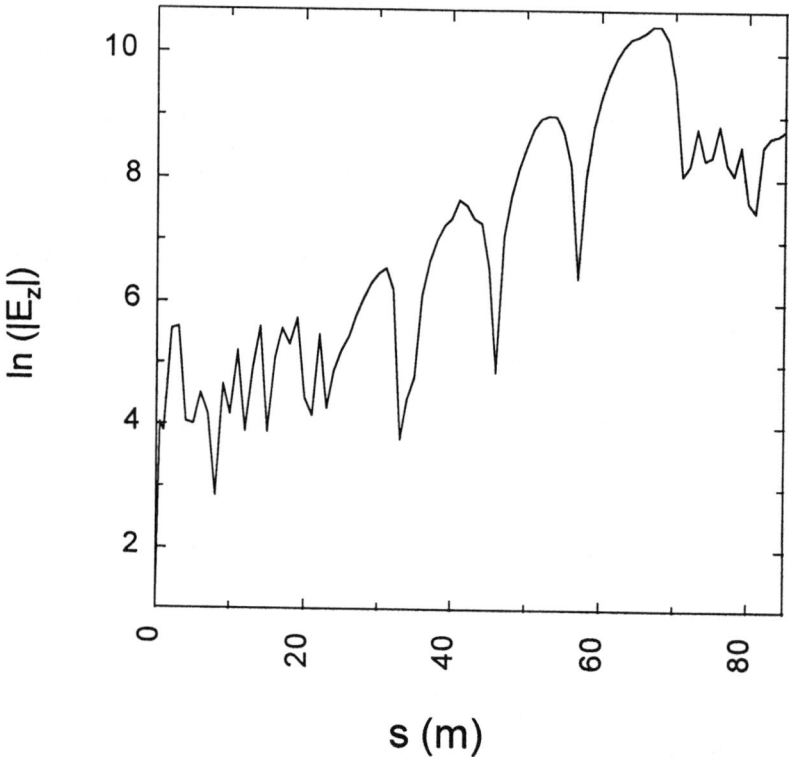

Figure 8. Evolution of the natural logarithm of the absolute value of the longitudinal self-electric field E_z, at a typical off-axis location, as the beam propagates. Exponential growth in the unstable fields is apparent.

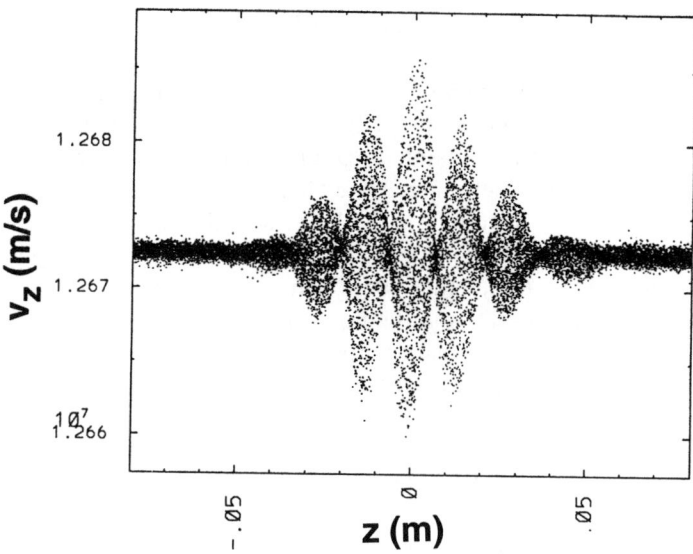

Figure 9. Plot of the z-v_z phase space after the beam has propagated 35 m, for a simulation in which the instability is seeded by perturbing the envelope radius near z=0. Note that the growth is largest at the site of the initial perturbation, and is spreading outward.

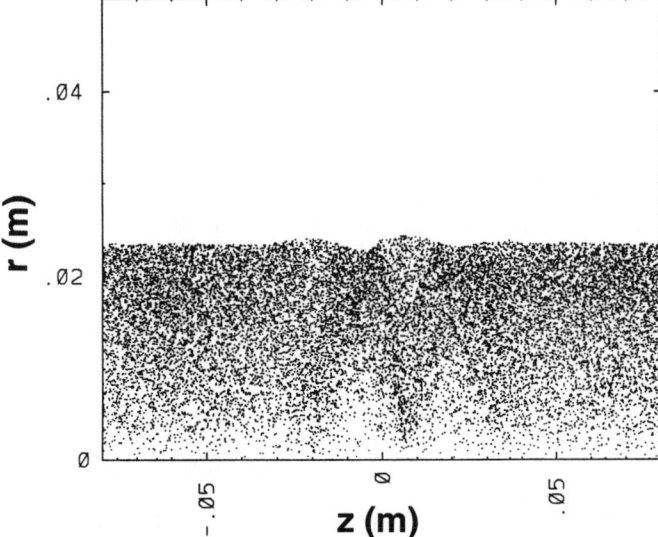

Figure 10. Plot of the r-z configuration space of the beam after 42.5 m of propagation showing how the beam envelope has been perturbed. It is also possible to see the density structure that results from the inherently oblique self-electric fields that characterize the unstable eigenfunctions.

SUMMARY

Two examples have been presented to show how the accuracy of simulations of beam dynamics can be limited by the finiteness of the gridded representation in a PIC code. It is not always feasible to conduct simulations, especially using a multi-dimensional code, which can resolve phenomena in the transverse direction on the dimensions of the sheath or in the longitudinal direction on a scale of the beam radius, so that it would be easy to incorrectly model this physics. In performing simulations, it may therefore be important, to check, sometimes on a scaled basis, whether increases in the numerical resolution strongly affect the beam dynamics. One of the primary purposes of this report is to show how such numerical exploration may reveal interesting phenomena which might otherwise be missed.

ACKNOWLEDGMENT

The authors acknowledge support for the research reported here by the United States Department of Energy at the Naval Research Laboratory under contracts DE-AI02-93ER40700 and DE-AI02-94ER54232, at the Lawrence Berkeley National Laboratory under contract AT2510000, and at the Lawrence Livermore National Laboratory under contract W-7405-ENG-48.

REFERENCES

1. Charles K. Birdsall and A. Bruce Langdon, *Plasma Physics via Computer Simulation*, McGraw Hill, New York, 1985.
2. I. Haber,"High Current Simulation Codes," *High Current, High Brightness, and High Duty Factor Ion Injectors*, Ed. George H. Gillespie, Yu-Yun Kuo, Denis Keefe, Thomas P. Wangler, AIP Conf. Proc. 139, (AIP, New York, 1986) p. 107.
3. I. Haber and H. Rudd, "Numerical Limits on P.I.C. Simulation of Low Emittance Transport,"*Linear Accelerator and Beam Optics Codes*, AIP Conf. Proc. 177, p. 161 (AIP, New York, 1988).
4. I. Hofmann, L. J. Laslett, L. Smith and I. Haber, "Stability of the Kapchinskij-Vladimirskij (K-V) Distribution in Long Periodic Transport Systems", *Particle Accelerators* **13**, 145 (1983).
5. *Induction Linac Systems Experiments, Conceptual Engineering Design Study*, p. 28, PUB-5219 (LBL, March, 1989).
6. C. M. Celata, W. M. Fawley, L. J. Laslett, I. Haber, "Simulation Studies of Space-Charge-Dominated Beam Transport in Large Aperture Ratio Magnetic Quadrupoles," *Il Nuovo Cimento*, **106 A**, N. 11, 1637 (Nov. 1993). Also published in *Proc. of the 1993 Particle Accel. Conf.*, 724, (IEEE, NJ 1993).
7. William M. Fawley, "Space-charge Dominated Transport in Magnetic Quadrupoles with Large Aperture Ratios," LBL-33608 (LBL, Aug. 1993).
8. L. Jackson Laslett, "On the Aperture Ratio and Related Transport Limitations" HIFAN-427 (LBL, April,1989).

9. I. Haber and H. Rudd, "Emittance Growth and Particle Loss During Intense Beam Propagation in a Nonlinear Periodic Transport System," *Nucl. Instr. and Methods in Phys. Res.* **A278**, 1, 174 (1989).
10. M. Reiser, "Free Energy and Emittance Growth in Nonstationary Charged Particle Beams," *J. Appl. Phys.* **70** 1919 (1991).
11. E. G. Harris, "Unstable Plasma Oscillations in a Magnetic Field," *Phys. Rev. Letters*, **2**, 34 (1959).
12. Debra A. Callahan, A. Bruce Langdon, Alex Friedman and Irving Haber, "Longitudinal Beam Dynamics for Heavy Ion Fusion," *Proc. 1993 Particle Accel. Conf.*, 730 (IEEE, 1993).
13. I. Haber, D. A. Callahan, A. Friedman, and A. B. Langdon, "Transverse-Longitudinal Energy Equilibration in a Long Uniform Beam," to be published in the *Proc. of the 1995 Particle Accel. Conf.*
14. I. Haber, D. A. Callahan, A. Friedman, D. P. Grote, and A. B. Langdon, "Transverse-Longitudinal Temperature Equilibration in a Long Uniform Beam," to be published in *Journal of Fusion Engineering Design*.

Injection into a Circular Machine with a KV Distribution*

E. Crosbie and K. Symon[1]

Argonne National Laboratory, 9700 South Cass Avenue, Argonne, IL 60439 USA

Abstract. In order to achieve a maximum space charge limit in the IPNS-II synchrotron it is desirable to inject a Kapchinskij-Vladimirskij (KV) distribution (1). We rederive the KV distribution, first starting from a smoothed Hamiltonian and then for the full alternating gradient case. The microcanonical distribution can be generalized slightly so as to allow one to alter the aspect ratio of the beam ellipse.

The KV distribution requires that the injected particles all have the same total transverse oscillation energy, and also that they are distributed uniformly throughout the entire energy shell. This requires painting the injected beam uniformly in the three independent dimensions of the energy shell. We have devised two scenarios for doing this, one involving a suitable variation of the x and y injected amplitudes during the injection process, and the second involving introducing a small coupling between the x and y motions.

We have written a program to simulate the injection process which includes the turn-to-turn forces between the (500) injected turns. If we omit the turn-to-turn forces then the resulting space charge density distributions are indeed very nearly uniform within a circular beam cross section for either KV injection scenario, but are neither uniform nor circular for other plausible scenarios. With turn-to-turn forces included, the interturn scattering can be fairly important and the resulting density distributions tend to develop lower density halos.

If we add a gradient bump to simulate magnetic quadrupole errors in the lattice, then the effects of half-integral resonances can be clearly seen. When the space charge forces between turns depress the tune to a resonance, beam growth keeps the tunes constant at the edge of the stop band, unless the resonance is crossed quickly. The resultant growth of the beam can be seen in the density distribution if resonant effects are dominant, i.e. starting with tunes near the resonance. If we start farther from the resonance, in which case we inject higher intensity beams, the turn-to-turn forces dominate the final density distribution. In that case the final distribution is nearly the same whether the resonance is present or not, though the effect of the resonance on the final tune can still be clearly seen.

---

* Work supportred by the U.S. Department of Energy, Office of Basic Energy Sciences, under Contract No. W-31-109-ENG-38.

[1] permanent address: Dept. of Physics, University of Wisconsin-Madison, 1150 University Ave., Madison, WI 53706 USA.

I. INTRODUCTION

The Kapchinskij-Vladimirskij (KV) distribution (1) was originally of mathematical interest as a many particle problem which could be solved analytically. It gave an idea of the effects of space charge forces in high intensity accelerators, but it was not expected that any real accelerator would contain a microcanonical distribution.

Recently it has been suggested that the KV distribution may be of practical interest for high intensity machines in that it may provide the maximum space charge limit for such a machine. One can make a plausible argument that the maximum beam intensity is obtained for a distribution for which all particles have the same tune, at least when the resonance is approached. One should therefore first reduce the chromaticity of the accelerator ring as much as possible, and second, make the betatron frequencies independent of amplitude, i.e., make the focussing forces linear. One way to make the focussing forces linear is to start with external focussing forces which are linear, and then make the space charge forces also linear by using a KV distribution.

Chapter II reviews the theory of the KV distribution, generalizing it slightly to include an elliptical beam cross section. We give first a simplified treatment based on treating the betatron oscillations as simple harmonic motions. We then treat the alternating gradient case. Finally two injection scenarios are described which produce a KV distribution (if beam-beam interactions are neglected during the injection process). In the first the injected x and y amplitudes follow a prescribed schedule. In the second there is a coupling between the x and y motions.

We have written a simulation code for the injection process which includes the space charge interactions between the injected turns. It also provides a gradient bump to simulate the effect of quadrupole errors which can drive a half-integral resonance. The code was used to study the injection process for the proposed IPNS upgrade (2), a 2 GeV rapidly cycling synchrotron designed to deliver a 1 MW proton beam. During the injection process 500 turns are injected. Chapter III presents the results of simulating the injection process, without turn-to-turn space charge forces, for a KV scenario and for a typical injection scenario which is not specifically designed to produce a KV distribution. The KV scenario indeed produces a circular beam cross section of uniform density. Non-KV scenarios produce a beam cross section which is neither circular nor of uniform density.

In describing the injection process we will use the following terminology. We will specify a time by the number of turns (revolutions of the beam around the accelerator) since injection started. The beam at any time consists of a number of beamlets. By a beamlet we mean that part of the beam which was injected during a particular previous turn. A beamlet will be specified by the number of the turn during which it was injected. By a turn-to-turn force we mean the electric force exerted by one beamlet on another.

The turn-to-turn forces are included in Chapter IV. They have a substantial effect on the resulting distribution. The cross section still has a rough circular symmetry, but the beam has a low density halo. The program provides a means of calculating the betatron oscillation frequencies of any beamlet during any turn. Plots of horizontal and vertical tunes versus time for selected beamlets, as well as the average tunes for all beamlets present in the machine clearly show the depression of tune due to the increasing space charge forces as injection proceeds. The proposed KV scenario produces a more uniform beam density and smaller tune shifts than the other injection scenario. When the gradient bump is added the effect of the half-integral resonance is clearly seen in the tune plots. In cases where the tunes would otherwise cross the resonance at $\nu = 5.5$ during injection, the gradient bump causes the beam to expand when the resonance is reached and the tune levels off at a value corresponding to the edge of the stop band. The optimum parameters correspond to arranging that horizontal and vertical tunes both just reach the edge of the stop band at the end of injection. This gives the maximum injected beam for a given beam cross section without an expanded halo.

There appear to be two regimes when beams exceeding the space charge limit are injected. If one or both tunes start at a value not too far from the resonance stop band, then the resonance dominates the process and the beam has a halo which expands to keep the tune at the edge of the stop band. If we start with tunes far enough from the resonance, which requires injecting a more intense beam in order to reach the stop band, then the fluctuating turn-to-turn forces dominate the process and cause the beam to expand because of the resulting diffusion in betatron amplitudes. The resulting beam density profiles do not depend very much on whether the resonance driving bump is present or not, although even in this case the effect of the bump can clearly be seen in the plots of tunes versus time.

II. THEORY

In Section 1 we will derive the KV distribution for the smooth case when the betatron motion is a simple harmonic oscillator and the parameters are arranged to produce a circular beam cross section. Section 2 treats the full alternating gradient case and allows the beam cross section to be elliptical. Section 3 presents an injection scenario in which the x and y amplitudes during injection are programmed to paint the energy shell uniformly and so produce a KV distribution. Section 4 discusses an alternative way of producing a KV distribution using a small xy coupling.

1. A Simple KV Distribution.

We start from the simple Hamiltonian

$$H = \frac{1}{2}p_x^2 + \frac{1}{2}k_x x^2 + \frac{1}{2}p_y^2 + \frac{1}{2}k_y y^2 \quad , \tag{1.1}$$

where the independent variable is the azimuthal distance s. The moments are

$$p_x = \frac{dx}{ds} \quad , \quad p_y = \frac{dy}{ds} \quad . \tag{1.2}$$

We are assuming that the focussing forces are linear. The space charge forces for the KV distribution will also be linear, and are included in the constants k_x, k_y. The solution of the equations of motion is a simple harmonic oscillation in both dimensions of frequencies

$$v_x = Rk_x^{1/2} \quad , \quad v_y = Rk_y^{1/2} \quad , \tag{1.3}$$

in oscillations per revolution, where $2\pi R$ is the circumference.

Introduce angle-action variables:

$$x = (2RJ_x / v_x)^{1/2} \sin \gamma_x \quad , \quad p_x = (2v_x J_x / R)^{1/2} \cos \gamma_x \quad ,$$
$$y = (2RJ_y / v_y)^{1/2} \sin \gamma_y \quad , \quad p_y = (2v_y J_y / R)^{1/2} \cos \gamma_y \quad . \tag{1.4}$$

The Hamiltonian becomes

$$H = \frac{v_x}{R} J_x + \frac{v_y}{R} J_y \quad . \tag{1.5}$$

The action variables

$$J_x = \frac{R p_x^2}{2 v_x} + \frac{v_x x^2}{2R} \quad , \quad J_y = \frac{R p_y^2}{2 v_y} + \frac{v_y y^2}{2R} \tag{1.6}$$

are constants of the motion and are each equal to the area of the corresponding phase ellipse divided by 2π. The angle variables are

$$\gamma_x = v_x s / R + \vartheta_x \quad , \quad \gamma_y = v_y s / R + \vartheta_y \quad , \tag{1.7}$$

where ϑ_x, ϑ_y are arbitrary constants. If we substitute from Eqs.(1.7) into Eqs.(1.4), we get the general solution of the equations of motion.

Now write the distribution function for the beam in the form

$$D(J_x, J_y, \gamma_x, \gamma_y) = A\delta(2J_x \cos^2 \zeta + 2J_y \sin^2 \zeta - J_0) \quad . \tag{1.8}$$

This is a slight generalization of the standard microcanonical distribution, in that it allows J_x and J_y to appear with arbitrary factors in the total action J_0. It becomes a standard microcanonical distribution if we set $\zeta = \pi/4$. The advantage of introducing the parameter ζ is that it allows us to adjust the shape of the beam in x,y space; for example, we can change an ellipse into a circle, even when $k_x \neq k_y$.

The particle density in physical space is

$$\rho(x,y) = \int dp_x \int dp_y D(x, p_x, y, p_y)$$

$$= A \int dp_x \int \delta(2J_x \cos^2 \zeta + w - J_0) \frac{dw}{2\sin^2 \zeta \frac{\partial J_y}{\partial p_y}}$$

$$= \frac{A v_y^{1/2}}{2R^{1/2} \sin \zeta} \int_{-P_{x1}}^{P_{x1}} \frac{dp_x}{\left[J_0 - \left(\frac{R p_x^2}{v_x} + \frac{v_x x^2}{R} \right) \cos^2 \zeta - \frac{v_y y^2}{R} \sin^2 \zeta \right]^{\frac{1}{2}}}$$

$$= \frac{(v_x v_y)^{1/2} A}{2R \sin \zeta \cos \zeta} \int_{-\pi/2}^{\pi/2} \frac{\cos \vartheta \, d\vartheta}{\left[1 - \sin^2 \vartheta \right]^{\frac{1}{2}}}$$

$$= \frac{\pi (v_x v_y)^{1/2} A}{R \sin 2\zeta} \, , \qquad (1.9)$$

where we have put

$$w = 2J_y \sin^2 \zeta \, ,$$

$$p_{x1} = \frac{v_x^{1/2}}{R^{1/2} \cos \zeta} \left[J_0 - \frac{v_x x^2 \cos^2 \zeta}{R} - \frac{v_y y^2 \sin^2 \zeta}{R} \right]^{\frac{1}{2}} \, , \qquad (1.10)$$

$$p_x = p_{x1} \sin \vartheta \, .$$

The density is constant (within the beam), which implies that the space charge force is linear, as we will see below. In evaluating the second line of Eq.(1.9), we have assumed that there are values of p_x, p_y for which the argument of the δ-function in Eq.(1.8) vanishes. This will be true provided the point x,y lies within the ellipse

$$\frac{v_x x^2 \cos^2 \zeta}{R} + \frac{v_y y^2 \sin^2 \zeta}{R} = J_o \, . \qquad (1.11)$$

Outside this ellipse, the particle density is zero. The area of this ellipse is

$$A_e = \frac{2\pi R J_0}{\left[v_x v_y \right]^{1/2} \sin 2\zeta} \, . \qquad (1.12)$$

Since the density (1.9) inside the ellipse is uniform, it is just the number N of particles per unit length divided by the area of the ellipse:

$$\rho(x,y) = \rho_0 = \frac{N[v_x v_y]^{1/2} \sin 2\zeta}{2\pi R J_0} \ . \qquad (1.13)$$

The electric field due to this particle density can be written

$$\mathbf{E} = -\nabla \phi \ , \qquad (1.14)$$

where the electric potential satisfies, inside the ellipse (1.11), the equation

$$\nabla^2 \phi = \frac{\partial^2 \phi}{\partial x^2} + \frac{\partial^2 \phi}{\partial y^2} = -\frac{e\rho_0}{\varepsilon_0} \ , \qquad (1.15)$$

and variations in the azimuthal direction are neglected. A solution of Eq.(1.15) is

$$\phi(x,y) = \phi_0 - \frac{1}{2}\kappa_x^2 x^2 - \frac{1}{2}\kappa_y^2 y^2 \ , \qquad (1.16)$$

where the constants κ_x and κ_y must satisfy

$$\kappa_x^2 + \kappa_y^2 = \frac{e\rho_0}{\varepsilon_0} \ . \qquad (1.17)$$

This condition can be satisfied by setting

$$\kappa_x = \kappa \cos \xi \ , \quad \kappa_y = \kappa \sin \xi \ , \qquad (1.18)$$

where ξ is arbitrary and

$$\kappa^2 = \frac{e\rho_0}{\varepsilon_0} = \frac{Ne[v_x v_y]^{1/2} \sin 2\zeta}{2\pi \varepsilon_0 R J_0} \ . \qquad (1.19)$$

The general solution of Eq.(1.15) is obtained by adding to the solution (1.16) the general solution of the homogeneous equation

$$\frac{\partial^2 \phi}{\partial x^2} + \frac{\partial^2 \phi}{\partial y^2} = 0 \qquad (1.20)$$

which is well-behaved at the origin. This must then be matched to a solution of Eq.(1.20) between the beam and the vacuum chamber. Since we are only looking for some self-consistent solution of the problem, we simplify the problem by taking a round beam:

$$\tan^2 \zeta = \frac{v_x}{v_y} \quad , \tag{1.21}$$

which makes Eq.(1.11) the equation for a circle:

$$x^2 + y^2 = a^2 = \frac{v_x + v_y}{v_x v_y} RJ_0 \quad , \tag{1.22}$$

where a is the radius of the beam. Note that we have made the beam round without assuming that $v_x = v_y$. If they are equal then $\zeta = \pi/4$. In any case, we set $\xi = \pi/4$, so that the potential has circular symmetry. Inside the beam, the potential (1.16) is then

$$\phi(x,y) = \phi(r) = \phi_0 - \frac{1}{4}\kappa^2(x^2 + y^2) = \phi_0 - \frac{1}{4}\kappa^2 r^2 \quad . \tag{1.23}$$

Outside, the potential is

$$\phi(r) = C \ln \frac{r}{b} \quad , \tag{1.24}$$

where the additive constant is chosen to make the potential vanish at the vacuum chamber radius b. The potential and its radial derivative have to be continuous across the boundary. The potential is then

$$\phi(x,y) = \begin{cases} \dfrac{ea^2 \rho_0}{4\varepsilon_0}\left(1 + 2\ln\dfrac{b}{a}\right) - \dfrac{e\rho_0}{4\varepsilon_0}(x^2 + y^2) & \text{inside the beam} \quad , \\ -\dfrac{ea^2 \rho_0}{2\varepsilon_0} \ln \dfrac{r}{b} & \text{outside} \quad , \end{cases} \tag{1.25}$$

where

$$\rho_0 = \frac{N v_x v_y}{\pi R J_0 (v_x + v_y)} \quad . \tag{1.26}$$

The space charge forces in the non-relativistic limit (i.e., neglecting magnetic self-forces) are

$$F_x = -e\frac{\partial \phi}{\partial x} = \frac{e^2 \rho_0}{2\varepsilon_0} x \quad ,$$
$$F_y = -e\frac{\partial \phi}{\partial y} = \frac{e^2 \rho_0}{2\varepsilon_0} y \quad . \tag{1.27}$$

The constants k_x, k_y in Eq.(1.1) can now be written in terms of the external focussing force constants k_{ex}, k_{ey}:

$$k_x = k_{ex} - \frac{e^2 \rho_o}{2\varepsilon_0 \beta^2 \gamma^2 mc^2} ,$$

$$k_y = k_{ey} - \frac{e^2 \rho_o}{2\varepsilon_0 \beta^2 \gamma^2 mc^2} ,$$

(1.28)

where β and γ are the relativistic parameters, whose variation with s are neglected, m is the mass, and an extra factor γ is added to the denominators to include the effects of the magnetic forces.

We now have a complete, self-consistent solution of the equations of motion for a round beam, including the effects of space charge forces.

2. The KV Solution for an AG Ring.

We will derive the general Kapchinskij-Vladimirskij solution for an elliptical beam in an alternating gradient ring following the same steps as in the treatment of the simpler problem above. Our treatment is a generalization of the KV paper (1), since they eventually assume a round beam. We start with the Hamiltonian (1.1), but we allow the force coefficients to depend (periodically) on s:

$$H(x, p_x, y, p_y) = \frac{1}{2} p_x^2 + \frac{1}{2} p_y^2 + \frac{1}{2} k_x(s) x^2 + \frac{1}{2} k_y(s) y^2 .$$ (2.1)

We will assume that the wavelengths for the variations of the functions $k_x(s)$, $k_y(s)$ are much longer than the cross sectional dimensions of the beam, so that the fields can be calculated treating the beam as a uniform elliptical cylinder at each azimuth s. The action variables are the Courant-Snyder invariants:

$$J_x = \frac{(\beta_x(s) p_x + \alpha_x(s) x)^2 + x^2}{\beta_x(s)} ,$$

$$J_y = \frac{(\beta_y(s) p_y + \alpha_y(s) y)^2 + y^2}{\beta_y(s)} ,$$

(2.2)

where the parameters $\alpha(s)$ and $\beta(s)$ are periodic functions of s.

We again write the generalized microcanonical distribution in the form (1.8). The calculation of the spatial density proceeds just as in the preceding section, and we get

$$\rho(x,y) = \int dp_x \int dp_y \, D(x,p_x,y,p_y)$$
$$= \frac{\pi A}{\left[\beta_x(s)\beta_y(s)\right]^{1/2} \sin 2\zeta} \quad , \tag{2.3}$$

within the ellipse

$$\frac{x^2 \cos^2 \zeta}{\beta_x(s)} + \frac{y^2 \sin^2 \zeta}{\beta_y(s)} = J_0 \quad , \tag{2.4}$$

and zero outside. The density is again uniform within the ellipse, but varies periodically in s, as does the area of the ellipse which is

$$A_e = \frac{2\pi J_0 \left[\beta_x(s)\beta_y(s)\right]^{\frac{1}{2}}}{\sin 2\zeta} \quad . \tag{2.5}$$

We will neglect any variation in the azimuthal velocity so that the linear density N (particles per unit length along s) is a constant of the motion. For a bunched beam, N may vary along the bunch, but remains constant at the location of any given particle at least for many revolutions, so its variation may be neglected in studying the betatron oscillations. The spatial density is then

$$\rho(x,y,s) = \rho_0(s) = \frac{N \sin 2\zeta}{2\pi J_0 \left[\beta_x(s)\beta_y(s)\right]^{\frac{1}{2}}} \quad . \tag{2.6}$$

We have to solve Eq.(1.15) which will be written in the form

$$\frac{\partial^2 \phi}{\partial x^2} + \frac{\partial^2 \phi}{\partial y^2} = \begin{cases} -\kappa^2(s) & \text{inside the beam} \\ 0 & \text{outside} \end{cases} \quad , \tag{2.7}$$

where $\kappa^2(s)$ is given by Eq.(1.17), with $\rho_0(s)$ given by Eq.(2.6), and the beam boundary is given by Eq.(2.4). We are assuming that the dependence on s is slow, so we neglect derivatives with respect to s.

In order to solve Eq.(2.7), one could write a solution in the form (1.16) or (1.23) inside the beam and try to fit the boundary condition at the wall and at the beam boundary by adding suitable solutions of the homogeneous equation inside and outside. Instead, since the beam boundary is an ellipse, we will use confocal elliptic coordinates (3, p.1195):

$$\begin{aligned} x &= h \cosh \mu \cos \lambda \quad , \\ y &= h \sinh \mu \sin \lambda \quad , \end{aligned} \tag{2.8}$$

which gives ellipses of constant μ and hyperbolas of constant λ with foci at $x = \pm h(s)$, $y = 0$. The coordinate μ runs from 0 to ∞. The coordinate λ is an angle from 0 to 2π and is roughly equal to the polar angle θ. The (positive, negative) x-axis is given (outside the foci) by $\lambda = 0, \pi$; the y-axis is given by $\lambda = \pm \pi/2$. We choose as coordinate foci the foci of the ellipse (2.4):

$$h(s) = \left[\frac{\beta_x(s)J_0}{\cos^2 \zeta} - \frac{\beta_y(s)J_0}{\sin^2 \zeta}\right]^{\frac{1}{2}}, \qquad (2.9)$$

so that the ellipse (2.4) is an ellipse of constant $\mu = \mu_b$:

$$\frac{x^2}{h^2 \cosh^2 \mu_b} + \frac{y^2}{h^2 \sinh^2 \mu_b} = \frac{x^2 \cos^2 \zeta}{\beta_x J_0} + \frac{y^2 \sin^2 \zeta}{\beta_y J_0} = 1, \qquad (2.10)$$

from which Eq.(2.9) follows. The elliptic coordinate of the beam ellipse is given by

$$\tanh \mu_b(s) = \left[\frac{\beta_y(s)}{\beta_x(s)}\right]^{\frac{1}{2}} \cot \zeta . \qquad (2.11)$$

We will usually omit explicit dependences on s, except when introducing a new quantity. Note that the ellipse of constant μ approaches a circle as μ becomes large (so that $\sinh \mu \doteq \cosh \mu$), and that the eccentricity approaches 1 for small μ. For $\mu = 0$ the ellipse shrinks to the line segment connecting the foci. The major axis of the beam ellipse is taken to be horizontal.

We will assume that the conducting vacuum chamber wall is also elliptical and confocal with the beam. The potential vanishes at the wall. In elliptic coordinates Eq.(2.7) becomes (3, p.504)

$$\frac{1}{h^2(\cosh^2 \mu - \cos^2 \lambda)}\left(\frac{\partial^2 \phi}{\partial \mu^2} + \frac{\partial^2 \phi}{\partial \lambda^2}\right) = \begin{cases} -\kappa^2 & \text{inside the beam} \\ 0 & \text{outside} \end{cases}, \qquad (2.12)$$

This equation is to be solved keeping ϕ and its normal derivative continuous across the beam boundary, and with $\phi = 0$ at the wall which we take to be the ellipse $\mu = \mu_w$. A particular solution inside can be found either by solving Eq.(2.12) by separation of variables or by taking the solution (1.23) and substituting from Eq.(2.8). The result is

$$\phi = -\frac{\kappa^2 h^2}{4}(\cosh 2\mu + \cos 2\lambda) . \qquad (2.13)$$

To this we add a solution of the homogeneous equation inside, and another outside. A set of solutions of the homogeneous equation periodic in λ is

$$\phi = 1, \quad \phi = \mu,$$
$$\phi = \cosh m\mu \cos m\lambda,$$
$$\phi = \sinh m\mu \sin m\lambda, \qquad (2.14)$$
$$\phi = \sinh m\mu \cos m\lambda,$$
$$\phi = \cosh m\mu \sin m\lambda,$$

where m is any positive integer. The second and the last two solutions are not well behaved at the origin, due in part to the peculiar behavior of the coordinate system near $\mu = 0$. The rest are polynomials of order m in x and y. It is clear from Eq.(2.13) that we need the solutions for $m = 0$ (the first two) and $m = 2$. We therefore write

$$\phi(\mu,\lambda) =$$

$$\begin{cases} \dfrac{\kappa^2 h^2}{4}\left[\cosh 2\mu + \cos 2\lambda - A + B\cosh 2\mu \cos 2\lambda\right] \text{ inside,} \\ \dfrac{\kappa^2 h^2}{4}\left[C(\mu_w - \mu_b) - D(\cosh 2\mu \sinh 2\mu_w - \sinh 2\mu \cosh 2\mu_w)\cos 2\lambda\right] \text{ outside,} \end{cases} \qquad (2.15)$$

where the coefficients are already adjusted to satisfy the boundary condition at the wall. We have to require that ϕ and $\partial\phi/\partial\mu$ be continuous at the beam ellipse; the result is

$$A = 2(\mu_w - \mu_b)\sinh 2\mu_b + \cosh 2\mu_b,$$
$$B = \frac{\cosh 2\mu_b \cosh 2\mu_w - \sinh 2\mu_b \sinh 2\mu_w}{\cosh 2\mu_w},$$
$$C = 2\sinh 2\mu_b, \qquad (2.16)$$
$$D = \frac{\sinh 2\mu_b}{\cosh 2\mu_w}.$$

For comparison with the development in the previous section we would need the circular limit of the above equations where $h \to 0$ and $\mu \gg 1$ almost everywhere. In that limit $\cosh \mu$ and $\sinh \mu$ approach $(e^\mu)/2$, the ellipses become circles, and λ becomes the polar angle θ. In that limit, $\mu \to \ln 2r/h$.

The solution (2.15) inside the beam can be written

$$\phi = -\frac{\kappa^2 h^2}{4}(A+B) + \frac{1}{2}\kappa^2(x^2+y^2) + \frac{B}{2}\kappa^2(x^2-y^2), \qquad (2.17)$$

from which the electric space charge forces follow:

$$F_x = -e\kappa^2(1+B)x \ ,$$
$$F_y = -e\kappa^2(1-B)y \ . \tag{2.18}$$

The coefficients in Eq.(2.1) can now be written in terms of the external focussing coefficients and the space charge forces:

$$k_x(s) = k_{ex}(s) - \frac{e\kappa^2(s)(1+B(s))}{\beta^2\gamma^2 mc^2} \ ,$$
$$k_y(s) = k_{ey}(s) - \frac{e\kappa^2(s)(1-B(s))}{\beta^2\gamma^2 mc^2} \ . \tag{2.19}$$

We now have a complete, self-consistent solution of the general KV problem. If the vacuum chamber wall is not an ellipse or is not confocal with the beam boundary then matching boundary conditions becomes more difficult. It may be necessary to add terms with $m > 2$ to the solution, in which case terms in x and y of order higher than two will appear in the solution (2.17). There will then be nonlinear terms in the forces (2.18) and our solution is no longer self-consistent. However for a reasonable wall shape one would expect such terms to be small, especially inside the beam. In any case, if beam and vacuum chamber are circular, the distribution (2.4) will result in linear space charge forces. For a circular beam in a concentric circular vacuum chamber, the KV distribution (2.4) always leads to linear focussing forces.

3. The Painting Scenario.

The KV distribution is essentially a microcanonical distribution with the beam distributed uniformly over a three-dimensional energy shell corresponding to a fixed total energy in the four-dimensional phase space of the x and y betatron oscillations. We need to construct a scenario which allows us to paint the energy shell uniformly. To simplify the treatment, our discussion will be based on the treatment in Section 1 which starts from the smoothed Hamiltonian (1.1).

If we inject at a fixed point in the phase space, the betatron oscillations will spread the beam over the γ_x, γ_y phase plane. In order to spread it over the three-dimensional surface defined by Eq.(1.8), we need to vary the action variables in an appropriate way. To this end introduce the variables

$$J_0 = 2\cos^2\zeta J_x + 2\sin^2\zeta J_y \ ,$$
$$J_m = 2\cos^2\zeta J_x - 2\sin^2\zeta J_y \ . \tag{3.1}$$

The Jacobian of this transformation is constant. Therefore if area is conserved in the J_x, J_y phase plane then it is also conserved in the J_0, J_m phase plane.

The total action J_0 is to be held constant and J_m is to be varied slowly. If the variation of J_m is slow compared with the betatron frequencies, then near each value of J_m the betatron motion will distribute the injected beam uniformly over the γ_x, γ_y phase plane, provided there is no rational relation with small denominator between ν_x and ν_y. The J_0 shell must be painted uniformly, so we require that dJ_m/dt be constant:

$$J_m = -J_0\left(1 - \frac{2t}{T}\right), \qquad (3.2)$$

where T is the total injection time. Note that we want to paint both positive and negative values of J_m. Equation (3.2) is adjusted for the case in which $J_m = -J_0$ initially, i.e., the y amplitude is maximum and the x amplitude is zero. The injected x, y actions are given by

$$J_x = \frac{1}{2\cos^2\zeta} J_0 \frac{t}{T},$$
$$J_y = \frac{1}{2\sin^2\zeta} J_0 \left(1 - \frac{t}{T}\right). \qquad (3.3)$$

The painting scenario can be achieved in the IPNS upgrade by using H(-) injection with a stripping foil, an internal horizontal orbit bump, and an external vertical deflection of the injected beam.

4. The Coupling Scenario.

Yanglai Cho (4) has pointed out that coupling the x and y betatron motions may allow us to achieve a KV distribution. He proposes to make the x and y betatron tunes equal and provide a small coupling between them. Then inject with zero y amplitude and a large fixed x amplitude. The coupling causes the y oscillation energy to increase at the expense of the x energy. This has two effects. First, it causes the previously injected beamlet to move away from the inflector and remain away for one beat period, thus facilitating multi-turn injection. Second, it results in a distribution in which all particles have the same total oscillation energy.

Unfortunately this procedure does not result in a microcanonical distribution, since it does not fill the energy shell uniformly. It fills only a two-dimensional torus scanned by the phases of the two coupled normal modes. Filling the three-dimensional energy shell requires also sweeping a suitably chosen variable analogous to J_m in Eq.(3.1). We have carried out the analysis [(5), Section 4] and have carried out corresponding simulations for this scenario. The results are similar to those presented later for the painting scenario. Since the coupling scenario seems to have no advantages over the painting scenario, we omit further discussion in this paper.

III. DOES IT WORK? — SIMULATION

5. Injecting with Painting Scenario.

We have written a program to simulate the scenario (3.3) as applied to the IPNS Upgrade (2). The injection time T corresponds to 500 injected turns. The maximum injected x-amplitude is 50 mm   Figure (5.1) shows the x and y amplitudes of each beamlet as it is injected. The points lie on a circle beginning with zero x amplitude and maximum y amplitude at turn 1 and ending with zero y amplitude and maximum x amplitude at turn 500.

Figure (5.2) shows the resulting density in xy space at the end of injection. Each of the small circles represents one injected beamlet. The spatial density is fairly uniform within a circle. Figure (5.3) shows the final space charge shifted horizontal tunes of the 500 beamlets. The vertical tunes are similar. In these calculations, the space charge forces between beamlets are omitted, except at the end of the injection process when we turn on the interaction forces for one turn in order to calculate the space charge shifted tunes resulting from the density shown in Fig.(5.2).

6. Injecting with Non-KV Scenarios.

Figure (6.1) shows the injected amplitudes for a non-KV scenario. It differs from that shown in Fig.(5.1) in that the sum of the amplitudes is held constant instead of the sum of the actions (proportional to amplitudes squared). Although Figs.(5.1) and (6.1) are not much different, the resulting density distribution shown in Fig.(6.2), in contrast to that in Fig.(5.2), is neither circular nor uniform. Likewise the space-charge shifted tunes after injection, shown in Fig.(6.3) are not all equal as in Fig.(5.3).

IV. DOES IT REALLY WORK? — SIMULATION WITH SPACE CHARGE FORCES.

7. Simulation and Tune Measurement.

In order to include the effect of space charge forces, we calculate at each integration step the total force on each beamlet due to each of the other beamlets. In this way we include not just the Vlasov term, containing the smoothed out space charge force, but also the fluctuating beamlet-beamlet forces. In addition the equations of motion include for each beamlet terms like that on the right side of Eq.(7.2) below, to drive the resonance $v = 5.5$ for both x and y motions. The force between two beamlets is inversely proportional to the distance between them unless they overlap, in which case, the force drops linearly to zero as their centers approach one another.

In order to find the tune of a simulated beamlet, we find the average space charge force over one turn in the following way. We assume that we may

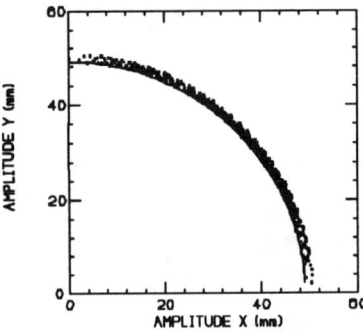

FIGURE 5.1 Painting scenario - Injection amplitudes.

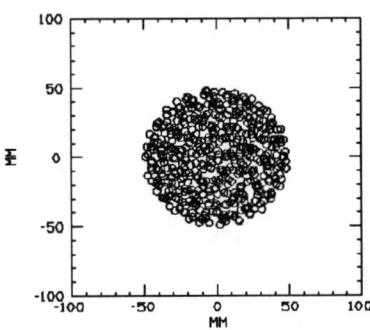

FIGURE 5.2 Painting scenario - Final density.

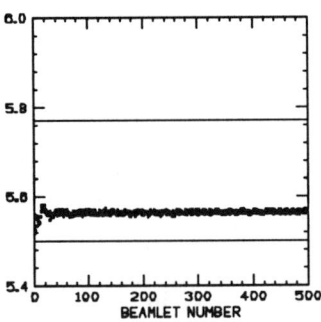

FIGURE 5.3 Painting scenario - Final x tunes.

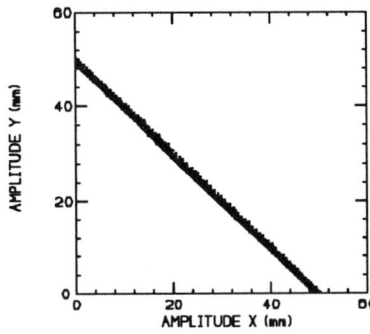

FIGURE 6.1 Non-KV scenario - Injection amplitudes.

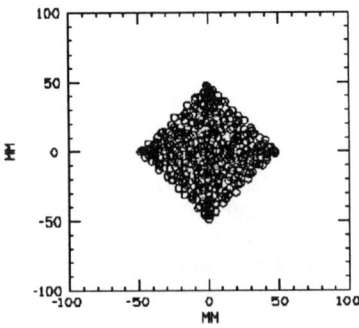

FIGURE 6.2 Non-KV scenario - Final density.

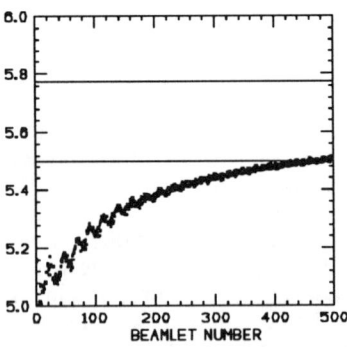

FIGURE 6.3 Non-KV scenario - Final x tunes.

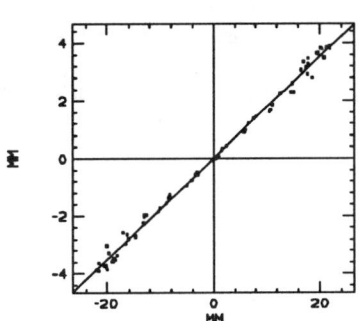
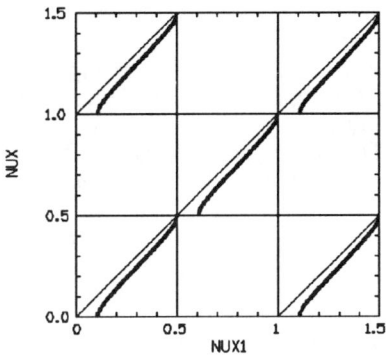

FIGURE 7.1 Force vs position for a beamlet during one turn.

FIGURE 7.2 Tune vs. tune parameter.

approximate the average space charge force on any beamlet by a linear function of the displacement x (horizontal or vertical). The integration time step is Δs. At each integration step, we calculate the total momentum increment Δp of any given beamlet due to the forces from all other beamlets. Figure (7.1) shows a plot of $R \Delta p_x$ vs x for a typical beamlet at each integration step during one turn. The least squares linear fit to the data is also shown in the figure and is written in the form

$$\frac{\Delta p}{\Delta s} = Ax + B \ . \tag{7.1}$$

We then assume we may approximate the equation of motion by the linear equation

$$x'' + R^{-2}v_0^2 x - Ax = B + ax\delta(s - \pi R) \ , \tag{7.2}$$

where $R^{-2}v_0^2 x$ is the mean focussing from the lattice structure and the last term represents a quadrupole error term which is introduced to drive a possible half-integral resonance. The delta function is periodic with period $2\pi R$. The quadrupole bump is placed half way around the ring so that the reference point $s = 0$ is a symmetry point for the bumped lattice. Since the equilibrium orbit $x=x_e$ must satisfy Eq.(7.2), the deviation from the equilibrium orbit satisfies the homogeneous linear equation

$$(x - x_e)'' + R^{-2}(v_0^2 - S)(x - x_e) = a(x - x_e)\delta(s - \pi R) \ , \tag{7.3}$$

where

$$S = AR^2 \qquad (7.4)$$

is the space charge defocussing coefficient. For algebraic convenience we will henceforth take x to be the deviation from the bumped equilibrium orbit and replace x-x_e by x.

The phase vector is carried from $s=0$ to $s=\pi R$, (from the reference point to just before the bump), via a matrix **A**:

$$\begin{pmatrix} x \\ p \end{pmatrix}_{\pi R} = \mathbf{A} \begin{pmatrix} x \\ p \end{pmatrix}_0 . \qquad (7.5)$$

The matrix **A** is given by

$$\mathbf{A} = \begin{pmatrix} \cos\frac{\sigma_1}{2} & \sin\frac{\sigma_1}{2} \\ -\sin\frac{\sigma_1}{2} & \cos\frac{\sigma_1}{2} \end{pmatrix}, \qquad (7.6)$$

where

$$\sigma_1 = 2\pi\left(v_0^2 - S\right)^{1/2} \qquad (7.7)$$

is the phase advance for the normal lattice plus space charge but without the gradient bump. We will call the quantity $(v_0^2 - S)^{1/2}$ the (horizontal or vertical) *tune parameter*. It is the space charge shifted tune in the absence of any resonance driving term. The matrix which carries the phase vector across the bump at $s=\pi R$ is

$$\mathbf{B} = \begin{pmatrix} 1 & 0 \\ -a & 1 \end{pmatrix} . \qquad (7.8)$$

The matrix which carries the phase vector once around the ring is then

$$\mathbf{M} = \mathbf{ABA} = \begin{pmatrix} \cos\sigma_1 - \frac{a}{2}\sin\sigma_1 & \sin\sigma_1 - \frac{a}{2}(1-\cos\sigma_1) \\ -\sin\sigma_1 - \frac{a}{2}(1+\cos\sigma_1) & \cos\sigma_1 - \frac{a}{2}\sin\sigma_1 \end{pmatrix} . \qquad (7.9)$$

The trace of **M** gives the phase advance σ around the ring:

$$\cos\sigma = \cos\sigma_1 - \frac{a\sin\sigma_1}{2} . \qquad (7.10)$$

If we consider σ as a function of σ_1 (or of the tune parameter) there will be unstable stop bands at integer and half integer resonances, i.e. at $\sigma=2n\pi$, where n is an integer or half-integer. Let us assume that a is small and neglect all but the lowest order terms in a. If $a=0$, a solution of Eq.(7.10) is $\sigma=\sigma_1$. For small a Eq.(7.10) may be written in the form

$$\cos\sigma_1\left(1-\frac{(\sigma-\sigma_1)^2}{2}\right)-(\sigma-\sigma_1)\sin\sigma_1 = \cos\sigma_1 - \frac{a\sin\sigma_1}{2} \quad , \qquad (7.11)$$

where we have kept terms of order $(\sigma-\sigma_1)^2$ since near the stop bands $\sin\sigma_1$ is small of order a and all terms in Eq.(7.11) are second order. The solution of Eq.(7.11) is

$$\sigma = \sigma_1 - \tan\sigma_1\left(1\pm\left[1+\frac{a}{\tan\sigma_1}\right]^{\frac{1}{2}}\right) . \qquad (7.12)$$

Away from the resonance (i.e., $\tan\sigma_1 \gg a$) the solution (7.12) is, to lowest order in a,

$$\sigma = \sigma_1 + \frac{a}{2} \quad . \qquad (7.13)$$

This solution is valid away from the integer and half-integer resonances. There is a second solution but it is not valid since it corresponds to $\sigma-\sigma_1 \gg a$. The edges of the stop bands occur where the solution of Eq.(7.10) is $\cos\sigma = \pm 1$. One edge will be at $\sigma_1 = 2n\pi$. The other edge, to first order in a, occurs where the square root in Eq.(7.12) vanishes, at

$$\sigma_1 \doteq 2n\pi - a \quad . \qquad (7.14)$$

In the stop band the solution of the equation of motion has the form

$$x = e^{\pm\Gamma}e^{\pm 2\pi i n s/R} \quad , \qquad (7.15)$$

with a growth rate approximately

$$\Gamma \doteq \left[-(\sigma_1-2n\pi)(\sigma_1-2n\pi+a)\right]^{1/2} \doteq a/2 \quad , \qquad (7.16)$$

where the last member is the growth rate at the center of the stop band. Since the growth rate has a vertical slope as a function of σ_1 at the edges of the stop band, it is roughly equal to $a/2$ throughout most of the stop band.

In Fig.(7.2) $v=\sigma/2\pi$ is plotted as a function of the tune parameter $v_1=\sigma_1/2\pi=(v_0^2-S)^{1/2}$. Note that according to Eq.(7.10) σ is a periodic function of σ_1. Figure (7.3) is a typical plot of the calculated shifted tune v_y of a beamlet as a function of time. In this case a number of the calculated values lie in the stop band and are plotted at the top of the figure. In order to include values which lie in the

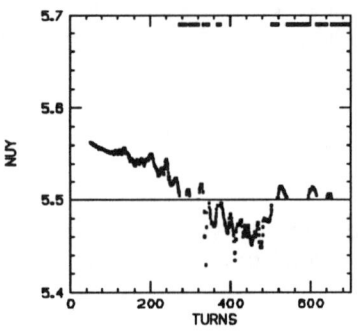
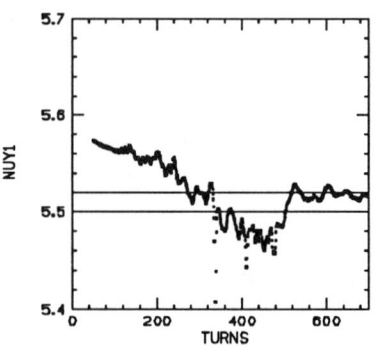

FIGURE 7.3 Tune vs. time for a typical beamlet.

FIGURE 7.4 Tune parameter vs. time for the beamlet of Fig.(7.3).

stop band, we will generally plot the tune parameter $v_1 = (v_0^2-S)^{1/2}$ as in Fig.(7.4), which also shows the edges of the stop band. Outside the stop band the tune parameter is nearly equal to the actual tune. Inside the stop band the motion is unstable with a growth rate given by Eq.(7.16). Note that there are substantial fluctuations of the calculated tunes. These are due to the fluctuating character of the turn-to-turn space charge forces. In the tune calculation the actual space charge forces are replaced with mean linear approximations which are also subject to fluctuations from turn to turn.

8. Effect of Beamlet-Beamlet Forces.

The beam density for the KV scenario [Fig.(5.1)] with the space charge forces included is shown in Fig.(8.1). The beam is still roughly circular, but is not as

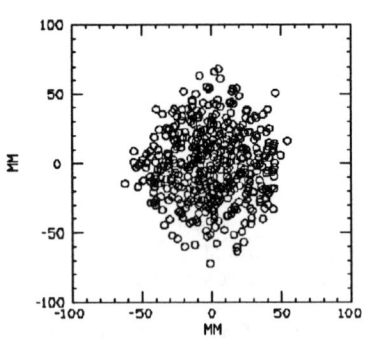
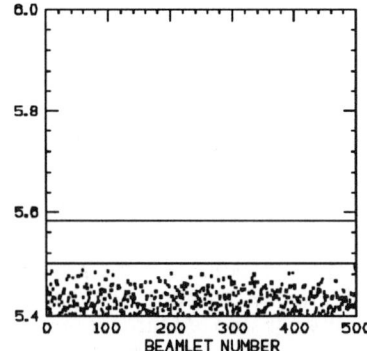

FIGURE 8.1 Final density for KV scenario with space charge.

FIGURE 8.2 Final x tunes for KV scenario with space charge.

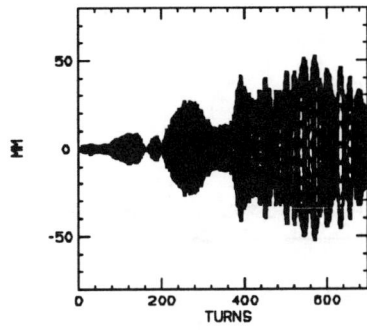
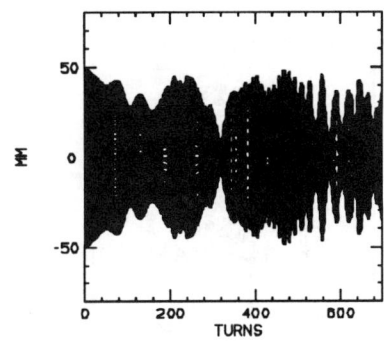

FIGURE 8.3 x coordinate during KV injection with space charge.

FIGURE 8.4 y coordinate during KV injection with space charge.

uniform as in Fig.(5.2). As a result, the final tunes shown in Fig.(8.2) are not as constant as in Fig.(5.3). During injection, the beam does not yet have a KV distribution, so there are nonlinear space charge forces. There are also coupling forces between beamlets whose effects can be seen in Figs.(8.3) and (8.4) which show the x and y coordinates of the first injected beamlet as it passes the reference point during injection. Either the nonlinear space charge forces or diffusion due to beamlet-beamlet forces may be responsible for the non-uniformity of the beam in Fig.(8.1).

Figures (8.5) and (8.6) show the average x and y tunes, averaged over all injected beamlets, vs. turn number during the injection process. The two outer curves in these figures are the rms deviations from the average tunes. The increasing depression of the tunes due to the increasing space charge forces is evident. The tunes are depressed below the resonance at $\nu = 5.5$ because there is as yet no term in the simulation to drive the resonance.

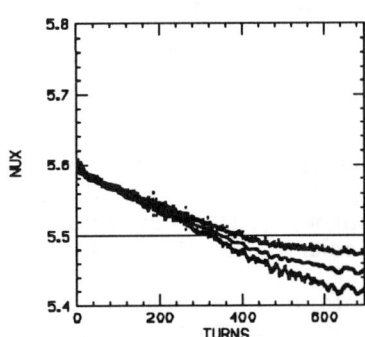
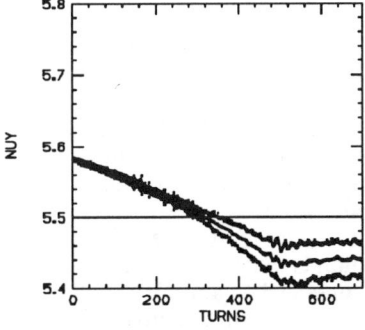

FIGURE 8.5 x tune during KV injection with space charge.

FIGURE 8.6 y tune during KV injection with space charge.

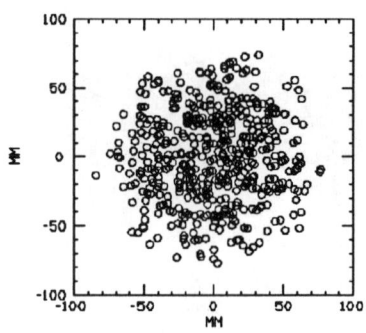 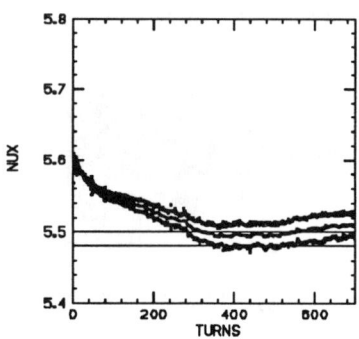

FIGURE 9.1 Final density for KV injection with resonance included.

FIGURE 9.2 x tune parameters during KV injection with resonance included.

9. Effect of an Imperfection Resonance.

Using the same injection scenario with a gradient bump included to drive the resonance $\nu=5.5$, the final density distribution is shown in Fig.(9.1) The x tunes during injection are shown in Fig.(9.2). The effect of the resonance on the tune history can be clearly seen. The resonance causes the beam density to expand to keep the tunes out of the stop band. The total injected current for this case is 27 A, with a bunching factor of 0.75. This is greater than required to depress the tune to the resonance and hence exceeds the conventionally defined space charge limit. We have also seen cases with large injected beam currents where the tune changes so rapidly that it can cross the resonance before the beam has time to expand.

10. The Space Charge Limited Case.

In a realistic case where we wish to inject the maximum possible beam without seriously increasing the beam size, we would choose an initial tune as far from the half-integral resonance as possible, and inject just enough beam to reduce the tunes to the edges of the stop bands. This corresponds to the conventional definition of the space charge limit. Figures (10.1), (10.2) and (10.3) show the final density and the tune history for this case, following the KV scenario (5.1). The total injected current is 54 A. The simulated tune shifts in Figs.(10.2) and (10.3) are equal to those calculated from the Laslett formula, as they should be if the simulation is done correctly.

Figures (10.4) and (10.5) show the same case for injection with the non-KV scenario. The final density is not much different, although the approach to resonance is more rapid in this case. For this case which starts far from the resonance and with a large injected beam, it would appear that the final density distribution is not dominated by the resonance, but instead is dominated by either the beamlet-beamlet collisions or the nonlinearities in the space charge forces or both. To illustrate this, we show in Fig.(10.6) the final density for the same case

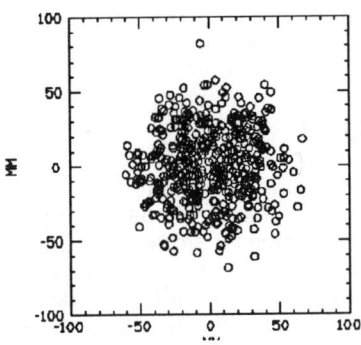

FIGURE 10.1 Final density, KV injection, space charge limited case.

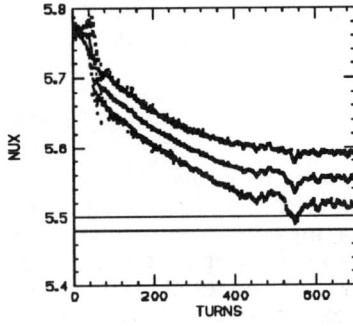

FIGURE 10.2 x tune parameter during KV injection, space charge limited case.

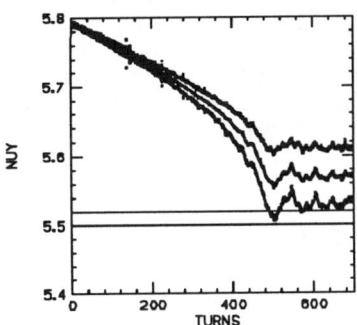

FIGURE 10.3 y tune parameter during KV injection, space charge limited case.

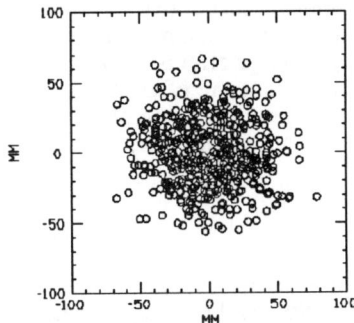

FIGURE 10.4 Final density, non-KV injection, space charge limited case.

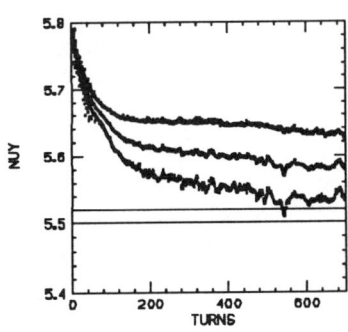

FIGURE 10.5 x tune parameter during non-KV injection.

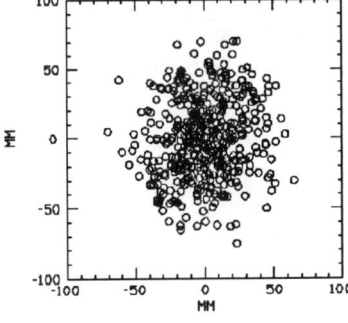

FIGURE 10.6 Final density, same case as Fig.(10.4), without resonance.

but with the bump that drives the resonance turned off. There is little difference in the final density in Figs.(10.4) and (10.6).

V. CONCLUSIONS.

We have presented the theory of the KV distribution, including alternating gradient effects, and including the case of an elliptical beam. We have presented practical injection scenarios which lead to KV distributions if space charge forces are neglected during injection. The resulting distributions are uniform and circular (or elliptical), and result in uniform space charge shifted tunes for all particles.

When the effects of space charge and of beamlet-beamlet forces are included, injection with a KV scenario may have some advantage, but the resulting distribution is not exactly a KV distribution and the density is not exactly uniform. Two regimes may be distinguished. If the initial tunes are close to the resonance, the final density distribution is dominated by the amplitude growth of particles in the resonance stop band. This growth limits the space charge detuning so that the final tunes lie just above the stop band. If the initial tunes are far from resonance, and the injected beam intensity is large, the final density distribution is dominated by space charge effects — nonlinear forces and/or beamlet-beamlet collisions. This may mean that the effective space charge limit may sometimes occur at a beam intensity where the beam blow-up reaches the maximum acceptable value before the maximum acceptable tune shift is reached.

REFERENCES.

1. Kapchinskij,I. M. and Vladimirskij,V. V., "Limitations of Proton Beam Current in a Strong Focusing Linear Accelerator Associated with the Beam Space Charge," in *Proc. CERN Symposium on High Energy Accelerators, I* , 1959, pp. 274–288.
2. "IPNS Upgrade - A Feasibility Study," ANL-95/13, April 1995.
3. Morse, P.M. and Feshbach, H., *Methods of Theoretical Physics,* New York: McGraw-Hill, 1953.
4. Cho,Y., private communication.
5. Symon, K. R. "Kapshinskij-Vladimirskij Distribution," Argonne National Laboratory Neutron Source Accelerator Note NSA-95-5 (available on request).

Reducing space charge tune shift with a barrier cavity [1]

M. Blaskiewicz

AGS Dept. Brookhaven National Lab

Implementation of a barrier cavity rf system appears to be a straightforward and relatively inexpensive way to increase the output current of second stage synchrotrons. This note serves as a general introduction to the relevant beam dynamics and addresses the problem of driving a cavity in the short burst mode.

INTRODUCTION

A barrier cavity rf system (1,2) is proposed as a tool in the continuing effort to obtain higher beam currents. Unlike a single frequency rf system a barrier cavity rf system employs a single turn voltage waveform similar to that shown in Figure 1. Each of the pulses making up the waveform tends to repel particles resulting in two bunches separated by narrow gaps. By varying the amplitude and phase of the pulses some novel rf manipulations become possible.

SINGLE PARTICLE BEAM DYNAMICS

Before proceeding it will be useful to define some relevant coordinates. The azimuth θ refers to longitudinal position with respect to the physical machine and increases by 2π each turn. I assume the RF field is limited to a narrow range around $\theta = 0$, several gaps will be similar as long as the time delays are correct. The RF system creates an electric field whose component along the beam velocity is given by $V(t)\delta_p(\theta)/R$ where $V(t)$ is the gap voltage as a function of time, R is the machine radius and $\delta_p(\theta)$ is the periodic delta function. I assume a constant magnetic field and an ideal energy E_0 with corresponding angular frequency ω_0. A comoving coordinate is defined by the relative phase $\phi = \theta - \omega_0 t$. Let ϕ_n and E_n correspond to the relative phase and the total energy a particle has just after traversing the cavity for the nth time. To leading order and neglecting collective effects,

$$\phi_{n+1} - \phi_n = -\frac{2\pi\eta(E_n - E_0)}{\beta^2 E_0} \quad (1)$$

---

[1] Work performed under the auspices of the U.S. Department of Energy.

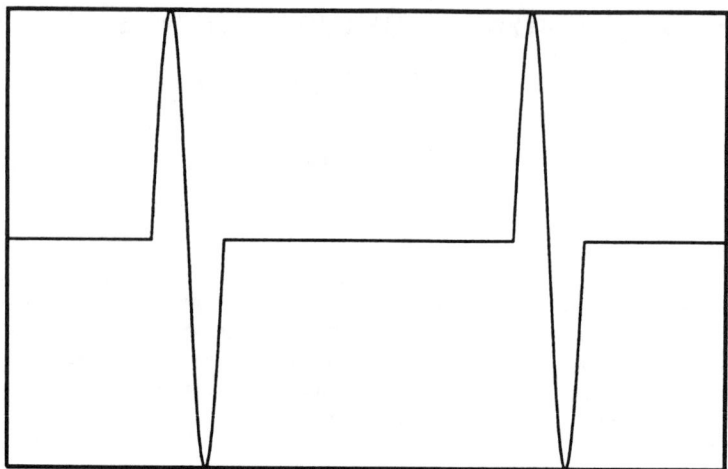

FIG. 1. Voltage .vs. time for one revolution period. The phases and amplitudes of the two components may vary slowly from one turn to the next.

$$E_{n+1} - E_n = qV(nT_0 - \phi_{n+1}/\omega_0) \tag{2}$$

where q is the charge of the particle, η is the frequency slip factor, $\beta = v/c$ and T_0 is the revolution period. The difference equations can be approximated by differential equations. With n as the smooth time-like variable the differential equations for coordinate ϕ with conjugate momentum E can be obtained from the Hamiltonian,

$$H(\phi, E) = -\frac{\pi\eta(E - E_0)^2}{\beta^2 E_0} + U(\phi). \tag{3}$$

The potential is given by

$$U(\phi) = -q \int_0^\phi V(x/\omega_0) dx. \tag{4}$$

Slow changes in the confining potential can be included in the usual way to produce $U(t, \phi)$. As long as $U(t, \phi)$ varies slowly enough in t, the adiabatic invariant given by $J = \oint E d\phi$, where the integral is over one period of the motion, will be nearly conserved. Figure 2 shows simulation results for a possible scenario using the AGS and AGS Booster which have a 4 : 1 circumference ratio. To make the connection with adiabatic invariants visible $\sqrt{H}$ and $\sqrt{U}$ are plotted. Figure 2a shows four Booster cycles worth of particles outside the barrier bucket and another Booster cycle has just been injected. In Figure 2b the injected particles have been squeezed yielding a similar energy distribution for particles inside and outside the bucket. In Figure 2c the second voltage component has collapsed. In 2d this component has returned in phase with the first. In Figures 2e and f the voltage components separate and grow making room for the next Booster cycle. In plot units, the initial peak momentum is 4 while the final peak momentum is 5 and the number of particles outside the barrier increases in the same ratio. There is very little emittance dilution during the process.

MULTIPARTICLE BEAM DYNAMICS

From a multiparticle standpoint a barrier cavity rf system has several features which distinguish it from a single frequency rf system. By optimizing the injection process it is possible to reduce the peak current without significantly diluting the longitudinal emittance. Additionally, the peak current in a barrier cavity system is time dependent, growing with each transfer. Reducing the peak current results in a smaller space charge tune shift which should reduce stopband losses. As an example consider a single bunch in the Booster having a parabolic line density of full length equal to half the Booster circumference. The gap in the AGS (cf. Figure 2) is taken to be the same length. Figure 3 shows the peak current in the AGS in units of Booster peak current assuming that the initial filamentation of the parabolic bunch in the gap is essentially instantaneous. The initial filamentation causes a fractional emittance increase of $4/\pi$ and reduces the peak current to 2/3 of the Booster value. Notice that the peak current for the stored bunch does not reach the initial injected intensity until the sixth Booster transfer and that the peak current for eight transfers always remains below the peak current in the Booster. After the injection process is completed the beam could be captured in a conventional rf system and quickly accelerated. The important point is that the time average space charge tune depression is greatly reduced with the barrier cavity system.

Coherent instabilities with a barrier cavity rf system are different from the sort found with a single frequency rf system. When the barrier half width is small compared to the bunch length (ϕ_b), the synchrotron tune (ν_s) is proportional to the fractional energy deviation,

$$\nu_s \approx \frac{\pi}{\phi_b \beta^2} \left| \eta \frac{E - E_0}{E_0} \right|. \tag{5}$$

The large spread in synchrotron tune implies a lot of Landau damping and in a first approximation the coasting beam stability criteria with the same peak current should give reasonable limits.

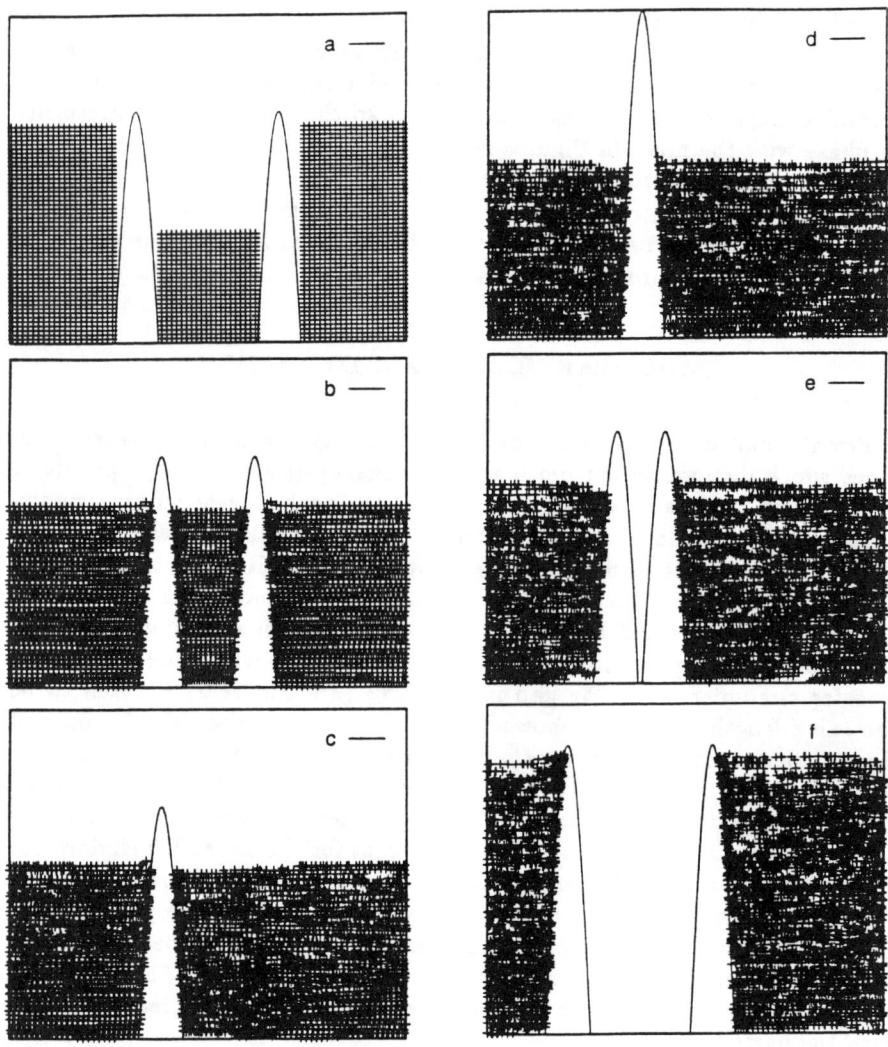

FIG. 2. Simulation of an injection scenario using a barrier cavity rf system.

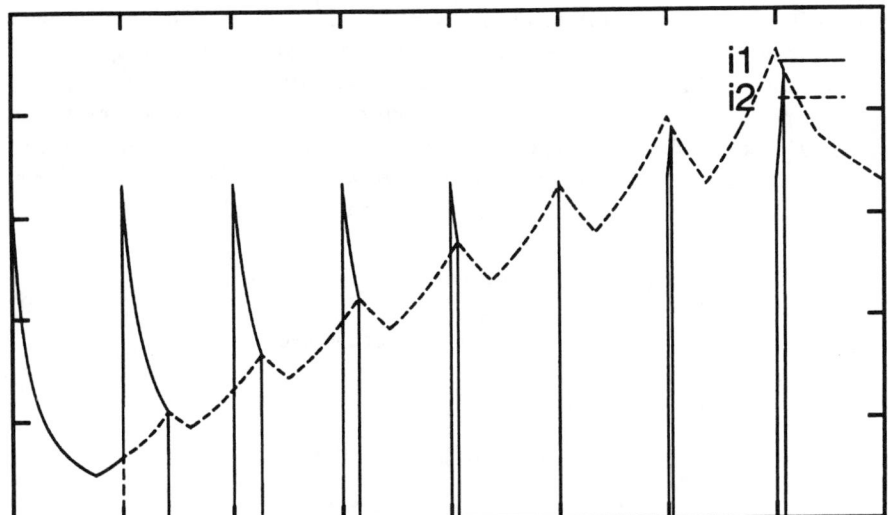

FIG. 3. (Peak current in AGS)/(Peak current in Booster) for eight Booster transfers.

RF CONSIDERATIONS

Generating an rf waveform like that shown in Figure 1 can be done in at least two ways. The first way is to approximate the waveform by a truncated Fourier series and to create the net voltage per turn using several single frequency rf systems operating at several revolution frequency harmonics. This appears to be a fairly complicated solution and will not be considered further.

Another way to generate the voltage waveform in Figure 1 is to assume that only a single type of rf cavity is available and to tailor the current pulse driving the cavity to create the desired voltage. For simplicity assume that the cavity impedance can be characterized by a single resonator

$$Z(\omega) = \frac{R_s}{1 + iQ(\omega_r/\omega - \omega/\omega_r)}, \tag{6}$$

where R_s is the shunt impedance, Q is the quality factor, and ω_r is the resonant frequency. For a current pulse $I(t)$ the voltage across the cavity is given by

$$V(t) = \int_0^\infty I(t-\tau)W(\tau)d\tau, \tag{7}$$

$$W(t) = \frac{\omega_r R_s}{\tilde{\omega} Q} \frac{d}{dt} \sin(\tilde{\omega} t) e^{-\omega_r t/2Q}, \tag{8}$$

where $W(t)$ is the wake potential and $\tilde{\omega} = \omega_r\sqrt{1 - 1/4Q^2}$. The cavity need not ring after the current pulse is turned off. Let $I_0(t)$ be some easily created current pulse of finite duration. Driving the cavity with a current $I(t) = I_0(t) + \exp[-\pi/\sqrt{(4Q^2 - 1)}]I_0(t - \pi/\tilde{\omega})$ results in no cavity voltage after the current $I(t)$ stops flowing. Another way to create a voltage pulse of finite duration is to define the voltage pulse required and then, via Fourier analysis, calculate the necessary current pulse. A particularly simple result is (4)

$$I(t) = \begin{cases} I_0\left(1 + \dfrac{\sin(\omega_r t)}{Q}\right), & 0 < t < 2\pi/\omega_r \\ 0, & \text{otherwise} \end{cases} \quad (9)$$

$$V(t) = \begin{cases} \dfrac{I_0 R_s}{Q}\sin(\omega_r t), & 0 < t < 2\pi/\omega_r \\ 0, & \text{otherwise.} \end{cases} \quad (10)$$

Note that the discontinuity of the current in equation 9 is not a fundamental problem. Convolving the current pulse with a Bartlett window (isosceles triangle function) yields a current pulse that is continuous with continuous derivative and smooths out the voltage without creating any ringing.

The results of a hardware test are shown in Figure 4 (3). The persistent high frequency ringing after the main voltage pulse is due to a parasitic resonance and would be removed by active feedback in an operational mode.

CONCLUSIONS

A barrier cavity rf system appears feasible on both theoretical and practical grounds. The main effect of such a system is to reduce the time average of the peak current. The benefits involve a reduction in space charge tune shift and an increase in synchrotron frequency spread. Longitudinal phase space dilution is small. Since the rf system generally contributes a small fraction to the total cost of a synchrotron the installation of barrier buckets is relatively inexpensive as well.

REFERENCES

1. J.E. Griffin, C. Ankenbrandt, J.A. MacLachlan, A. Moretti, IEEE TNS, 30, p. 3502, 1983.
2. V.K. Bharadwaj, J.E. Griffin, D.J. Harding, J.A. MacLachlan, IEEE TNS, 34, p. 1025, 1987.
3. J.M. Brennan, private communication.
4. M. Meth, private communication.

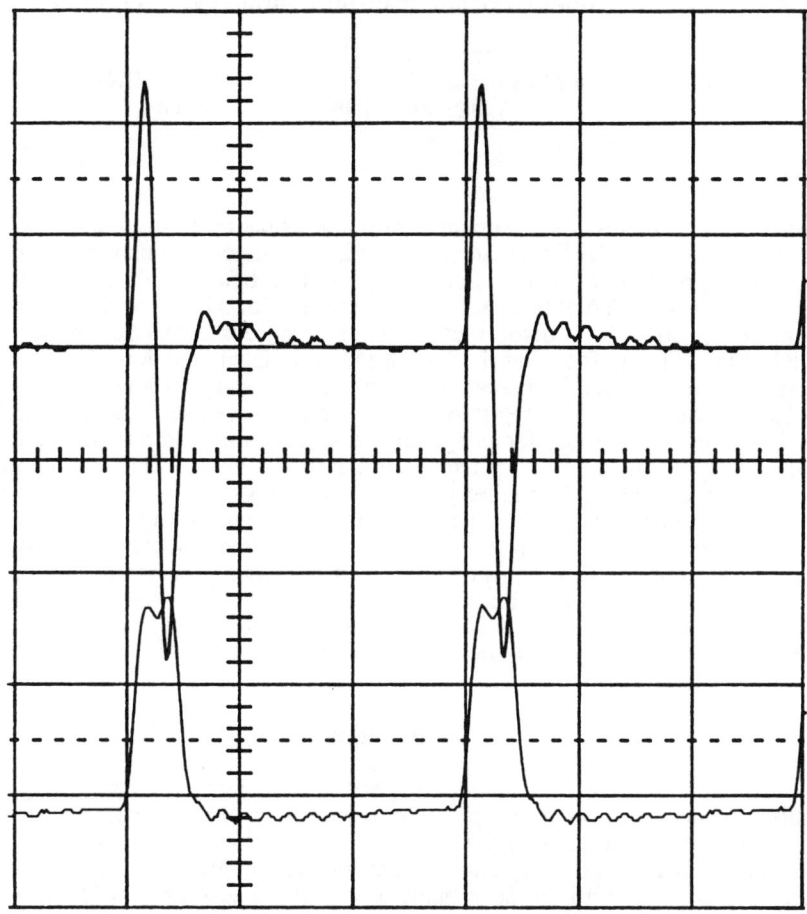

FIG. 4. Experimental confirmation of pulsed operation using existing hardware at the AGS. The time increment is 1μs per box. The top trace is the gap voltage (1160 volts per box), while the bottom trace is the tube current (40 amps per box).

On Space Charge Dominated Beam Transport without Emittance Growth

Yuri K. Batygin
The Institute of Physical and Chemical Research (RIKEN),
Hirosawa, 2-1, Wako-shi, Saitama, 351-01, Japan

Abstract

Possible mechanisms of beam emittance conservation in high current transport lines and in RF linacs are discussed. Invariance of beam emittance is treated as a problem of proper matching of the beam with focusing and accelerating channels. To obtain matching conditions for a beam with an arbitrary distribution function in transport channel, it is necessary to accept that the potential of the external focusing field is a highly nonlinear function of radius. The solution for external potential is obtained from the stationary Vlasov's equation for beam distribution function and Poisson's equation for electrostatic beam potential. Gradual change of nonlinear focusing field results in adiabatic transformation of the beam with initial nonlinear distribution into the beam, matched with the linear focusing channel. Alternating-gradient focusing structure with higher order multipole component creates better matching of the beam with the channel than pure quadrupole structure. Beam emittance growth in linear accelerator is studied for a drift tube linac with solenoid focusing. It is shown that beam phase space distortion in RF linac can be eliminated by the appropriate choice of the value of focusing field with respect to acceleration gradient. An analytical approach is illustrated by results of a particle-in-cell simulation.

1. Introduction

Distortion of beam emittance in particle accelerators result in serious limitations in achieved value of phase space density of the beam and particle losses. Prevention of emittance growth and halo formation in high intensity beams is a key problem for next generation of particle accelerators. Importance of the problem is connected with the development of particle accelerators for heavy ion fusion , spallation neutron sources, radioactive waste transmutation and other applications. In most of the cases the beam emittance growth have to be suppressed to provide small beam losses in accelerator or to produce small beam size at the target. Most of the emittance growth is observed in the low energy part of accelerator facility where particles are slow and space charge forces are significant. Two main reasons contribute to space charge dominated beam emittance growth : nonlinear space charge forces and RF beam defocusing in linear accelerators. Nonlinear space charge forces of laboratory beam produce strong emittance growth in linear focusing channel due to mismatching of the beam profile with focusing field (see ref. [1-12]). Recently [13-14] it was shown that emittance of the high brightness beam is conserved in highly nonlinear focusing field. The required focusing field have to be linear function of radius near axis and drops nonlinearly from linear function far away from axis. Ideal way to create required potential distribution is plasma lens with specific distribution of the opposite charged particles. Approximate matched conditions for the beam with nonlinear distribution function can be obtained in an alternative-gradient quadrupole structure with higher order multipole components. Adiabatic change of focusing field along the structure

© 1996 American Institute of Physics

results in gradual transformation of initially nonlinear distribution into the distribution, matched with the linear focusing channel.

One of the main source of emittance growth in RF linac is a dependence of transverse oscillation frequency on RF phase. This effect is the most pronounced at the stage of beam bunching, where phase length of the beam is large. This phenomena is studied for the case of drift tube linac with solenoid focusing. It is shown, that beam emittance growth can be controlled by adjusting of the value of longitudinal magnetic field with respect to RF field.

2. Matching of the arbitrary beam the uniform focusing channel

Let us consider the beam with arbitrary distribution function in a uniform focusing transport channel. Assume that the beam is matched with the channel, hence the Hamiltonian is a constant of motion but no assumptions about linearity of focusing forces are adopted:

$$H = \frac{p_x^2 + p_y^2}{2m} + qU(x,y) = \text{const}. \tag{1}$$

Four-dimensional beam distribution function obey Vlasov's' equation:

$$\frac{df}{dt} = \frac{\partial f}{\partial x} v_x + \frac{\partial f}{\partial y} v_y - q\left(\frac{\partial f}{\partial p_x}\frac{\partial U}{\partial x} + \frac{\partial f}{\partial p_y}\frac{\partial U}{\partial y}\right) = 0. \tag{2}$$

Substitution of distribution function into Vlasov's equation gives expression for total potential of the structure which is a combination of the external focusing potential, U_{ext}, and the space charge potential U_b of the beam, $U = U_{ext} + U_b$. Self potential of the beam U_b is a known function derived from Poisson's equation:

$$\frac{1}{r}\frac{\partial}{\partial r}\left(r\frac{\partial U_b}{\partial r}\right) = -\frac{\rho(r)}{\varepsilon_0}, \tag{3}$$

where $\rho(r)$ is the space charge density of the beam. Combining solutions of Vlasov's equation for total potential of the structure, U, and space charge potential of the beam, U_b, the external potential of the focusing structure can be found $U_{ext} = U - U_b$. The solution of this problem has to be defined for every specific particle distribution.

3. Gaussian Beam Matched with the Channel

Let us consider a z-uniform beam with a Gaussian distribution function in four-dimensional phase space, which is close to the experimentally observed beam distribution:

$$f = f_0 \exp\left(-2\frac{x^2 + y^2}{R^2} - 2\frac{p_x^2 + p_y^2}{p_0^2}\right). \tag{4}$$

This distribution makes an elliptical phase space projection at every phase plane with normalized root-mean-square (RMS) beam emittance $\varepsilon=(4/mc)$ $\sqrt{<x^2><p_x^2>-<xp_x>^2}$ = Rp_o/mc. Substituting the distribution function (4) into Vlasov's equation yields an expression for the total unknown potential of the structure:

$$\frac{mc^2}{q}(x\,p_x + y\,p_y) = \frac{R^4}{\varepsilon^2}\left(p_x\frac{\partial U}{\partial x} + p_y\frac{\partial U}{\partial y}\right). \tag{5}$$

The solution of the equation (5) is a quadratic function of coordinates which creates linear focusing:

$$U(x,y) = \frac{mc^2}{q}\frac{\varepsilon^2}{R^4}\left(\frac{x^2+y^2}{2}\right). \tag{6}$$

Appearance of quadratic terms in the total potential of the structure is quite clear because phase space projections of the beam have elliptical shape and an ellipse is conserved in a linear field. The space charge field of the beam E_b is calculated from Poisson's equation using a known space charge density of the beam ρ_b:

$$\rho_b = \frac{2I}{\pi c\beta R^2}\exp\left(-2\frac{r^2}{R^2}\right), \quad E_b = -\frac{\partial U_b}{\partial r} = \frac{I}{2\pi\varepsilon_0\beta c}\frac{1}{r}\left[1-\exp\left(-2\frac{r^2}{R^2}\right)\right], \tag{7}$$

where I is the beam current and β is the longitudinal velocity of particles. Subtraction of the space charge field from the total field of the structure gives the expression for the external focusing potential of the structure which is required for conservation of beam emittance:

$$E_{ext} = -\frac{mc^2}{qR}\left[\frac{\varepsilon^2 r}{R^3} + 2\frac{I}{I_c\beta}\frac{R}{r}\left(1-\exp\left(-2\frac{r^2}{R^2}\right)\right)\right], \tag{8}$$

where $I_c = 4\pi\varepsilon_0\,m\,c^3/q = A/Z\cdot 3.13\cdot 10^7$ amp is a characteristic value of the beam current. The relevant potential of the focusing field is given by the expression:

$$U_G(r) = \frac{mc^2}{q}\left[\left(\frac{\varepsilon^2}{2R^4} + \frac{2I}{I_c\beta R^2}\right)r^2 + \right.$$

$$\left. + \frac{2I}{I_c\beta}\left(-\frac{r^4}{2R^4} + \frac{2}{9}\frac{r^6}{R^6} +...+ \frac{(-1)^{k+1}\,2^k}{2k\,k!}\frac{r^{2k}}{R^{2k}}\right)\right]. \tag{9}$$

Let us note that the external potential of the structure consists of two parts: quadratic (which produces linear focusing) and higher order terms which describe nonlinear focusing. The linear part depends on the values of beam emittance and beam current while the nonlinear part depends on beam current only. This means

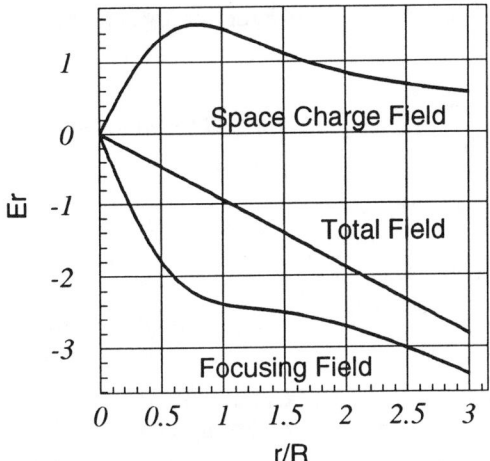

Fig. 1. Space charge field of the Gaussian beam, total field of the structure and required external focusing field to maintain stationary beam distribution.

that the external field has to compensate the nonlinearity of self-field of the beam and produce required linear focusing of the beam to keep the elliptical beam phase space distribution. Fig. 1 illustrates the relationships between space charge field of the beam, total field, and focusing field of the structure. The external focusing field obtained from the above consideration is a complicated function of radius which is linear near the axis and becomes nonlinear far from the axis. One of the ways to create the required focusing potential is to introduce inside the transport channel an opposite charged cloud of particles (plasma lens) with the space charge density:

$$\rho_{ext} = \frac{I_c}{2\pi c\, R^2}\left[\frac{\varepsilon^2}{R^2} + \frac{4\,I}{\beta\, I_c}\exp(-2\frac{r^2}{R^2})\right] . \qquad (10)$$

Another possibility of creation of nonlinear focusing field is connected with the utilization of higher order multipole components in alternating-gradient quadrupole structure (see chapter 7).

4. Particle-in-Cell Simulation of Beam Emittance Conservation

To verify the possibility of conservation of beam emittance in a nonlinear focusing field, a beam dynamics simulation using general-purpose particle-in-cell code BEAMPATH [15] has been performed. A beam of particles is represented as a collection of large number of modeling particles (typically 30000). Equations of motion are integrated using a second order integrator with constant time step Δt:

1) $\vec{p}_i^* = \vec{p}_{i-1/2} + \vec{E}_i\, \Delta t$,   3) $\vec{p}_{i+1/2} = \vec{p}_i^* + \vec{E}_i\, \Delta t$

2) $\vec{p}_i^* = \vec{p}_i + \Delta t\, \dfrac{(\vec{p}_i^* + \vec{p}_i)}{2\gamma_i} \times \vec{B}_i$ 4) $\vec{r}_{i+1} = \vec{r}_i + \dfrac{\vec{p}_{i-1/2} + \vec{p}_{i+1/2}}{2\gamma_i}\Delta t$. (11)

The space charge field of a beam in different coordinate systems is found from Poisson's equation:

$$\frac{\partial^2 U_b}{\partial x^2} + \frac{\partial^2 U_b}{\partial y^2} = -\frac{\rho(x,y)}{\varepsilon_o} \quad \text{for z-uniform beam;} \tag{12}$$

$$\frac{1}{r}\frac{\partial}{\partial r}\left(r\frac{\partial U_b}{\partial r}\right) + \frac{\partial^2 U_b}{\partial z^2} = -\frac{\rho(r,z)}{\varepsilon_o} \quad \text{for axial-symmetric bunches;} \tag{13}$$

$$\frac{\partial^2 U_c}{\partial x^2} + \frac{\partial^2 U_c}{\partial y^2} + \frac{\partial^2 U_c}{\partial z^2} = -\frac{\rho(x,y,z)}{\varepsilon_o} \quad \text{for quadrupole-symmetric bunches} \tag{14}$$

The Dirichlet boundary condition $U_b(x,y,z) = 0$ for potential is imposed on the surface of an infinite pipe and periodic boundary conditions in longitudinal direction $U(x,y,z) = U(x,y,z+\beta\lambda)$ are adopted for bunched beam. The region occupied by an ensemble of particles is divided into uniform rectangular meshes of dimension NX x NY, NR x NZ or NX x NY x NZ. The charge of every particle is distributed among the nearest four nodes (2D problem) or eight nodes (3D problem) of the grid inversely proportional to the distance of the particle from each node. Poisson's equation is solved by combination of decomposition method and Fast Fourier Transform.

In fig. 2 the results of the beam dynamics with initial Gaussian distribution in linear and nonlinear focusing channel are presented. Parameters of the beam were chosen as follows: A/Z=1, I = 2 amp, ε = 0.12 π cm mrad, R = 0.15 cm, β = 0.0178. The external focusing potential for the linear focusing channel was taken as

$$U_L(r) = \frac{mc^2}{q}\left(\frac{\varepsilon^2}{2R^4} + \frac{I}{I_c \beta R^2}\right) r^2, \tag{15}$$

which corresponds to the matched conditions for an equivalent KV beam with the same RMS beam emittance ε, and RMS beam size, R. In the case of nonlinear focusing, the external potential is represented by eq. (9). Let us note that quadratic terms in potentials (9) and (15) are different.

From results of simulations, it is seen that in both cases the sizes of the beam in real space are close to constant which is typical for matching of the beam, taking into account RMS beam sizes. But in the case of linear focusing, the beam is mismatched in the phase plane which results in 50% emittance growth accompanied by halo formation. At the same time, the beam is completely matched with the nonlinear focusing channel, and this results in conservation of all beam characteristics and does not suffer any emittance growth.

5. "Water Bag" and Parabolic Beam Matched with the Channel

The analogous results can be obtained for a beam with the other distributions with elliptical symmetry. Let us consider "water bag" distribution in four-

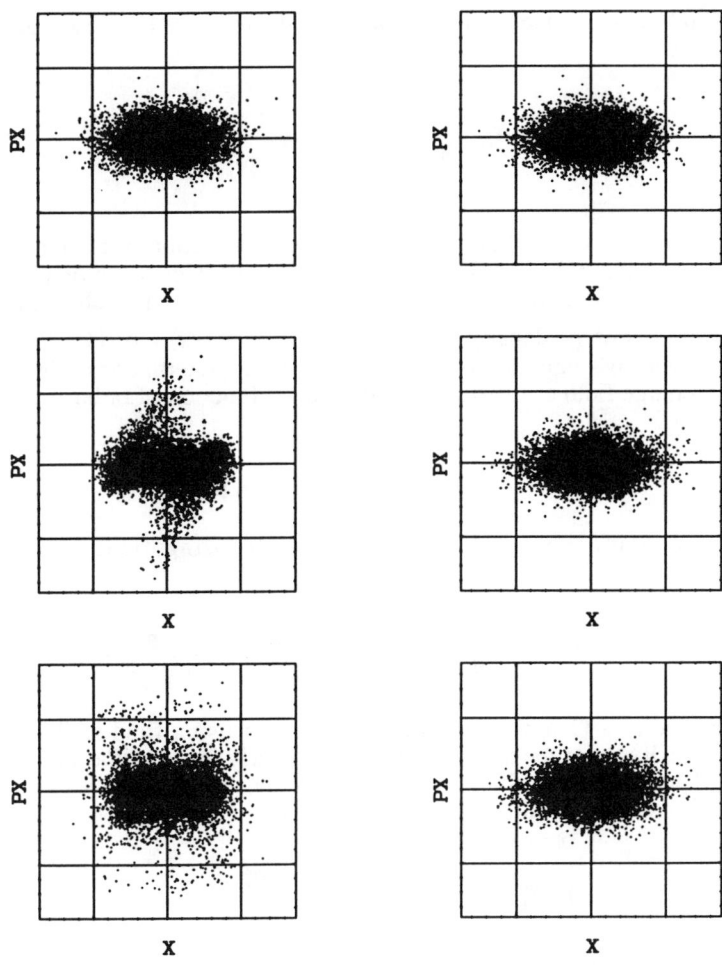

Fig. 2. Emittance growth of the Gaussian beam in linear focusing channel (left column) and emittance conservation in non-linear focusing channel (right column).

dimensional phase space which corresponds to a uniformly populated 4D hypervolume:

$$f = f_0, \; \frac{2}{3} \left(\frac{x^2 + y^2}{R^2} + \frac{p_x^2 + p_y^2}{p_0^2} \right) \leq 1 . \tag{16}$$

The coefficient 2/3 in eq.(16) is chosen from normalization of the distribution and reflects the fact that the maximum beam sizes for such a distribution are $\sqrt{3/2}$ larger than RMS beam parameters R, and p_0. This distribution is characterized by a parabolic space charge density function in real space $\rho(r) = \rho_0 [1 - (2r/3R)^2]$. The solution of Vlasov's equation is the same as for the potential described by eq. (6). The space charge field of the beam is a two-terms function of radius:

$$U_b = - \frac{I}{3\pi \varepsilon_0 \beta c} \frac{r^2}{R^2} \left(1 - \frac{r^2}{6R^2} \right) . \tag{17}$$

The corresponding potential of the external focusing field is given by the expression:

$$U_{WB}(r) = \frac{mc^2}{q} \left[\frac{\varepsilon^2}{2R^4} r^2 + \frac{4 I r^2}{3 I_c \beta R^2} \left(1 - \frac{r^2}{6R^2} \right) \right] . \tag{18}$$

As in the case of a Gaussian beam the required focusing field is close to a linear function of radius near the axis and drops nonlinearly far from the axis.
For "parabolic" distribution phase space density of particles monotonically decreases from center of the beam to the boundary of 4-dimensional hypervolume:

$$f = f_0 \left(1 - \frac{x^2 + y^2}{2 R^2} - \frac{p_x^2 + p_y^2}{2 p_0^2} \right) . \tag{19}$$

Maximum beam sizes for such a distribution are $\sqrt{2}$ larger than RMS beam parameters R, and p_0 which is reflected in coefficient 2 in denominator of distribution in eq. (19). Space charge density function of the beam ρ_b and electrical field of the beam E_b are defined by expressions:

$$\rho_b = \frac{3 I}{2 \pi c \beta R^2} \left(1 - \frac{r^2}{2 R^2} \right)^2 ; \; E_b = \frac{3 I r}{4 \pi \varepsilon_0 \beta c R^2} \left[1 - \left(\frac{r}{\sqrt{2} R} \right)^2 + \frac{1}{3} \left(\frac{r}{\sqrt{2} R} \right)^4 \right] . \tag{20}$$

The relevant potential of the focusing field which is required to conserve beam emittance is given by the expression:

$$U_P(r) = \frac{m c^2}{q} \left[\frac{r^2}{2 R^2} \left(\frac{\varepsilon^2}{R^2} + \frac{3 I}{I_c \beta} \right) + \frac{3 I}{8 I_c \beta} \left(- \frac{r^4}{R^4} + \frac{r^6}{9 R^6} \right) \right] . \tag{21}$$

6. Adiabatic Transformation of Nonlinear Beam Distribution

Conservation of the beam distribution function requires the focusing field to be essentially nonlinear. Conventional focusing structures employ quadrupole lenses and axial - symmetric lenses (both electrostatic and magnetostatic) where linear focusing component is usually dominate. It is interesting to verify whether it is possible to transform initially nonlinear distribution into distribution matched with the linear focusing channel. Suppose, the focusing field provides perfect matching of the initial beam with nonlinear distribution function. If the focusing field is changed adiabatically, beam emittance is a constant of motion:

$$\varepsilon = \int dx\, dp_x = inv \qquad (22)$$

analogously for y, p_y. The restriction of adiabatic change of parameters means that the system should changed much slower, than the period of oscillation of particles, i.e. total change of the focusing structure requires at least a few transverse particle oscillations. The adiabatic invariant (22) is not conserved exactly during the field transformation [16], but we can expect that change of the value of invariant (beam emittance) will be small.

The final expected beam distribution is a distribution matched with the linear focusing structure. This class of matched beams was studied in detail in ref. [12, 17-19]. The distribution function of the matched beam is a function of the Hamiltonian containing linear focusing term:

$$f = f[\frac{p_x^2 + p_y^2}{2m} + \frac{k^2}{2}(x^2 + y^2) + q\, U_b(x,y)] \ . \qquad (23)$$

In eq.(23) parameter k^2 describes the focusing of particles in solenoids or a smoothed external focusing in alternative-gradient structure and $U_b(x,y)$ is the space charge potential of the beam. A general property of the solution is that with increasing beam current, the profile of the matched beam has to be more and more flat while the phase space projection (beam emittance) has to be more and more close to a rectangle.

In computer simulations which were performed for the beam with parabolic distribution (19), the potential of the focusing field was gradually transformed from the field (21) required by nonlinear matched conditions to the linear focusing field (15):

$$U(r,z) = U_P(r) + [U_L(r) - U_P(r)]\frac{(z - z_0)}{L}, \qquad (24)$$

where z -longitudinal coordinate, L - drift space of the matcher. Parameters of the numerical problem were chosen the same as for the case presented at fig. 2. Distance L = 114 cm was selected to perform 5 transverse oscillations of particles along the matcher. After adiabatic transformation, beam was transported in linear focusing channel to check the result of adiabatic matching. For comparison, the beam dynamics in pure linear focusing channel was calculated. The results of simulation in both cases are presented at Table 1 and at figs. 3,4.

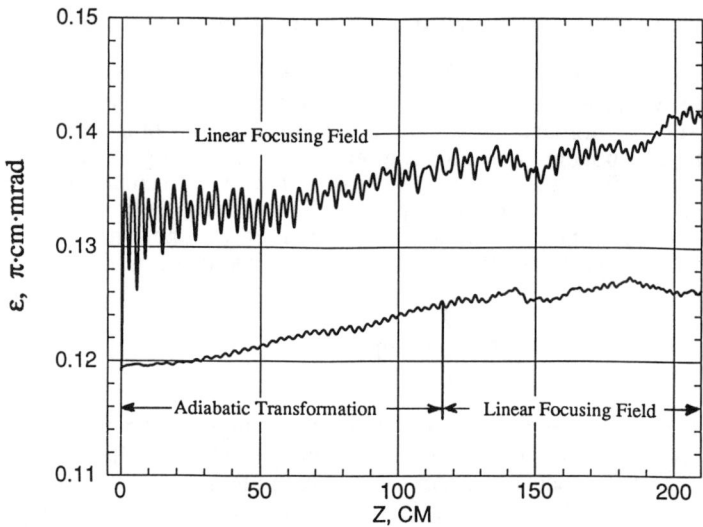

Fig. 3. Beam emittance growth in the uniform focusing channel with linear focusing field (up) and in matching section with nonlinear field (bottom).

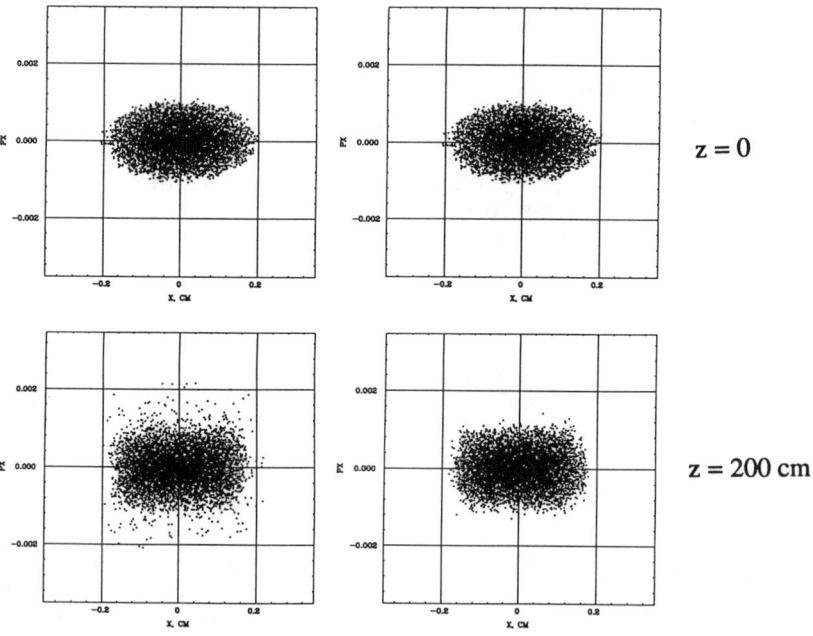

Fig. 4. Phase space projections of the beam in the uniform focusing channel with linear focusing field (left) and in matching section with nonlinear field (right).

Table 1. Results of Beam Emittance Growth Simulation.

	Initial Emittance	Final Emittance
Linear Focusing Field	0.12 π cm mrad	0.14 π cm mrad
Nonlinear Adiabatic Matching	0.12 π cm mrad	0.126 π cm mrad

Calculations show that in pure linear focusing channel beam experienced strong emittance growth while during adiabatic transformation the emittance growth is substantially smaller. Phase space projection of the beam in matching section is changed according to changes of focusing field and final beam distribution is completely matched with the channel. The performed study shows that nonlinear beam distribution can be transformed into distribution matched with linear focusing channel using relatively short adiabatic matching section with nonlinear focusing field.

7. Focusing by a Static Field

The required external focusing field obtained from the above consideration is a complicated function of radius which is linear one near the axis and drops nonlinearly far from the axis. This specific feature of the focusing field restricts the possible ways to produce the appropriate potential distribution. Axial-symmetric electrostatic and magnetic lenses have aberrations which increase the focusing of charged particles with radius as compared with linear focusing [12]. Time-independent field provides focusing effect which can be described by a linear term as well as higher order terms. The paraxial equation of radial motion of a particle in the electrostatic lens with field distribution along the axis $E_z(z)$ is given by

$$\frac{d^2 r}{dt^2} = \frac{P_\theta^2}{m^2 r^3} - \frac{q}{m}[\frac{r}{2}\frac{\partial E_z}{\partial z} - \frac{r^3}{16}\frac{\partial^3 E_z}{\partial z^3} + ...] \ . \qquad (25)$$

where P_θ is an azimuth component of canonical momentum of particle. After passing through the lens the slope of the particle trajectory is changed as $\Delta r' = -(r/f)(1 + C_\alpha r^2)$, where f is a focal length of the lens and C_α is a spherical aberration coefficient. From eq. (25) it follows that the changing of slope of trajectory is larger for particles with larger radius. Spherical aberrations of axial-symmetric lenses result in hollow beam profile formation and emittance growth [12].

Most of the focusing channels are based on alternative-gradient principle employing alternating focusing - defocusing quadrupole lenses with linear focusing field distribution across the aperture. The higher order multipole lenses (sextupoles, octupoles, etc.) create essentially nonlinear field due to azimuth variation of potential $U_{ext} = U_0 r^n \cos n\theta$. Focusing and defocusing directions are repeated after azimuth angle shift $\Delta\theta=\pi/n$. Potential of the quadrupole alternating gradient focusing channel is presented as follows:

$$U(r,\varphi,z) = \frac{G_2(z)}{2} r^2 \sin 2\varphi + \frac{G_6(z)}{6} r^6 \sin 6\varphi + ... \qquad (26)$$

where $G_2(z)$ is a quadrupole gradient and $G_6(z)$ is a duodecapole component. To create required nonlinear compensation of space charge field in two orthogonal x - y directions the sign of duodecapole component $G_6(z)$ should be opposite to the sign of quadrupole component $G_2(z)$. Let us consider one dimensional problem for particle oscillating in the field (26):

$$m\frac{d^2 x}{dt^2} = q[G_2(z) x + G_6(z) x^5] \quad . \qquad (27)$$

We restrict our consideration by FD (focusing-defocusing) structure where all lenses have the same length D and the period of the structure is S = 2D. Solution of the problem can be represented as a combination of slow varying deviation of particle from axis X and fast oscillating variable ξ. Employing averaging method [16] one can obtain the following equation for slow variable X which is essential for definition of the focusing properties of the channel:

$$\frac{d^2 X}{d\tau^2} + \mu_0^2 [X + 6\frac{G_6}{G_2} X^5] = 0 \quad , \qquad (28)$$

where $\tau = z/S$ is a dimensionless longitudinal coordinate and μ_0 is a frequency of the smoothed oscillations in FD structure:

$$\mu_0 = \frac{G_2 D^2}{\sqrt{2} \frac{mc^2}{q} \beta^2} \quad . \qquad (29)$$

Let us consider beam with parabolic distribution. Required field to conserve beam emittance for parabolic particle distributions is given by differentiating of potential (21):

$$E_{ext} = -\frac{mc^2 r}{q R^2} [\frac{\varepsilon^2}{R^2} + \frac{3 I}{I_c \beta} (1 - \frac{r^2}{2 R^2} + \frac{r^4}{12 R^4})] \quad . \qquad (30)$$

It consists of terms proportional to r, r^3, r^5, while the field of FD structure consists of terms x, x^5. Let us choose the parameters of quadrupole structure from two conditions: (i) linear parts of fields (21), (30) are the same, (ii) fields are equal each other at the boundary of the beam at $r = \sqrt{2}R$. It gives the following expressions for the field gradient and for the duodecapole component:

$$G_2 = \sqrt{8} \frac{mc^2 \beta}{q RD} (\frac{\varepsilon^2}{R^2} + \frac{3 I}{I_c \beta})^{1/2} \quad ; \quad G_6 = -\frac{1}{3} \frac{I}{I_c} G_2 \frac{D^2}{R^6} \frac{1}{\mu_0^2 \beta^3} \qquad (31)$$

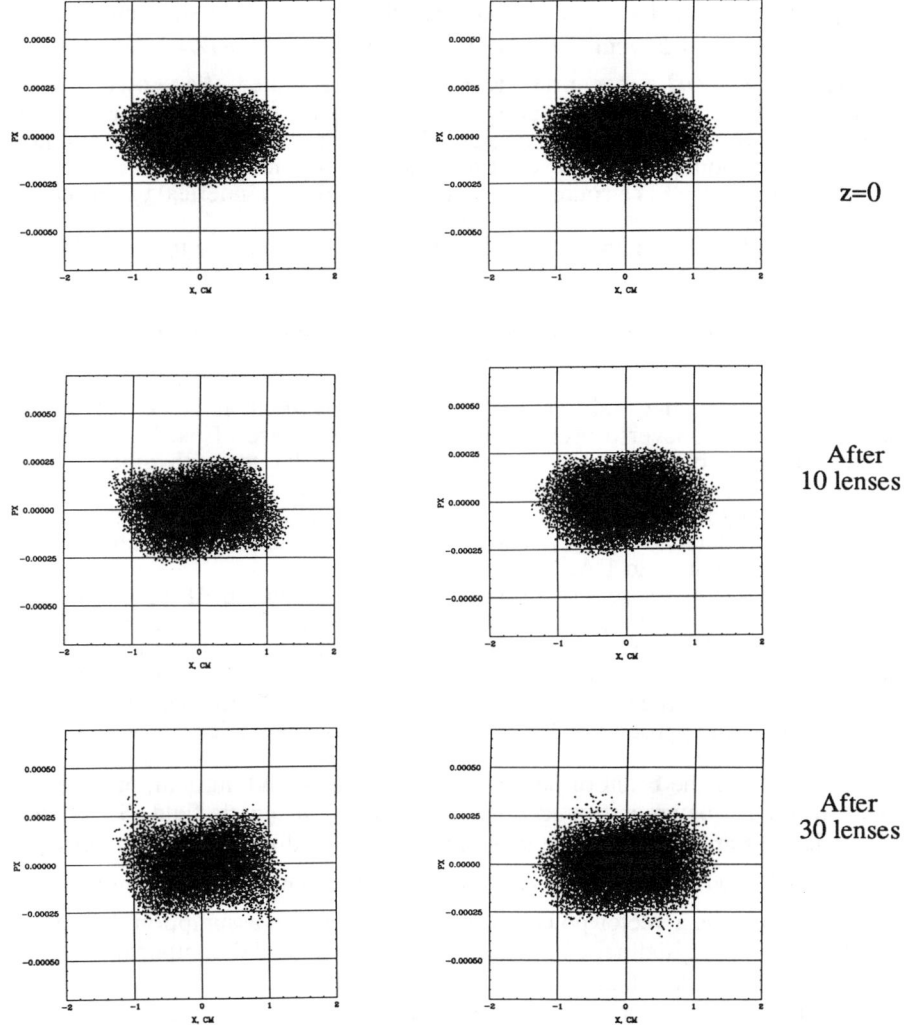

Fig. 5. Beam dynamics in quadrupole FD structure with field gradient $G_2 = 36 \frac{kV}{cm^2}$.

Fig. 6. Beam dynamics in quadrupole FD structure with field gradient $G_2 = 36 \frac{kV}{cm^2}$ and duodecapole component $G_6 = -1 \frac{kV}{cm^6}$.

At figs. 5, 6 the results of beam dynamics simulation in FD structure with $G_2 = 36 kV/cm^2$, $G_6 = -1 kV/cm^2$; D = 1 cm for the beam with A/Z =1, W =150 keV, I = 100 mA, $\varepsilon = 0.2\ \pi$ cm·mrad, R = 1cm are presented. As shown, the beam emittance shape is better conserved in the focusing channel with duodecapole component. It confirms the assumption that nonlinear focusing field component compensates nonlinear space charge field, but in the same time creates x-y coupling which itself is a source of emittance growth. In numerical experiments the effect of nonlinear space charge field compensation and therefore, beam matching, was observed for beams with the value of phase space density up to the 0.6 A/cm mrad.

8. Suppression of the Beam Emittance Growth in RF Linac due to RF Defocusing

In RF linac the main source of emittance growth is RF defocusing, i.e. dependence of transverse oscillation frequency on phase of particle in RF field. Here this effect is studied for drift tube linear accelerator with superconducting solenoid focusing [20]. The advantages of such type of accelerator are high energy gain, weak dependence of beam dynamics on misalignments and large value of transverse acceptance which allows to use this structure for acceleration of the beam with current up to 1 A.

At fig. 7 the phase space trajectory of the particle in RF accelerating field and solenoid focusing field is presented. It is clear, that phase space ellipse in phase space x-Px is deformed due to RF defocusing. The effective emittance of the beam after many particle oscillations in RF field can be significantly larger than the initial beam emittance. As it is shown below, this emittance growth can be controlled by an appropriate choice of longitudinal magnetic field with respect to RF field.

Consider the beam of particles with charge q and mass m, propagating in RF drift tube structure surrounded by solenoid with magnetic field $B_z=B$. We use Cartesian system of coordinate (x,y,z) with canonical-conjugate variables (x,P_x), (y,P_y), (z,P_z) where $\vec{P}=\vec{p}+q\vec{A}$ is a canonical momentum, $\vec{p}$ is a momentum of particle and $\vec{A}$ is a vector potential of the field. Due to assumption that magnetic field has only longitudinal component, the vector potential components are $A_x = -By/2$, $A_y = Bx/2$, $A_z = 0$ and $p_z = P_z$.

Hamiltonian for a non relativistic particle in this system is given by

$$H = \frac{1}{2m}[(P_x - qA_x)^2 + (P_y - qA_y)^2 + p_z^2] + q[\sum_{n=1}^{N} V_n I_0(k_n r)\cos k_n z \sin \omega t + U] \quad (32)$$

where electrostatic potential of RF structure is described by Fourier-Bessel series with expansion coefficients V_n, wave numbers $k_n = 2\pi n/\beta\lambda$, ω is a RF frequency and U=U(x,y,z,t) is a space charge potential of the beam. Let us substitute the new canonical variables $\eta = p_z - p_s$, $\xi = z - z_s$ for the old variables p_z, z. It corresponds to transform from laboratory coordinate system to the system moving with the velocity of synchronous particle $v_s = p_s/m$. Generating function of transformation is $F = -(\xi + z_s)(p_z - p_s) - xP_x - yP_y$ and a new Hamiltonian is given by $K = H + \partial F/\partial t$ [16].

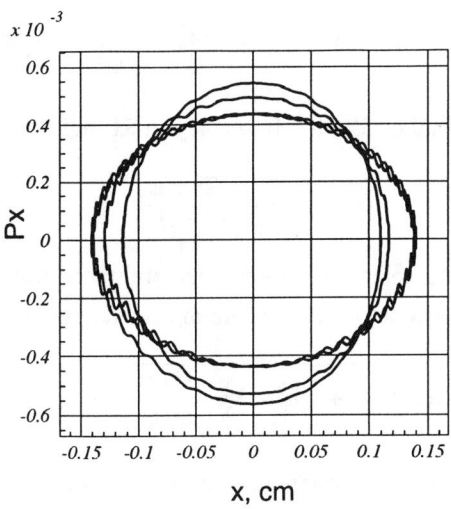

Fig. 7. Phase space trajectory of particle in standing wave RF accelerator.

Expression for a new Hamiltonian contains constant of motion $M = P_y x - P_x y$ and terms p_s^2, $z_s \dot{p}_s$ which do not depend on new variables, therefore they can be omitted. Besides in potential function of RF structure we take into account only first term $V_1 = 2V_0 T/\pi$ which describes the acceleration with synchronous phase $\varphi_s = \omega t - kz$ where V_0 is a potential applied to a gap g between drift tubes and T is a transit time factor for a structure of a radius a:

$$T = \frac{1}{I_0(ka)} \frac{\sin \pi g/\beta\lambda}{\pi g/\beta\lambda} \qquad (33)$$

The final expression for a new Hamiltonian is

$$K' = \frac{1}{2m}(P_x^2 + P_y^2 + \eta^2) + \frac{m\omega_L^2}{2}(x^2 + y^2) + \frac{qTV_0}{\pi}[I_0(kr)\sin(\varphi_s - k\xi) + k\xi\cos\varphi_s] + qU \qquad (34)$$

where notation of Larmor frequency is used $\omega_L = qB/2m$. The Hamiltonian (34) is a total energy of particle which oscillates around synchronous particle of the beam in the moving coordinate system [17]. If parameters of the structure are changing adiabatically along the channel, the Hamiltonian is a constant of motion and any function $f = f(K')$ gives the distribution function for a beam matched with the channel. Real beam in accelerating structure is usually mismatched.

For near-axis particles in a long bunch the transverse motion of particles can be considered separately from the longitudinal one and the corresponding Hamiltonian of transverse motion of particle is [17]

$$K_r = \frac{1}{2m}(P_x^2 + P_y^2) + \frac{m}{2}[\omega_L^2 - \frac{\Omega^2}{2}\frac{\sin\varphi}{\sin\varphi_s}](x^2+y^2) + qU , \qquad (35)$$

where Ω is a longitudinal oscillation frequency in RF field:

$$\Omega^2 = \omega^2 q \frac{V_0 T \sin\varphi_s}{mc^2 \pi\beta^2} . \qquad (36)$$

Matched condition for a beam with normalized emittance ε and radius R in uniform focusing channel is $\varepsilon = \omega_r R^2/c$ [17], where ω_r is a depressed transverse oscillation frequency

$$\omega_r = \sqrt{\omega_1^2 + \omega_L^2 - \frac{\Omega^2}{2}\frac{\sin\varphi}{\sin\varphi_s}} - \omega_1 , \qquad (37)$$

and $\omega_1 = cI/(\beta\gamma^2 I_c \varepsilon)$ is a frequency, defined by space charge of the beam. For beam in longitudinal magnetic field the beam emittance $\varepsilon = (4/mc)(<x^2><P_x^2> - <xP_x>^2)^{1/2}$ has to be defined at the phase plane of canonical conjugate variables (x,P_x) instead of usually taken variables (x,p_x).

Dispersion of transverse oscillation frequencies (37) due to dependence on RF phase results in distortion of phase space trajectories and finally in beam emittance growth. Beam emittance growth can be reduced by the appropriate choice of the value of Larmor frequency with respect to longitudinal oscillation frequency. The lower value of magnetic field is limited by transverse stability constraint $\omega_L > \Omega$. With the increasing of ω_L/Ω the dispersion of transverse oscillation frequency (37) is damped.

At fig. 8 the results of beam dynamics study in RF proton linac for energy 3 MeV and beam current 250 mA are presented. Drift tube accelerator structure consists of prebuncher, buncher and acceleration section. Synchronous phase is changing monotonously from 90° to 30°. The final value of accelerating gradient is 2.5 Mev/m. Total transmission efficiency obtained in simulation is 90%.

From results of simulation it is seen that the emittance growth is occurred mainly at the stage of beam bunching, where the amplitude of phase oscillation is large. When the beam is bunched, the emittance growth is saturated. For the value of magnetic field, close to transverse stability limit $\omega_L/\Omega = 1.25$, the strong emittance growth (up to 100%) was observed. With the increasing of magnetic field, the emittance growth was seriously reduced and finally can be made close to zero (see case $\omega_L/\Omega = 2.5$).

9. Conclusions

Conservation of beam emittance was treated as a problem of proper matching of the beam with a uniform focusing channel. Matched conditions for the beam with elliptical phase space projections but nonlinear space charge forces in a uniform focusing channel require the focusing field to include nonlinear terms of higher order than quadratic. The solution for the external potential is attained from the stationary Vlasov's equation for beam distribution function and Poisson's

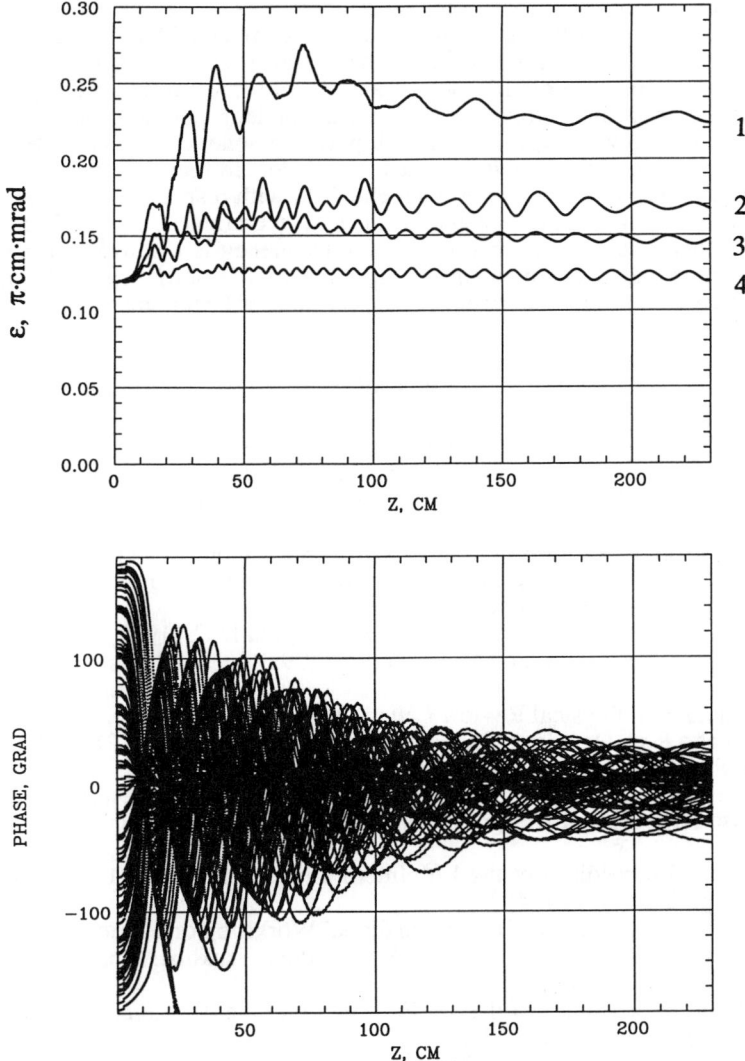

Fig. 8. Beam emittance (up) and phase trajectories (bottom) in RF linear accelerator with solenoid focusing: 1) $\omega_L/\Omega = 1.25$; 2) $\omega_L/\Omega = 1.75$; 3) $\omega_L/\Omega = 2$; 4) $\omega_L/\Omega = 2.5$.

equation for electrostatic beam potential. Different examples of Gaussian, "water bag" and parabolic distributions in 4D phase space are considered. Gradual change of nonlinear focusing field result in adiabatic transformation of initial nonlinear particle distribution into distribution, matched with the linear focusing channel. Two cases of nonlinear focusing field: plasma lens and alternating gradient quadrupole channel with higher order multipole components were considered. High current beam dynamics issues in RF linac with solenoid focusing were considered. Appropriate choice of the value of magnetic field with respect to acceleration field gives small emittance growth of the beam in accelerating structure. The ratio of Larmor frequency to longitudinal oscillation frequency is a convenient parameter to operate transverse beam mismatching due to RF defocusing. Results of a particle-in-cell simulation confirm the conservation of beam emittance in transport line and RF linac.

References

1. P.M.Lapostolle, IEEE Trans. Nucl. Sci. NS-18, 1101 (1971).
2. J.Struckmeier, J.Klabunde, and M.Reiser, Part. Accel. 15 (1984) 47.
3. T.P.Wangler, K.R.Crandall, R.S.Mills, and M.Reiser, IEEE Trans. Nucl. Sci. NS-32, (1985) 2196.
4. R.A.Jameson, Proceedings of the PAC93, Washington D.C. (1993) 3926.
5. J.S.O'Connell, T.P.Wangler, R.S.Mills and K.R.Crandall, Proceedings of the PAC93, Washington D.C. (1993) 3657.
6. R.D.Ryne, Proceedings of the PAC93, Washington D.C. (1993) 3229.
7. J.M.Lagniel, Nucl. Instr. and Meth. A 345 (1994) 46; Nucl. Instr. and Meth. A 345 (1994) 405.
8. R.L.Gluckstern, Physical Review Letters, Vol. 73, 1247 (1994).
9. C.Chen and R.C. Davidson, Physical Review Letters, Vol. 72, 2195 (1994).
10. C.L.Bohn, Physical Review Letters, Vol. 70, 932 (1993).
11. C.Chen and R.C. Davidson, Physical Review E, Vol. 49, 5679 (1994).
12. M.Reiser, Theory and Design of Charged Particle Beams, Wiley, New York, 1994.
13. Y.Batygin, Proceedings of the 17th Intern. Linac Conference (LINAC94), Tsukuba, Japan (1994) 487.
14. Y.Batygin, Proceedings of the International Workshop on Particle Dynamics in Accelerators: Emittance in Circular Accelerators, Tsukuba, Japan (1994), KEK Proceedings 95-7, 88.
15. Y.Batygin, Proceedings of the 3rd European Particle Accelerator Conference (EPAC92), Berlin (1992) 822.
16. L.Landau and E.Lifshitz, Mechanics, Pergamon Press, 1975.
17. I.M.Kapchinsky, Theory of Resonance Linear Accelerators, Moscow, Atomizdat, 1966, Harwood, 1985.
18. P.Lapostolle, YII Int. Conference on High Energy Accelerators, Yerevan (1969) 205.
19. J.Struckmeier and I.Hofmann, Part. Accel., 39 (1992) 219.
20. B.I.Bondarev, A.N.Kurmanov, A.V.Mishenko, V.M.Pirozhenko, O.V.Plink, V.A.Smirnov, Proc. BEAMS-90, Novosibirsk, 1990, Vol. 2, 860.

STATISTICAL AND FOKKER PLANCK APPROACH

Generalized Free-Energy Principle and Emittance Growth

Patrick G. O'Shea
Free-Electron Laser Laboratory
Department of Physics
Duke University
Durham NC 27708-0319

e-mail: oshea@fel.duke.edu

Abstract

The connection between entropy growth and rms emittance growth in linear accelerators is studied. Emittance growth is divided into two classes: reversible and irreversible depending on the corresponding entropy change. A generalized free-energy function is shown to act as a driver for phase space evolution and emittance growth.

1. Introduction

An important question in beam physics is: to what extent is beam phase-space degradation reversible? This topic has been addressed previously both theoretically [1,2] and experimentally [3,4] for special cases. In previous work [1,2] the onset of irreversible dynamics has been described in terms of the cessation of laminar particle motion as manifested by trajectory crossing in phase space. In this paper we study the underlying issues in beam degradation (i.e. emittance growth) in the presence of time dependent and nonlinear, space-charge and external forces from a statistical thermodynamic view point. We use the entropy change as the delimiter between reversible and irreversible emittance growth. As a consequence of this analysis we consider a generalized free-energy principle, and derive an expression for a generalized thermodynamic potential, analogous to the Helmholtz free-energy, that may act as a driver for emittance growth. A comprehensive source of information on the topic of emittance growth, and its relation to free-energy concepts, is the recent book by Reiser [5].

A practical, and widely used, figure of merit for beam optical quality is the normalized rms emittance [6,7] $\tilde{\varepsilon}_{n,x}(z) = \frac{1}{mc}\sqrt{\langle x^2 \rangle \langle p_x^2 \rangle - \langle xp_x \rangle^2}$, where z is the laboratory coordinate of the beam centroid, x and p_x are conjugate spatial and momentum coordinates respectively, the brackets represent averages over the particle distribution, m is the particle mass and c is the velocity of light. Similar expressions can be written for the y and z emittances. It is well known that $\tilde{\varepsilon}_{n,x}$ is invariant under time-independent linear symplectic transformations [8], but not necessarily in other deterministic processes such as when nonlinear forces, or forces that are correlated with the longitudinal position in the particle bunch, are present.

The general connection between emittance and entropy was made over two decades ago by Lawson, Lapostolle and Gluckstern [9]. More recently, the connection between emittance growth and entropy growth has been noted by a number of other authors [5,10-13], but has not been developed extensively.

Entropy is considered a macroscopic quantity in equilibrium thermodynamics, or a microscopic quantity of a statistical ensemble. In macroscopic thermodynamics, entropy cannot be defined for a system that is not in thermal equilibrium. In general, beams in linear accelerators are not in thermal equilibrium [14], and the non-equilibrium region may extend from the cathode to the beam dump. In most cases we must rely on the statistical interpretation of entropy as a measure of the information available about a distribution. In the most general sense the statistical entropy of a system can be written [15]

$$S = -k_B \sum_i f_i \ln(f_i) \qquad (1)$$

where k_B is Boltzmann's constant, i denotes a microcanonical state of the system, f_i is the statistical probability of that state, and $\sum_i f_i = 1$.

In the case of a bunched beam with a very large number of particles (N) per bunch we can write the sum as an integral by noting that a state corresponds to a six-dimensional volume element $A_6 = \delta x \delta p_x \delta y \delta p_y \delta \zeta \delta p_\zeta$ and the probability is equivalent to the product of A_6 times the distribution function $\rho_6 = \rho(x, p_x, y, p_y, \zeta, p_\zeta)$, so that:

$$S_6 = -k_B N \int \rho_6 \ln[A_6 \rho_6] dx dp_x dy dp_y d\zeta dp_\zeta \qquad (2)$$

where we choose the normalization $\int \rho_6(x)d^6x = 1$. The time derivative of the entropy is given by $\dot{S} = -k_B N \int \dot{\rho}_6 \ln[A_6 \rho_6] dx dp_x dy dp_y d\zeta dp_\zeta$. Therefore the entropy of a distribution is invariant under deterministic processes where Liouville's theorem is valid, *if the variation of ρ_6 across the cell is negligible.*

How should we choose the size of the volume element A_6? The quantum-mechanical lower bound is $\hbar^3$. In practice one should choose a larger value of A_6 based on the limits of our ability to make observations of the beam, i.e., on the resolution of our instrumentation, on the precision with which we can apply external forces, or on the limits of physical phenomena of interest. We assume that we have no knowledge of changes in the phase-space density on scales smaller than A_6. All techniques for measuring phase-space distributions involve segmentation of data into bins analogous to A_6, or its lower dimensional analogues, and ρ_6 is averaged over each cell. Whether or not a measured entropy change occurs will depend on our choice of A_6. Because of the coarse graining of the distribution function, it is possible for a process that is strictly Liouvillian in nature to result in an apparent (i.e. calculated) entropy change if a larger, rather than a smaller, value of A_6 is chosen [10]. This concept will be important in determining whether or not emittance growth is reversible in a particular situation.

We may also consider the cell size concept in relation to the trajectory-crossing condition for the onset of irreversible dynamics of previous work [1,2]. Trajectories with an initial separation ΔR that later have zero separation are said to have crossed. If we consider two trajectories with initial separation $\Delta R > \sqrt[6]{A_6}$, that later have a separation $\Delta R \leq \sqrt[6]{A_6}$, then the entropy will appear to have grown. Setting $A_6 = 0$ in the entropy formalism results in an irreversibility criterion equivalent to the non-laminarity condition.

2. Determination of the entropy

We now proceed to develop a useful relationship between entropy and rms emittance. When considering bunched beams it is useful to divide the bunch into a number of sub-bunches or slices of longitudinal length $\delta\zeta$ in the laboratory frame, where $\delta\zeta \ll \sigma_b$ the bunch length, and ζ_i is the longitudinal coordinate of the ith sub-bunch relative to the centroid of the bunch. Consider each sub-bunch to contain a large number of particles, N_s. The ensemble entropy of the bunch is the sum of the entropies of the sub-bunches, i.e., $S = \sum_i S_i$ where S_i is the entropy of the ith sub-bunch. Furthermore, we assume that the longitudinal momentum spread δp_ζ within each slice is small relative to the average momentum of the slice. If the longitudinal density distribution is $\rho(\zeta,p_\zeta)$, then $N_i = \delta\zeta\delta p_\zeta \rho(\zeta,p_\zeta)$. The slice entropy and the rms emittance can be readily evaluated for specific distribution functions. If we consider a specific class of distribution functions where the four dimensional distribution function of each slice can be written in separable form such that $\rho(x,p_x,y,p_y) = \rho_x(x,p_x)\rho_y(y,p_y)$, that the form of ρ_x and ρ_y is the same, and that $\tilde{\varepsilon}_{x,n} = \tilde{\varepsilon}_{y,n} = \tilde{\varepsilon}_{s,n}(\zeta)$, then from eq. (2) the relationship between the slice entropy, $S_s(\zeta,z)$, and the slice emittances can be written as

$$S_s(\zeta,z) = 2k_B N_s \ln(\frac{D\tilde{\varepsilon}_{s,n}}{A_2}) \qquad (3)$$

where $\delta x \delta p_x = \delta y \delta p_y = A_2$ and D is a unitless distribution-dependent parameter. We can evaluate D for specific distributions of interest as shown in Table 1. We expect D to be largest for a Gaussian distribution because the maximum entropy state corresponds to that of an equilibrium Maxwell-Boltzmann distribution, which will have a Gaussian distribution in phase space when space-charge forces are not important[16,17].

The slice entropy, and hence the total bunch entropy, does not depend on the orientation of the slice in transverse phase space. Therefore, it is useful to introduce a quantity called the entropy of the average slice $S_{<s>}$ as

$$S_{<s>}(z) = 2k_B N \ln(\frac{D\langle\tilde{\varepsilon}_{s,n}\rangle}{A_2}) \qquad (4)$$

where $\langle\tilde{\varepsilon}_{s,n}\rangle = \int \tilde{\varepsilon}_{s,n}\rho(\zeta)d\zeta$ is the slice emittance averaged over the bunch.

$S_{<s>}$ has similar properties to the total entropy of the bunch. In particular, $\Delta S_{<s>} > 0$ when $\Delta S > 0$, and $\Delta S_{<s>} = 0$ when $\Delta S = 0$.

Distribution	$(r = x^2 + x'^2)$	Distribution Parameter (D)
Gaussian	$e^{-\frac{r^2}{2\sigma^2}}$	$2e\pi \approx 5.44\pi$
Parabolic	$1 - \frac{r^2}{a^2} \quad r \le a$ $0 \quad r > a$	5π
Uniform	$1 \quad r \le a$ $0 \quad r > a$	4π
Hollow	$r^2 e^{-\frac{r^2}{2\sigma^2}}$	3.105π

Table 1. Distribution parameter (D) for some common distribution functions

3. Differential phase space rotation

The emittance growth may have two components. One involves differential rotation of the slices in phase space. Various effects such as non-uniform longitudinal space-charge distributions, transverse wakefields and phase dependent rf forces will result in the twisting of the phase space being correlated with position in the bunch. The second component results from growth in the emittance of individual slices. On scales shorter than the slice length $\delta\zeta$ we assume that the transverse phase space is uncorrelated with ζ. On scales longer than $\delta\zeta$ we assume that there may be correlations between the phase space and the location of the slice in the bunch.

The phase space orientation of the ith slice may be characterized by Twiss parameters $\hat{\alpha}_i = \hat{\alpha}(\zeta,z)$, $\hat{\beta}_i = \hat{\beta}(\zeta,z)$, and $\hat{\gamma}_i = \hat{\gamma}(\zeta,z)$, with $\hat{\alpha}_i^2 + 1 = \hat{\beta}_i \hat{\gamma}_i$. We envision processes where the phase space distribution of the beam evolves such that the Twiss parameters change in a fashion determined by ζ dependent forces. If the forces that determine the phase space evolution are correlated in ζ (i.e. non-stochastic), the phase space evolution will be correlated in ζ also. The emittance change of each sub-bunch is determined by nonlinear and stochastic transverse forces.

We can write the normalized rms emittance of the bunch for the x coordinate in terms of the slice emittances and slice Twiss parameters as

$$\tilde{\varepsilon}_n^2(z) = \int \tilde{\varepsilon}_{s,n}(\zeta,z)\hat{\beta}(\zeta,z)\rho(\zeta)d\zeta \int \tilde{\varepsilon}_{s,n}(\zeta,z)\hat{\gamma}(\zeta,z)\rho(\zeta)d\zeta$$
$$- (\int \tilde{\varepsilon}_{s,n}(\zeta,z)\hat{\alpha}(\zeta,z)\rho(\zeta)d\zeta)^2 \quad (5)$$
$$= \langle \tilde{\varepsilon}_{s,n}(\zeta,z)\hat{\beta}(\zeta,z)\rangle \langle \tilde{\varepsilon}_{s,n}(\zeta,z)\hat{\gamma}(\zeta,z)\rangle - \langle \tilde{\varepsilon}_{s,n}(\zeta,z)\hat{\alpha}(\zeta,z)\rangle^2$$

We define a normalized emittance correlation coefficient as $C^2(z) = \dfrac{\tilde{\varepsilon}_n^2}{\langle \tilde{\varepsilon}_{s,n}\rangle^2}$.

Since $\tilde{\varepsilon}_n(z) \geq \tilde{\varepsilon}_{s,n}$ always, $C(z) \geq 1$ always. When all the equivalent phase-space ellipses of the slices are aligned, $C = 1$, its minimum value. When the ellipses are not aligned, $C > 1$, because of the differential expansion of the slices and rotation of each phase-space slice with respect to its neighbor. Differentiating $C(z)$ with respect to z gives

$$\frac{d\tilde{\varepsilon}_n^2}{dz} = \langle \tilde{\varepsilon}_{s,n}\rangle^2 \frac{dC^2}{dz} + 2\tilde{\varepsilon}_n^2(z)\frac{d(\ln\langle \tilde{\varepsilon}_{s,n}\rangle)}{dz} \quad (6)$$

Changes in the total rms emittance of the bunch occur either from changes in the phase space correlations or from changes in the slice emittances. It is also possible for $\tilde{\varepsilon}_n$ to remain constant while C decreases and for the slice emittance to increase as the beam is transported through the accelerator. It is now appropriate to examine this in the context of entropy changes. Substituting eq. (4) into eq. (6) we obtain

$$\frac{d\tilde{\varepsilon}_n^2}{dz} = \langle \tilde{\varepsilon}_{s,n}\rangle^2 \frac{dC^2}{dz} - \frac{2\tilde{\varepsilon}_n^2}{D}\frac{dD}{dz} + \frac{\tilde{\varepsilon}_n^2}{k_B N}\frac{d(S_{(s)})}{dz} \quad (7)$$

Let us consider the physical interpretation of eq. (7). We will consider two cases, one where the entropy change is zero, the other where the entropy change is non-zero.

4. Emittance growth without entropy change

Zero entropy-change implies $\dfrac{\tilde{\varepsilon}_n^2}{D}\dfrac{dD}{dz} = -\dfrac{C^2}{2}\dfrac{d\langle\tilde{\varepsilon}_{s,n}\rangle^2}{dz}$, i.e., the distribution adjusts itself to compensate for emittance changes. We assume that the forces that drive the phase space evolution are smoothly varying in space and time. The value of C(z) is calculable from the beam dynamics.

We consider a general zero-entropy-change case where each slice is moving under transverse space-charge forces $qE_{sc}(x,\zeta,z)$, externally applied time-dependent forces $qE_x(x,\zeta,z)$ (e.g., from rf fields, wakefields, etc.), external time-independent magnetic or electric focusing forces represented by $qK_x(x,z)$, and longitudinal time-dependent electric forces $qE_z(x,\zeta,z)$, such that the motion of particles in each slice is given by

$$x'' + \frac{qE_z(\zeta,z)x'}{mc^2\beta^2\gamma} + \frac{qK_x(x,z)}{mc^2\beta^2\gamma} - \frac{qE_{sc}(x,\zeta,z)}{mc^2\beta^2\gamma^3} + \frac{qE_x(x,\zeta,z)}{mc^2\beta^2\gamma} = 0 \qquad (8)$$

where q is the particle charge, and $mc\beta\gamma$ is the longitudinal momentum. We can determine the change of emittance with longitudinal distance by differentiating eq. (7) to obtain

$$\begin{aligned}\frac{d\tilde{\varepsilon}_n^2(z)}{dz} &= \langle\tilde{\varepsilon}_{s,n}\hat{\beta}'\rangle\langle\tilde{\varepsilon}_{s,n}\hat{\gamma}\rangle + \langle\tilde{\varepsilon}_{s,n}\hat{\beta}\rangle\langle\tilde{\varepsilon}_{s,n}\hat{\gamma}'\rangle - 2\langle\tilde{\varepsilon}_{s,n}\hat{\alpha}\rangle\langle\tilde{\varepsilon}_{s,n}\hat{\alpha}'\rangle + \langle(\tilde{\varepsilon}_{s,n})'\hat{\beta}\rangle\langle\tilde{\varepsilon}_{s,n}\hat{\gamma}\rangle \\ &+ \langle\tilde{\varepsilon}_{s,n}\hat{\beta}\rangle\langle(\tilde{\varepsilon}_{s,n})'\hat{\gamma}\rangle - 2\langle\tilde{\varepsilon}_{s,n}\hat{\alpha}\rangle\langle(\tilde{\varepsilon}_{s,n})'\hat{\alpha}\rangle\end{aligned} \qquad (9)$$

where $\tilde{\varepsilon}_{s,n}$ and the Twiss parameters are functions of both ζ and z. Using equation (8) we may evaluate the Twiss parameters for each slice as

$$\hat{\alpha}' = \hat{\gamma} - \frac{q}{mc^2\beta\tilde{\varepsilon}_{s,n}}\langle xK_x\rangle_s + \frac{q}{mc^2\beta\gamma^2\tilde{\varepsilon}_{s,n}}\langle xE_{sc}\rangle_s - \frac{q}{mc^2\beta\tilde{\varepsilon}_{s,n}}\langle xE_x\rangle_s - \frac{(\tilde{\varepsilon}_{s,n})'}{\tilde{\varepsilon}_{s,n}}\hat{\alpha},$$

$$\hat{\beta}' = 2\alpha + \left(\frac{eE_z}{mc^2\beta^2\gamma} - \frac{(\tilde{\varepsilon}_{s,n})'}{\tilde{\varepsilon}_{s,n}}\right)\hat{\beta},$$

$$\hat{\gamma}' = -\frac{2}{mc^2\beta\tilde{\varepsilon}_{s,n}}\langle x'K_x\rangle_s + \frac{2q}{mc^2\beta\gamma^2\tilde{\varepsilon}_{s,n}}\langle x'E_{sc}\rangle_s$$

$$-\frac{2q}{mc^2\beta\tilde{\varepsilon}_{s,n}}\langle x'E_x\rangle_s - \left(\frac{qE_z}{mc^2\beta^2\gamma} + \frac{(\tilde{\varepsilon}_{s,n})'}{\tilde{\varepsilon}_{s,n}}\right)\hat{\gamma}$$

where $\langle\ \rangle_s$ indicates an average over a slice, and after some straightforward algebra, we can re-write eq. (9) as

$$\frac{d\tilde{\varepsilon}_n^2(z)}{dz} = \frac{2q}{mc^2\gamma}\left[\begin{array}{c}\langle x'E_{sc}\rangle\langle x^2\rangle - \langle xx'\rangle\langle xE_{sc}\rangle \\ -\gamma^2\left(\langle x'E_x\rangle\langle x^2\rangle - \langle xx'\rangle\langle xE_x\rangle + \langle x'K_x\rangle\langle x^2\rangle - \langle xx'\rangle\langle xK_x\rangle\right)\end{array}\right]$$

$$= \frac{-2\langle x^2\rangle\gamma}{mc^2N}\left[\frac{1}{\gamma^2}\frac{d[U(z)-U_o(z)]}{dz} + \frac{d[V(z)-V_o(z)]}{dz} + \frac{d[W(z)-W_o(z)]}{dz}\right]$$

$$= \langle\tilde{\varepsilon}_{s,n}\rangle^2\frac{dC^2}{dz} + C^2\frac{d\langle\tilde{\varepsilon}_{s,n}\rangle^2}{dz}$$

(10)

where $\langle\ \rangle$ represents the average over the entire bunch. The expressions in eq. (9) correspond to the emittance growth resulting from self-field and external-field energy changes, where U is the space-charge field energy of the beam and U_o is the space-charge field energy of the equivalent uniform beam; V is the transverse kinetic energy of the bunch induced by the external time-dependent forces, V_o is the kinetic energy that would have been induced if the transverse rf forces were linear in x and phase independent; W is the transverse kinetic energy of the bunch induced by the external time-independent focusing forces, and W_o is the kinetic energy that would have been induced if the external focusing forces were linear in x. The portion of the expression in eq. (10) involving the space-charge field energy has been derived previously [1,11,18,19] for the limited case of drifting beams with linear time-independent external focusing fields. We see that emittance growth in the zero entropy-change case can be characterized by a general energy principle in the form of eq. (10). In the case where no emittance growth occurs expansion and contraction of the bunch result in simple exchanges between potential and kinetic energy.

The emittance changes described by eq. (10) are deterministic, not stochastic, and can be positive or negative depending on the details of the beam dynamics. The time dependent forces give rise to changes in the correlated emittance coefficient C, and the forces that are nonlinear in x give rise to the slice emittance growth. Application of appropriate forces, such as the inverse of the forces that caused emittance growth, will result in removal of the emittance growth if no entropy growth has occurred.

A practical demonstration of emittance growth reversal can be found in rf electron photoinjectors where solenoidal emittance compensation is used [2]. In this case space-charge forces and rf forces combine to introduce correlations into the bunch, and result in large emittance growth over the first few centimeters of beam acceleration. Because of the rapid cooling of the bunch in the longitudinal direction during acceleration, initially there is negligible diffusion of the particles from slice to slice. The emittance growth may be removed by appropriate focusing of the bunch [2], resulting in electron beams of unprecedented brightness [20, 21], i.e., C(z) grows and can be brought back close to unity when appropriate focusing forces are applied.

5. Emittance growth with entropy change

We now consider cases where the entropy of the bunch changes. We note that both C and the slice emittance can increase or decrease. The entropy of the distribution, however, will tend to increase. Such irreversible events can be driven by such effects as Coulomb collisions and thermal diffusion [23], space-charge wave breaking [1,24,25], rf noise, or phase-space filamentation on the scale of the cell size A_2 [10].

We can include the entropy change in our expression for emittance growth by combining eq. (7) and eq. (10) and rearranging terms. This results in a generalized expression for the emittance growth that involves both the reversible and irreversible components

$$\frac{d\tilde{\varepsilon}_n^2}{dz} = \frac{-2\gamma\langle x^2\rangle}{mc^2 N}\left(\frac{1}{\gamma^2}\frac{d[U-U_o]}{dz} + \frac{d[V-V_o]}{dz} + \frac{d[W-W_o]}{dz} - T_{ef}\frac{dS_{\langle s\rangle}}{dz}\right) \quad (11)$$

where $T_{ef} = \dfrac{mc^2 \tilde{\varepsilon}_n^2}{k_B \gamma \langle x^2\rangle}$ is the effective transverse temperature of the beam as measured in the laboratory frame [5].

We can define a thermodynamic potential function, F, that is analogous to the Helmholtz free-energy, where $\dfrac{d\tilde{\varepsilon}_n^2}{dz} = \dfrac{-2\gamma\langle x^2\rangle}{mc^2 N}\dfrac{dF}{dz}$, with $\dfrac{dF}{dz}$ defined with reference to eq. (10). Analysis of the generalized free-energy provides insight into phenomena that have been noted in previous work. The system (in this case the phase space of the bunch) will evolve so that F moves toward a minimum value. Changes in T correspond to expansions and contractions of the bunch. In the case of a cold beam where the space charge dominates over the emittance, the U term can be much greater than the TS term, and the distribution will tend toward uniformity [1,11,18,19,22], its lowest energy state. (We note that for T = 0 the appropriate Maxwell-Boltzmann distribution is a uniform distribution [16]). In the case of a hot, i.e., emittance-dominated beam, the TS term dominates, leading to a Gaussian transverse distribution such as in high energy storage rings [16,17]. In the case of nonlinear external forces, the distribution function should relax to a form appropriate to that force, such that $\dfrac{d[W(z)-W_o(z)]}{dz} = 0$.

In practice we can consider emittance growth to have two components: the reversible part where $\Delta S = 0$ and the irreversible part where $\Delta S > 0$, such that $\Delta\tilde{\varepsilon}(z)^2 = \Delta\tilde{\varepsilon}(z)_I^2 \pm \Delta\tilde{\varepsilon}(z)_R^2$ where the subscripts R and I stand for reversible and irreversible respectively. A determination as to what portion of the emittance growth is reversible in a particular process can be made by evaluating the entropy change using eq. (2). Whether or not entropy growth has occurred will depend, among other things, on the choice of A and $\delta\zeta$.

There is no guarantee that reversible emittance growth can in fact be reversed. This is because effective entropy growth can occur in the absence of stochastic phenomena within the beam. The degree of achievable reversibility depends on our ability to apply corrective forces to the bunch. For example, it is possible to measure the phase-space distribution with a fine resolution and to conclude that no entropy growth has occurred. If, however, we are unable to apply corrective forces on this fine scale, we should conclude that effective entropy growth has occurred and recalculate the entropy using a coarser scale. Therefore, our choice of A_2 and $\delta\zeta$ should be made with reference to the spatial and temporal resolution of our corrective apparatus.

A question that is related to reversibility is as follows: When presented with a beam of given phase-space distribution and emittance, what are the criteria for determining whether or not the emittance may be reduced by the application of deterministic forces? Based on our previous discussion there appear to be two conditions under either of which the emittance may be reduced: 1) if the correlated emittance coefficient $C > 1$, or 2) if the distribution parameter $D < 2e\pi$,

where $D = \dfrac{A_2 \exp\left(\dfrac{S_{<s>}}{2k_B N}\right)}{\langle \tilde{\varepsilon}_{s,n} \rangle}$. Under condition 1) the emittance reduction may be accomplished by applying time dependent forces. Under condition 2) for emittance reduction to occur, D must increase while the emittance decreases in order to keep $S_{<s>}$ constant. For example, in the latter case, take an space-charge dominated beam that has an initially uniform distribution and focus it nonlinearly so that it becomes more Gaussian in form.

Entropy growth does not necessarily imply emittance growth. We see from eq. (10) that changes in entropy can, in principle, be offset or overcome by the application of appropriate nonlinear focusing forces.

A detailed analysis of the time scales for the onset of irreversible dynamics and entropy growth in particular cases is beyond the scope of this paper. An approach using the Fokker-Planck equation [5,11,24,25,26] may be adapted to determine the stochastic entropy growth rate in certain cases. A useful approach for future work will be to use particle simulations to directly evaluate the entropy changes in various situations of interest, so as to determine whether or not phase space evolution can be described by an equation such a eq. (11), and to access the validity of other predictions of the theory presented herein.

Acknowledgments

This work was supported by the Office of Naval Research under contract N00014-91-c-0226. The author would like to acknowledge stimulating discussions with C.L. Bohn, B.E. Carlsten, M. Reiser, J. Struckmeier, and T.P. Wangler, and the encouragement and support of J.M.J. Madey.

References

[1] O.A. Anderson, Part. Accel. **21**, 197 (1987).
[2] B.E. Carlsten, Part. Accel., **49**, 27(1995).
[3] D.X. Wang et al., Phys. Rev. Lett., **73**, 66 (1994).
[4] P.G. O'Shea et al., Nucl. Instr. Meth., **A331**, 62 (1993)
[5] M. Reiser, *Theory and Design of Charged Particle Beams*, Wiley-Interscience, New York, (1994).
[6] P.M. Lapostolle, IEEE Trans. Nucl. Sci. **NS-18**, 1101 (1971)
[7] F.J. Sacherer, ibid., 1105.
[8] A.J. Dragt, F. Neri and G. Rangarajan, Phys. Rev. A, **45**, 2572 (1992).
[9] J.D. Lawson, P.M. Lapostolle and R.L. Gluckstern, *Part. Accel.*, **5**, 61 (1973).
[10] J.D. Lawson, *The Physics of Charged Particle Beams*, Oxford University Press, Oxford, p. 193, 1988.
[11] J. Struckmeier, *Part. Accel.*, **45**, 229 (1994).
[12] K.J. Kim and R.G. Littlejohn, to appear in the proceeding of the 1995 IEEE Particle Accelerator Conference, Dallas TX, May 1-5, 1995.
[13] P.G. O'Shea, Nucl. Instr Meth **A358**, 36, (1995).
[14] N. Brown and M. Reiser, Phys. Plasmas **2**, 965 (1995).
[15] R.T. Cox, *The Statistical Mechanics of Irreversible Change*, The Johns Hopkins Press, Baltimore, 1955.
[16] M. Reiser and N. Brown, Phys. Rev. Lett. **71**, 2911 (1993).
[17] I. Hoffmann et al., Phys. Rev. Lett. **75**, 3842 (1995).
[18] T.P. Wangler et al., IEEE Trans. Nucl. Sci., **NS-32** (1985).
[19] I. Hofmann and J. Struckmeier, Part. Accel. **21**, 69 (1987).
[20] P.G. O'Shea et al., Phys. Rev. Lett. **71**, 3661 (1993).
[21] P.G. O'Shea et al., Nucl. Instr. Meth. **A341**, 7 (1994).
[22] J. Struckmeier, J. Klabunde, M. Reiser, Part Accel. **15**, 47 (1984).

[23] M. Reiser and N. Brown, Phys. Rev. Lett. **74**, 1111 (1995).
[24] C. L. Bohn, Phys. Rev. Lett., **70,** 932 (1993).
[25] C.L. Bohn and J.R. Delayen, Phys. Rev E, **50**, 1516 (1994).
[26] G.H. Jansen, *Coulomb Interactions in Particle Beams*, Academic Press, New York, 1990.

Conceptual Foundation of the Fokker-Planck Approach to Space-Charge Effects

Courtlandt L. Bohn

Continuous Electron Beam Accelerator Facility[1]
Newport News, VA 23606

An rms-mismatched beam can evolve rapidly to a configuration of quasiequilibrium under the influence of space-charge forces. As it evolves, its emittance grows and a diffuse halo forms. The beam's distribution function accounts for all of the complicated dynamics. Unfortunately, the distribution function is difficult to calculate inasmuch as the physics lies at the interface between classical mechanics and thermodynamics. This paper presents the foundation for a statistical theory of the dynamics of nonequilibrium space-charge-dominated beams. Within certain approximations, the theory takes on a Fokker-Planck form. Key questions arise concerning the nature of the dynamical friction and diffusion in the beam's phase space and of the quasiequilibrium configuration that ensues.

INTRODUCTION

A charged-particle beam in an accelerator can be significantly away from thermal equilibrium. This is particularly true following transitions in the effective external focusing force. In an intense beam, space charge drives the subsequent evolution. These beams are generally "cool", and charge redistribution in response to the change in external focusing leads to wavebreaking in the beam's phase space, an event which marks the onset of turbulence. Excess free energy goes toward establishing a mode spectrum. The mode spectrum interacts with particle orbits to generate phase mixing, a process which generally decreases the coarse-grained phase-space density near the phase point of each particle. The fluctuating space-charge potential also widens the range of energies of the particles, a process known in the astrophysical community in connection with a fluctuating gravitational potential in a large stellar system as "violent relaxation".

Phase mixing and violent relaxation can drive the beam toward a quasiequilibrium state on a time scale much shorter than the Coulomb relaxation time. Several numerical simulations show relaxation of rms-mismatched

---

[1] CEBAF is operated by the Southeastern Universities Research Association for the Department of Energy under Contract DE-AC05-84ER40150.

© 1996 American Institute of Physics

space-charge-dominated beams to quasiequilibrium within a few plasma periods, corresponding to only a few lattice periods (1). One might suspect this to be due to numerical noise. However, the process has also been unambiguously observed in laboratory experiments (2).

Inasmuch as they behave like nonneutral plasmas, three-dimensional beams which are strongly turbulent are characterized by an average "collision frequency" $\sim g^{-1}$ times larger than in a quiescent beam, where $g \equiv 1/n\lambda_D^3$ is the plasma parameter (3), with n representing the number density and λ_D representing the Debye length. In space-charge-dominated beams g^{-1} is large, and interactions between particles and localized fluctuations are important. In weak turbulence the average collision frequency is $\sim g^{1/2}$ times smaller than in strong turbulence.

Jean Delayen and I have recently advanced and applied a statistical theory of nonequilibrium beams (4,5). It includes the evolving mode spectrum to account self-consistently for rapid changes in the electrostatic self-field of the beam. Readers interested in a detailed account are referred to these papers. The following section, which is largely extracted from Ref. (4), presents the fundamental conceptual foundation of the theory, outlining how an equation of the Fokker-Planck type can materialize from a "reduction" of Liouville's equation for the distribution function. Subsequent discussion presents key questions that arise in the context of this approach.

FOKKER-PLANCK MODEL

A precise statistical treatment of the beam dynamics involves the microscopic Klimontovich density distribution. This distribution consists of a self-consistent superposition of the orbits of all constituent particles of the beam in the Cartesian phase space $(\mathbf{x}, \mathbf{u})$ of a single particle, and it satisfies Liouville's theorem. The microscopic distribution is

$$f(\mathbf{x}, \mathbf{u}, t) = \frac{1}{N} \sum_{i=1}^{N} \delta[\mathbf{x} - \mathbf{x}_i(t)] \delta[\mathbf{u} - \mathbf{u}_i(t)] , \tag{1}$$

where $[\mathbf{x}_i(t), \mathbf{u}_i(t)]$ denotes the orbit of the ith particle, and N is the total number of particles. The ultimate goal of a numerical experiment is to specify the Klimontovich distribution as a function of time.

To develop an analytic formalism, it is more practical to work with a macroscopic distribution function. We thus consider the macroscopic, coarse-grained distribution function $\bar{f}(\mathbf{x}, \mathbf{u}, t)$ which is found by averaging the Klimontovich distribution over scales substantially greater than those associated with localized turbulent fluctuations:

$$\bar{f}(\mathbf{x}, \mathbf{u}, t) = \frac{1}{\Delta V(\mathbf{x}, \mathbf{u})} \int_{\Delta V(\mathbf{x}, \mathbf{u})} d\mathbf{x}' \, d\mathbf{u}' \, f(\mathbf{x}', \mathbf{u}', t) , \tag{2}$$

where $\Delta V(\mathbf{x}, \mathbf{u})$ is a phase-space volume element centered on the coordinates $(\mathbf{x}, \mathbf{u})$ which is large compared to the size of the turbulent fluctuations but small compared to the size of the beam. In what follows, the "bar" is a signature of quantities calculated from the coarse-grained distribution $\bar{f}$.

After resolving the Klimontovich distribution into two components, $\bar{f}$ and fluctuations about $\bar{f}$, and averaging Liouville's equation, we are left with a "collision" term involving the fluctuations. Working with a coarse-grained distribution function is tantamount to neglecting nonlinear coupling between fluctuations in the particle distribution and fluctuations in the electromagnetic field. This approach results in the reduction of Liouville's equation to an equation of the Fokker-Planck type (6):

$$(\partial_t + \mathbf{u} \cdot \nabla_x + \bar{\mathbf{K}} \cdot \nabla_u)\bar{f} = \nabla_u \cdot (\mathbf{F}\bar{f}) + \nabla_u \cdot (\mathbf{D} \cdot \nabla_u \bar{f}), \tag{3}$$

where $\bar{\mathbf{K}}$ is the net acceleration of a particle in the comoving frame found from the potentials Φ_f and $\bar{\Phi}_s$ associated with the external focusing force and coarse-grained internal space-charge force, respectively, i.e.,

$$\bar{\mathbf{K}} = -QM^{-1}\nabla_x(\Phi_f + \bar{\Phi}_s); \tag{4}$$

the vector $\mathbf{F}$ and tensor $\mathbf{D}$ are coefficients of friction and diffusion, respectively, and Q and M are the charge and mass, respectively, of the beam particles. According to Poisson's equation, $\bar{\Phi}_s$ is determined from the coarse-grained density, which is in turn determined from $\bar{f}$:

$$\nabla_x^2 \bar{\Phi}_s(\mathbf{x}, t) = -\frac{NQ}{\varepsilon_0} \int d\mathbf{u}\, \bar{f}(\mathbf{x}, \mathbf{u}, t)., \tag{5}$$

in which ε_0 is the permittivity of free space.

If the coarse-grained beam is regarded to be uniform so that it can be Fourier transformed using the periodic boundary conditions of a homogeneous cube of volume V, then the transport coefficients are

$$\mathbf{F} = \frac{1}{V}\frac{2\pi Q^2}{M\varepsilon_0} \sum_{\mathbf{k}} \frac{\mathbf{k}}{k^2} \frac{\delta(\omega_\mathbf{k} - \mathbf{k}\cdot\mathbf{u})}{\epsilon'(\mathbf{k}, \omega_\mathbf{k})}, \tag{6}$$

and

$$\mathbf{D} = \frac{1}{V}\frac{2\pi Q^2}{M^2\varepsilon_0} \sum_{\mathbf{k}} \frac{\mathcal{E}_\mathbf{k}}{\omega_\mathbf{k}} \frac{\mathbf{k}\mathbf{k}}{k^2} \frac{\delta(\omega_\mathbf{k} - \mathbf{k}\cdot\mathbf{u})}{\epsilon'(\mathbf{k}, \omega_\mathbf{k})}. \tag{7}$$

In these expressions, $\mathcal{E}_\mathbf{k}$ is the energy contained in the fluctuation with wavevector $\mathbf{k}$ and angular frequency $\omega_\mathbf{k}$, and in this quasilinear formulation it evolves in the manner

$$\frac{\partial \mathcal{E}_\mathbf{k}}{\partial t} = 2\gamma_\mathbf{k}\mathcal{E}_\mathbf{k} + \frac{\pi M\omega_p^2 \omega_\mathbf{k}}{k^2 \epsilon'(\mathbf{k}, \omega_\mathbf{k})} \int d\mathbf{u}\, \bar{f}(\mathbf{u})\delta(\omega_\mathbf{k} - \mathbf{k}\cdot\mathbf{u}), \tag{8}$$

where

$$\gamma_{\mathbf{k}} \simeq \left[\frac{\partial \Re\left[\epsilon(\mathbf{k},\omega)\right]}{\partial \omega}\right]^{-1} \Im\left[\epsilon(\mathbf{k},\omega_{\mathbf{k}})\right] . \qquad (9)$$

The first term on the right-hand side of Eq. (8) accounts for Landau damping or growth at the rate $\gamma_{\mathbf{k}}$ from absorption or induced emission of mode energy by the particles, respectively, and the second term accounts for spontaneous Cherenkov emission of mode energy by the particles. The dielectric response function is

$$\epsilon(\mathbf{k},\omega) = 1 + \frac{\omega_p^2}{k^2} \int d\mathbf{u} \, \frac{\mathbf{k} \cdot \nabla_u \bar{f}(\mathbf{u})}{\omega - \mathbf{k} \cdot \mathbf{u}} , \qquad (10)$$

$\epsilon'(\mathbf{k},\omega_{\mathbf{k}})$ denotes $[\partial\epsilon/\partial\omega]_{\omega=\omega_{\mathbf{k}}}$, $\omega = \omega_{\mathbf{k}} + i\gamma_{\mathbf{k}}$ is the solution of $\epsilon(\mathbf{k},\omega) = 0$, and $\omega_p^2 = nQ^2/\varepsilon_0 M$ is the square of the plasma frequency. The $\mathbf{k}$-summation in Eqs. (6) and (7) is over the growing (unstable) modes obtained from these zeroes, for which $\gamma_{\mathbf{k}} > 0$. The friction $\mathbf{F}$ arises from particles losing energy to unstable modes via spontaneous Cherenkov emission, and the diffusion $\mathbf{D}$ results from particles recoiling in response to absorption and induced emission. In turn, diffusion gives rise to turbulent heating of the particles.

The effect of fluctuations is to change the shape of $\bar{f}$ continuously until there are no more growing modes and $\gamma_{\mathbf{k}} < 0$ for all $\mathbf{k}$. When this has occurred, the modes quickly dissipate by linear Landau damping, and the turbulence vanishes. For example, in the presence of a background isotropic Maxwellian velocity distribution, the fluctuation spectrum evolves as

$$\frac{\partial \mathcal{E}_{\mathbf{k}}}{\partial t} = \frac{\pi M \omega_p^2 \omega_{\mathbf{k}}}{k^2 \epsilon'(\mathbf{k},\omega_{\mathbf{k}})} \left(1 - \frac{\mathcal{E}_{\mathbf{k}}}{k_B T}\right) \int d\mathbf{u} \, \bar{f}(\mathbf{u}) \delta(\omega_{\mathbf{k}} - \mathbf{k} \cdot \mathbf{u}) , \qquad (11)$$

in which k_B is Boltzmann's constant and T is the beam's thermal-equilibrium temperature. This shows explicitly that $\mathcal{E}_{\mathbf{k}}$ dissipates if $\mathcal{E}_{\mathbf{k}} > k_B T$, and the energy spectrum thermalizes toward the equipartition value $\mathcal{E}_{\mathbf{k}} = k_B T$.

In the Fokker-Planck equation (3), the left-hand side accounts for systematic effects arising from the external focusing field and the coarse-grained space-charge field, and it therefore includes resonances between global space-charge modes and the focusing force if any are present. The process of phase mixing is included there. The right-hand side accounts for stochastic effects of the collective-mode spectrum. Fast collisionless relaxation and energy redistribution between modes and particles are included there.

This quasilinear formalism highlights the ingredients of a self-consistent solution for the dynamics of a nonequilibrium beam. As the fluctuations evolve, they change the shape of the coarse-grained distribution function, which in turn modifies the evolution of the fluctuations. As the mode spectrum dissipates, the friction and diffusion coefficients eventually settle down toward their thermal values associated with Coulomb collisions, and the beam then slowly proceeds to thermal equilibrium. A word of caution , however: with respect to charged-particle beams, "thermal equilibrium" has meaning only in the approximation that the frequency of the external focusing forces is high enough that there is no significant beam evolution between focusing lenses or acceleration gaps.

KEY QUESTIONS

The statistical theory is obviously very complicated. It is also only approximate, particularly with regard to its application to an inhomogeneous beam, for in this case it is very difficult to calculate the normal modes. Yet, at the same time, it simplifies matters by operating over the 6-dimensional phase space of a single particle rather than the $6N$-dimensional phase space of the N particles comprising a beam bunch.

To date, applications have incorporated a simple model of the turbulence and Brownian particle motion to specify the Fokker-Planck coefficients. Practical applications of the statistical theory therefore require more accurate models of these coefficients. Establishing viable models will undoubtedly require carefully planned numerical experiments. A fundamental approach would be to calculate the fluctuation spectrum at each position and time step during the course of a numerical experiment and then construct the coefficients from the spectrum using Eqs. (6) and (7) as guides. A second approach might be to infer them from the orbits of test particles in a sequence of numerical experiments. It may also prove fruitful to try to measure these quantities in laboratory beams.

Mismatch-induced excitation and subsequent relaxation of a turbulent mode spectrum constitutes a transient phase in the beam's evolution. As the fluctuations damp, the beam will evolve more slowly. It is not obvious that violent relaxation drives the beam all the way to thermodynamic equilibrium. The process stops when the potential ceases to change and there is no guarantee that the excess free energy is fully equilibrated at that time.

Numerical simulations and laboratory experiments suggest the beam assumes a configuration of quasiequilibrium after a few plasma periods. At this stage the microscopic fluctuations are probably weak, and they will eventually damp to thermal levels. Subsequent evolution is very slow and, except in circular machines, will take much longer than the transit time of the beam through the remainder of the accelerator. Thus, it is important to establish the properties of the quasiequilibrium beam, especially in machines where there is concern about radioactivation due to scraping of a diffuse halo. For example, if the quasiequilibrium beam were to comprise only a few low-order global modes of oscillation, then models describing the oscillating-core-single-particle interaction (7) would probably be sufficient to compute the bulk of the halo over most of the machine. One would need to be careful, though, because in the early transient stage a statistically few particles could conceivably interact resonantly with many modes and be launched into orbits of very large amplitude. This is a consideration, for example, in high-current proton machines envisioned as spallation neutron sources, for which losses less than a nA/m are of concern (8).

One possibly fruitful approach might be to formulate a "maximum-entropy" principle. It would take into account that not all "final" states are equally probable in the configuration of quasiequilibrium. For example, there is not enough time available to populate the largest-amplitude orbits in the Maxwell-Boltzmann distribution. Accordingly, one would expect the core to have progressed closer to thermal equilibrium than the halo, and this is consistent with observations from laboratory and numerical experiments. This approach has been applied to self-gravitating stellar systems with some success (9).

A third area of investigation of possible relevance to accelerators is to consider explicitly a periodic driving force in a frame comoving with the beam. Stochastic processes in the presence of periodic forcing may be susceptible to resonances, which in turn will influence the dynamics. This has been seen, for example, in bistable systems (10). It is also of interest to establish unambiguously the properties of a beam in thermal equilibrium subject to a periodic external force.

REFERENCES

1. See, for example, T. P. Wangler, in *High-Brightness Beams for Advanced Accelerator Applications*, edited by W. W. Destler and S. K. Guharay, AIP Conf. Proc. **253**, (AIP, New York, 1992), pp. 47-56; R. A. Jameson, Los Alamos Report No. LA-UR-93- 12, 1993 (unpublished), and also related articles in this Proceedings.
2. D. Kehne, M. Reiser, H. Rudd, in *High-Brightness Beams for Advanced Accelerator Applications, ibid.*, pp. 47-56, and related discussion in these Proceedings.
3. S. Ichimaru, *Basic Principles of Plasma Physics* (Benjamin, Reading, MA, 1973), Chap. 11.
4. For a comprehensive account, see C. L. Bohn and J. R. Delayen, Phys. Rev. E **50**, 1516 (1990).
5. The theory was introduced in C. L. Bohn, Phys. Rev. Lett. **70**, 932 (1990), and it has been applied in C. L. Bohn and J. R. Delayen, Proc. 1994 Linac Conf., KEK Report, K. Takata, ed., 505 (1995), and in *International Conference on Accelerator-Driven Transmutation Technologies and Applications*, edited by E. D. Arthur, A. Rodriguez, and S. O. Schriber, AIP Conf. Proc. **346**, pp. 371-375 (1995).
6. S. Ichimaru's textbooks are good starting references: Ref. (3) and *Statistical Plasma Physics, Vol. 1* (Addison-Wesley, Redwood City, CA, 1992). The quasilinear formalism described herein is based largely on his treatment. D. Pesme, Phys. Scr. **T50**, 7 (1994) argues that the Fokker-Planck equation may also apply to systems in which nonlinear coupling predominates, provided the transport coefficients are judiciously chosen.

7. See, for example, R. L. Gluckstern, Phys. Rev. Lett. **73**, 1247 (1994), and related articles in this Proceedings.
8. J. R. Delayen and C. L. Bohn, in *International Conference on Accelerator-Driven Transmutation Technologies and Applications, op. cit.*, pp. 460-465.
9. D. N. Spergel and L. Hernquist, Astrophys. J. Lett. **397**, L75 (1992).
10. P. Jung, Phys. Reports **234**, 175 (1993).

RF Linac Designs with Beams in Thermal Equilibrium

Martin Reiser and Nathan Brown*
Institute for Plasma Research
University of Maryland
College Park, MD 20742

Abstract

Beams in conventional radio-frequency linear accelerators (rf linacs) usually have a transverse temperature which is much larger than the longitudinal temperature. With high currents, space charge forces couple the transverse and longitudinal particle motions, driving the beam toward thermal equilibrium, which leads to emittance growth and halo formation. A design strategy is proposed in which the beam has equal transverse and longitudinal temperatures through the entire linac, avoiding these undesireable effects. For such equipartitioned linac beams, simple analytical relationships can be derived for the bunch size, tune depression, and other parameters as a function of beam intensity, emittance and external focusing. These relations were used to develop three conceptual designs for a 938 MeV, 100 mA proton linac with different tune depressions, which are presented in this paper.

---

*Current address: G. H. Gillespie Associates, Inc., P.O. Box 2961, Del Mar, CA 92014.

© 1996 American Institute of Physics

1. Introduction

Space charge forces play a dominant role in high-current radio-frequency linear accelerators (rf linacs) for advanced accelerator applications such as injector linacs for high-energy colliders, free electrons lasers, spallation neutron sources, the transmutation of radioactive nuclear waste, tritium production and heavy-ion inertial fusion. In conventional rf linacs the beams typically have a transverse temperature, $T_\perp$, which is much larger than the longitudinal temperature, $T_\parallel$. Space charge forces couple the transverse and longitudinal particle motion and drive the beam toward thermal equilibrium. This equipartitioning effect can lead to emittance growth and halo formation, which is unacceptable for high-power, high-quality beams in advanced accelerator applications.

The applied transverse and longitudinal focusing forces acting on an axially symmetric bunched beam in an rf linac are defined by

$$F_x = \gamma_0 m \beta_0^2 c^2 k_{x0}^2 x, \; F_z = \gamma_0^3 m \beta_0^2 c^2 k_{z0}^2 z. \tag{1}$$

The transverse focusing wave number is

$$k_{x0} = \frac{\sigma_{x0}}{n\beta_0 \lambda}. \tag{2}$$

Here, σ_{x0} is the transverse phase advance per cell without space charge and $n\beta_0\lambda$ is the cell length, where n is an integer ≥ 1, β_0 is the beam velocity in units of the speed of light, c, and $\lambda = c/f$ is the rf wavelength, where f is the linac frequency. The longitudinal focusing wave number is

$$k_{z0} = \left(\frac{-2\pi q E_m \sin \phi_0}{\lambda m c^2 \beta_0^3 \gamma_0^3}\right)^{1/2}, \tag{3}$$

where q and m are, respectively, the particle charge and mass, E_m is the electric field amplitude, ϕ_0 is the synchronous phase, and $\gamma_0 = (1-\beta_0^2)^{-1/2}$ is the relativistic energy factor.

The ratio of the transverse and longitudinal focusing strengths, k_{x0}/k_{z0}, determines the bunch shape and the resulting beam physics, including the ratio of the transverse and longitudinal temperatures.

A design strategy is proposed in which the beam in an rf linac is equipartitioned through the entire system. The emittance growth, halo formation and beam losses which can occur due to the equipartitioning process in existing machines are avoided. To illustrate the flexibility in designing such a linac, three design examples are presented of a 938 MeV, 100 mA proton linac with different tune depressions and beam radii.

2. Conventional Linac Design

In conventional rf linacs the transverse phase advance per cell, σ_{x0}, is kept constant along the channel. The product of the accelerating gradient, E_m, and $\sin\phi_0$ is also kept constant along the channel. This yields the scaling

$$\begin{aligned} k_{x0} &\propto 1/\beta_0, \\ k_{z0} &\propto 1/(\beta_0\gamma_0)^{3/2}, \\ k_{x0}/k_{z0} &\propto \beta_0^{1/2}\gamma_0^{3/2}. \end{aligned} \quad (4)$$

In conventional linacs the ratio of the transverse and longitudinal focusing strengths, k_{x0}/k_{z0}, increases with energy, so that the beam radius and the ratio of the radius and the bunch length decrease. This leads to a temperature anisotropy, i.e. the ratio of the transverse and longitudinal temperatures also increases with distance along the channel. When space charge forces play a significant role in the beam physics, as is the case in many advanced applications of rf linacs, the two-temperature distribution does not satisfy the stationary Vlasov equation. Coupling between the transverse and longitudinal space charge forces drives the beam toward thermal equilibrium. This equipartitioning effect, first demonstrated in computer simulation studies by Jameson [1], and, more recently, by Wangler et al. [2], leads to emittance growth, halo formation and beam losses. A review of efforts to design rf linacs in which these detrimental effects are reduced has also been given by Jameson [3].

3. Conditions and Design Relations for Thermal Equilibrium

The condition that a beam be equipartitioned ($T_\perp = T_\parallel$) with constant transverse and longitudinal emittance requires that the bunch aspect ratio remain constant as the energy changes. The effective transverse and longitudinal emittances are defined, respectively, as [4]

$$\epsilon_{nx} = 5\tilde{\epsilon}_{nx} = \tilde{x}\left(\frac{\gamma_0 k_B T_\perp}{mc^2}\right)^{1/2},$$

and
$$\epsilon_{nz} = 5\tilde{\epsilon}_{nz} = \tilde{z}\left(\frac{\gamma_0^3 k_B T_\|}{mc^2}\right)^{1/2},$$

where $\tilde{\epsilon}_{nx}$ and $\tilde{\epsilon}_{nz}$ are, respectively, the rms transverse and longitudinal emittances, $\tilde{x}$ and $\tilde{z}$ are, respectively, the rms values for the bunch size in the x and z directions, and k_B is Boltzmann's constant. It is assumed here that the beam is symmetric about the axis of propagation.

In terms of the emittances, the condition for equipartitioning is

$$\frac{\epsilon_{nx}}{\epsilon_{nz}} \frac{\gamma_0 z_m}{a} = 1. \tag{5}$$

Here, $a = \sqrt{5}\tilde{x}$ is the radius and z_m is the half-length of the equivalent uniform ellipsoid in the laboratory frame; $\gamma_0 z_m$ is the half-length in the beam frame where the factor γ_0 is due to the relativistic length contraction. Note that the aspect ratio $(\gamma_0 z_m/a)$ in the beam frame remains constant during acceleration while the aspect ratio in the laboratory frame decreases as γ_0^{-1}.

The coupled envelope equations for a matched beam are [4]

$$k_{x0}^2 a - \frac{3Nr_c}{2\beta_0^2 \gamma_0^3 a z_m}\left(1 - \frac{ga^2}{2\gamma_0^2 z_m^2}\right) - \frac{\epsilon_{nx}^2}{\beta_0^2 \gamma_0^2 a^3} = 0, \tag{6}$$

$$k_{z0}^2 z_m - \frac{3Nr_c g}{2\beta_0^2 \gamma_0^5 z_m^2} - \frac{\epsilon_{nz}^2}{\beta_0^2 \gamma_0^6 z_m^3} = 0. \tag{7}$$

The geometry factor, g, depends on the ratios $\gamma_0 z_m/a$ and b/a, where b is the radius of the drift tube [4, 5]. In most light-ion and low-energy electron rf linacs, where γ_0 is in the range $1 \leq \gamma_0 \lesssim 5$, the half-length $\gamma_0 z_m$ of the bunches in the beam frame is small compared to the tube radius b (i.e., $\gamma_0 z_m < b$). In this case, image effects can be neglected and the geometry factor g can be approximated by the "free-space" factor g_0 which depends only on the ratio $\gamma_0 z_m/a$ and is given by [4, 5]

$$g_0 = \frac{2}{\xi^2}\left[\frac{1}{2\xi}\ell n\left(\frac{1+\xi}{1-\xi}\right) - 1\right], \tag{8a}$$

$$\xi = \left(1 - \frac{a^2}{\gamma_0^2 z_m^2}\right)^{1/2} \quad \text{for } \gamma_0 z_m > a, \tag{9a}$$

and

$$g_0 = \frac{2}{\xi^2}\left[1 - \frac{\tan^{-1}\xi}{\xi}\right], \tag{8b}$$

$$\xi = \left(\frac{a^2}{\gamma_0^2 z_m^2} - 1\right)^{1/2} \quad \text{for } \gamma_0 z_m < a. \tag{9b}$$

If the beam is in thermal equilibrium the ratio $\gamma_0 z_m/a$ remains constant during acceleration, depending only on the emittance ratio $\epsilon_{nz}/\epsilon_{nx}$ according to Eq. (5); the geometry factor g_0 is then also constant and can be calculated from Eqs. (8), (9). Substituting $\gamma_0 z_m/a = \epsilon_{nz}/\epsilon_{nx}$ from Eq. (5) into Eq. (6) and treating g_0 as a constant, we obtain from Eqs. (6), (7) the two uncoupled fourth-order algebraic equations in a and z_m for an equipartitioned beam:

$$k_{x0}^2 a - \frac{3Nr_c}{2\beta_0^2\gamma_0^2 a^2}\frac{\epsilon_{nx}}{\epsilon_{nz}}\left(1 - \frac{g_0}{2}\frac{\epsilon_{nx}^2}{\epsilon_{nz}^2}\right) - \frac{\epsilon_{nz}^2}{\beta_0^2\gamma_0^2 a^3} = 0, \tag{10}$$

$$k_{z0}^2 z_m - \frac{3Nr_c g_0}{2\beta_0^2\gamma_0^5 z_m^2} - \frac{\epsilon_{nz}^2}{\beta_0^2\gamma_0^6 z_m^3} = 0. \tag{11}$$

Using our approximation for fourth-order algebraic equations of this type [6] we find the approximate solutions for the bunch radius a and half-length z_m:

$$a \simeq \left[\frac{3Nr_c}{2\beta_0^2\gamma_0^2 k_{x0}^2}\frac{\epsilon_{nx}}{\epsilon_{nz}}\left(1 - \frac{g_0}{2}\frac{\epsilon_{nx}^2}{\epsilon_{nz}^2}\right) + \left(\frac{\epsilon_{nx}}{\beta_0\gamma_0 k_{x0}}\right)^{3/2}\right]^{1/3}, \tag{12}$$

$$z_m \simeq \left[\frac{3Nr_c g_0}{2\beta_0^2\gamma_0^5 z_m^2} + \left(\frac{\epsilon_{nz}}{\beta_0\gamma_0^3 k_{z0}}\right)^{3/2}\right]^{1/3}. \tag{13}$$

These relations exhibit the scaling with space charge and emittance (first and second terms, respectively, in the brackets on the right-hand side). They are exact in the space-charge or emittance-dominated limits, and the accuracy is always within 3.4%.

The factor g_0 can be readily calculated for any particular aspect ratio $\gamma_0 z_m/a$ from the two emittances. As an example, suppose that $\epsilon_{nz}/\epsilon_{nx} = 4$, hence $\gamma_0 z_m/a = 4$. Then $\xi = 0.968$ from Eq. (9a) and $g_0 = 2.413$ from Eq. (8a). In the region $0.7 \leq \gamma_0 z_m/a \leq 4$, the factor g_0 can be approximated by the linear relation $g_0 \simeq 2\gamma_0 z_m/3a$. This can be used to obtain simpler solutions for either an equipartitioned beam [7] or for the conventional rf linac where the beam is not in thermal equilibrium (Eqs. (A.4.27) and (A.4.28) in Ref. 4). In the latter case, the emittances do not remain constant due to the equipartitioning effect, and one needs to know how ϵ_{nz} and ϵ_{nx} vary with energy if one wants to calculate the bunch parameters a and z_m along the linac. For the above example of $\gamma_0 z_m/a = 4$, the linear approximation yields $g_0 = 2.667$, which is about 9.5% higher than the exact value of 2.413.

The ratio between the transverse and longitudinal focusing constants for an equipartitioned beam can be derived from the envelope equations as

$$\frac{k_{x0}}{k_{z0}} = \left[\frac{\frac{3}{2}(Nr_c a/\epsilon_{nx}^2)(\epsilon_{nz}/\epsilon_{nx})g_r + \epsilon_{nz}^2/\epsilon_{nx}^2}{\frac{3}{2}(Nr_c a/\epsilon_{nx}^2)(\epsilon_{nx}/\epsilon_{nz})g + 1}\right]^{1/2}, \quad (14)$$

where g was replaced by g_0 and g_r is defined as

$$g_r = 1 - \left(\frac{g_0}{2}\frac{\epsilon_{nx}^2}{\epsilon_{nz}^2}\right). \quad (15)$$

In conventional rf linacs, k_{x0}/k_{z0} is increased as the beam propagates along the channel. In an equipartitioned rf linac, k_{x0}/k_{z0} varies only slightly as the bunch changes its size and its space charge tune depression as the energy changes. The coupled envelope equations can also be used to obtain the transverse and longitudinal space charge tune depressions,

$$\frac{k_x}{k_{x0}} = \left(1 - \frac{3Nr_c g_r}{2\beta_0^2 \gamma_0^3 a^2 z_m k_{x0}^2}\right)^{1/2}, \quad (16)$$

$$\frac{k_z}{k_{z0}} = \left(1 - \frac{3Nr_c g_0}{2\beta_0^2 \gamma_0^5 z_m^3 k_{z0}^2}\right)^{1/2}. \quad (17)$$

4. Three Designs with Different Space Charge Tune Depressions

To demonstrate the versatility of the design of an equipartitioned rf linac, we give three examples with different bunch aspect ratios and different space charge tune depressions. All three examples are of a 2 MeV to 938 MeV proton linac, with an average current of 100 mA. The transverse normalized

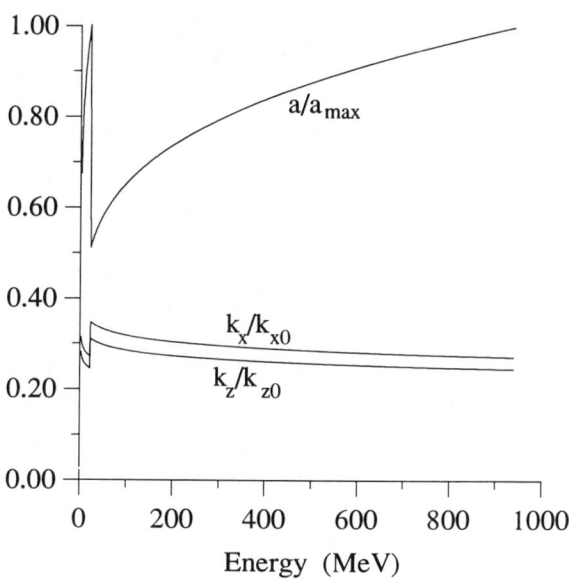

Figure 1: Variation of beam radius and tune depression with energy for Linac Design 1 of equipartitioned linac.

emittance is $\epsilon_{nx} = 6.85 \times 10^{-7}$ m-rad, and the synchronous phase is $\phi_0 = -40°$. The frequency is 200 MHz, ($\lambda = 1.5$ m) and the field gradient is $E_m = 1.6$ MV/m at low energy; the frequency is 800 MHz ($\lambda = 0.375$ m) and the field gradient is E_m 3.2 MV/m at high energy. This frequency change from 200 MHz to 800 MHz makes it possible to increase the field gradient E_m, and hence the focusing strength in both directions (k_{z0} and k_{x0}), which reduces the beam size. The transition point from the low energy structure to the high energy structure, which in all three designs occurs at about 22 MeV, is chosen so that the beam reaches the same peak radius at the end of each structure (i.e., at 22 MeV and at 938 MeV).

Figure 1 shows the peak radius of the equivalent ellipsoid as a function of energy, normalized to its maximum value; this is the same curve for all three design examples. Also in Figure 1 are the transverse and longitudinal space charge tune depressions for Design 1, from Eqs. (13) and (14). Design 1 has an aspect ratio in the beam frame of 2, and therefore, according to the

Table 1: Beam and channel parameters at 2 MeV, 22 MeV (after transition to the high energy structure), and 938 MeV, for three design examples of equipartitioned beams.

Design	Parameter	2 MeV	22 MeV	938 MeV
1	a	2.7 mm	2.0 mm	4.0 mm
	z_m	5.3 mm	3.9 mm	4.0 mm
	$\Delta\phi_m$	20°	18°	4.4°
	k_{x0}	6.1 m^{-1}	3.0 m^{-1}	0.13 m^{-1}
2	a	1.6 mm	1.2 mm	2.4 mm
	z_m	6.5 mm	4.8 mm	4.8 mm
	$\Delta\phi_m$	24°	22°	5.3°
	k_{x0}	11 m^{-1}	5.0 m^{-1}	0.22 m^{-1}
3	a	0.97 mm	0.74 mm	1.4 mm
	z_m	7.7 mm	5.8 mm	5.8 mm
	$\Delta\phi_m$	28°	26°	6.3°
	k_{x0}	20 m^{-1}	9.4 m^{-1}	0.39 m^{-1}
	k_{z0}	4.1 m^{-1}	1.9 m^{-1}	0.084 m^{-1}

equipartitioning condition [Eq. (5)], a longitudinal emittance of 1.27 ×10^{-6} m-rad. The thermal equilibrium distribution for the final state of this design (at 938 MeV) is shown in Fig. 4; the aspect ratio in the lab frame is 1 due to the relativistic length contraction. The fact that this beam is space charge dominated can be seen by the near uniformity of the density over most of the bunch, and the sharp drop-off at the edges of the bunch.

Several parameters are shown in Table 1, including the phase half-width of the equivalent uniform ellipsoid, which is defined as $\Delta\phi_m = 2\pi z_m/\beta_0 \lambda$.

Figure 2 shows the space charge tune depressions of an equipartitioned beam with an aspect ratio in the beam frame of $\gamma_0 z_m/a = 4$ (Design 2); the channel parameters are the same as in the previous case except that the transverse focusing strength is increased to get a larger aspect ratio. The thermal equilibrium distribution for the final state of this design (at 938 MeV) is shown in Fig. 5. The tune depressions are higher than in Design 1 due to the increased focusing of the beam with the same transverse emittance. The longitudinal emittance in this case is $\epsilon_{nz} = 4\epsilon_{nx} = 2.74 \times 10^{-6}$ m-rad.

Figure 3 shows the space charge tune depressions of an equipartitioned

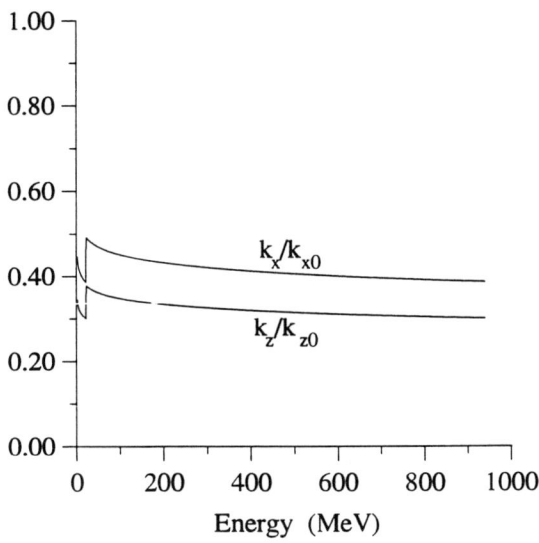

Figure 2: Tune depression vs. energy for Linac Design 2.

beam with an aspect ratio in the beam frame of $\gamma_0 z_m/a = 8$ (Design 3), representing a further increase in the transverse focusing strength. The thermal equilibrium distribution for the final state of this beam is shown in Fig. 6, which reveals the more peaked profile (compared with that of Design 1) corresponding to the higher values of the tune-depression parameters (k_x/k_{x0}, k_z/k_{z0}) in this emittance-dominated beam. Table 1 shows some parameters for Designs 1, 2, and 3.

In the first design, the phase width is small enough compared to the width of the rf bucket that the focusing nonlinearities of the rf bucket are insignificant. In Design 2 and Design 3 these focusing nonlinearities are included in the calculation of the thermal equilibrium distributions [8], but the symmetry of Figs. 5 and 6 about $z = 0$ shows that these nonlinearities do not have a significant effect on the beam distribution.

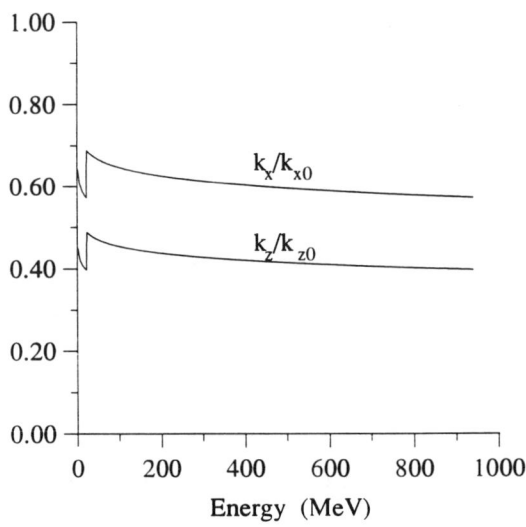

Figure 3: Tune depression vs. energy for Linac Design 3.

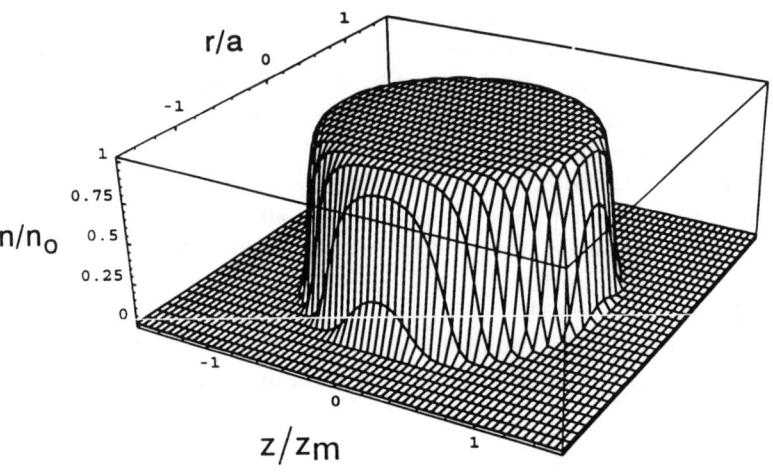

Figure 4: Bunch density profile for Linac Design 1 at full energy (938 MeV).

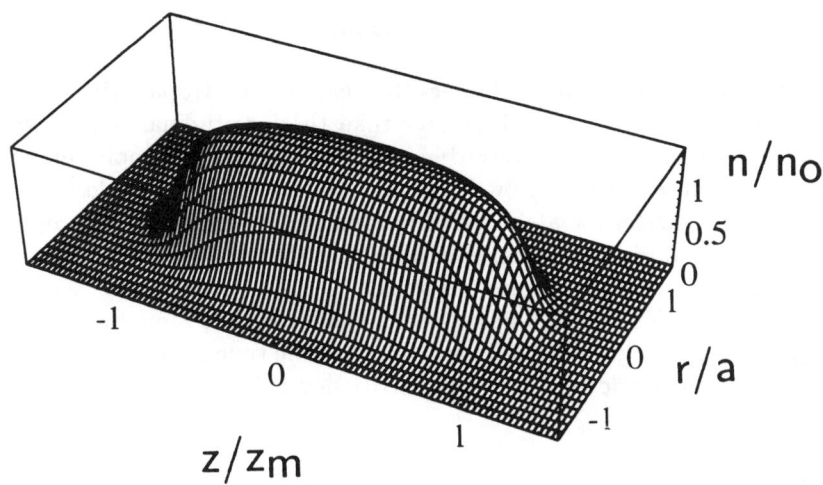

Figure 5: Bunch density profile for Linac Design 2 at 938 MeV.

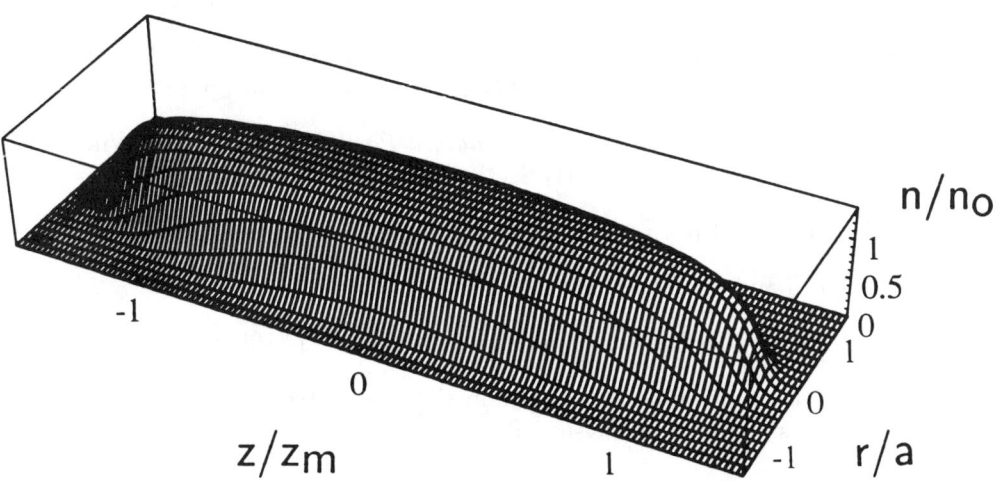

Figure 6: Bunch density profile for Linac Design 3 at 938 MeV.

5. Conclusion

In the prevailing design for rf linacs the beam is anisotropic, with a transverse temperature that is much greater than the longitudinal temperature. The relaxation of the beam toward equilibrium with an isotropic temperature can lead to emittance growth, halo formation and particle losses even if the beam is carefully matched. An alternative design has been proposed in which the beam is in thermal equilibrium so that emittance growth, halo formation and particle losses are minimized.

Analytic relations for the beam parameters have been found, and three examples have been given for different values of the tune depression, showing the flexibility in the design of an equipartitioned rf linac.

References

[1] R. A. Jameson, IEEE Trans. Nucl. Sci. **28**, 2408 (1981).

[2] T. P. Wangler, T. S. Bhatia, G. H. Neuschaefer, and M. Pabst, Conf. Record of the 1989 IEEE Trans. Nucl Sci. NS-28, 2399 (1989).

[3] R. A. Jameson, in *Advanced Accelerator Concepts*, ed. J. S. Wurtele, AIP Conference Proceedings **279** (AIP, New York, 1993), p. 969.

[4] M. Reiser, *Theory and Design of Charged Particle Beams* (John Wiley & Sons, Inc., New York, 1994) Section 5.4 and Appendix 4.

[5] C.K. Allen, N. Brown and M. Reiser, Part. Accel. **45**, 149 (1994).

[6] N. Brown and M. Reiser, Part. Accel. **43**, 231 (1994).

[7] M. Reiser and N. Brown, Phys. Rev. Lett. **74**, 1111 (1994).

[8] N. Brown and M. Reiser, "Longitudinal current losses in rf linear accelerators," submitted to Physical Review E.

NEUTRON SPALLATION SOURCES

Critical Beam Dynamical Issues in Neutron Spallation Sources

M. Pabst and K. Bongardt

Forschungszentrum Jülich GmbH, 52425 Jülich, Germany

and

A.P. Letchford

Rutherford Appleton Laboratory, Chilton, Didcot, UK

Abstract

The accelerator part of proposed neutron spallation sources consists of a high intensity linac and compressor ring or rapid cycling synchrotron. The most critical part of such a high current machine is to keep activation caused by particle loss along the linac or at ring injection down to an acceptable limit. Sources of particle loss along the linac can be beam mismatch, resonances of any kind, temperature transfer within a bunch and/or nonlinear internal or external forces. In addition machine errors like misalignments, tolerances and rf errors have to be considered. All these sources cause emittance growth. The common way of setting up the beam dynamics of high intensity linacs is governed by avoiding these sources and testing it by Monte-Carlo simulations. To get information on the possible loss mechanism, the only way is to increase the particle number of the Monte-Carlo simulations and to study phase space distributions in detail.

Monte-Carlo simulations with 50000 particles for the 1.334 GeV coupled cavity linac of the European Spallation Source (ESS) are presented. It is shown that it is possible to design a non-space charge dominated linac for 200 mA bunch current with almost constant emittances. A detailed study of the phase space distribution along the linac shows a small number of halo particles nearby the bunch core. This halo is acceptable for ring injection. Some information related to particle loss in the linac and in the compressor ring afterwards is extracted and comments for positioning scrapers are made.

Introduction

All proposed pulsed spallation source projects consist of a high intensity H^- – linac followed by one (or more) compressor rings or rapid cycling synchrotrons [1]. In Fig. 1, as a typical example, the layout of the ESS linac is shown [2]. The low energy part consists of two H^- – ion sources with 70 mA peak current each, a 2 MeV bunched beam transfer line between two RFQs for installing a fast chopping device and a 5 MeV funnelling line afterwards. The drift tube linac (DTL) operates at 350 MHz, the coupled cavity linac (CCL) at 700

© 1996 American Institute of Physics

MHz. The transition energy is 70 MeV. In the 1.334 GeV high energy transfer line, a 4 m long 700 MHz cavity is positioned after 75 m, acting as a bunch rotator [4]. The linac operates at 50 Hz with 6% duty cycle. All the mentioned parameters are more or less typical for high intensity H^-- injector linacs.

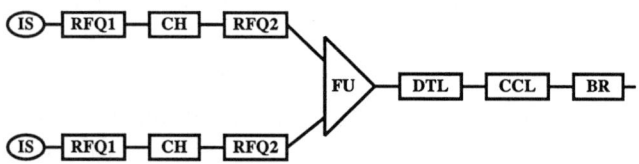

Fig. 1 ESS linac layout: IS: ion source, CH: chopper, FU: funneling, BR: bunch rotator

The main task is to reduce losses at ring injection. Therefore the linac pulse has to be chopped at the ring revolution frequency. In addition at the high energy end the energy spread has to be limited by the bunch rotation cavity. The use of a funnelling scheme implies a second bunched beam transfer line but it relaxes the constraints of the chopping line and the H^-– ion sources considerably [10]. Beside the loss problem at ring injection particle losses along the linac should be about 1 nA/GeV/m in order to allow hands on maintenaince.

Layout and beam dynamical design of the coupled cavity linac section

Because particle loss below 120 MeV causes negligible activation, results are presented for the technically designed ESS 700 MHz CCL. The main features of the technical linac design are the following items.

Costs are minimized by choosing an accelerating gradient $E_oT = 2.8$ MV/m. These costs include structure, rf, ten years of operation and buildings without extensive shielding.

Concerning losses one has to be aware of 'matching' losses in the transverse and longitudinal direction resulting from the change of the transverse focusing period and the accelerating gradient between the preceding 350 MHz DTL and the 700 MHz CCL. The matching losses occur mainly after injection into the CCL. Therefore the input energy of the CCL has to be as low as possible, here 70 MeV, to be far below the neutron production threshold of about 120 MeV.

To reduce the phase slip at injection into the CCL the tanks have to be short enough. For the ESS–CCL the number of cells with constant length per tank decreases from 16 to 10. This allows the choice of a constant synchronous phase of — 25 ° and gives a phase slip of ± 4° in the first tank.

Transverse focusing is provided by doublets located after every second tank in $5\beta\lambda/2$ long intertank sections. Doublets are favoured over singlets giving a more round beam with smaller average diameter and beam envelope oscillations. The quadrupole gradients vary between 25 and 15 T/m.

Diagnostic equipment and steering elements are placed in short intertank sections. The length of these is $3\beta\lambda/2$. The increase of the intertank sections with β allows the installation of scrapers at the high β end and the use of less compact doublets which can be supplied by stable dc power supplies.

For setting up the beam dynamics several constraints have to be fulfilled.

First, a constant transverse tune is not possible along the CCL. Due to a decreasing beam radius in case of a constant transverse tune, space charge increases at higher energies. Therefore the effective longitudinal focusing force goes to zero. This problem can be solved by decreasing the transverse tune σ_t with β.

Second, to avoid temperature exchange between transverse and longitudinal direction the beam should be close to equipartitioning.

Third, to avoid resonances the transverse and the longitudinal tunes should not be too large.

Fourth, the tune depression should be high in order to reduce the number of possible resonances.

As an example results are shown for the ESS CCL. The bunch current is 214 mA, the normalized transverse rms emittance is 0.6 π mm mrad and the longitudinal rms emittance is 1.2 π ° MeV. In Fig. 2 the tunes for full current are shown. The transverse tune decreases like $\sigma_t = \sigma_{t0}\,(\gamma_0/\gamma)^{-2.5}$. γ is the relativistic factor and γ_0, σ_0 are the values at 70 MeV. Not shown are the tune depressions and the zero current tune. The tune depressions are constant around 0.8 in the transverse and the longitudinal direction indicating a non-space charge dominated beam. This results in a quite large value of 102 ° for the initial zero current transverse tune. At 105 MeV this tune has decreased to values below 90 °. The initial longitudinal tune starts below 90 °.

In Fig. 3 the ratio of the transverse to longitudinal temperature (equipartition ratio) is shown. Two curves are presented which differ by a factor γ^2. In the bunch system the beam is almost equipartitioned along the linac but not in the laboratory system.

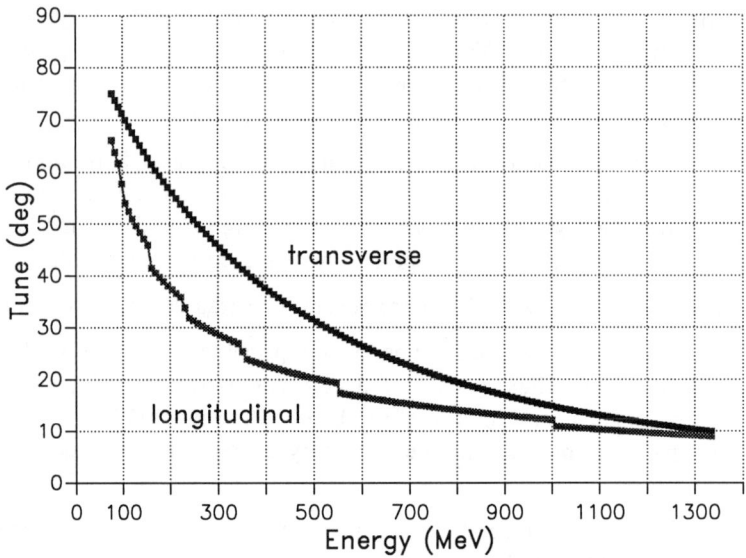

Fig. 2 Transverse and longitudinal full current tunes along the linac

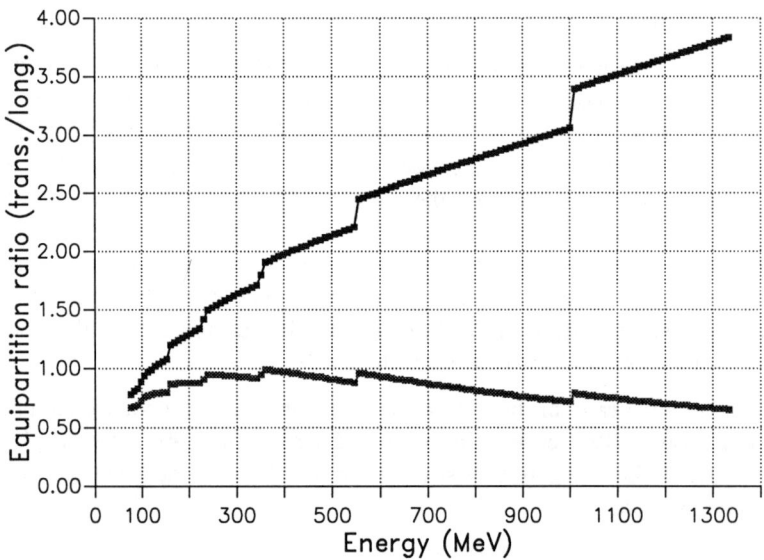

Fig. 3 Ratio of transverse to longitudinal temperatures. The upper curve is calculated in the laboratory system, the lower in the bunch system.

Results from Monte-Carlo simulations

In order to check the above design, Monte-Carlo simulation have been done with 50000 interacting particles. The three-dimensional space charge forces are calculated every $\beta\lambda/2$ in order to take into account the variation of the transverse beam radius. The initial phase space distribution is of a Waterbag type in the four transverse coordinates and, independently, also in the two longitudinal coordinates.

In Fig. 4 the rms emittances are shown along the CCL. Less than 10% rms emittance growth is observed. This indicates that no dangerous resonances are present and no temperature exchange takes place between the transverse and longitudinal plane.

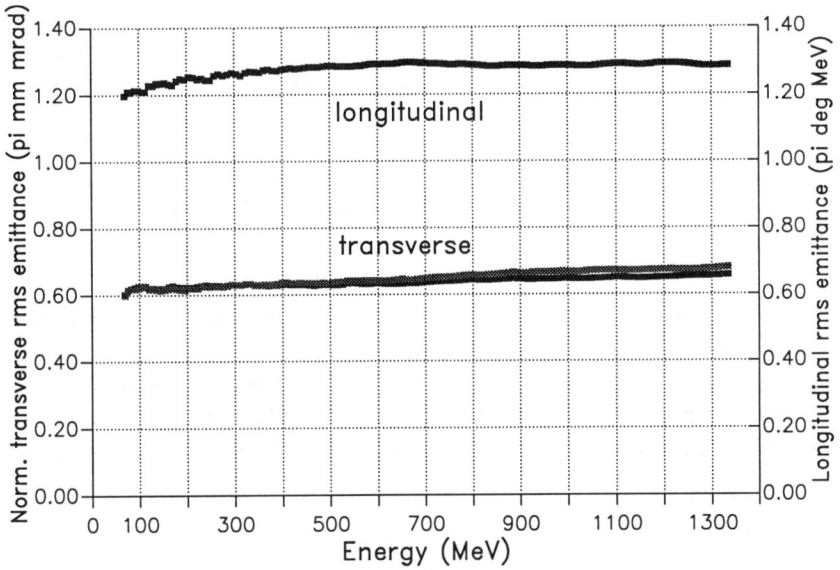

Fig. 4 Normalized rms emittances along the CCL

The transverse mismatch factors [5] are presented in Fig. 5. The observed oscillation of this factor for a perfectly matched beam at injection is probably due to the change of the cell length and the number of cells per tank. Due to the small tune depression of about 0.8 the relatively large average mismatch of 0.3 causes no emittance growth, consistent with previous numerical results [6]. A similar argument holds for the initial transverse zero current tune of about 102 ° because the dangerous 90 °– resonance is passed quite fast.

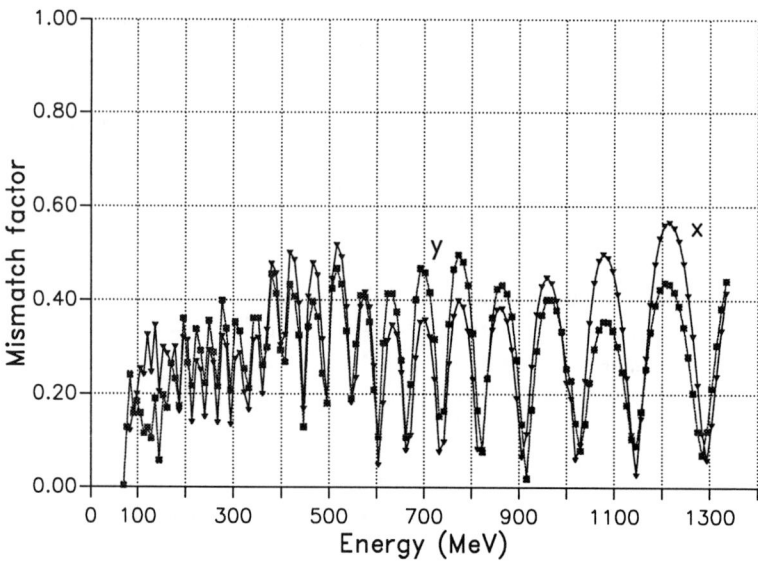

Fig. 5 Transverse mismatch factors along the CCL. The beam is almost matched at injection.

Rms quantities, e.g. radius, emittance, matching parameters, are useful to check the beam dynamical design of the linac. However, these quantities give no information about particles loss. To get some information one has to consider the outermost particles of the phase space distribution. The question arises as to what can be evaluated and is it possible to extract a quantity related to particle loss.

Fig. 6 and 7 show projections of the phase space distribution at the injection energy of 70 MeV and at the linac end at about 1.334 GeV. The projections considered are the real transverse space (x,y) and the longitudinal phase space ($\Delta\phi$, ΔW). The initial distributions have 'sharp' edges due to the fact that it presents an analytical one. At the end the distributions have no sharp edges anymore. The boundaries have become 'soft'. We can even recognize some satellite particles, called halo. Looking at the output distributions we count between 40 and 50 particles, corresponding to 0.1% out of 50000 particles. The transverse halo particles are not necessarily longitudinal ones and vice versa. Usually they are different.

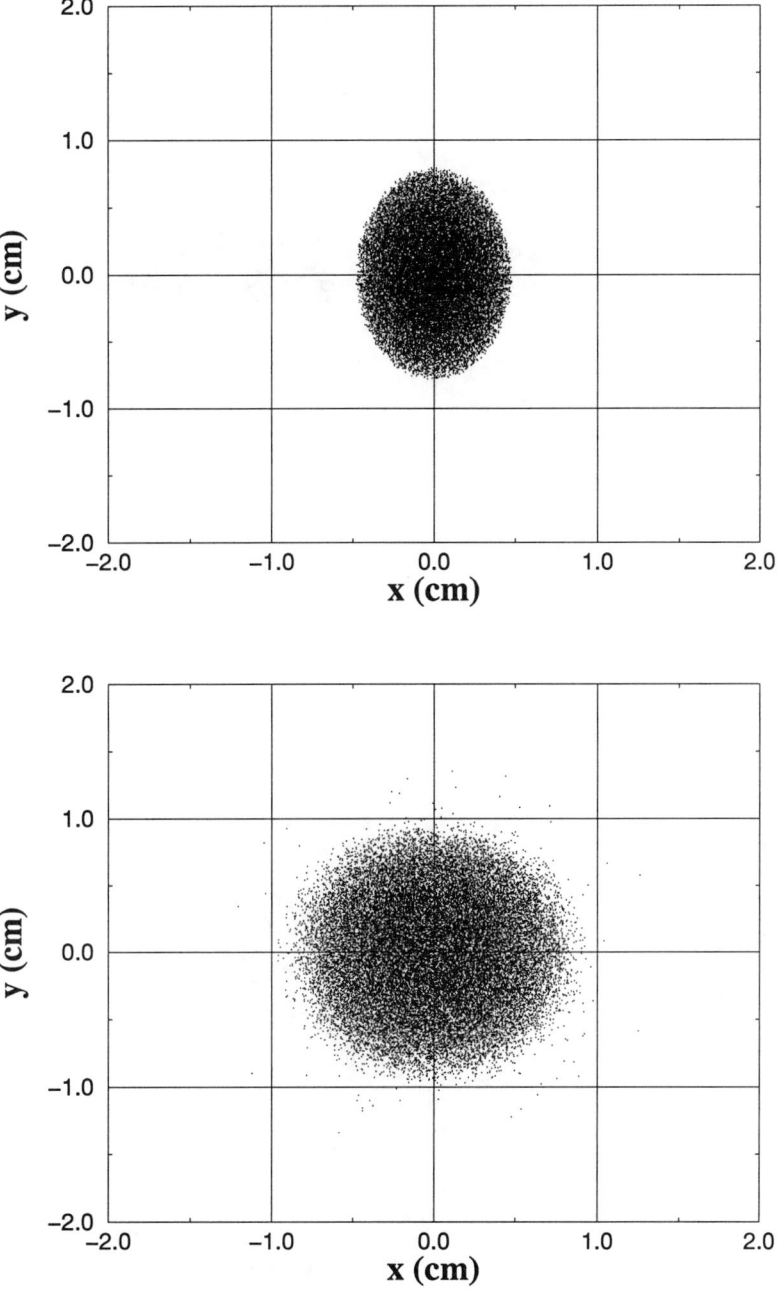

Fig. 6 Real space (x,y) at the input (top) and output of the CCL

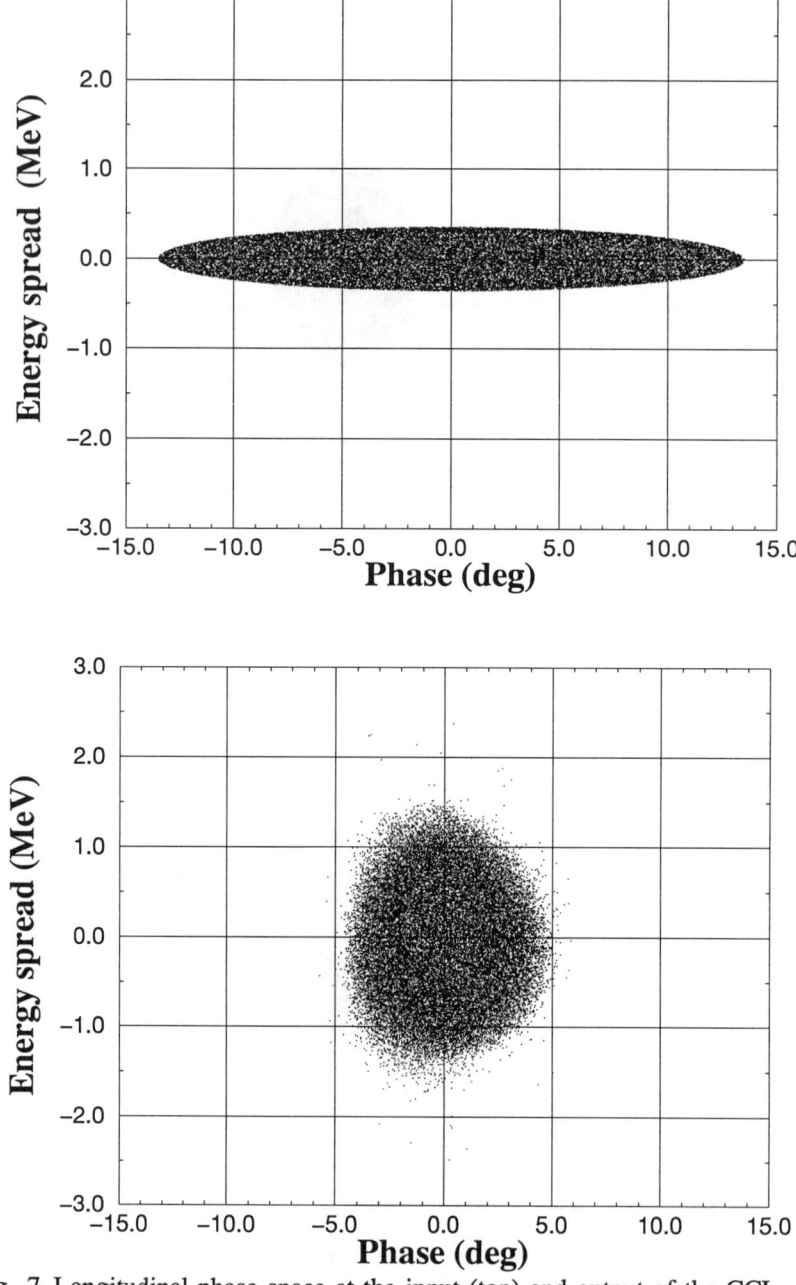

Fig. 7 Longitudinal phase space at the input (top) and output of the CCL

In order to quantify the properties of the halo we evaluate the phase space distribution in more detail. For the 90%, 99%, 99.9%, 99.99% and 100%–emittances, results are shown in Fig. 8. Here we calculate a single particle emittance by choosing the same matching parameters α and β as for the rms emittance. By ordering these emittances the curves of Fig. 8 are obtained. Both graphs show a strong oscillation for the 100% and the 99.99%-emittances. This is due to the fact that the 100% emittance is defined by one particle only while the 99.99%–emittance contains 5 particles. The outermost particles are not the same along the linac. These emittances are oscillating because the α and β of the rms emittances are mismatched. This is obvious by comparing the oscillations with the oscillating mismatch factor in Fig. 5. For a simulation with 50000 particles the 99.9%–emittance shows little oscillation and therefore is a useful quantity. In the x-direction this emittance stabilizes around 10 times the rms emittance, in the longitudinal direction around the value 7. A strong increase is seen at the beginning.

The presented results for the ESS linac with a tune depression of 0.8 only differs markedly from numerical results for linacs with large tune depressions up to 0.1. For the non-space charge dominated beam in the CCL our results show a small number of halo particles nearby the beam core. In the case of space charge dominated beams a large production of halo particles far outside the core has been observed analytically [7] and numerically [8,9].

For the ring injection some useful conclusions can be drawn from the simulation with 50000 particles. One requirement of the injection into the compressor ring is that 10^{-4} particles of the beam are allowed to be outside ±2 MeV. The longitudinal output distribution in Fig. 7 shows about twice as many particles outside the limit. Between the linac and compressor ring is a 130 m long transfer line where the energy spread increases further. To improve the situation a bunch rotator is definitely needed [4]. Transversely the situation is much more favorable for the ring injection. As shown in Fig. 8, 1% of the beam is outside 8 times the rms emittance and 0.1% outside 11 times the rms emittance.

Seeing losses along the linac in relation to acceptances and emittances, allows some general remarks to be made. The longitudinal acceptances increases due to acceleration, e.g. for the CCL from 12.5 up to 160 π ° MeV. This gives an increasing ratio from 10 to 130. At high energies, even for the outermost particles, the longitudinal emittance only is 40 times the rms emittance, see Fig. 8. The transverse acceptance of the CCL is about 30 π mm mrad giving a ratio between acceptance and rms emittance of about 50. To improve the situation the transverse emittance has to be decreased.

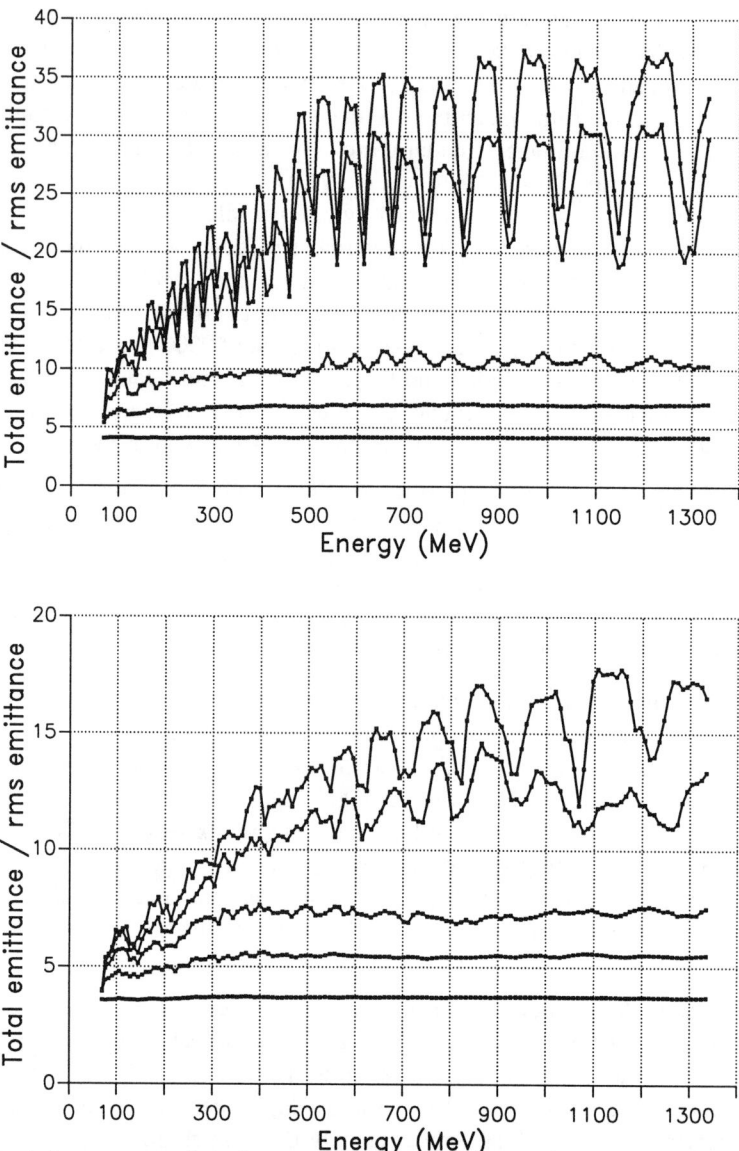

Fig. 8 Ratio between total emittances and rms emittances for the transverse (upper graph) and the longitudinal direction (lower graph). Curves (top to bottom) correspond to 100%, 99.99%, 99.9%, 99%, 90% emittances.

Concerning scraping particles transversly the following comments can be made. In general the beam will be mismatched by a certain amount resulting in an additional oscillation of the beam envelope. By localizing the maxima of the envelope, positions for scrapers can be defined. From Fig. 8 it is obvious that even for the outermost particles their radii are not increasing continuously. It seems to be possible to scrape the beam at medium energies and thereby improving the condition at the high energy end.

References

[1] H. Lengeler, "Proposals for Spallation Sources in Europe", Fourth European Particle Accelerator Conference, London, 1994, World Scientific, p.249–253

[2] H. Klein, "Spallation Neutron Sources", Proceedings of the 1994 International Linac Conference, Tsukuba, 1994, p. 322–327

[3] M. Pabst and K. Bongardt, "Beam Dynamics in the 1.3 GeV High Intensity ESS Coupled Cavity Linac", Proceedings of the Particle Accelerator and International Conference on High Energy Accelerators, Dallas, 1995

[4] K. Bongardt, M. Pabst and A. Letchford, "Halo Containment in the ESS Linac to Ring Transfer Line", these conference proceedings

[5] J. Guyard and M. Weiss, "Use of Beam Emittance Measurements in Matching Problems", Proton Linear Accelerator Conference, Chalk River, 1976

[6] C. Chen and R. A. Jameson, "Self-Consistent Simulation Studies of Periodically Focused Intense Charged Particle Beams", Phys. Rev. E, vol. 52, 1995, p. 3034

[7] R. Gluckstern, "Survey of Halo Formation Studies in High Intensity Proton Linacs, see ref. 2, p.333

[8] J. M. Langiel and A. C. Piquemal, "On the Dynamics of Space Charged Dominated Beams", see ref. 2, p.529

[9] R. Ryan, "Halos of Intense Proton Beams", see ref. 3

[10] K. Bongardt and M. Pabst, "Design Criteria for High Intensity H^- – Injector linacs", see ref. 3

Halo Containment in the ESS Linac to Ring Transfer Line

K. Bongardt and M. Pabst

Forschungszentrum Jülich GmbH, 52425 Jülich, Germany

and

A.P. Letchford

Rutherford Appleton Laboratory, Chilton, Didcot, UK

Abstract

The key issue of the accelerator part of the European Spallation Source ESS is the loss free ring injection. For achieving this goal, the linac beam has to be collimated in both transverse planes and limited in energy spread. By a bunch rotation cavity, positioned 75 m after the linac, the energy spread is reduced by a factor 3. In addition the mean kinetic energy has to be ramped during the injection pulse. For a bunch current of 214 mA, space charge forces are present in all 3 planes. They act quite drastically longitudinally due to the lack of focusing. After the bunch rotation cavity, where the bunch length is larger than the pipe radius, image forces in all 3 planes can no longer be neglected.

In order to fulfill the collimation requirements for loss free ring injection under various beam current conditions, transverse beam scraping and a longitudinal two stage collimation system are foreseen. There are two independently phased cavities, one for energy ramping and another for bunch rotation. The bunch rotator reduces the number of particles outside the wanted ±1 MeV value considerably. The remaining particles are stripped and collected in an achromatic collimation system.

Monte-Carlo simulations with 50000 particles are presented for the 130 m long straight transfer line after the linac end. They show an increase of the energy spread before the bunch rotation cavity and a strong filamentation in the longitudinal phase space afterwards. Comments are made about the influence of image forces.

Introduction

The most critical part of the accelerator of the European Spallation Source [1] is the loss free injection into the compressor rings. Due to the 5.1 MW average beam power at 1.334 GeV, particle loss above 10^{-7}/m forbids unconstrained hands on maintenance of accelerator components. At the injection of the 1.2 msec long linac pulse into the two compressor rings the loss must be at the 10^{-5} level. To achieve this with a 1000 turn H$^-$–injection scheme the linac beam has to be truncated in both transverse planes. There should be less than 1% of particles outside a geometrical emittance of 5.4 π mm mrad, corresponding to 20 times the normalized rms linac emittance of 0.6 π mm mrad. The mean kinetic energy has to be varied by 4 MeV during injection, corresponding to 2×10^{-3} Δp/p ramping. Less than 10^{-4} particles outside an energy spread of ± 2 MeV can be accepted by the stripping foil. The linac pulse has to be chopped at the 1.67 MHz revolution frequency with 60% chopping efficiency [2].

At the end of the 700 MHz ESS coupled cavity linac, we have the following situation. Transversely there are about 10^{-3} particles outside a geometrical emittance of 5.4 π mm mrad. Longitudinally one has about 2×10^{-4} particles outside an energy spread of ± 2 MeV [4]. Uncorrelated randomly distributed field errors in the coupled cavity linac shift the mean energy of the bunch center by ± 0.6 MeV at the linac end [12]. Therefore the energy spread has to be reduced at least by a factor of 2; either by dephasing the last linac section or by placing a bunch rotation cavity afterwards. By dephasing the last linac section a short transfer line between linac and compressor rings fulfills the collimation conditions. But a short transfer line is excluded for 2 reasons: (i) the doublet focusing scheme of the linac does not match the triplet focusing scheme of the compressor ring and (ii) a vertical switchyard magnet has to be installed in order to give a final vertical separation of 2 m for the two accumulator rings.

Without dephasing the last linac section there will be a long transfer line with a bunch rotation cavity afterwards. Space charge forces are present in all 3 planes due to the 214 mA bunch current. As a consequence, all injection requirements can be fulfilled, but the high β transfer line is quite different than expected.

Final bunch rotation

The beam motion is studied for a 130 m straight transfer line including a bunch rotation cavity. The rms values of the energy spread and the longitudinal emittance are shown in Fig. 1 and 2 for the ESS linac design parameters: 214 mA bunch current at 700 MHz and 1.334 GeV [3]. Shown are the results for a longitudinal drifting beam compared with bunched beam transfer line where constant longitudinal focusing is applied. For the multiparticle calculation the linac output phase space distribution is used as the input distribution for the transfer line simulation [4]. The simulation is done with 50000 interacting particles. The three dimensional space charge forces are calculated about every 40 cm taking into account the beam envelope oscillations. The same focusing scheme is adopted as at the end of coupled cavity linac. A quadrupole doublet is positioned every 5.4 m resulting in an average transverse radius of 3.5 mm.

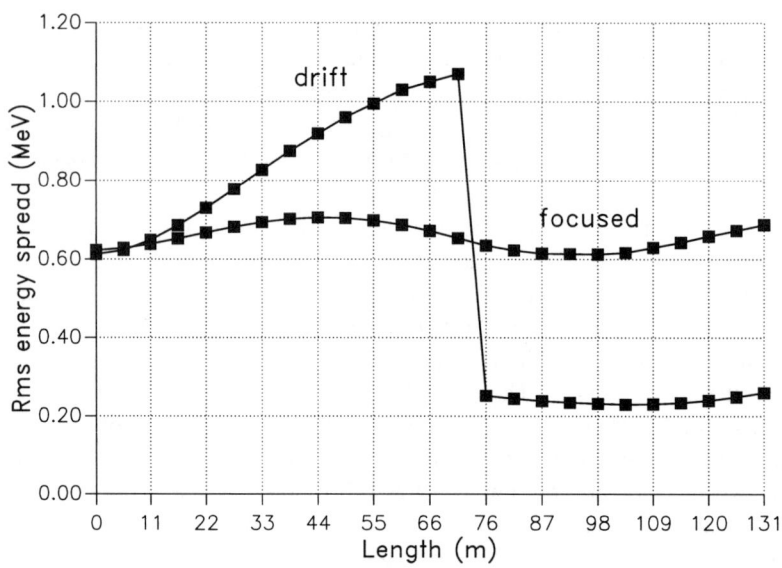

Fig. 1 Rms energy spread along the transfer line for longitudinally drifting and focused beams

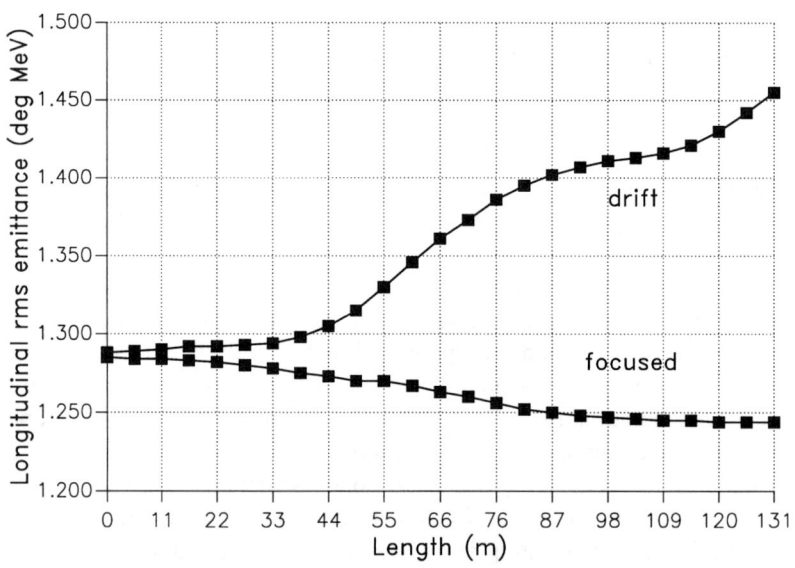

Fig. 2 Longitudinal rms emittance along the transfer line for longitudinally drifting and focused beams

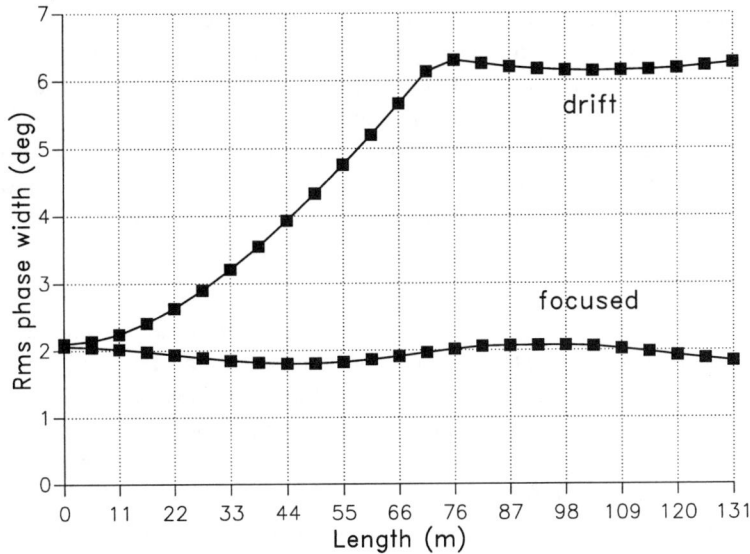

Fig. 3 Rms phase width along the transfer line for longitudinally drifting and focused beams

The rms energy spread and the longitudinal rms emittance are not constant along the transfer line as expected in a simple drift space. Due to small, but still not negligible, space charge forces the particles are not 'drifting' longitudinally. Instead, they are moving in a 'plasma channel' with decreasing strength. The envelope equation for the dense beam core with its mainly linear space charge forces is space charge dominated and not emittance dominated in spite of the high kinetic energy. The asymptotic value of the rms energy spread can be predicted quite well by the linear part of the space charge forces [5].

The rms phase width changes from 2° to 6° (10° without bunching cavities), see Fig. 3. For the bunched beam transfer line the phase width is constant. Along the ESS transfer line, the longitudinal rms emittance increases by the same amount as along the whole coupled cavity linac [4]. With a bunch rotation cavity, placed after 75 m corresponding to 6° rms phase width the rms energy spread is reduced to 0.2 MeV, see Fig. 1. About 11 MV rotation voltage is applied to a 4 m long 700 MHz coupled cavity.

Filamentation in the longitudinal phase space

In Fig. 4 the longitudinal phase space is plotted at the linac end, before and after the bunch rotation cavity and 60 m behind. As input, the particle distribution at the linac end is chosen.

An increase of the energy spread is recognized before the bunch rotation cavity. This is expected from linear space charge forces. Unexpected is the strong filamentation behind the bunch rotation cavity. This filamentation is probably due to 'mismatch' caused by the lack of longitudinal focusing in the transfer line. But even for the strong filamented phase space distribution at 130 m, there are less than 10^{-4} particles outside an energy spread of ±1 MeV. We get almost the same filamentation effect and longitudinal rms emittance increase if we start at the beginning of the transfer line with an analytical distribution, but with the same rms parameters.

Fig. 5 shows for the same initial distribution the output distribution after 130 m for a bunched beam transfer line. This bunched beam transfer line keeps all the rms parameters constant as shown in Fig. 1, 2 and 3. There is a no longitudinal rms emittance increase and filamentation. Transversely the rms radii are constant to about 3.5 mm. The total synchrotron tune is about $90°$ in the bunched beam transfer line. It is evident from Figs. 2, 4 and 5 that the phase space filamentation is connected with rms emittance increase. Detailed calculations are going on to detect the driving term.

Image charges along the high β transfer line

All the results presented here are calculated with direct Coulomb forces but without the interaction with the metallic vacuum chamber. This assumption is correct at the beginning of the transfer line, where $\gamma z_m < b$. γz_m is the bunch length in the lab system and b is the beam pipe radius of about 4 cm for the ESS high β line.

For short bunches, particles at the head and the tail can interact with each other. As the bunch length is increasing, particles at the head and the tail are shielded from each other due to the Lorentz contracted field lines with an angular spread of the order $1/\gamma$. For very long cylindrical bunches the total space charge potential becomes proportional to the local longitudinal line density [6].

For the ESS high β line, γz_m is greater than the pipe radius after the bunch rotation cavity. Transversely, the particles are filling up about 1/3 of the beam pipe. We have an approximately cylindrical beam with a line density proportional to $(z_m^2 - z^2)^{1/2}$ due to the filling of the 700 MHz coupled cavity linac. This line density results in non-linear longitudinal space charge forces, even for particles close to the bunch center.

In the intermediate regime between the short bunch and the long bunch limit, no analytical expression for the total space charge potential is available. On the

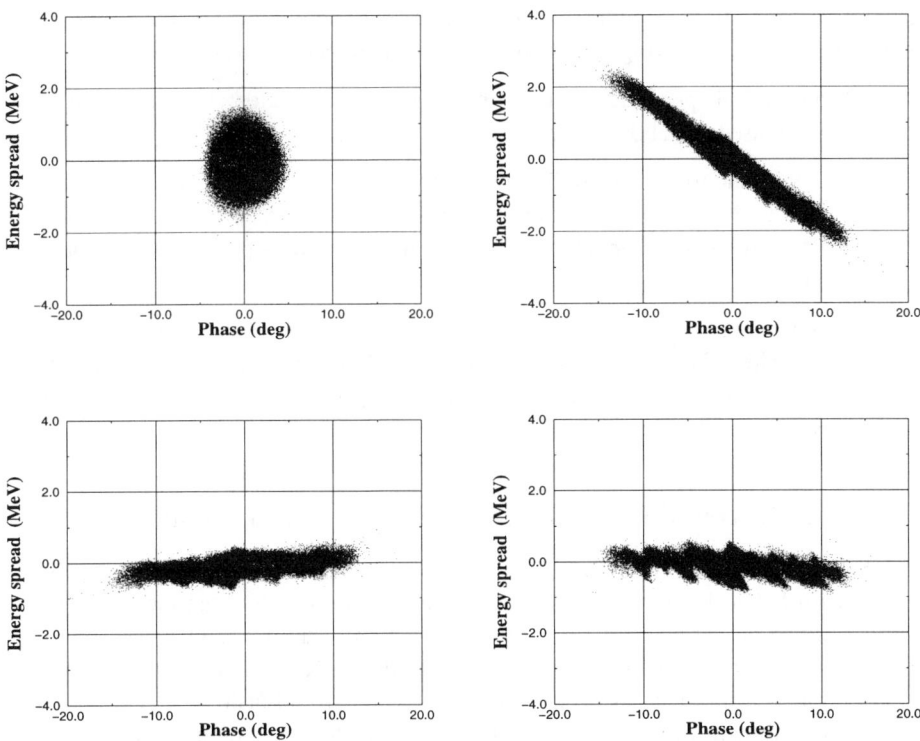

Fig. 4 Longitudinal phase space projections at the end of the coupled cavity linac, before and after the buncher and 70 m behind (upper left, upper right, lower left, lower right)

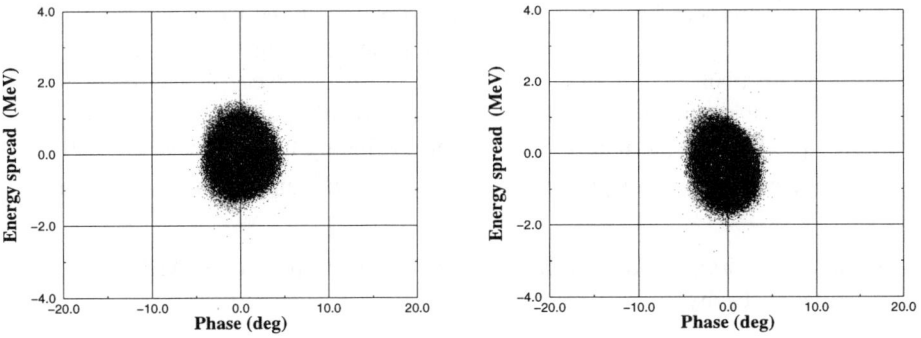

Fig. 5 Longitudinal phase space projections at the beginning and end for a bunched beam transfer line of same length as in Fig. 4

other hand, analytical formulas for both cylindrical and elliptical distributions [7, 8, 9] indicate the following results for the ESS parameters given above:

- head and tail particles see non-linear longitudinal fields: the linear part is decreased by about 30 %.

- for constant transverse radii, head and tail particles see a radial field, significantly smaller than the field at the bunch center. The radial field at the bunch center is about 20 % higher than the corresponding Coulomb field.

to study image charges accurately in the intermediate regime between short and long bunches, the total forces on each particle have to be calculated by double series over Bessel functions. Based on the method given in [10] a fast converging formula will be implemented soon.

Halo containment in the ESS high β transfer line

In order to allow loss free injection into the ESS compressor rings, the linac beam has to be truncated in all 3 planes and in addition, the mean kinetic energy has to be varied by 4 MeV during injection. Due to the increased phase width, image forces cannot be neglected after the bunch rotation cavity. If the distribution is truncated, the ring injection scheme is insensitive to the actual particle distribution in the tail [11]. This important feature simplifies drastically the design of the ESS high β transfer line in the longitudinal and the transverse direction.

The quadrupole strength has to be adjusted dependent on beam current in order to assure correct matching. For example, if the quadrupole gradients are kept constant along the transfer line, we have a decreasing oscillation of the rms radii between 3.5 and 2.5 mm. This is caused by the direct Coulomb forces for 214 mA bunch current because the increasing phase width causes a transverse mismatch. No filamentation or emittance increase is observed. Image forces will enhance the mismatch and in addition cause deformation of the transverse phase space ellipse because particles are no longer oscillating between head and tail.

To overcome all these problems the ESS high β transfer line consists of 3 parts:

- a straight transfer line 75 m from the linac end up to a 700 MHz bunch rotation cavity

- a 42.5 π m circumference, 180° achromatic collimation system

- a 75 m final matching section up to the H^--stripping foils

The transverse focusing scheme is flexible enough to handle beam currents down to 50 % of the design value during the start up period. In order to simplify the operation of the compressor rings during the start up period and to guarantee a collimated beam under various beam currents and mismatch conditions, horizontal scraping is foreseen before the rotation cavity. Vertical collimation is applied after the achromatic system. All collectors are designed for 5 kW average beam power corresponding to 0.1% of the full beam power.

Longitudinal halo scraping is obtained by a two stage collimation system. There are two independently phased cavities for energy ramping and bunch rotation. This reduces the particle number outside the desired ±1 MeV value, considerably. The remaining particles are stripped and collected in an achromatic collimation system [13]. This two stage longitudinal beam collimation is quite superior to all other possibilities considered. By varying the amplitude and phase of both rf cavities, there are four independent knobs for achieving the longitudinal halo collimation under varying beam current and ramping conditions including the influence of field errors and mismatch effects.

The achromatic collimation system can be avoided by dephasing the last linac section and keeping the beam bunched afterwards. This system is less flexible concerning varying beam current and energy ramping conditions. It requires bunching cavities all along the transfer line.

Acknowledgment

The authors thank G. H. Rees for many clarifying discussions about the longitudinal halo collimation system and very useful hints about the image forces. They express their gratitude to V. G. Vaccaro for providing a fast convergent series presentation for the image forces.

References

[1] H. Lengeler, "Proposals for Spallation Sources in Europe", Fourth European Particle Accelerator Conference, London, 1994, World Scientific, p.249–253

[2] G. H. Rees, "Important Design Issues of High Output Current Proton Rings", see ref. 1, p 241

[3] H. Klein, "Spallation Neutron Sources", Proceedings of the 1994 International Linac Conference, Tsukuba, 1994, p. 322–327

[4] M. Pabst K. Bongardt and A. P. Letchford, "Critical Beam Dynamical Issues in Neutron Spallation Sources", these conference proceedings

[5] K. Bongardt, "Matching Problems between Linac and Compressor Ring", Proceedings ICANS XII, 1993, RAL-report 94–025, p. A-155

[6] A. W. Chao, "Physics of Collective Beam Instabilities in High Energy Accelerators", Wiley, New York, 1993

[7] C. K. Allen, N. Brown and M. Reiser, "Image Effects for Bunched Beams in Axis Symmetric Systems", Particle Accelerators, vol. 45, 1995, p. 149

[8] M. Reiser, "Theory and Design of Charged Particle Beams", Wiley, New York, 1994

[9] G. H. Rees and J. V. Trotman, "Longitudinal Envelope Equation", ESS-95-25-R report, Feb. 1995

[10] G. Miano and L. Verolino," Some Integral Involving Bessel Functions", Il Nuovo Cimento, vol. 110 B, 1995, p. 421

[11] C. R. Prior, "Transverse and Longitudinal Tracking Studies for ESS and ISIS", this conference proceedings

[12] M. Pabst and K. Bongardt, "Beam Dynamics in the 1.3 GeV High Intensity ESS Coupled Cavity Linac", Proceedings of the Particle Accelerator and International Conference on High Energy Accelerators, Dallas, 1995

[13] V. Lebedev and G. H. Rees, "Transfer Line from Linac to Accumulator", ESS-95–18–R report, Feb 1995

SPACE CHARGE BEAM DYNAMICS STUDIES FOR A PULSED SPALLATION SOURCE ACCELERATOR[*]

Y. Cho and E. Lessner

Argonne National Laboratory
Argonne, IL 60439
U. S. A.

Abstract. Feasibility studies for 2-GeV, 1-MW and 10-GeV, 5-MW rapid cycling synchrotrons (RCS) for spallation neutron sources have been completed. Both synchrotrons operate at a repetition rate of 30 Hz, and accelerate 1.04×10^{14} protons per pulse. The injection energy of the 2-GeV ring is 400 MeV, and the 10-GeV RCS accepts the beam from the 2-GeV machine. Work performed to-date includes calculation of the longitudinal space charge effects in the 400-MeV beam transfer line, and of both longitudinal and transverse space charge effects during the injection, capture and acceleration processes in the two rings. Results of space charge calculations in the rings led to proper choices of the working points and of rf voltage programs that prevents beam loss. Space charge effects in the 2-GeV synchrotron, in both transverse and longitudinal phase space, have major impact on the design due to the fact that the injection energy is 400 MeV. The design achieves the required performance while alleviating harmful effects due to space charge.

I. INTRODUCTION

A study of the feasibility of upgrading the Intense Pulsed Neutron Source (IPNS) at ANL to a 1-MW spallation source, based on a 2-GeV RCS, has been completed [1]. An additional design study for a further upgrade from a 1-MW source to a 5-MW source, using a 10-GeV RCS, has also been completed [2]. Since transverse space charge effects are inversely proportional to $\beta^2\gamma^3$ and longitudinal space charge effects are inversely proportional to γ^2, space charge effects are largest at or near the 400-MeV injection energy of the 2-GeV ring.

What follows is a brief description of accelerator configurations for the 1-MW and 5-MW proton sources, a short summary of the beam dynamics work performed to enable us to tolerate the space charge effects, and an introduction of the detailed work presented in this workshop by the other team members.

Figure 1 shows the layout of the 1-MW pulsed neutron source. It utilizes existing buildings and other infra-structure from the former 12-GeV Zero Gradient Synchrotron (ZGS). The total space being re-used for the pulsed source is about 50,000 m². The 2-GeV RCS is housed in the ZGS tunnel, and two neutron producing target stations are placed in former experimental area buildings. The only

---

[*] Work supported by the U. S. Department of Energy, Office of Basic Energy Sciences under the Contract W-31-109-ENG-38.

new construction required for the 1-MW facility is a shielded enclosure for the 400-MeV linac and a building to house the linac klystrons and related equipment.

The former ZGS tunnel geometry can accommodate a 2-GeV RCS. That constraint fixes the extraction energy at 2 GeV, thus 0.5 mA of beam current is required to produce a 1-MW proton source. An RCS repetition rate of 30 Hz is preferred by the neutron source users. A 30-Hz rate enables the users to avoid accidental events in which neutrons from previous pulses mimic those of the current pulse, referred to as frame-overlap. Low repetition rates imply that in order to get a time-averaged current of 0.5 mA, the number of protons per pulse has to be large, in this case 1.04×10^{14}. The requirement for this large number of protons per pulse leads to space charge limit considerations.

The best injection energy to achieve 1.04×10^{14} protons per pulse was determined to be 400-MeV after consideration of the space charge limit dependence on $\beta^2\gamma^3$, and the ring acceptance. This will be discussed in Section III.

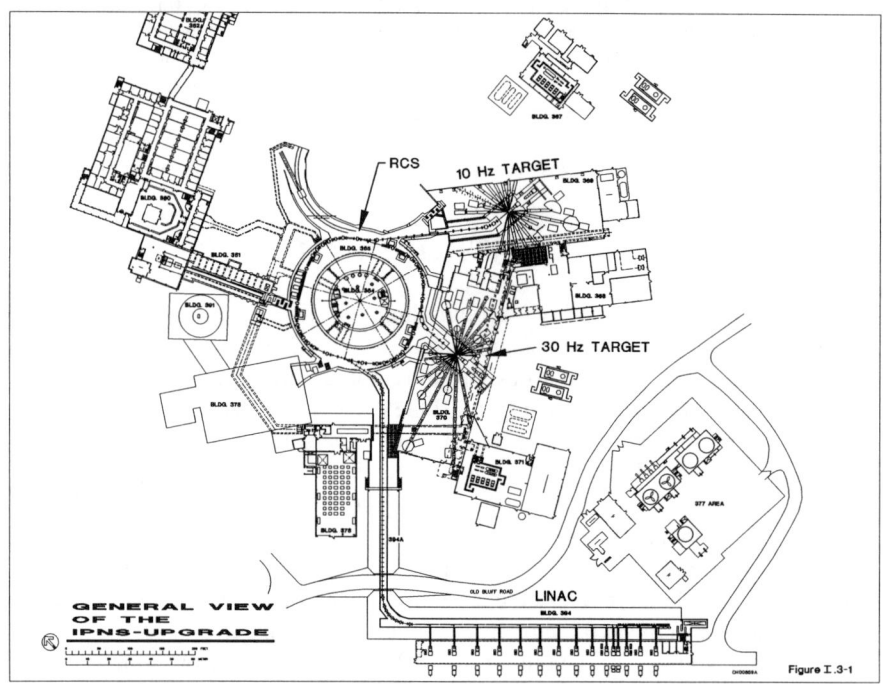

FIGURE 1. 1-MW Proton Source Machine Layout.

Figure 2 shows a general layout of the 5-MW facility based on a 10-GeV RCS that accelerates the 2-GeV (0.5-mA) beam from the 2-GeV RCS to 10 GeV (0.5 mA). In this case space charge effects are expected to be less because the energy of particles is higher. However, the bunch configuration in the ring is uneven, so space charge considerations should still be taken into account. The bunch configuration is uneven for geometrical reasons, as explained below.

The rf harmonic number of the 2-GeV ring must be changed from h = 1 to h = 2 to upgrade the 1-MW facility to a 5-MW facility, thus its h=2 rf frequency is 3 MHz. Changing the harmonic number does not change any design or performance parameters, and is done in order to raise the synchrotron frequency in the 10-GeV ring. The circumference of the 10-GeV RCS is 4 times that of the 2-GeV ring. In order to further increase the synchrotron frequency of the 10-GeV ring near its extraction energy, it is planned to inject the 2-GeV beam into a waiting 6-MHz bucket in the 10-GeV ring. When two 2-GeV beam bunches are transferred to the 10-GeV ring, only two of 16 buckets in the large ring are occupied.

Both transverse tune shifts and rf bucket distortions due to space charge must be studied during beam transfer from the low energy ring to the high energy ring.

Some of the areas in which space charge dominated beam dynamics was a consideration in designing this facility are:

1) Longitudinal space charge forces in the linac and 400-MeV LET may alter the energy spread of the beam being injected into the 2-GeV RCS.
2) Transverse tune shift and spread may occur during injection, capture and acceleration in a 400-MeV to 2-GeV RCS.
3) Longitudinal space charge suppression of bucket area may occur during injection, capture and acceleration in a 400-MeV to 2-GeV RCS.
4) Longitudinal and transverse effects of the 2-GeV beam in the 10-GeV RCS must be evaluated.

The first two topics are discussed in detail in this report, and the last two topics are described in detail by E. S. Lessner elsewhere in these proceedings [3]. The possibility of injecting a Kapchinskij-Vladimirskij (KV) distribution into a circular machine is also presented in these proceedings by E. Crosbie and K. Symon [4].

II. SPACE CHARGE EFFECTS IN THE LOW ENERGY TRANSFER LINE

The 400-MeV linac design included space charge effects, and the linac is described in reference [1]. The linac system consists of an H^- ion source, a low energy beam transport with a chopper, a 2-MeV radio-frequency quadrupole operating at 425 MHz, a 70-MeV drift tube linac (DTL) operating at 425 MHz, and a coupled cavity linac (CCL) operating at 1275 MHz. There is a frequency jump of a factor of three between the DTL and CCL, resulting in two empty buckets for every occupied one.

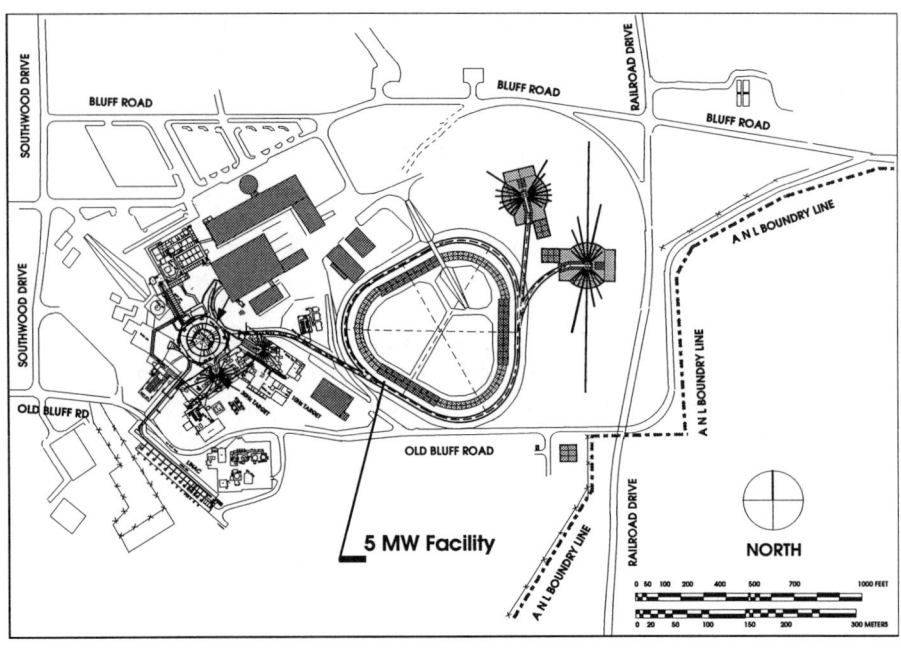

FIGURE 2. Site Layout Showing the Location of the 5-MW Facility.

The output linac micro-bunch has an energy spread of ± 0.8 MeV and a phase spread of ± 8° (in the 1275 MHz system) measured at the base line. This corresponds to a micro-bunch phase space area of 1.4×10^{-5} eV sec. Each of the 0.5-msec-long linac macro-pulses contains 1.04×10^{14} protons. The macro-pulse is chopped at 1.1 MHz to facilitate injection into the RCS. The chopping operation places 75 % of the beam into a waiting synchrotron bucket. The remaining 25 % of the beam is discarded upstream of the RFQ to avoid later losses, since it may not be captured in the RCS. These numbers can be expressed in terms of the linac pulse current. The linac average pulse current with 25 % chopping is 44 mA. The number of protons in a micro-bunch is, then, 6.3×10^{8}, and the bunch length at the base line is ± 17 psec at the exit of the linac.

Longitudinal space charge effects of the 6.3×10^{8} protons initially confined in the ± 17 psec bucket and drifting through the low energy transport line were investigated using TRACE-3D [5], a six-dimensional tracking program that calculates the envelopes of a bunched beam while taking into account linear space charge forces.

The 157.3-m-long 400-MeV transfer line is constructed using FODO cells with the following three features: 1) a 90° horizontal achromatic bend near the linac, 2) a

3-m vertical achromatic elevation change, and 3) a 72° achromatic bend near the synchrotron for injection matching.

Figure 3 shows the energy spread of the beam being transported as a function of distance from the linac exit point with and without space charge effects. It shows that when the space charge effects are taken into account, the energy spread changes very rapidly (within 20 m) from the initial spread of 0.8 MeV to a plateau of 1.6 MeV after some 60 m. When the space charge potential is not taken into account, the energy spread, as expected, remains constant.

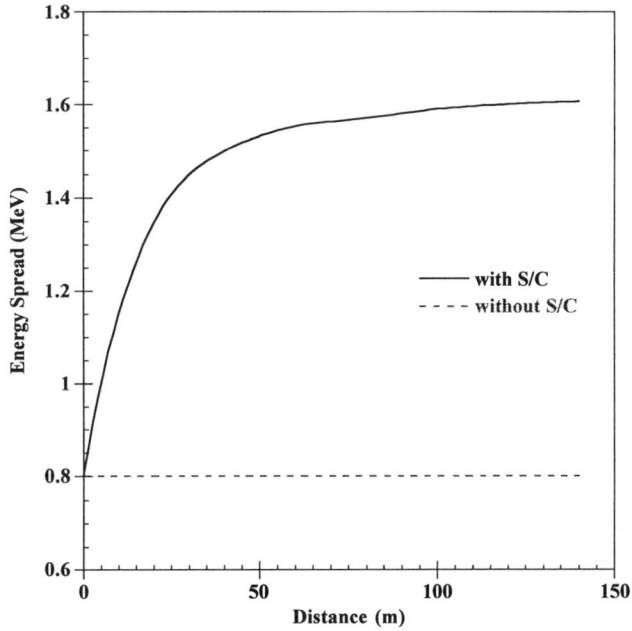

FIGURE 3. Variation of Energy Spread as a Function of Drift Distance, as Calculated with TRACE-3D.

Figure 4 shows the evolution of the micro-bunch length expressed in units of 1.275 GHz rf phase angle. The bunch length, $\Delta\phi$, increases from ± 8.0° to ± 320° when space charge effects are taken into account. When space charge effects are turned off in the calculation, the bunch length increases to ± 150°. The non-linear expansion of the bunch length, particularly in the early part of the transport shown in the figure, comes from the fact that the head and tail of the bunch both contain particles with both $+\Delta E$ and $-\Delta E$ at the exit of the linac. Therefore, the time expansion of the bunch is a quadrature sum of linear expansions due to $|\Delta E|$ together with travel time and the initial bunch length.

In addition to the TRACE-3D tracking study, the order of magnitude effect was estimated by analytic considerations. Calculations are further simplified by the assumption that the particles are distributed uniformly in radial space and parabolic in longitudinal space. The density is zero at the end of the bunch with the line density, $\lambda(z)$, given by:

$$\lambda(z) = \frac{3N}{4z_m}(1 - \frac{z^2}{z_m^2})$$

where, N is number of particles in the distribution, $\pm z_m$ is the maximum bunch length at a given point in time, and z is the longitudinal coordinate measured from the center of the bunch.

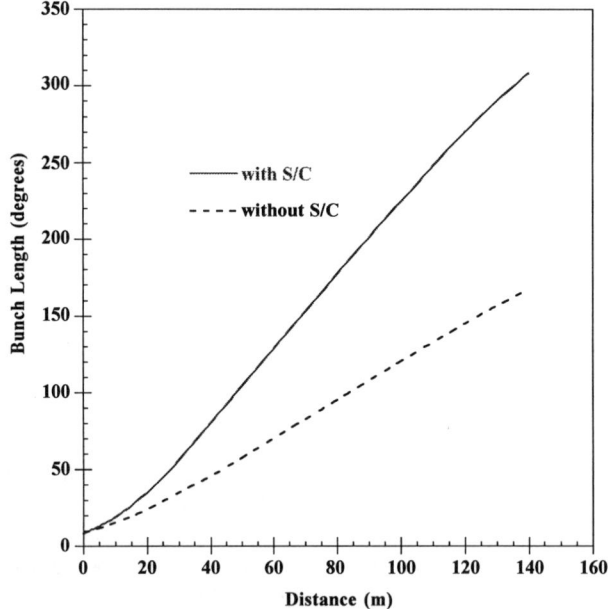

FIGURE 4. Variation of Bunch Length as a Function of Drift Distance, as Calculated with TRACE-3D.

It is expected that during the drift, the bunch length will change due to the momentum spread in the beam, and the momentum spread of the bunch will change due to acceleration and deceleration coming from the self-field of the bunch, which varies as the bunch length changes. The coupled equations of motion for a particle in the distribution are:

$$\frac{d\delta z}{dt} = \frac{c\beta}{\gamma^2}\frac{\delta p}{p},$$

$$\frac{d\delta p}{dt} = \frac{3e^2 N g_o}{8\pi\varepsilon_o \gamma^2 z_m^3}\delta z$$

where δz is the longitudinal position deviation and δp is the momentum deviation from the central particle coordinates, respectively; ε is the vacuum permittivity; c, β and γ are the speed of light and relativistic parameters; and $g_o = 1 + 2\ln(b/a)$ is a capacitive factor for a beam of radius a traversing a beampipe of radius b.

From the above single particle equations of motion, one can obtain the time development of the bunch envelope coordinates δz_m and δp_m using the WKB approximation. The envelope equations are better suited to the time development of the bunch.

$$\frac{d\delta z_m}{dt} = \frac{c\beta}{\gamma^2 p}\delta p_m,$$

$$\frac{d\delta p_m}{dt} = \frac{c\beta\varepsilon^2}{\gamma^2 p \delta z_m^3} + \frac{3e^2 N g_o}{8\pi\varepsilon_o \gamma^2 \delta z_m^2}$$

where ε is the longitudinal emittance of the micro-bunch (1.4 × 10^{-5} eV sec). Numerical calculation of the evolution of the envelope equations is performed using a leap-frog algorithm. This calculation showed that evolution of beam energy spread, ΔE, as the macro-pulse travels through the transport line varied from 0.8 MeV to an asymptotic value of 1.7 MeV.

The two different computations give similar results of ± 1.6 and ± 1.7 MeV. The 2-GeV RCS is designed to accept an energy spread of ± 2.5 MeV, and this simple analysis shows that by the time the 400-MeV beam reaches the injection point, the energy spread has grown to ± 1.7 MeV. More detailed study is under way in order to confirm these preliminary conclusions.

III. TRANSVERSE SPACE CHARGE EFFECTS IN THE 2-GEV RCS
(Space Charge Limit and the Choice of Linac Energy)

It is customary to use the Laslett tune shift due to space charge to calculate the number of protons that can be accelerated by a synchrotron. Knowledge of the longitudinal and transverse beam distributions is required for these calculations. These distributions come either from assumptions based on previous experience or from tracking studies. The initial estimate of the space charge limit is obtained by using elliptic beam distributions and a bunching factor obtained from assumptions. A refined study of the space charge limit was performed after a longitudinal tracking study from injection to full-energy acceleration was made.

We now discuss the initial estimate of the space charge limit and choice of the injection energy to the 2-GeV ring. The discussion presented here closely follows that described in *Report of the ISIS Project Group* [6]. It is reasonable to assume the local elliptical energy distribution of Hofmann and Pederson [7] for longitudinal motion of protons in the bunch. For this distribution, self-forces caused by longitudinal space charges are directly proportional to the external focusing forces. This allows bunching factors and rf bucket area reductions to be calculated analytically. The bunching factor, B_f, is defined to be the ratio of mean to peak number of particles per unit length, and is given by

$$B_f = \frac{1}{2\pi} \frac{\sin\phi_2 - \sin\phi_1 - (\phi_2 - \phi_1)\cos\phi_2 - 0.5(\phi_2 - \phi_1)^2 \sin\phi_s}{\cos\phi_s - \cos\phi_2 + (\phi_s - \phi_2)\sin\phi_s}, \quad (1)$$

where ϕ_2 is the maximum angle, ϕ_1 is the minimum angle, and ϕ_s is the synchronous angle.

If we assume a small synchronous phase angle of 5° for early acceleration, the bunching factor from Equation (1) is about 0.4. For the transverse charge distribution, it is reasonable to assume that the protons are distributed in an elliptical geometry of semi-axes a (horizontal) and b (vertical) with the normalized two-dimensional elliptic density distribution, $\rho(x,y)$:

$$\rho(x,y) = \frac{3}{2\pi ab}\left[1 - \left(\frac{x}{a}\right)^2 - \left(\frac{y}{b}\right)^2\right]^{1/2}.$$

The density at the center of such a distribution is 1.5 times that of a beam with uniform-filled, elliptical cross section. It is believed that the density at the core of the beam soon after the injection process will not exceed the assumed elliptic distribution. The Laslett transverse space charge limit, N, in a circular machine is directly proportional to the transverse emittance of the stacked beam and inversely proportional to the central density. Therefore, the space charge limit for an elliptic distribution is 1.5 times less than that of a uniformly distributed beam of similar geometry.

The Laslett incoherent transverse space charge limit can be written

$$N = -\Delta v_y \varepsilon_y \left(1+\frac{a}{b}\right) \beta^2 \gamma^3 \frac{B_f}{(r_p G_v F_v)}, \qquad (2)$$

where

$$F_v = 1 + \left(1+\frac{a}{b}\right)\left[\varepsilon_1 \frac{b^2}{h^2} + \beta^2 \gamma^2 B_f \left[\varepsilon_1 \frac{b^2}{h^2} + \varepsilon_2 \frac{b^2}{g^2}\frac{\rho_m}{R}\right]\right] \bigg/ G_v,$$

- h = vacuum chamber half-height;
- g = bending magnet gap half-height;
- ε_1 = electrostatic image coefficient = 0.1;
- ε_2 = magnetostatic image coefficient = 0.411;
- ρ_m = bending radius of the machine;
- R = average radius of the machine;
- r_p = classical radius of proton = 1.5347×10^{-18} m;
- Δv_y = Allowed vertical tune shift due to space charge;
- G_v = form factor = 1.2 for elliptic distributions and 1.0 for uniform distributions; and
- ε_y = vertical emittance of stacked beam at injection, or acceptance of the machine not taking "beam-stay-clear" into account.

As Equation (2) indicates, the space charge limit is proportional to the phase space area of the stacked beam. Therefore the phase space area must be decided while taking into account its implications for such machine parameters as magnet apertures and vacuum chamber dimensions. It is prudent to iterate the values of the choices of vertical and horizontal emittances, ε_y and ε_x, with consideration given to other parameters.

A detailed study led to the following conclusions:

1. Although it is customary to have ε_y smaller than ε_x, so that the ring dipole magnet gap height can be made smaller, a very high-intensity machine like the one being described may not benefit from having $\varepsilon_y < \varepsilon_x$. The important point is to minimize the beam losses and to control them. There are some advantages to having equal acceptances, because coupling resonances can be used to stack the beam in a KV-like distribution [4]. Having equal acceptances in both planes alleviates concerns about the coupling resonances.

2. Using Equation (2) and assuming a stacked beam emittance of $375\pi \times 10^{-6}$ m rad in each plane, the space charge limit of the machine under these conditions

gives the number of particles per pulse as a function of the injection energy. This number can be converted to the time-averaged beam current at a repetition rate of 30 Hz. The allowed tune shift, Δv_y, used is 0.15. Figure 5 shows the average current versus the injection energy.

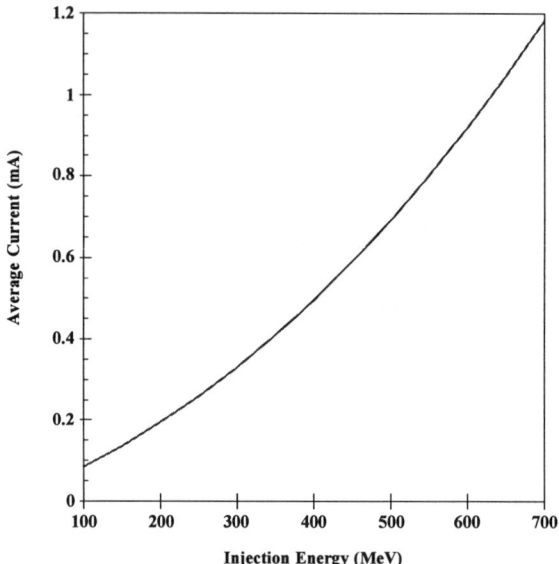

FIGURE 5. Average Current from Space Charge Limit.
$\Delta v_y = 0.15$, $\varepsilon_y = \varepsilon_x = 375\,\pi \times 10^{-6}$ m.

From this figure, one can conclude that an injector linac of 400 MeV allows 0.5-mA operation of the synchrotron. The current of 0.5 mA at a 30-Hz repetition rate implies that the number of protons per pulse is 1.04×10^{14}.

In this preliminary estimate of the space charge limit, the use of $\Delta v_y = 0.15$ is quite conservative, and there is a safety margin associated with the assumption. However, the calculation was based on the assumption of an elliptic charge distribution. In order to obtain refined information on the particle distribution, an extensive tracking study was performed to simulate both capture and acceleration processes and to minimize beam loss during these processes. A detailed discussion of the tracking is given in Reference [3]. The longitudinal charge distribution as a function of time from injection to extraction is obtained from the tracking study. This gives the time-varying bunching factor.

Because the goal is to accelerate 1.04×10^{14} protons per pulse, we use the time-varying bunching factor, the emittance of 375×10^{-6} m, and Equation (2) a time-varying, incoherent, space charge tune shift. Figure 6 shows both the time-varying bunching factor and the space charge tune shift during the first 6 msec of the acceleration cycle.

Note that the space charge tune shift is maximal around 1 to 2 msec after injection, and the maximum tune shift is about 0.19.

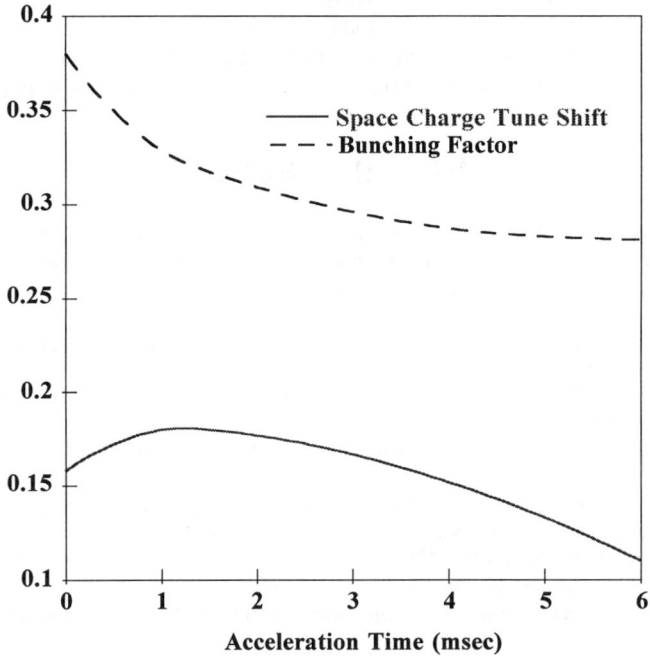

FIGURE 6. Time Variation of the Bunching Factor (top curve) and Time Variation of the Space Charge Tune Shift for the First 6 msec.

IV. LONGITUDINAL SPACE CHARGE EFFECTS IN THE 2-GEV RCS

As expected, the longitudinal space charge effect is largest during the injection and capture processes because the effect varies as $1/\gamma^2$. To study the effect in detail, we have developed an extensive longitudinal tracking study program based on a new computing code developed for this purpose [8].

The discussion on the result of the study is presented by Lessner, Cho, Harkay and Symon in these proceedings in a paper titled "Longitudinal Tracking Studies for a High Intensity Proton Synchrotron" [3].

V. SPACE CHARGE EFFECTS IN THE 10-GEV RCS

The maximum space charge effect in the 10-GeV ring is at the injection energy of 2 GeV. The transverse Laslett tune shift at injection into 2 buckets from the 2-GeV RCS is estimated to be 0.19 using Equation (2).

Longitudinal effects concern bucket area distortion due to space charge and matching of 2-GeV bunches from the lower energy ring to waiting buckets of the higher energy ring. Details of this work are also discussed in Reference [3].

VI. REFERENCES

[1] "IPNS Upgrade - A Feasibility Study", ANL Report ANL-95/13 (April, 1995).
[2] Y. Cho et al., "A 10-GeV, 5-MW Proton Source for a Pulsed Spallation Source," *Proc. of 13th Meeting of the International Collaborations on Advanced Neutron Sources*, PSI, Villigen, Switzerland, to be published (October 11-14, 1995).
[3] E. Lessner, Y. Cho, K. Harkay and K. Symon, "Longitudinal Tracking Studies for a High Intensity Proton Synchrotron," in these proceedings.
[4] E. Crosbie and K. Symon, "Injection into a Circular Machine with a KV Distribution," in these proceedings.
[5] K. R. Crandall, "TRACE-3D Documentation," LANL Report LA-UR-90-4146, (August 1987).
[6] *Report of the ISIS Project Group,* Rutherford Laboratory (November 1986).
[7] A. Hofmann and F. Pederson, "Bunches with Local Elliptic Energy Distributions," IEEE Trans. Nucl. Sci. **NS-26** (3), 3526 (1979).
[8] Y. Cho, E. Lessner and K. Symon, "Injection and Capture Simulations for a High Intensity Proton Synchrotron," *Proc. of the European Particle Accelerator Conference*, London, page 1228 (1994).

Longitudinal Tracking Studies for a High Intensity Proton Synchrotron [*]

E. Lessner, Y. Cho, K. Harkay, K. Symon[†]

Argonne National Laboratory
9700 Cass Ave., Argonne, IL 60439

Abstract. Results from longitudinal tracking studies for a high intensity proton synchrotron designed for a 1-MW spallation source are presented. The machine delivers a proton beam of 0.5 mA time-averaged current at a repetition rate of 30 Hz. The accelerator is designed to have radiation levels that allow hands-on-maintenance. However, the high beam intensity causes strong space charge fields whose effects may lead to particle loss and longitudinal instabilities. The space charge fields modify the particle distribution, distort the stable bucket area and reduce the rf linear restoring force. Tracking simulations were conducted to analyze the space charge effects on the dynamics of the injection and acceleration processes and means to circumvent them. The tracking studies led to the establishment of the injected beam parameters and rf voltage program that minimized beam loss and longitudinal instabilities. Similar studies for a 10-GeV synchrotron that uses the 2-GeV synchrotron as its injector are also discussed.

Introduction

Feasibility studies of a proton source for a 1-MW spallation source were performed at Argonne National Laboratory [1]. The machine is designed to deliver a proton beam of 0.5 mA time-averaged current at a repetition rate of 30 Hz. The 1-MW proton source is based on a 2-GeV, 30-Hz rapid cycling synchrotron. A key design feature of the accelerator system is to minimize beam losses during the whole injection to extraction cycle, reducing activation to levels consistent with hands-on maintenance. The time-averaged current in

---

[*]Work supported by U. S. Department of Energy, Office of Basic Energy Sciences under Contract No. W-31-109-ENG-38.
[†]Also Dept. of Physics, University of Wisconsin-Madison, Madison, WI 53706

the synchrotron is 0.5 mA, corresponding to 1.04×10^{14} protons per pulse. This intensity is about two times higher than that of existing machines and causes strong space charge fields that affect the dynamics of the injection, capture and acceleration processes.

The space charge fields modify the particle distribution, distort the stable bucket area, and reduce the rf linear restoring force, causing loss of focussing or even defocussing. These effects may lead to particle loss. Another factor that affects the longitudinal dynamics is the tune depression introduced by the space charge forces. Space charge can potentially lead to collective instabilities which may introduce emittance growth and/or beam loss.

Tracking simulations were conducted to study space-charge-related effects in the accelerator and ways to circumvent them. The results were used to design the injected beam chopping requirements, introduced to facilitate lossless capture, the injected beam energy spread and an rf voltage program that prevented losses and longitudinal instabilities.

Similar studies were also conducted for the 10-GeV rapid cycling synchrotron of a 5-MW proton source [2]. The 10-GeV synchrotron uses the 2-GeV machine as its injector.

The bucket area reduction due to the space charge is an important consideration in the matching process between the extracted beam from the 2-GeV machine and the waiting bucket of the 10-GeV machine. In this case, neglect of space charge contributions in the matching calculations leads to a badly mismatched beam and consequent dilution as the beam is accelerated.

In the following, we discuss first the relevant features of the simulation code that we developed for the studies. Then, we discuss space-charge-related issues in the 2-GeV synchrotron. Finally, we discuss those issues that are specific to the 10-GeV synchrotron.

Tracking Algorithm

A Monte-Carlo-based tracking computer program was developed to study space charge effects [3][4] and to determine the injected beam parameters and rf voltage profile for the synchrotron.

The equations of motion including space charge forces are derived in a Hamiltonian form. We assume that there is a synchronous particle that remains in the equilibrium orbit and whose energy is denoted by E_s. The Hamiltonian for a particle of energy E and phase Φ is:

$$H = \frac{h\eta_s\omega_s^2}{\beta_s^2 E_s}W^2 + \frac{eV}{2\pi}(cos\Phi + \Phi sin\Phi_s - cos\Phi_s - \Phi_s sin\Phi_s) - \frac{e^2 g_0 h^2}{4\pi\varepsilon_0 \gamma_s^2 R}\lambda(\Phi), \qquad (1)$$

where:
the subscript s refers to the synchronous particle,
$\lambda(\Phi)$ is the linear particle density,
ε_0 is the vacuum permittivity,
R is the average ring radius,
and g_0 is a capacitive geometrical factor $= 1 + 2ln(b/a)$, where a is the beam radius and b is the vacuum chamber radius. We define $W = (E - E_s)/\omega_s$, such that $dW \cdot d\Phi$ is conserved.

The differential equations derived from this Hamiltonian describe the time-evolution of a particle subject to an accelerating force that depends on the rf phase and to the space charge forces that act on the particle. These differential equations are solved numerically by using leapfrog difference equations that are suitable for rapid numerical calculations. The coordinates of each particle, W and Φ are advanced by a specified time-step, τ, according to:

$$W_{n+1/2} = W_{n-1/2} + \frac{eV_n\tau}{2\pi}(sin\Phi_n - sin\Phi_{s,n}) + \frac{e^2 g_0}{4\pi\varepsilon_0}\frac{h^2\tau}{R\gamma_{s,n}^2}(\frac{d\lambda(\Phi)}{d\Phi})_n, \quad (2)$$

$$\Phi_{n+1} = \Phi_n + h\tau(\frac{\eta_s\omega_s^2 W}{\beta_s^2 E_s})_{n+1/2} + \Phi_{s,n+1/2} - \Phi_{s,n-1/2}, \quad (3)$$

where the subscript n indicates that a quantity is evaluated at a time $t = n\tau$.

The values of the Φ coordinates are calculated at integer time-steps and the values of W, at half-integer time-steps. The last two terms in Eq.(3) account for the changes in Φ_s. These equations have first order simplicity but second-order accuracy, because the derivatives are evaluated at the mid-point of each interval. The numerical simulation of the injection, capture and acceleration processes follow an ensemble of macro-particles, whose initial coordinates in phase and energy are chosen according to a specified distribution.

The turn-by-turn injection process is simulated by randomly placing N_p/N_t macro-particles into the machine. Here, N_p is the total number of macro-particles, and N_t is the total number of injection turns. The particles are distributed uniformly in phase within the maximum and minimum values of the injected beam. The energy distribution follows a cosine distribution, where the base is equal to the maximum energy spread of the beam.

During injection, the time-step is equal to the revolution period. During the acceleration process, the time-step can be increased to reduce computer time, but must, of course, be kept much smaller than the synchrotron period, τ_s, or other relevant period of the motion. In the simulations, the time-step size is kept smaller than $\tau_s/30$.

The initial phase space coordinates for each particle are saved. The analysis of lost particles using these parameters permits the design of a more efficient injection scheme by depopulating the undesirable phase space. The vacuum chamber wall is represented by a momentum cutoff that depends on the maximum betatron oscillation and dispersion function. Particles of equal, or higher, momenta than the cutoff are removed from the calculations, so as not to contribute to the space charge forces.

The variation of the capacitive geometrical factor during acceleration is also taken into account [6]. The geometrical factor depends on the ratio of the vacuum chamber radius and the beam radius. Since the latter decreases as the beam is accelerated, the ratio increases.

The space charge fields are calculated by using binning and data smoothing techniques that alleviate the statistical fluctuations introduced by the relatively small number of macro-particles used in the simulations.

The macro-particles are binned first into an appropriate number of bins. The number of bins must be chosen so that the average number of macro-particles per bin is not too small and the bin length not too large so as to reveal the bunch structure. A large number of bins reduces the bin occupancy, introducing higher frequency components. This can be seen in Figure 1, where 32 bins and 128 bins were used to estimate the derivative of the particle linear density distribution. A beam of length 180° and 10^4 macro particles were used in the simulation. The mean bin occupancy is 312 for the coarser binning and is 78 for the finer binning. For the latter, the ringing at the ends of the bunch that results from the Fourier series behaviour at a simple discontinuity (Gibbs phenomenon) is quite noticeable.

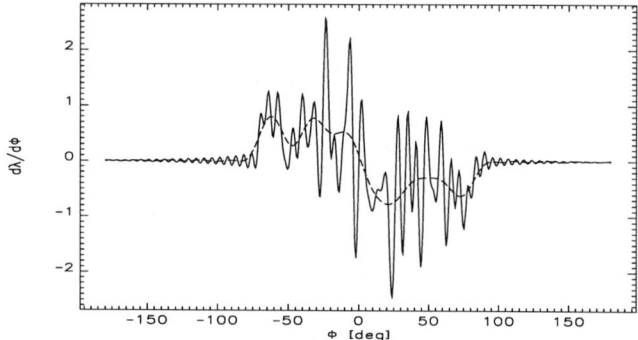

Figure 1: Gradient of the Linear Particle Density Distribution for a Beam of Length 180° Obtained with 32 Bins (Continuous Line) and 128 Bins (Dashed Line). The mean bin occupancy is 312 for the coarse binning and 78 for the finer binning.

The projected phase distribution is binned by the cloud-in-cell method [5]. For each particle, a weighted contribution is assigned to the two closest grid points, according to how far the particle is from these points.

The data is then fast-Fourier-transformed and convolved with a $sink/k$

kernel, where k is a harmonic number. This smoothens the binned density and provides a finer grid, from which the fields can be obtained by interpolation.

Finally, the Fourier components are multiplied by a Lanczos convergence factor:
$$\frac{sin\pi k}{k_c} / \frac{\pi k}{k_c},$$
where k_c is a cutoff harmonic number. The filtering process eliminates the ringing at the ends of the bunch, but has the disadvantage of introducing an overestimation of the density at the ends of the bunch and an underestimation of its peak. Figure 2 shows the linear particle density distributions from the example used in Figure 1 obtained after application of the same kernel interpolation but with cutoff harmonics of 9 and 17. The coarsely-binned data is obtained by using 32 bins and the interpolated data is obtained with 512 bins. As shown in the figure, the cutoff at the 9th harmonic underestimates the peak of the distribution by 8% when compared with the peak of the distribution obtained from the 32-binned unfiltered data. The same cutoff overestimates the density at the ends of the bunch by 10%.

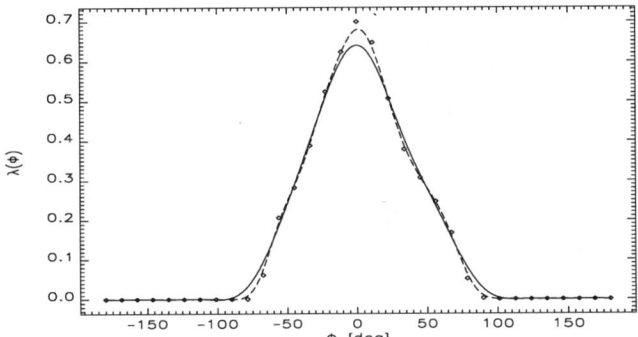

Figure 2: Linear Particle Density Distribution Obtained by Using Different Filter Cutoff Frequencies. The continuous line represents the density from a 9th harmonic cutoff and the dotted line that from a 17th harmonic cutoff. The squares represent the data before smoothening and filtering is applied.

The choice of the number of bins and cutoff frequency is based on an "optimal" filter method that gives the best estimate of the actual standard probability distribution from the computed density of the simulation particles [7].

Space Charge Issues in the 2-GeV Synchrotron

The accelerator system of the 1-MW source consists of an injector linac and a rapid cycling synchrotron (RCS). The 400-MeV linac delivers H$^-$ ions to RCS, where they are injected after stripping by a thin carbon foil. RCS accelerates the beam from 400 MeV to 2 GeV and its rf system is operated with a harmonic number equal to one. The rf frequency varies from 1.1 MHz to 1.5 MHz. The synchrotron magnets are powered by a dual-frequency resonant circuit that excites the magnets at 20 Hz and de-excites them at 60 Hz, resulting in an effective rate of 30 Hz and a reduction of 1/3 in the required peak rf voltage value. The ring circumference is 190.4 m and the synchrotron fits in the former Zero Gradient Synchrotron (ZGS) enclosure. A layout of the accelerator system is shown in Figure 3.

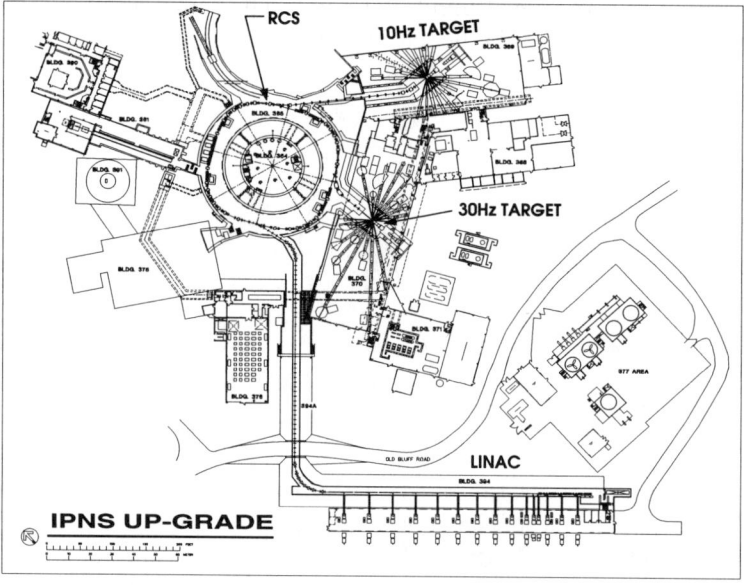

Figure 3: Schematic Layout of the IPNS Upgrade.

The space charge self-fields in the longitudinal direction vary as the inverse of the energy squared. In RCS, γ varies from 1.4 to 3.1 and the space charge self-fields affect the whole machine cycle. The fields change the beam dynamics and distort the rf bucket, thus enhancing the probability of beam losses. Since the main goal of the rf program design was to minimize losses, space charge effects in the synchrotron were analyzed in detail.

Some of the issues related to the prevention of beam losses in RCS are, at injection, choice of the injected-beam parameters and minimum initial bucket necessary to contain the beam while avoiding excessive dilution. During the injection period, the rf voltage must be increased to overcome the space-charge-induced bucket area reduction. During acceleration, the voltage needs to provide sufficient rf linear restoring force and the beam needs to have large momentum spread to prevent the microwave instability. During the whole cycle, the impedance due the space-charge self-fields must be minimized. In the following, we examine these issues separately.

The linac beam macro pulse is injected into a waiting bucket of RCS. Injection occurs in 0.5 msec, or 561 turns, under a flatbottom guide field. The choice of the initial particle distribution in phase space is guided by the requirements of high capture efficiency and longitudinal stability. The first requirement determines how much the linac macro pulse has to be chopped, since adiabatic capture is not pratical in RCS. The second requirement determines the bunch area of the beam. The simulation studies indicated that the best capture efficiency is obtained by discarding 25% of the linac beam and that the microwave instability is avoided when the bunch area after injection is 7.3 eV sec. Allowing a dilution factor of about two, an injected beam area of 3.3 eV sec satisfies the requirement. This area implies, in turn, that the the energy spread of the injected beam be equal to ± 2.5 MeV.

An important consideration in the studies was to avoid the need for large values of the rf voltage. The initial voltage was chosen so as to provide a bucket area sufficient to contain the incoming beam but not so large as to allow excessive dilution and the need for high rf voltage values at later times. The minimum voltage required to contain the beam injected with an energy spread of ±2.5 MeV, cut by 25%, is 40 kV. Figure 4 depicts the bucket formed by this voltage and the particle distribution of the first injected turn. The voltage provides the needed bucket height at the edges of injected the bunch, as shown in the figure.

The rf voltage was raised rapidly during injection to avoid deleterious bucket area reduction caused by the increasing particle density as multiple turns are accumulated in the ring. In Figure 5(a) we show the phase space distribution at the end of injection, obtained by maintaining the voltage constant at 40 kV during the injection period. The initial bucket area of 7.0 eV sec is adequate to contain the injected beam of $\Delta E = \pm$ 2.5 MeV and $\Delta \Phi = \pm$ 135° as it undergoes synchrotron oscillations, but not enough to account for the bucket area reduction caused by the space charge forces. The bucket is 100% full and particles spill out of the bucket as it shortens during acceleration. For comparison, Figure 5(b) shows the phase space distribution at the end of injection when the rf voltage is raised by 43%, from 40 kV to 70 kV. In this case, the bucket area is 9.0 eV sec and the particles remain in the bucket

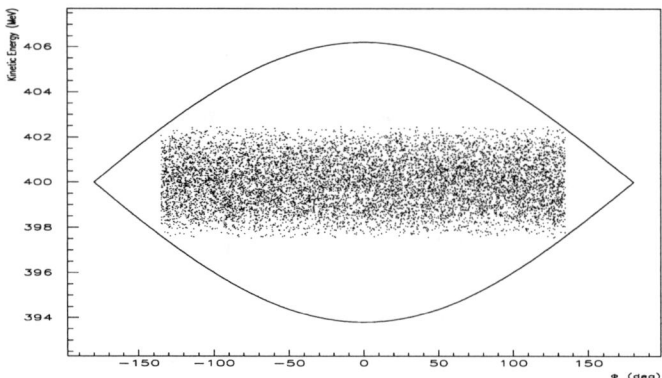

Figure 4: Bucket and Phase Space Distribution of the First Injected Turn in RCS.

as it is accelerated.

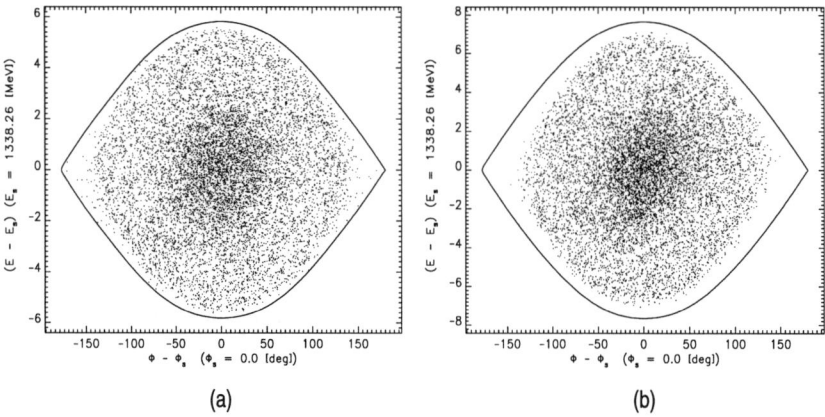

Figure 5: Particle Distribution at the End of Injection when (a) the Rf Voltage Is Maintained Constant During Injection and (b) the Rf Voltage Is Increased by 43%. Note the different scales used in the plots.

During the early stages of acceleration, the space charge forces in RCS are large enough to cause strong defocussing. An example of loss of focussing is shown in Figure 6(a). The figure depicts the instantaneous energy gain per turn at 1 msec of acceleration of a beam of initial energy spread ± 1.0 MeV chopped at 34% and injected into a 4.5 eV sec bucket. In the figure, the dashed line represents the energy gain due to the space charge, the short-dashed, that due to the rf voltage and the continuous line depicts the sum total of the energy gains. The acceleration provided by the rf is overcome by the deceleration caused by the space charge, as indicated by the negative slope of the total energy gain near the synchronous phase. Figure 6(b) shows the bucket and the phase space distribution of the beam at 1 msec, where the bucket distortion is clearly seen.

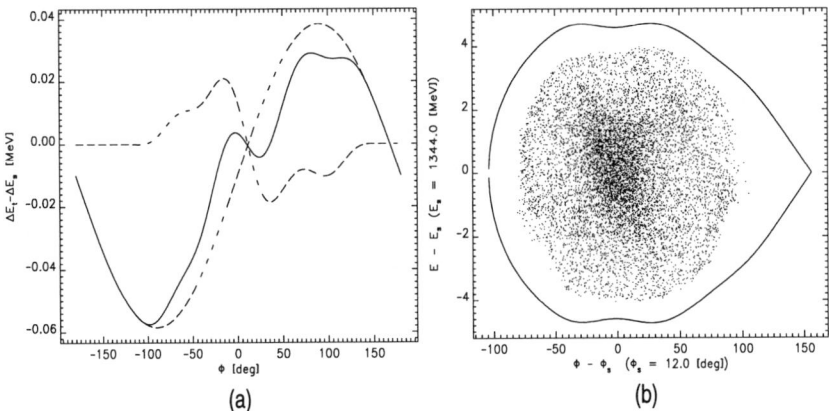

Figure 6: (a) Instantaneous Energy gain per Turn at 1 msec of Acceleration in RCS. The dashed line depicts the energy gain per turn due to the space charge alone, the short-dashed line represents the gain due to the rf voltage alone, and the continuous line is the sum total of the energy gains. (b) Bucket and Phase Space Distribution of the Beam whose Energy Gain per Turn is Shown in (a).

In general, loss of focussing may not be a concern, provided the bucket area is large compared to the bunch area and the beam is well matched to the bucket. In our studies, we opted for an adequate but not large bucket area in the ealier stages of acceleration while maintaining a well focussed beam.

Another factor that affects the longitudinal dynamics is the tune depression introduced by the space charge. In the presence of space charge, the synchrotron frequency does not decrease continuously with amplitude, as in the no-space-charge case. This can be seen as follows. The synchrotron frequency in presence of space charge produced by a beam of a given length can be estimated by the usual computation of the area under the Hamiltonian contour between two turning points of the motion:

$$\tau = \frac{1}{f_s} = 2\int_{\Phi_1}^{\Phi_2}\{2B[\Lambda(\Phi_2) - \Lambda(\Phi) + \frac{eV}{2\pi}(cos\Phi + \Phi sin\Phi_s - cos\Phi_2 - \Phi_2 sin\Phi_s)]\}^{-1/2}d\Phi, \qquad (4)$$

where:
τ and f_s are the synchrotron period and frequency, respectively;

Φ is the rf phase; Φ_s is the synchronous phase;
Φ_1 and Φ_2 are the turning points of motion;
Λ is the space charge potential;
V is the peak rf voltage; and the constant

$$B = -\frac{h\eta\omega_s^2}{\beta_s^2 E_s},$$

with h the harmonic number, ω_s the angular revolution frequency, E_s the synchronous energy, β_s and γ_s the relativistic factors, and $\eta = 1/\gamma_t^2 - 1/\gamma_s^2$. γ_t is the transition energy. The contribution due to the space charge potential is given by the first two terms in the integrand of Equation (4).

In Figure 7(a) the synchrotron frequency is plotted as a function of amplitude for a bunch of 79° half phase spread, at the mid-point of the acceleration cycle in RCS.

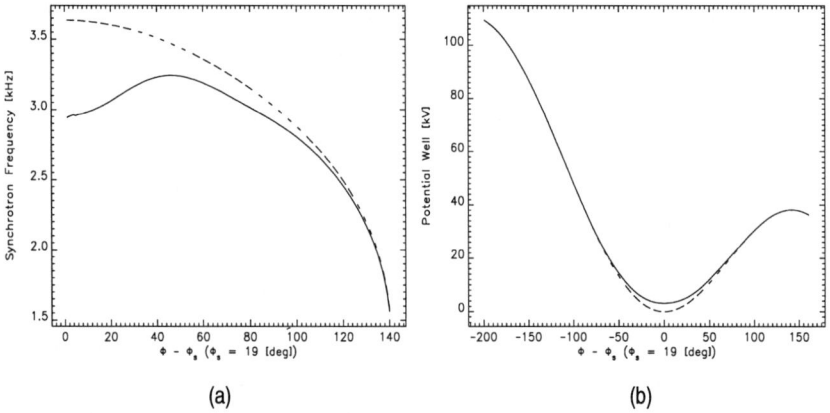

(a)   (b)

Figure 7: (a) Synchrotron Frequency as a Function of Amplitude, for a Beam of ±79° Phase at t=12.5 msec of the Acceleration Cycle in RCS. The continuous line depicts the synchrotron frequency including space charge. The synchrotron frequency without space charge is represented by the dashed line. (b) Potential Well Including Space Charge (continuous line) and Without Space Charge (dotted line).

In the figure, the frequency variation calculated without the space charge contribution is also shown, for comparison. The corresponding potential distributions (with and without space charge) are plotted in Figure 7(b). Figure 7(a) shows that when the space charge potential is included, the particles near the synchronous phase have the largest frequency depression from the bare frequency, about 21%. As the particle density decreases near the end of the bunch, the depression decreases and tends asymptotically to zero towards the unstable fixed point.

During the acceleration cycle, the beam is kept above the microwave instability threshold by maintaining a large momentum spread. Longitudinal instability thresholds were obtained from the estimated coupling impedances [1]. The contribution to the longitudinal impedance from space charge dominates in RCS, where it is -220j ohms at injection energy and -50j ohms at extraction energy. RCS operates below the transition energy, thus the longitudinal microwave instability is not expected to occur unless there is a large resistive component. However, a conservative approach was adopted to ensure that the momentum spread is sufficient to satisfy the Keil-Schnell (KS) stability criterion, which is given by [8]:

$$|\frac{Z}{n}| \leq \frac{F|\eta|\beta^2 E/e}{I_{pk}}(\frac{\Delta p_{fwhm}}{p})^2 \qquad (5)$$

Studies were performed to choose an rf voltage profile that provides adequate bucket area and momentum spread. Tracking studies show that the RCS beam remains in the stable region defined by the KS criterion over the duration of the acceleration cycle [9]. Figure 8 depicts the momentum spread of the beam obtained from tracking together with the longitudinal instability threshold according to the K-S criterion.

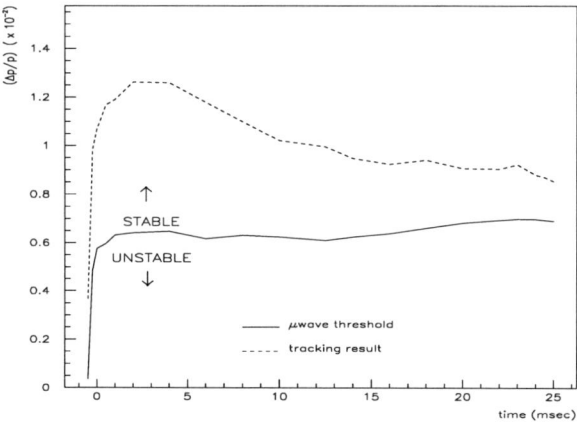

Figure 8: Longitudinal Instability Threshold and Momentum Spread of the Beam in RCS Obtained from Tracking Simulations.

The contribution to the longitudinal impedance from the space charge self-fields dominates in RCS. To minimize the space charge self-field impedance, the ceramic vacuum chamber is constructed with a special rf shield, such as that used at the ISIS facility of Rutherford-Appleton Laboratory [10]. The shield follows the beam envelope at an aperture equal to the beam-stay-clear (BSC), defined as twice the emittance and the ±1% momentum spread of the injected beam. This reduces the longitudinal space charge impedance by 30% at injection and 20% at extraction.

Simulation studies addressing the aforementioned issues lead to the rf voltage profile and beam parameters for RCS that minimized the instabilities and beam losses. The rf voltage program and the time variation of the bucket and bunch areas in are shown in Figures 9 and 10, respectively. In the figures, t=0 indicates the start of the acceleration clock. The initial voltage is 40 kV, which is the minimum necessary to contain the beam injected with an energy spread of ±2.5 MeV, cut by 25%. The bucket area is 7 eV sec and the bunch area is 3.4 eV sec. During the injection period, the voltage is raised from 40 kV to 70 kV to overcome bucket area reductions due to space charges. At the end of injection, the bucket and the bunch area are 9.0 eV sec and 7.3 eV sec, respectively. From the end of injection to 7.5 msec the voltage is designed to maintain a constant bucket area of 9.0 eV sec. The voltage reaches a maximum of 169 kV at 7.5 msec. For the remaining part of the cycle the energy spread of the beam is controlled by manipulating the bucket area. The voltage is kept constant from 7.5 msec to 12.5 msec, while the bucket area grows from 9.0 to 11.3 eV sec. After 12.5 msec, the voltage is decreased from 169 kV to 114 kV. The rf voltage is kept high during the later part of the cycle so as to provide a high enough synchrotron frequency that the particles in the bunch can follow the rapidly changing synchronous phase.

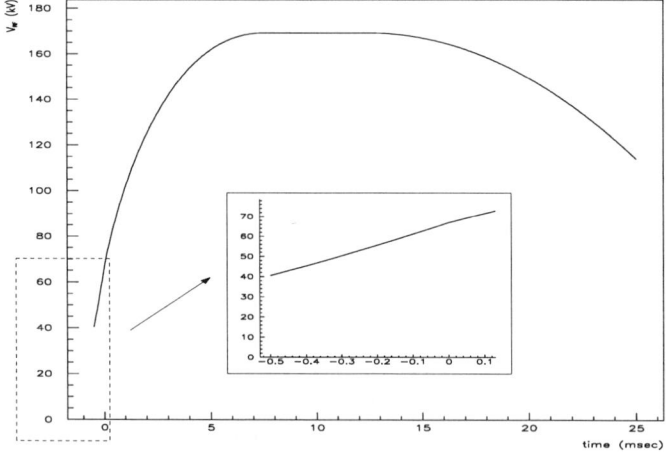

Figure 9: Rf Voltage Program for the Injection and Acceleration Cycles in RCS. The beginning of the acceleration clock starts at t=0.

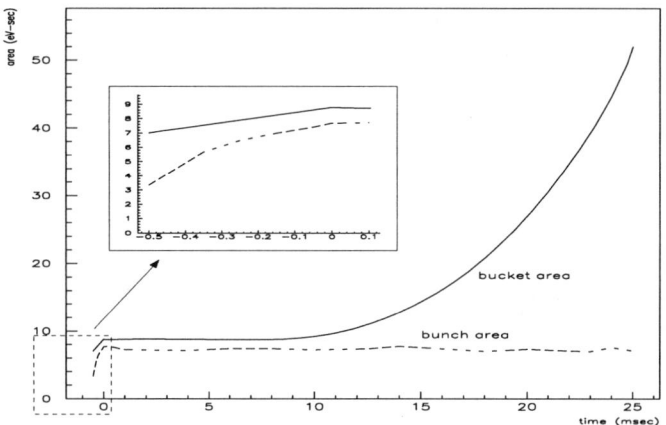

Figure 10: Bucket and Bunch Areas, Obtained from the Simulations under the rf Voltage Program Shown in Figure 9.

Space Charge Issues in the 10-GeV Synchrotron

The 10-GeV synchrotron (RCS-II) uses the 2-GeV synchrotron as its injector. In this case, the RCS rf system is run with a harmonic number equal to two, to provide two bunches for RCS-II. The circumference ratio between the 5-MW ring and the 1-MW ring is four. The RCS bunches at 3 MHz are transferred into waiting buckets of the 6-MHz RCS-II rf system which operates with a harmonic number of 16. RCS-II magnet are powered by a dual-frequency resonant circuit similar to the RCS circuit.

In RCS-II, the beam energy is high (γ varies from 3.1 to 11.7). However, space charge effects cannot be neglected in this machine, especially at injection and at extraction. At injection, space charge fields affect the matching between the incoming beam from RCS and the RCS-II waiting bucket. At extraction, the bunch length is short, about 22 nsec long, causing a high peak current.

The two bunches from RCS are injected into two buckets of RCS-II which are separated by an empty bucket. The bunch-to-bucket transfer between RCS and RCS-II is done by the following algorithm. The rf voltage at extraction of the 2-GeV beam determines the energy spread, ΔE, and bunch length, Δt, of the bunch. The energy spread and bunch length of the incoming beam are matched to a phase space area in the waiting bucket determined by the Hamiltonian contour that has the same ΔE and Δt. This determines the required initial voltage and the waiting bucket area. Figure 11(a) shows the rf bucket and the bunch population in RCS at extraction and Figure 11(b) shows the bunch injected into the waiting bucket of RCS-II. The dashed line in Figure 11(b) represents the Hamiltonian contour whose height, ΔE, and whose enclosed area are equal to those of the injected beam.

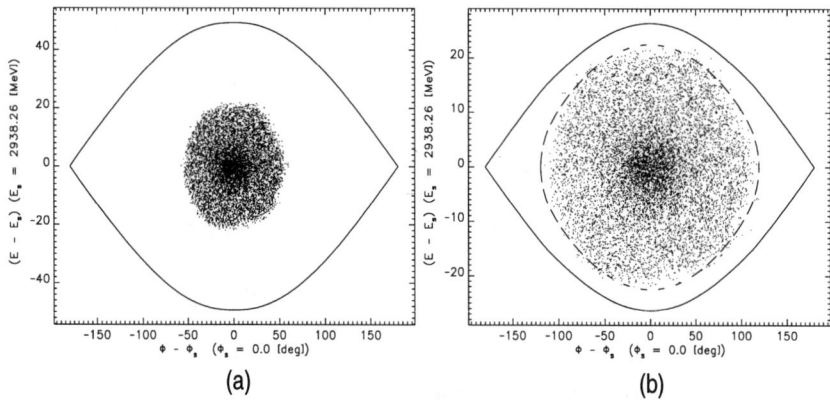

Figure 11: (a) Rf Bucket and Particle Distribution at Extraction in RCS. The single bunch area is 3.7 eV sec. (b) Rf Bucket and Phase Space Distribution at Injection in RCS-II. The dotted line indicates the contour enclosing an area of 3.7 eV sec.

Neglect of the space charge potential in the calculation of the Hamiltonian contour leads to underestimation of the necessary bucket area at injection and to consequent dilution of the beam as it is accelerated, due to mismatching. This can be seen in Figure 12(a), where we plot the phase space distribution of the beam at 6 msec into the acceleration cycle. In this case, the space charge contributions were not included in the bunch-to-bucket-transfer algorithm described above. The figure shows considerable dilution of the beam. For comparison, Figure 12(b) depicts the beam also at 6 msec, when the space charge contributions were included.

In RCS-II, the space charge impedance is smaller than in RCS by a factor of 5 at injection and 10 at extraction. The peak current in RCS-II is 160 A at injection and less than 800 A at extraction. The voltage has to be kept high at the later stages of acceleration not only to provide a high enough synchrotron frequency, as in RCS, but also to avoid collapse of the bucket due to the high peak current.

Figure 13 shows the rf voltage and bucket and bunch areas from injection to extraction in RCS-II obtained from simulations. The initial voltage determined by the matching procedure aforementioned is 0.7 MV and the corresponding single bucket area is 5.8 eV sec. This area is maintained at this initial value for the first 8.0 msec, when the voltage reaches a maximum value of 1.9 MV. The voltage is decreased to 1.7 MV at extraction.For the latter part of the cycle, the voltage is maintained high as explained above. The rf program for RCS-II

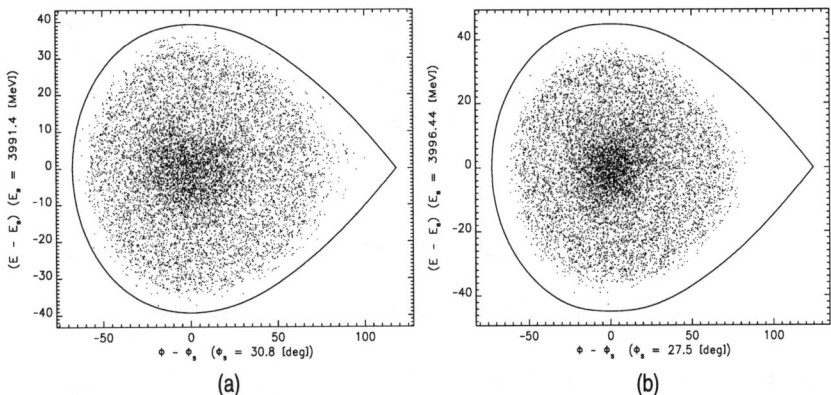

Figure 12: Phase Space Distribution of the Beam at 6 msec of the Acceleration Cycle in RCS-II when: (a) the Waiting Bucket Area Is Calculated without Space Charge and (b) the Waiting Bucket Area Is Calculated with Space Charge.

satisfies the no-loss criterion and satisfies the KS criterion through most of the acceleration.

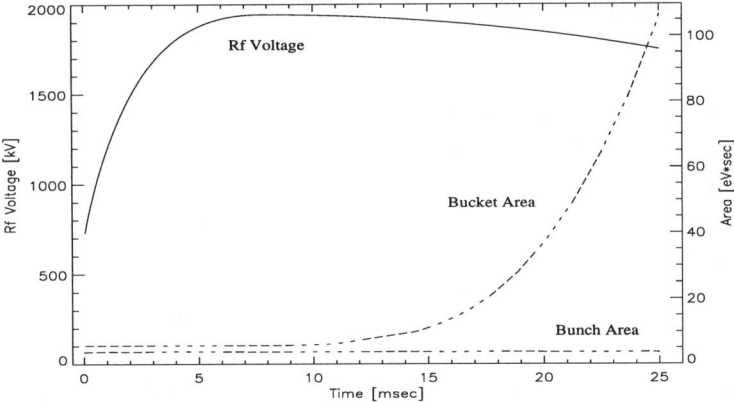

Figure 13: Rf Voltage Program for RCS-II, Showing the Time Evolution of the Bucket (solid line) and Bunch (dotted line) Over the Complete Cycle.

Summary

The space charge self-fields are large in the synchrotrons of a 1-MW and a 5-MW spallation neutron sources designed to deliver 1.04×10^{14} protons per pulse at a repetition rate of 30 Hz. Longitudinal space charge effects were examined in detail by tracking simulations and means to circumvent these effects were investigated. The tracking studies were used to formulate rf voltage profiles and beam parameters that facilitate keeping beam losses in both synchrotrons to radiation levels that allow hands-on maintainance.

References

[1] "IPNS Upgrade – A Feasibility Study," ANL Report ANL-95/13 (April, 1995).

[2] Y. Cho, et. al. "A 10-GeV, 5-MW Proton Source for a Pulsed Spallation Source," *Proc. of the 13th Meeting of the Int'l. Collaboration on Advanced Neutron Sources*, PSI, Villigen, to be published (Oct. 11-14, 1995).

[3] K. R. Symon, "Synchrotron Motion with Space Charge," ANL Report NSA-94-3 (April 4, 1994).

[4] Y. Cho, E. Lessner and K. Symon, "Injection and Capture Simulations for a High Intensity Proton Synchrotron," *Proc. of the European Particle Accelerator Conference*, London, page 1228 (1994).

[5] C. K. Bird and A. B. Langdon, "Plasma Physics Via Computer Simulations," *Adam Higler*, Bristol, England, 1991, page 19.

[6] K. C. Harkay, "Transverse Beam Size and Ratio (b/a) in Space Charge Geometry Factor for RCS-II," ANL Report NSA-94-1 (April 4, 1994).

[7] K. R. Symon, "Optimal Filtering in Space Charge Simulations," ANL Report NSA-94-4 (May 24,1994).

[8] V. K. Neil and A. M. Sessler, "Longitudinal Resistive Instabilities of Intense Coasting Beams in Particle Accelerators," Rev. Sci. Instrum. **36**(4), 429 (1965).

[9] K. Harkay, Y. Cho and E. Lessner, "Longitudinal Instability Analysis for the IPNS Upgrade," *Proc. of the 1995 Particle Accelerator Conference*, Dallas, Texas, to be published (May 1995).

[10] G. H. Rees, "Status Report on ISIS," *Proc. of the IEEE Particle Accelerator Conference*, London, page 1228 (1994).

Longitudinal and Transverse Tracking Studies for ESS

C.R. Prior

Rutherford Appleton Laboratory, Chilton, Oxon, UK
(December 1, 1995)

> The techniques currently being employed to model beam transport and injection into the ESS accumulator rings are described. These use a combination of one dimensional (longitudinal) and two- dimensional (transverse) particle tracking codes, incorporating a variety of methods of simulating self-field effects. A description of the proposed mechanism for painting longitudinal and transverse phase-space for ESS is given, and the results of the modelling and subsequent optimisation are discussed.

INTRODUCTION

The design of the European Spallation Source raises difficult theoretical and constructional problems which must be overcome in order to satisfy the highly stringent specifications. Only extremely low loss levels can be tolerated in the accumulator rings, and the method of injection and trapping needs to be examined in close detail if these requirements are to be met. From an initial framework, this entails computer modelling with a series of codes looking at different aspects of the process, taking into account the effects of space-charge in the beam.

In this paper, the ESS injection scheme is outlined and the manner in which the various parameters have been optimised is described. The procedure is, first, to treat the motion as decoupled and simulate injection into longitudinal phase-space. The partially optimised system is then modelled in two-dimensional (transverse) space. The whole process has, inevitably, to be repeated before a suitable scheme is decided upon. Finally, a three-dimensional particle tracking code can, in theory, be used to examine coupling between all three phase planes; however, given the time-scale involved (the order of years of CPU), this is unlikely to be a practical proposition and other methods of estimation may need to be found.

The various stages and results to date are described here.

© 1996 American Institute of Physics

THE ESS INJECTION SCHEME

For the ESS, a total of 2.34×10^{14} H^- ions are injected at 1.334GeV successively into each of two accumulator rings and are converted to protons by means of a stripping foil. At the point of injection, the dispersion of the rings is non-zero and this is used as a means of simultaneously painting longitudinal and horizontal phase-spaces by ramping the momentum of the incoming beam. Vertical painting is achieved by programming the vertical closed orbit via four bump magnets. The resulting oscillation amplitudes vary as injection proceeds from large to small in the horizontal plane and from small to large in the vertical plane. This correlation is the most favourable, as it helps to minimise the acceptances of the injection dipole and reduces the number of times the protons intercept the stripping foil [1,2].

COMPUTER MODELLING OF LONGITUDINAL INJECTION

A study of injection into longitudinal phase-space has been carried out using the computer code TRACK1D. This is a particle tracking code based on the standard equations of motion in a synchrotron, which, given in terms of phase, ϕ, and energy, E, are

$$\frac{d}{dt}\frac{\Delta E}{\omega_0} = \frac{e}{2\pi}[V(\phi) - V(\phi_s) + V_{sc}(\phi)] \tag{1}$$

$$\frac{d}{dt}\Delta\phi = \frac{h\omega_0 \eta}{\beta^2 E_s}\Delta E. \tag{2}$$

Here Δ refers to individual particle variations from synchronous values (suffix s), $\eta = 1/\gamma_t^2 - 1/\gamma^2$, $V(\phi)$ is the total applied cavity voltage and h is the harmonic number. ω_0 is the revolution frequency and is evaluated for the synchronous particle to the order of accuracy of the equations. The space-charge forces are given by

$$V_{sc}(\phi) = -e\frac{d\lambda(\phi)}{d\phi}[\frac{Rg_0}{2\epsilon_0\gamma^2} - L\beta^2 c^2], \tag{3}$$

where $\lambda(\phi)$ is the line density of the bunch, L is the total inductance per turn of the reactive wall, R is the radius of the machine and, for a beam with assumed circular cross-section of radius a in a circular pipe of radius b, $g_0 = 1 + 2\ln(b/a)$. The plan for ESS is to use a dual harmonic radio-frequency voltage for harmonic numbers h and $2h$ of the form

$$V(\phi, t) = V_0(t)[\sin\phi - \delta\sin 2\phi]. \tag{4}$$

A more general formulation has previously been analysed in relation to the ISIS synchrotron [3] and it has been shown how the stable regions of phase-space are increased, allowing more particles to be trapped with good bunching factor.

A feature of TRACK1D is the way in which it treats the problem of simulating a large number of injection turns (1000 for ESS) within the confines of a manageable number of beam modelling particles, usually of the order of ten to fifty thousand. This is achieved by allowing the simulation particles to carry variable charge. During injection, a Eulerian approach is adopted from the theory of fluid mechanics. Attention is focused on beam properties at the nodes of an imaginary grid in phase-space (the "particles"). During a revolution of the machine, charges at these points move away from the nodes under the equations of motion, (1) and (2), but are weighted back onto the gridpoints, using a scheme such as area weighting or the triangular–shaped density cloud (see, for example, [4]). Newly injected beam is similarly weighted onto the nodes. In this way a charge density is built up but the number of particles is equal to the number of nodes with non-zero charge and so bounded by the scaling of the grid. (Note that the same grid does not have to be used throughout, nor does it have to be regular.) At the end of the injection, the nodal charges are regarded as the macro-particles for the beam and tracking subsequently continues under the normal Lagrangian approach. The method has become feasible only since the advent of high-speed colour plotters, which provide an added dimension for the output of information from the code. It has proved highly successful for modelling behaviour in ISIS and in indicating possible future developments [3].

The remaining features of the code include standard leap-frog methods for integrating the equations of motion (1) and (2) and various routines to calculate space-charge forces from the line density via equation (3), incorporating smoothing techniques described in [4] to avoid spurious statistical effects.

Repeated use of the code has suggested that the scheme most suited to the ESS requirements demands a 60% chopped beam from the linac. During the injection process, the beam is ramped linearly from an initial momentum spread, $\Delta P/P$, of $[0,2] \times 10^{-3}$ to $[2,4] \times 10^{-3}$ over a period of 0.6 msec while 1000 turns are injected. R.F. steering is also operated, with the frequency held at 1.67178 MHz during injection - corresponding to a particle circulating naturally with momentum $\Delta P/P = 2 \times 10^{-3}$ - which then falls linearly to the synchronous revolution frequency 1.67138 MHz during a further 0.6 msec while the second ring is filled. This is shown schematically in figure 1. The effect is to fill the lower half of the phase-space bucket initially but, under the momentum ramping, to inject into the upper region towards the end of the cycle. The r.f. steering then directs the beam towards the centre of the ring.

Other parameters for ESS are $R = 26$m, $L = 0$, $h = 1$ and $b/a = 1.8$. The optimised δ is found to be 0.49, which is associated with a voltage pattern $V_0(t)$ programmed in five stages, from an initial 8 kV to 22 kV during injection, then a slower rise to a peak of 26 kV at 0.9 msec, followed by a linear decrease to 24 kV at 1.2 msec.

Results of the final simulation are shown in figure 2 (see also [5]). The shading is a manifestation of the different charges carried by the particles, the boundaries being effectively contour lines of charge density. The effect of the

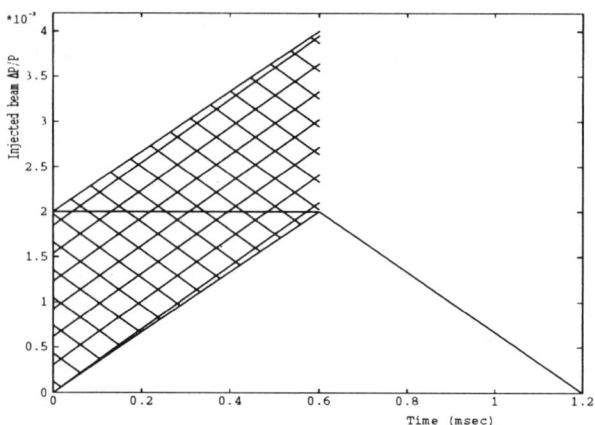

FIG. 1. Momentum ramping and r.f. steering for ESS

r.f. steering is a displaced phase-space bucket, centred on $\Delta P/P = 2 \times 10^{-3}$ during injection, then linearly progressing to $\Delta P/P = 0$. The momentum ramping varies the position of the incoming turns (whose superposition is clearly visible in the first three plots) relative to the r.f. bucket. An additional requirement for usable beam is that the final bunch remains between phases $\pm 124°$. This is achieved in the model and no beam loss is predicted. The final bunching factor is 0.464 (comparing well with the best value of 0.29 that could be found for the single harmonic voltage, $\delta = 0$).

COMPUTER MODELLING OF TRANSVERSE INJECTION

In the transverse plane, computer modelling has been carried out using the code TRACK2D, which is a development of earlier codes described in, for example, [6–8]. This program now includes the method of variable charge described above [1]. The layout at the foil is shown in figure 3. The incoming beam has 3σ unnormalised emittances of $2.45(\pi)\mu$rad.m in each plane and generates a beam of $150(\pi)\mu$rad.m, horizontally through the position of the foil and the momentum painting, and vertically through the collapse of the vertical closed orbit. The turns are mismatched to the ring, with $\beta_x = 2.4$m and $\beta_y = 1.8$m, and corresponding values of $\beta_x = 7.877$m and $\beta_y = 4.869$m in the machine. The dispersion at the injection point is $\alpha_p = 5.732$m. The first step has been to optimise the way the orbit is decreased to minimise heating

---

[1] Initial runs, however, have relied on the more straightforward approach of 1000 turns each of 250 particles of equal charge.

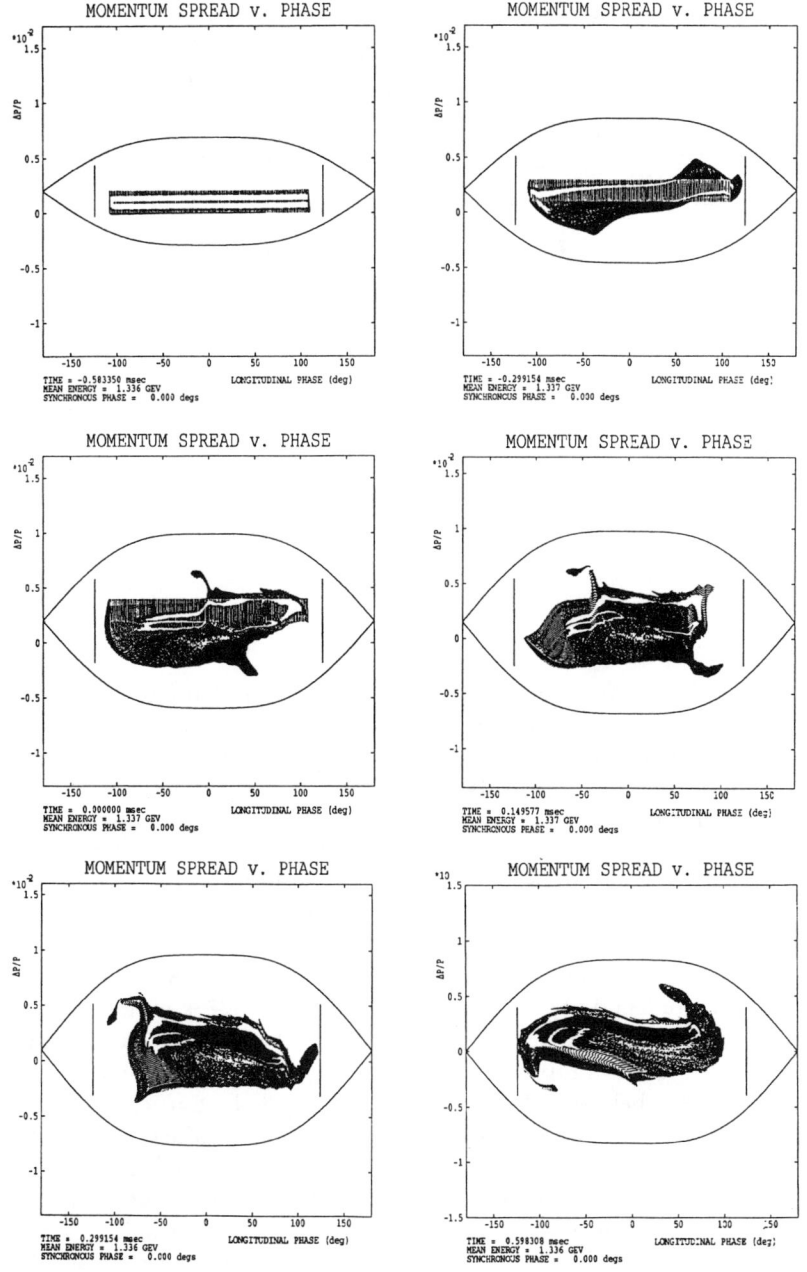

FIG. 2. Phase-space plots for the injection of 2.34×10^{14} protons per pulse at 1.334GeV into the ESS accumulator rings.

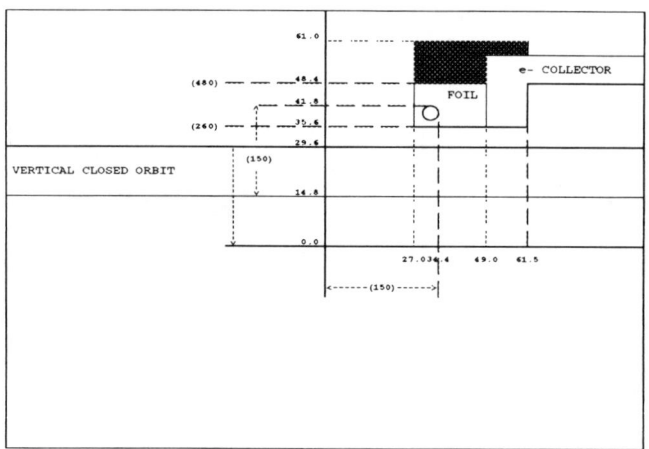

FIG. 3. ESS H⁻ transverse injection parameters at foil.
(Units: mm. Figures in brackets represent equivalent emittances in μrad.m.)

of the foil: this has been achieved using an analytical program neglecting space-charge effects, and indicates the bulk of the fall should be between the injection of turns 250 and 600, with the orbit held constant otherwise. The predicted number of foil traversals is between 6 and 7 per particle.

Results from tracking in the absence of space-charge, particularly the effects of ramping the momentum of the beam, are shown in figures 4 and 5. The vertical distribution is somewhat hollow while the horizontal distribution has a sharp peak towards the positive side of the beam. At the time of writing, attempts are being made to see how this is effected by small modifications to the method of ramping the longitudinal momentum. Space-charge also needs to be included, but, with a total distance of 327 km through which to track - and from which reliable results must be obtained - a preliminary method which regards individual turns as "blobs" of charge and calculates the coulomb force between them is being investigated. Earlier work [6] did suggest, however, that the internal space-charge effects of each of the turns cannot be ignored, and an attempt at a fully self-consistent run must be made at some future stage. It would also be of interest to carry out a full three-dimensional simulation, to look at, for example, equipartitioning of energy and coupling between the phase-planes, but, for such a long process, this is undoubtedly not a practical proposition at the present time.

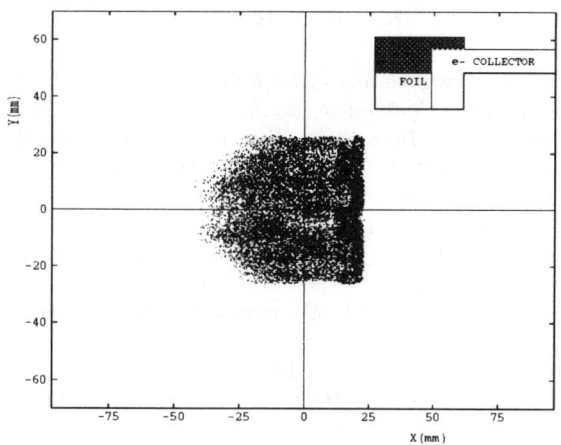

FIG. 4. ESS H⁻ injection at foil.

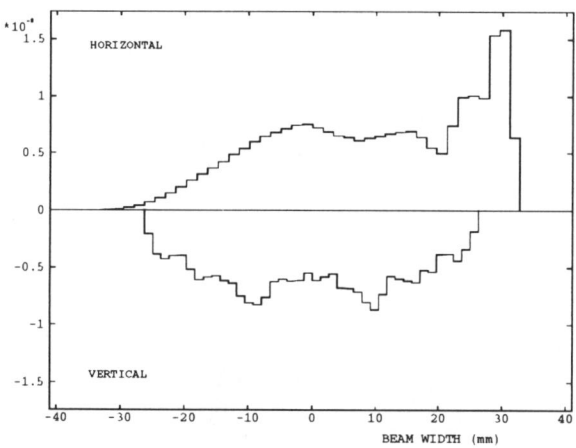

FIG. 5. ESS Transverse charge distribution of final beam.

REFERENCES

1. G.H. Rees: *Important Design Issues of High Output Current Proton Rings.* Proceedings of the 4th EPAC Conference, London, June 1994, p. 241.
2. G.H. Rees: *Space Charge Tune Shift, Fast Resonance Traversal and Current Limits in Circular Accelerators.* Proccedings of the 8th ICFA Advanced Beam Dynamics Workshop, Bloomington, Indiana, October 1995.
3. C.R. Prior: *Studies of Dual Harmonic Acceleration in ISIS.* Proceedings of the Twelfth Meeting of the International Collaboration on Advanced Neutron Sources, ICANS-XII, 1993, A-11.
4. R.W. Hocking and J.W. Eastwood: *Computer Simulation Using Particles.* McGraw Hill, 1981.
5. C.R. Prior: *Longitudinal Injection for the ESS.* ESS-94-7-R, November 1994.
6. C.R. Prior: *Multiturn Injection for Heavy Ion Fusion.* Proceedings of the Symposium on Accelerator Aspects of Heavy Ion Fusion, GSI, Darmstadt, March 1982, p. 290.
7. C.R. Prior: *Final Focusing of 10 GeV Bismuth Ions. ibid.* p. 376.
8. C.R. Prior: *Particle Tracking in Solenoid Channels.* AIP Conference Proceedings 152 on Heavy Ion Inertial Fusion, Washington, D.C., 1986.

HEAVY ION INERTIAL FUSION

INDUCTION-ACCELERATOR HEAVY-ION FUSION: STATUS AND BEAM PHYSICS ISSUES*

Alex Friedman
Lawrence Livermore National Laboratory
University of California
Livermore, CA 94550 USA

ABSTRACT

Inertial confinement fusion driven by beams of heavy ions is an attractive route to controlled fusion. In the U.S., induction accelerators are being developed as "drivers" for this process. This paper is divided into two main sections. In the first section, the concept of induction-accelerator driven heavy-ion fusion is briefly reviewed, and the U.S. program of experiments and theoretical investigations is described. In the second, a "taxonomy" of space-charge-dominated beam physics issues is presented, accompanied by a brief discussion of each area.

INDUCTION-ACCELERATOR HIF CONCEPT AND STATUS

1. Overview

The U.S. is developing the physics and technology of ion induction accelerators, with the goal of electric power production by means of heavy-ion beam–driven inertial fusion (heavy-ion fusion, or HIF). In addition, heavy-ion induction accelerators are an attractive driver option for a high-yield microfusion facility for defense research.

Accelerators for high-energy physics research typically have long lifetimes and good availability; an accelerator designed as a fusion driver should share these attributes. Since the beams will be focused onto the target by magnetic fields, the final lens will not be subject to serious damage from the exploding targets. Induction accelerators in particular are efficient, low-impedance devices that naturally accelerate a high-current beam, can readily amplify the current of that beam, and can easily meet repetition rate requirements of a few pulses per second. For these reasons, the heavy-ion induction accelerator is the driver being pursued for an inertial confinement fusion (ICF) power plant in the U.S. Inertial Fusion Energy (IFE) program, in the Office of Fusion Energy of the U.S. Department of Energy. However, significant technology development must be carried out before an HIF driver is realized. Because driver development is deemed the long lead-time item in the IFE development sequence, much of the programmatic emphasis is currently being placed on high-current ion accelerator research.

For a fusion power technology to be viable, it must offer an economically attractive power plant at a reasonable size (of order 1 GW electric output, a smaller plant size being preferable), and must have favorable properties with respect to environmental considerations. Recent studies indicate that the cost of electricity from an HIF power plant can compare favorably with those of tokamak fusion, future fission and coal, and suggest that the environmental attributes can be benign as well.[1,2,3]

Furthermore, the development program leading to a pilot power plant must be affordable. The HIF development cost will be minimal because the program builds upon a multi-billion dollar worldwide investment in particle accelerators, and upon a large and multi-faceted inertial fusion program based upon lasers and light-ion drivers. The planned National Ignition Facility (NIF), to be based upon a 1.8 MJ glass laser, can test the most important heavy-ion target physics issues: soft x-ray transport and drive symmetry, hohlraum plasma dynamics, capsule implosion hydrodynamics, and mixing of unwanted material into the central fuel. It will be possible to deploy experiments on the NIF which focus the laser beams into smaller hohlraums (filled with high-atomic-number doped gas at 1/10 the critical density) placed at each end of the larger hohlraum which contains the fusion capsule. These will emit their x-rays into the main hohlraum in the same way as the converters at the ends of a two-sided heavy-ion fusion hohlraum. Experiments using light-ion drivers will also yield information which is important to heavy-ion scenarios, including ion energy deposition and conversion to x-rays in foams, radiation transport, capsule symmetry requirements, and beam propagation in the target chamber.

The use of a non-solid first surface in the chamber obviates any need for a separate large-scale materials testing facility; experiments on the NIF and other lasers will provide much of the needed data. An Engineering Test Facility (ETF) accelerator built to explore target physics and target-chamber dynamics can be used in the subsequent demonstration power plant as well. In one scenario, such a program might be pursued as an international effort. In another scenario, the ETF driver might also serve as the driver for a high-yield microfusion capability for defense research.

These multiple-use attributes are made possible by the physical decoupling of the driver, the fusion chamber, and the target. A beam switchyard will enable one accelerator to drive several target chambers at the same time. The cost penalty associated with the requisite pulse repetition rate should be negligible. Reduced-yield targets can be used in scaled fusion chambers, enabling the resolution of nuclear engineering issues at minimal cost. Finally, an accelerator driver can be built in stages, thereby combining the accelerator development program with the construction of the eventual facility.

Research into HIF drivers outside the U.S. has emphasized radio-frequency (RF) linac technology in conjunction with storage rings. The U.S. induction-based research program and the RF-based efforts are complementary. The decision to pursue induction technology in the U.S. was made for well-defined reasons, including a lower predicted driver cost, and a belief that the critical issues arise at lower energy and so can be more readily resolved. RF linac technology is mature, but the required storage rings are challenging; the approach may prove entirely adequate to the fusion-driver task. Discussions of the advantages of each of these approaches can be found in Refs. [4,5].

In Subsection 2 below, we review the basic design of a radiatively-driven heavy-ion target using "two-sided" illumination (with beams in two clusters), and the requirements imposed by such a target upon the incoming beams.

In Subsection 3 we present a schematic of an induction linear accelerator designed around these requirements. The research program is working toward expansion of the usable design regime of a linear accelerator, toward a lower final particle energy. These considerations are also discussed in that subsection.

The recirculating induction accelerator promises cost reduction by re-using the same accelerating and focusing (confining) elements multiple times.[6] The recirculator concept is discussed in Subsection 4.

The goals and design of the Elise accelerator are briefly described in Subsection 5.[7] Proposed for construction at Lawrence Berkeley National Laboratory (LBNL), Elise is an outgrowth of the ILSE (Induction Linac Systems Experiments) concept, and represents the initial phase of that overall program.[8] Elise was designed to be a 5 MeV, multi-beam induction linac accelerating potassium (K^+, ~39 AMU) ions produced by LBNL's new electrostatic-quadrupole (ESQ) injector, which is also described briefly in Subsection 5. The ILSE accelerator is intended to serve as a testbed upon which a variety of experiments using driver-scale (in diameter and line-charge density) beams will be carried out. These will ultimately include beam merging, drift-compression, and focusing onto a small spot. The Elise project is currently "on hold."

A recently initiated experimental program in beam bending and recirculation at LLNL is described in Subsection 6.[9] The small prototype recirculator which is the ultimate goal of this work will be the first accelerator of its type. Because of the unique opportunity they will afford for the study of space-charge-dominated beam behavior on a long time scale, experiments on the small recirculator will be important to linear accelerators as well.

Finally, ongoing research is examining issues including beam propagation in the target chamber, long-time beam dynamics, and detailed beam optics in structures including injectors, beam combiners, plasma lenses, and magnetically confined transport of space-charge-dominated beams. A brief outline of these elements of the program is presented in Subsection 7.

2. Review of Requirements Imposed by the Target

The U.S. HIF program is examining a variety of target concepts. In one illumination geometry, the fusion capsule is enclosed in a hohlraum, which has a radiation converter of radius ~2-6 mm at each end. Each converter is illuminated by a cluster of beams with a narrow cone angle. This geometry is similar enough to that of an indirectly-driven laser fusion target that the design methods and tools used for the latter carry over directly. Furthermore, the illumination geometry is favorable for power-plant design, relative to concepts which require that the beams enter the fusion chamber from a wide multiplicity of directions. To achieve a gain of 10-100, such a target is projected to require a total incident beam energy of order 5 MJ in a suitably shaped pulse, with duration (for the main part of the pulse) of order 10 ns and a longer initial "foot" preceding. Other geometries include spherical or nearly-spherical targets, two-sided targets with the converter material distributed appropriately around the inside of the main hohlraum, single-sided targets with a single radiator, and "single-sided" targets with two radiators which are illuminated "sideways" (at 90° to the symmetry axis of the main hohlraum) by beam clusters coming from ports on the same side of the fusion chamber.

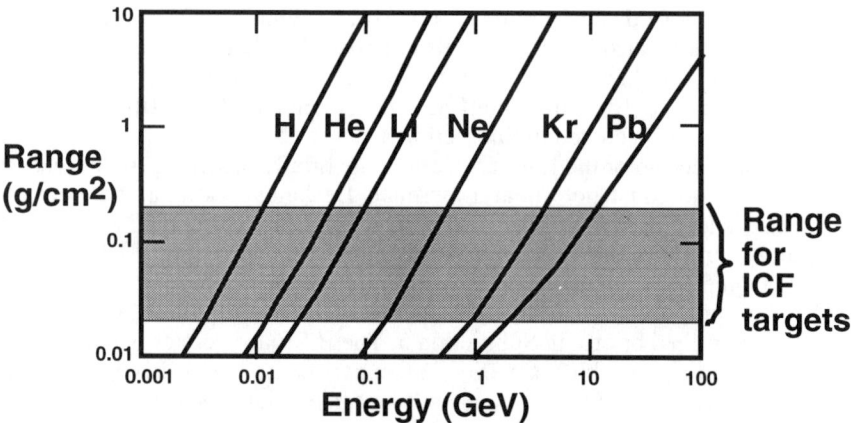

Fig. 1. Range-energy relation for various ions stopping in matter.

The energy-range relation of ions stopping in matter (Fig. 1) is an important factor in selecting the ion energy. It is necessary to stop the incident ions in a relatively small amount of matter (range $R \sim 0.02\text{-}0.2$ g/cm^2) if the amount of converter material heated by the incident beams is to be kept small enough for a specific energy deposition of order 10^8 J/g; the energy devoted to overcoming the specific heat of the converter material must be minimized if high conversion efficiency of ion energy to x-rays is to be obtained. For an ion of mass ~100-200 AMU, the required energy per ion is found to be of order 1-10 GeV. For a light ion of mass ~1-7 AMU, the corresponding energy is of order 3-100 MeV. The heavy-ion fusion program has emphasized drivers using singly-charged mass 100-200 ions because their relatively high allowed energy allows the requisite ~5 MJ to be achieved at a relatively low current (compared to that of a lighter ion scenario), which in turn allows non-neutralized ballistic focusing of the beam onto the target in near-vacuum. The achievable power on target is limited by several effects which add roughly in quadrature; these include beam space charge, beam emittance (thermal pressure), misalignments, and aberrations. The dependence of the focusable power upon ion energy as set by space charge is especially steep, varying as the 5/2 power.[5] Thus for non-neutralized focusing there is a strong incentive to operate at high ion energy and correspondingly large ion mass. However, as discussed in the next section, use of a lighter ion, or a higher charge state, may convey certain advantages, despite a requirement of at least partially neutralized transport to the target.

Most power plant concepts call for near-ballistic focus of the beams through a fusion chamber with radius 4-10 m. If one assumes a 10 m separation between the middle of the final magnetic optic and the target, and a 3 mm radius spot on target, the particles must be aimed with a "microdivergence" angle of $0.003/10 = 3\times10^{-4}$ radians. This same angle represents the ratio of transverse thermal velocity to directed beam velocity, and its square ($\sim 10^{-7}$) represents the ratio of transverse thermal to directed energies. For a 10 GeV beam, the transverse temperature must remain below ~1 keV. The transverse emittance, or projected phase space area, is the product of beam transverse dimension and velocity spread, the latter normalized to the directed velocity and so measured as an angle. The final optic does not produce a perfect image; geometric and chromatic aberrations limit the achievable spot size. Uncorrected optics are limited to a focusing angle of about 15 mrad. For a beam with a focal spot radius of 3 mm focused through a 15 mrad angle from a maximum radius 15 cm, one obtains an emittance requirement of ~ π(150 mm)(0.3 mrad) = π(3 mm)(15 mrad) = 45 π-mm-mrad (ignoring space charge). In addition, space charge causes beam spreading near the target, and the beam centroid must be precisely aimed and must not wander more than about 0.2 mm. These effects contribute jointly to the effective spot size; to compensate, the emittance must be somewhat smaller than that calculated above.

Similarly, the ability of the optical system to focus ions with a range of energies onto a small spot is limited. Chromatic aberrations limit the relative momentum variation $\delta p_z/p_z$ to about 0.3%. Optically-corrected focusing designs are employed in other accelerator applications, and have been studied for HIF [10], but remain to be shown to be practical.

Alternative modes of transporting the beams to the target include neutralized or partially neutralized ballistic transport, as well as self-pinched propagation in a channel. Neutralization may be important if operation at higher chamber pressures associated with a greater shot-to-shot repetition rate (for some chamber designs) is to be practical. By easing the effects of space charge, these modes open up a larger region in parameter space at lower ion energy and mass. Detailed simulation studies have shown that, for the nominal HYLIFE-II parameters,[1] even a relatively small degree (perhaps half a percent) of ionization of the background chamber vapor can result in excellent beam neutralization, and this nearly neutralized ballistic regime is well on its way toward replacing transport in near vacuum as the baseline model.[11,12] Pinch-mode propagation would offer a number of advantages, including smaller holes in the chamber wall, and acceptance of beams with a greater longitudinal velocity spread.[13] Recent progress in plasma lenses suggests that they may be useful in the introduction of a beam into such a channel.[14] However, these latter regimes remain less well understood.

3. Linear Accelerator Concepts

Induction accelerators apply electromotive force to the beams by passing them through a series of one-turn pulsed transformers (see Fig. 2). Pulsed power is routed to the transformer's "primary winding;" conducting surfaces are arranged so that an axial EMF appears across an "accelerating gap" in the beam pipe. The ferromagnetic core, which has a large permeability, gradually fills with magnetic flux. A short-circuit across the input is averted until the core's "Volt-seconds" have been consumed, affording almost-d.c., relatively flat pulses. Longitudinal confinement of the beams requires voltage "ears," or temporal ramps on the ends of the voltage waveform timed to coincide with the passage of the front and back tips of the beams through the accelerating gap (see Fig. 3). In contrast with RF accelerators, there is no requirement for constant frequency, so longitudinal

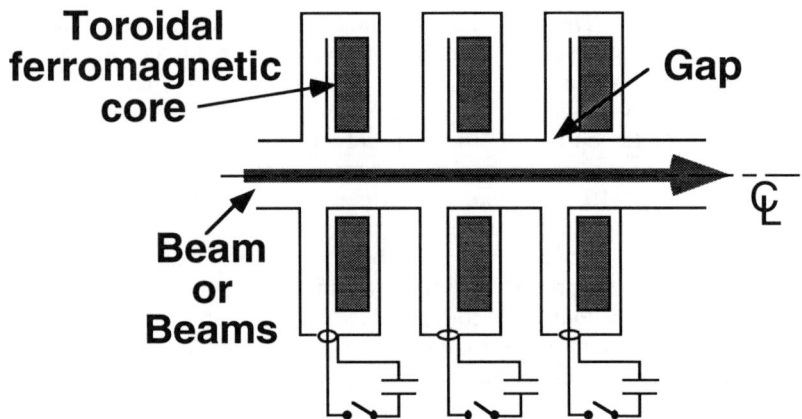

Fig. 2. Induction module geometry.

compression and increasing velocity can produce an increasing current, up to the limits of the transverse confining fields.

A "conventional" linear induction accelerator designed using ions of mass ~200 to meet the requirements imposed by the "conventional" target is depicted schematically in Fig. 4. Such an accelerator achieves economy by passing multiple beams through each induction core. Power amplification to the required 10^{14}-10^{15} W is achieved by beam combining, acceleration, and longitudinal bunching. The accelerator lattice must transversely "focus" the beam (confine it against its own space charge, and to a much lesser degree its own thermal pressure); in modern accelerators this is effected with alternating-gradient quadrupole lenses, which may be either electrostatic or magnetic. Each lens focuses the beam in one plane while defocusing it in the other; however, because the beam is larger (and thus experiences stronger fields) as it passes through a focusing element, the net effect is a strong focusing.

In such a "conventional" HIF accelerator, multi-beam injection with 16-64 beamlets is followed by acceleration with electrostatic transverse focusing. Four-into-one beam merging is effected at 10-100 MeV, with a transition to magnetic

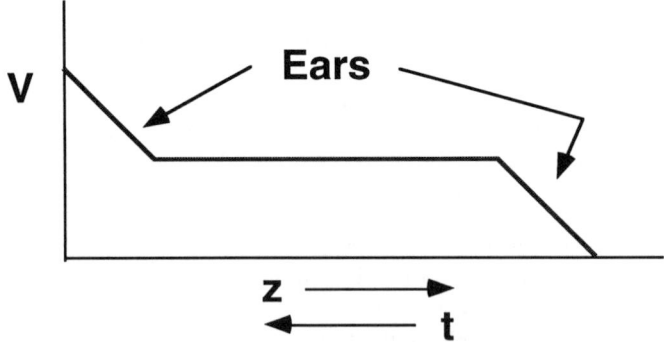

Fig. 3. Typical accelerating waveform.

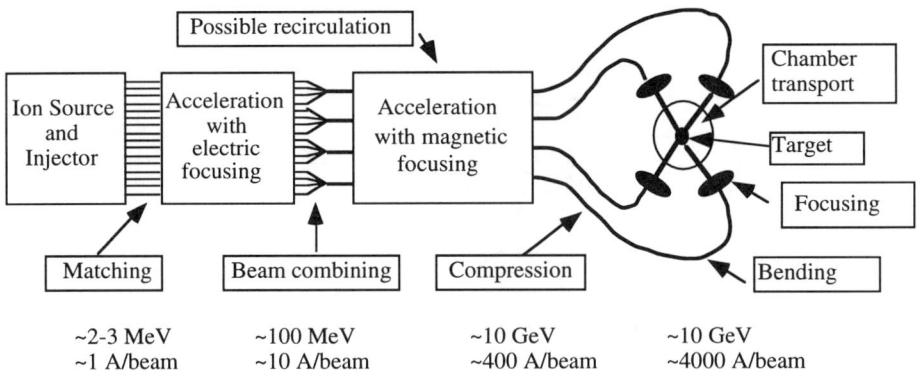

Fig. 4. Schematic of a "conventional" induction accelerator driver.

focusing in roughly the same energy range. The linear approach, in comparison with the recirculating approach discussed below, enjoys relatively simple timing and control, acceleration and focusing tailored to the beam at each point, no need for bends except near the target, and reduced vacuum requirements because debris from the wall is readily pumped away between shots and does not accumulate. However, the length and need for a high gradient ~1 MV/m establish the cost.

In present designs the major part of the accelerator employs magnetic focusing. It may well be possible to accelerate beams in an induction linac with an average gradient exceeding 1 MV/m, by suitable grading of insulators. The cost penalty of the larger cores needed for a higher gradient would be more than offset by the reduced length.

Beam merging[15] is likely to be an important cost-reduction technique because at low energy it is desirable to employ many beams, while at high energy economics drives the design toward fewer beams. Electrostatic quadrupoles are generally to be preferred at low energy because their focusing power scales more favorably with velocity at low energy. Also at low energy the quadrupoles must be closely spaced along the axis to provide sufficient confinement, and it is difficult to fit electromagnets in. Finally, electrostatic elements are cheaper to fabricate. The limits imposed by voltage breakdown imply that a small aperture, and thus a relatively large number of beams, must be used. In contrast, at higher energy focusing should use magnetic quadrupoles. Longitudinal compression at constant radius is possible with magnets because the confinable line charge density rises with the beam energy. Also, if small-aperture magnets were used, most of the induction core's bore would be filled with superconducting magnet insulation and coils instead of beam. Thus, at high energy, efficient core use dictates a large aperture, and a smaller number of thicker beams.

However, beam merging inevitably leads to an emittance increase because empty phase space is entrained in the combining process. This can be minimized by bringing the beams together in a "Stonehenge" geometry, which gives good packing. Such a geometry is depicted in Fig. 5.

As explained above, a high ion mass (of order 200) allows rigid beams with high energy, of order 10 GeV, and implies that relatively few ions are needed,

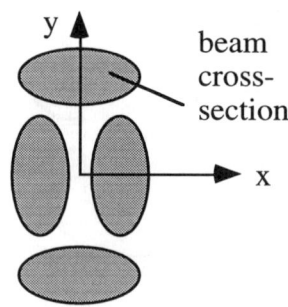

Fig. 5. "Stonehenge" geometry

reducing collective effects and easing focusing requirements. The baseline induction linac concept assumes a kinetic energy ~10 GeV. At lower masses, preservation of a small ion range in the target implies that the ion energy must be reduced. Use of a reduced ion energy would permit a shorter accelerator, and a smaller ion stopping range would enhance the target gain. However, the required current to produce ~5 MJ on target is considerably higher, and so a detailed optimization must be carried out. Smaller ion masses (as low as 39) are routinely used in present-day and near-term experiments. LBNL, LLNL, and Sandia National Laboratories are working together to understand these tradeoffs as they apply to the various accelerator technologies being considered for ion-beam driven fusion.

Substantial improvements in cost and efficiency of induction linacs may well be possible through the use of lower-energy ions. The energy gain in a core is given by the product IVτ, where I is the beam current, V the voltage gain, and τ the pulse duration. Because cost and losses increase with increasing Vτ, a large current and small voltage would be preferable. The minimum beam energy is set by the scaling of focusable current onto the target (assuming unneutralized focusing), with the beam energy varying as the inverse 2/5 power of the number of beams. It is also set by the "Maschke current limit." In practical terms, for magnetic focusing, $I_{max} \sim 4\times10^5 \, B \, \beta^2 \, a$ Amperes, where B is the quadrupole field at the pole tip (T), a the beam radius (cm), and β the ratio of beam velocity to the speed of light (the expression for electric focusing is similar). Note that I_{max} varies as the beam radius, not the beam area. One maximizes the current through a given core by using a large number of small beams. Thus, the ion energy can be lowered (while still delivering enough total energy to the target) by neutralizing the beam in the chamber and/or using a large number of small beams. The ultimate limits are set by alignment and fabrication precision; a general prejudice against using many beams must be overcome by experimental experience.[5]

"Nominal" beam and pipe radii in an electrostatic-focusing accelerator have generally been assumed to be a ~1.6 cm, b ~ 3 cm. Recent experiments [16] have refined our understanding of quadrupole breakdown voltage as a function of size, and at present $a \sim 1$ cm, $b \sim 2.3$ cm appears to be an optimal design point. (The calculated transportable current through a given pipe in a multi-beam array has not changed as a result of this new information, but the beams are smaller and there are more of them). Future improvements in technology may allow smaller apertures, with correspondingly larger multi-beam currents at the low-energy end.

4. Recirculating Induction Accelerators

A recirculating induction accelerator potentially offers reduced cost relative to a "conventional" linac because the accelerating and focusing elements are re-used many times in a single target shot. The overall accelerator length is reduced (to about 3.6 km in the "C-design" recirculator of Ref. [6], and possibly less), and the accelerating cores are smaller and are not driven so close to saturation because there is no need to accelerate the beam at the maximum possible rate. The recirculator designs considered to date employ greater axial compression than is typically assumed for linac designs, with a smaller number of longer beams used initially, and do not employ beam merging.

The beam dynamics issues which must be resolved before a recirculating driver can be built include centroid control, longitudinal control, avoidance of phase-space dilution in bends, and insertion/extraction of the beam into/out of the rings. As described in Subsection 6 below, these can be addressed at reduced scale in a small prototype recirculator. The waveform generators in a driver must supply variable accelerating pulses at high repetition frequencies of order 100 kHz, and accurate time-varying dipole fields with good energy recovery. These requirements are challenging, but advances in solid-state power electronics should make it possible to meet them through a technology development program; LLNL has already achieved 200 kHz bursts at 5 kV and 800 A, with pulse widths of 0.5-2 µs, but with a non-variable format.[17] Because of its long path length, a recirculator driver will require a high vacuum of $\sim 10^{-10}$ to 10^{-11} torr. Collisional interactions (intrabeam charge exchange, ionization of background gas, stripping of beam ions by gas collisions) can drive beam or gas ions into the walls of the beam pipe, and so cause the desorption of material off the walls. This material will interact with the beam on its next pass. Thus, a high vacuum is especially important. There remain uncertainties in some of the relevant cross sections; many of these can be resolved through experiments on existing accelerator facilities.

Current research on recirculator drivers has centered on multi-ring designs, with each ring augmenting the beam's energy by an order of magnitude over 50 to 100 laps. Relative to a "conventional" linac, the length is reduced by a factor of order 2-3, but the beam path length lengthened to perhaps ~200 km. Here too, the research program aims to develop the necessary physics and technology. Hybrid designs (with a recirculator at the low-energy end) are also possible.

5. ILSE Program, Elise Accelerator, and ESQ Injector

Early experiments with intense space-charge-dominated ion beams were very successful. These included large-aperture source development tests using cesium, as well as experiments performed on the SBTE and MBE-4 facilities at LBNL.[18,19] However, these experiments tested only a few driver elements: sources, injectors, beam matching, and acceleration with electric focusing. For economic reasons, the beams in SBTE and MBE-4 were small in comparison with those currently envisioned for a driver.

To begin addressing the physics and technology issues associated with full-scale beams, LBNL, in collaboration with LLNL, formulated an accelerator project and suite of experiments known as ILSE (Induction Linac Systems Experiments). The full ILSE accelerator concept consisted of a four-beam electric-focusing accelerator bringing a potassium beam to 5 MeV, followed by a one-beam magnetic-focusing accelerator to 10 MeV.[8] The beam would have the same line-charge density as that of a driver, and will be of the same diameter. However, to reduce cost it would use a shorter pulse and lower energy. The critical dimensionless parameters relating betatron wavelengths to the accelerator lattice period would be the same as in a driver. Experiments to be performed as part of the ILSE sequence include beam merging, drift-compression and pulse-shaping, beam bending, focusing onto a spot, and recirculation.

At the request of the Department of Energy, a proposal for construction at LBNL of the four-beam electrostatic-focusing part of the ILSE accelerator (the elements up to about 5 MeV) was submitted. This electrostatic-focusing accelerator is known as "Elise."[7] In December of 1994, the Elise proposal received "Key Decision 1" approval, allowing engineering design to begin. However, funding for the U.S. HIF program has remained approximately level, and the ILSE/Elise project is currently "on hold."

In anticipation of the ILSE program of accelerator development and experiments, LBNL, in collaboration with LLNL, has developed, over the past two years, a single-beam injector which produces a full-scale beam suitable for injection into one channel of the ILSE/Elise accelerator. This injector is of a novel "ESQ" type, employing a sequence of electrostatic quadrupole lenses with a superposed voltage gradient along the axis. The net effect is to both confine and accelerate the beam. The system consists of a dished hot-plate source with an internal filament, an axisymmetric diode section taking the beam to 1 MeV, and the ESQ section. This section decouples the strict relationship between accelerating and focusing voltages inherent in earlier "electrostatic aperture column" designs, alleviates difficulties associated with breakdown limits, and produces a high-quality beam.[20].

6. Planned Experiments in Bending and Recirculation

A recirculating accelerator using the ILSE linac as its front end would represent an attractive facility. Before such an ILSE-scale recirculator can be credibly proposed, it will be necessary to develop the technology and physics on a smaller scale. Lawrence Livermore National Laboratory, in collaboration with Lawrence Berkeley National Laboratory, is currently developing a small prototype heavy-ion, intense beam, recirculating induction accelerator. This "small recirculator" is intended to explore, in a scaled manner, the physics and technology issues involved in constructing a full scale recirculating driver for application to inertial fusion energy. The small recirculator will be developed and operated as a series of experiments over several years' time. Figure 6 illustrates the overall design of the final small recirculator, and lists some of the elements which must all work together, both in it and in a full-scale fusion driver.

The small recirculator will have a circumference of 14.4 meters, a 3.5 cm aperture radius (pipe radius) for the beam focusing and bending elements, and a half-lattice period of 36 cm. The beam will be transversely focused with alternating-gradient permanent magnet quadrupoles with a field of ~0.3 T at the pipe wall, and will be bent with electric dipole deflector plates. These quadrupoles and dipoles will each physically occupy about 30% of the axial lattice length, and the full recirculator ring will consist of 40 half-lattice periods, including several special large-aperture quadrupole magnets through which the beam will be inserted and extracted. The beam ion is singly charged potassium, and the beam will be accelerated from an initial particle kinetic energy of 80 keV to 320 keV over 15 laps by 34 induction cores. (No induction cores will be present in the lattice periods where the beam is

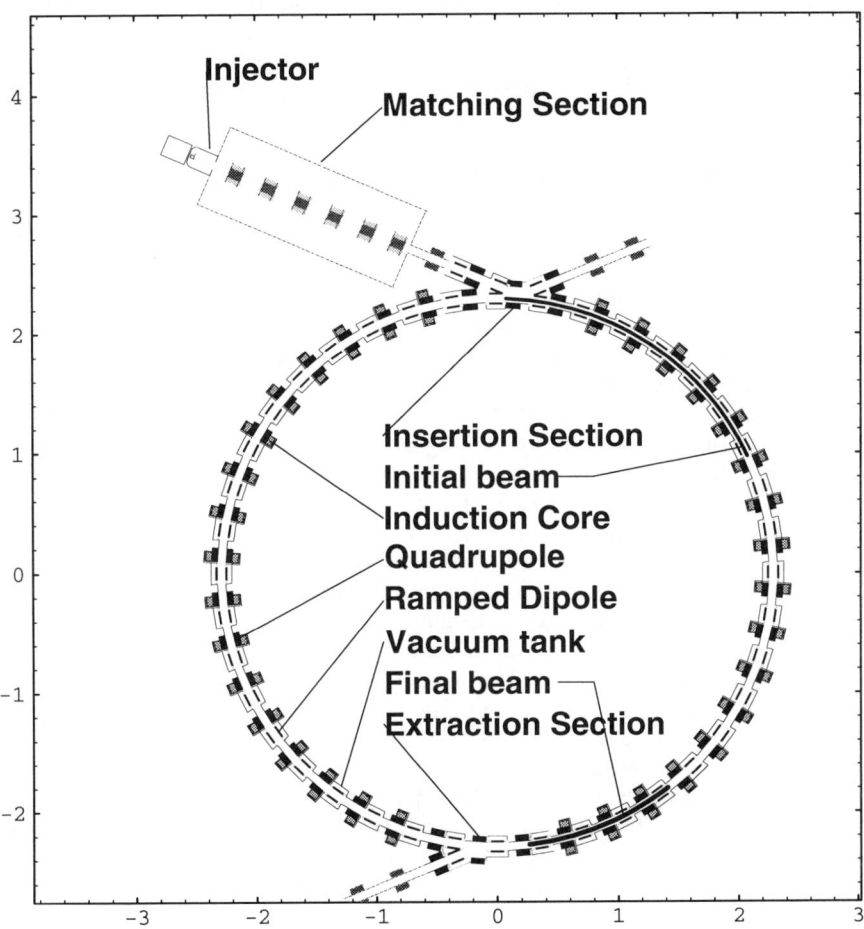

Fig. 6. Overview of final small recirculator configuration (scales in meters).

inserted and extracted.) The initial current of the beam on insertion into the ring will be 2 mA, corresponding to a line-charge density of 0.0036 µC/m and characteristic beam radius of 1.1 cm, and the initial pulse duration will be 4 µsec. Also, the initial undepressed phase advance per lattice period will be $\sigma_0 = 78°$, depressed by space charge to $\sigma = 16°$. After the full 15 laps of acceleration, the beam will have a final current of 8 mA, with a corresponding line-charge density and average beam radius of 0.00721 µC/m and 1.3 cm, and the final pulse duration will be 1 µsec. Also, at the final beam energy, the phase advances will decrease to $\sigma_0 = 45°$ and $\sigma = 12°$ per lattice period.

Because the beam in the small recirculator will be nonrelativistic and accelerating, the waveforms required to accelerate and bend the beam will be technologically challenging. They will require the accurate synthesis of a train of ~15 detailed voltage pulses with a repetition rate ranging from approximately 40 to 90 KHz at the initial and final beam energies. The voltage pulses for the electric bending dipoles must be correctly shaped and synchronized with respect to the pulses that power the induction cores. Furthermore, detailed "ear" pulse structures and lap-to-lap variation of the pulse duration must be added to the accelerating waveforms to maintain or decrease the beam length.

As of this writing, the source and injector diode are injecting a beam into the electrostatic-focusing matching section (a modified segment of LBNL's SBTE apparatus), which then injects the beam into a straight transport line. The matching section uses every second quadrupole of the original structure for focusing, with the quadrupole voltages individually tunable, and steering elements in some of the vacated quadrupole positions. The mechanical design of the recirculator "half-lattice period" is partially completed. Details of the experimental design [9,21] and status [9,22,23] were presented at the International Symposium, Princeton, 1995.

Linear experiments now under way are studying beam transport in a linear channel, using some of the quadrupole magnets destined for the recirculator. These experiments will afford phase-space measurements of the magnetic transport of space-charge-dominated ion beams. They will characterize the beam prior to injection into the recirculator, provide a test-bed for diagnostic development (especially the capacitive pickup centroid position monitors), and afford a preliminary assessment of the role of electrons in magnetic beam transport. Similar space-charge-dominated beam experiments using electromagnetic quadrupoles and longer pulses were carried out by the GSI group some years ago.[24]

Future experiments will study beam transport around a bend of order 90° (at first without any accelerating modules). The transition of the beam from a straight transport line into the ring will represent a change of curvature, and will allow study of the resulting emittance growth. Emittance growth can also result from imperfections in the focusing and bending fields; the small imperfections expected in the experiments will be well characterized by theory and measurement. Even over a short bend, detailed intercepting beam diagnostics (using a two-slit apparatus to measure both transverse position and velocity) should be able to detect relatively small changes in the distribution of beam particles as a result of the bend. An important goal of these initial experiments will be validation of the computer models and scaling laws used to predict the behavior of linear and recirculating drivers.

Later experiments will study insertion and extraction, acceleration, beam steering, bunch compression, and fully integrated operation of the recirculator. Until the ring is complete it will be possible to use intercepting diagnostics to characterize the beam and to calibrate the non intercepting diagnostics that will be critical to the successful operation of the full ring. As currently planned, the beam can be extracted and diagnosed with detailed intercepting diagnostics once or twice each lap. As with earlier linac experiments at LBNL, excellent shot-to-shot repeatability is anticipated and, so far, observed. The principal non intercepting diagnostic under development is a set of segmented capacitive pickups to be located inside the quadrupoles [22,23].

Theoretical calculations clearly show that bend-induced emittance growth in driver-scale recirculators will be minimal.[25] The parameters of the small recirculator, with its sharp bends, are such that a measurable amount of emittance growth is expected to take place over the full fifteen laps, with most of this growth occurring in the first two laps, as confirmed by detailed 3-D particle-in-cell simulations.[26,27]

The long beam path length in the machine (up to, and possibly exceeding, 300 full lattice periods) will provide a unique opportunity to observe and characterize the longitudinal propagation of space-charge waves along the beam. Such waves will be launched by mismatching the applied ear fields. Also to be explored are issues important to both recirculating and linear drivers, such as slow thermalization.

7. OTHER ELEMENTS OF THE U.S. HIF RESEARCH PROGRAM

The U.S. HIF program is also carrying out an intensive program of technology development and smaller-scale experiments. In addition to the injector and recirculator efforts described above, and the technology development program now underway, three other experimental efforts are being pursued at LBNL.

The first of these is an experimental study of beam merging, using the MBE-4 apparatus with new, angled sources feeding into one of the existing transport lines through a new combining section.[15] The final element through which the four beams pass must both bend and focus the beams. Because there is no room for conventional electrostatic quadrupole rods, the element is a "squirrel cage" made of individual wires, each held at the appropriate voltage. The beams are brought together in the "Stonehenge" configuration described above. The early stages of this experimental work have already had significant spin-off benefit in ion source improvements and understanding of beam current limits, which are proving larger than previously had been thought.

The second effort is a magnetic quadrupole development program. These "current dominated" quadrupoles are being designed with a geometry that will be suitable to the superconducting elements planned for a driver. Tests to date have focused on the mechanical and heat-transfer properties of the magnets; near-term tests will begin to consider their optical properties (the detailed winding design is still being optimized as of this writing). Experiments exploring beam transport through a set of such magnets will be carried out if funding is available, initially using the beam from LBNL's new ESQ injector.[28]

The third is a study of plasma lenses, building on recent work in Europe. This work is being carried out in collaboration with A. Tauschwitz, a visitor at LBNL from GSI. Such lenses may be attractive as final optics, either with or without a channel through which the focused beam is transported to the target.[14]

A smaller experimental program at the University of Maryland [29] is studying longitudinal beam dynamics, including wave propagation, and most recently is using a channel with a resistive wall to explore the longitudinal microwave instability as it applies to HIF drivers, where it is driven by the impedance of the accelerating modules. Detailed theoretical [30] and computational [31,32] studies indicate that this instability will have a moderate growth rate, and that careful beam control (and possibly feed-forward stabilization techniques) will be necessary to preserve beam quality. The Maryland experiments will serve to validate and normalize these calculations.

Because the space-charge-dominated beams in an induction accelerator are effectively non-neutral plasmas, theoretical and computational modeling of these beams is carried out using techniques related to those used in both the accelerator and plasma physics communities. Models employed range from simple zero-dimensional codes based upon analytically-derived scaling relations, through fluid- and moment-equation simulations, up to large and elaborate discrete-particle simulations. In addition to LBNL and LLNL, theoretical and simulation efforts are underway at the Naval Research Laboratory [33], the Princeton Plasma Physics Laboratory [34], and MIT.[35]

The CIRCE code [36] is a multi-dimensional model which solves an envelope equation (evolving moments such as centroid position and transverse extent) for each of a number (typically a hundred or greater) of "slices" of the beam. The longitudinal dynamics of the beam is modeled by evolving the positions and velocities of the slices using fluid equations. CIRCE is used to assess alignment tolerances, accelerating schedules, and steering techniques in both linacs and recirculators. It is useful for any application in which the evolution of the detailed internal degrees of freedom of the beam need not be resolved (*e.g.*, emittance growth processes); at present, the beam emittance is assumed constant in CIRCE.

Because the beam resides in the accelerator for relatively few plasma oscillation periods, particle-in-cell simulation techniques are especially effective and have proved invaluable to the design and analysis of ongoing experiments, and to the prediction of the behavior of future machines. The WARP code includes both fully three-dimensional (WARP3d) [26,27] and axisymmetric (WARPrz) [31,32] particle-in-cell simulation models. The former is used for "near first-principles" studies of MBE-4, the ESQ injector, the small recirculator, the Elise accelerator, beam bending, beam combining, and other accelerator experiments and elements. The axisymmetric model is used for long-term beam dynamics studies including the effects of accelerating module impedance. WARP3d uses a number of novel techniques to render it both accurate and efficient. These include a capability for subgrid-scale placement of internal conductor boundaries, a capability of simulating "bent" accelerator structures, and a technique for rapidly following particles through a sequence of sharp-edged accelerator "lattice elements," using a relatively small number of time steps while preserving accuracy. On some problems (such as the ESQ injector), the code is run in a quasi-steady state mode, whereby a 3-D run can be completed in just a few minutes of computer time; this makes it suitable for iterative design calculations. The ultimate goal of this code development is effective simulations of both the ILSE experiments and a driver, from the source through the final focus, into the codes used to model propagation in the fusion chamber, and ultimately into the target design codes. A number of other particle-in-cell codes employ a "slice" description of the beam (assuming slow variation of quantities along the beam); much application, as well as detailed studies of the properties of such particle-in-cell models of beams and plasmas, has been carried out.[33]

In addition to these accelerator development efforts, the program is exploring all of the proposed modes of propagating the beams through the fusion chamber to the target. The BIC (Beam-in-Chamber) code [11,12] is an axisymmetric electromagnetic code which allows the computational grid to taper down to small radius, preserving resolution as the beam converges toward the target. It is used to examine the near-ballistic propagation modes, and such processes as beam neutralization and stripping. Higher-density pinched propagation modes are being studied at LBNL, at Sandia National Laboratory, and at the Naval Research Laboratory.[13]

LLNL is involved in target design research, using essentially the same tools developed for target experiments with the Nova and NIF lasers and the Sandia ion-diode devices. The goal is to develop robust, high-gain targets which impose a minimal set of requirements on the driver. Recent progress has led to a much more detailed understanding of the radiation converters; integrated design efforts are well underway.[37]

Finally, research continues on improved HIF power-plant concepts. The most recent work, on an improved HYLIFE-II system with a liquid Flibe waterfall providing a renewable first wall, suggests an extremely competitive cost of electricity may be obtainable from an HIF power plant.[1] Recent studies of target injection and tracking [38], and planned experiments with model waterfalls [39], are important elements of the HIF power plant development program.

SPACE-CHARGE-DOMINATED BEAM PHYSICS ISSUES

The following "taxonomy" of issues was developed with induction-accelerator heavy-ion fusion in mind. It may have broader utility, but is not all-inclusive (even for HIF). The treatment is very brief, since much discussion might reasonably be devoted to each topic. No attempt is made to systematically cite all relevant references.

Note that, in general, U.S. HIF researchers generally use the term "space-charge-dominated" (with various hyphenations!) to refer to a situation wherein the transverse force balance is primarily between the space charge and the applied focusing forces. Such beams are typically space-charge-dominated longitudinally as well. In contrast, European researchers often refer to beams as space-charge-dominated when the longitudinal space charge force is greater the longitudinal thermal pressure, even when the beams are not transversely space-charge-dominated.

I. Ideal beam "equilibria"

A. Utility of the concept

Ideal equilibria have long been used as a tool for the analysis of transportable current, beam stability, etc. Since an accelerating beam in an alternating-gradient (A-G) lattice is not in true equilibrium, a question arises concerning the utility of approximate equilibrium as a goal for system design. Efforts to keep the beam "matched" to the transport channel minimize transverse particle excursions and keep more of the beam in the good-field region. Similarly, longitudinal matching minimizes the launching of waves on the beam. Thus, it is generally believed that beams which are "near equilibrium" will in general be well-behaved. However, it may be that in some cases the optimum design will involve a beam far from equilibrium, at some stage or stages of the acceleration process.

B. Transverse equilibria for infinitely long beams

For an axisymmetric system, or for a smooth-approximation treatment of an alternating-gradient system, the assumption of thermal equilibrium may be reasonable, especially if no time-varying external forces are applied. However, since the beams are practically collisionless, and instabilities etc. will in general not drive them all the way to thermal equilibrium, it is not clear that thermal states are reached on relevant time scales. An issue remains about the boundedness of thermal equilibria; in general they involve nonzero densities at infinity or on the pipe wall, and one must invoke exponential smallness of the error thus incurred. Furthermore, non-thermal equilibria might in principle be more appropriate in some cases.

For an A-G system, one asks whether there exists a nonsingular (i.e., non Kapchinskij-Vladimirskij) 4-D (or 5-D if longitudinal velocity spread is included) phase space density describing a continuous beam which is replicated after one lattice period. Formally such existence requires absence of particle loss. It remains unclear how best to find such an equilibrium. In principle one could use the Vlasov equation to numerically evolve densities on a 4-D or 5-D mesh over one lattice period, then vary parameters (via, *e.g.*, a quasi-Newton procedure) until a replicating state is found. In practice such a procedure would be extremely computationally intensive. Clever choice of a set of basis functions used to characterize the distribution may be the key to resolving this issue.

For one study of this area, see Ref. [46].

C. Longitudinal equilibria at beam ends, assuming axisymmetry

Questions to be resolved are: What are practical, achievable "ear" waveforms? What is the actual "equilibrium shape" of the beam tips (axial dependence of beam radius, line-charge density, emittance; ideally, the full 5-D particle distribution function), in the approximation that ear fields are continuously applied? Is there a particle distribution which replicates under periodic application of ear fields?

D. Three-dimensional equilibria

Little analytic work is likely to be done in full 3-D geometry. However, full 3-D simulations need appropriate initial conditions. In some cases the beam is best formed by simulating the injection process, but in others it is useful to start with an approximate equilibrium. In WARP simulations, specification of such initial states is usually carried out by assuming the beam has a semi-Gaussian distribution transversely (*i.e.*, a top-hat function in position and a Gaussian in velocity), with a variety of longitudinal dependences. Often, the longitudinal velocity in the main, flat-top part of the pulse is assumed Gaussian, with the line-charge density flat over the main pulse and tapering to zero at the ends with (*e.g.*) a parabolic falloff. The transverse beam size in the tips is set to have a specified dependence upon the local line-charge density, often assumed (in an *ad hoc* manner) to preserve a constant charge density and depressed phase advance within the beam envelope. When such a beam is followed over many lattice periods, it is seen to evolve toward a blunter shape, with charge density decreasing toward the tips.

II. Beams formed by injection

A. Beam ends

Questions to be resolved include: What are the shapes (in the sense described above) of the beam head and tail as the beam exits the diode? What are the shapes downstream, at the first practical location for application of time-dependent accelerating fields? How best to "catch" or "patch" the beam as it emerges from the injector and encourage it toward equilibrium (or some other desired state)?

For discrete-particle simulations, efficient and accurate modeling of time-dependent space-charge limited flow is a challenging problem, and the ideal algorithm has not yet (in this author's opinion) been identified.

B. Transverse profile

Questions to be resolved include: What is the transverse distribution function? What design strategy leads to a beam with minimal hollowing and/or distortion? What are the downstream effects of source nonuniformities?

C. Longitudinal behavior

Questions to be resolved include: What waves are launched from the beam ends as a result of the diode voltage rise and fall (for achievable rise and fall waveforms)? What waves are launched as a result of ripple or droop in the injection voltage? How do such wave motions thermalize?

As a result of acceleration, the longitudinal thermal velocity (in the beam frame) can be expected to decrease, in both the injector and the main accelerator (under certain assumptions).[40] However, effects such as rapid transfer of thermal energy from transverse to longitudinal motions (described below, in Subsection VI) mitigate the cooling. One must assess how much "accelerative cooling" actually occurs in the injection process for each circumstance, perhaps by detailed simulation, ideally by experiment.

III. Transverse dynamics

A. Dynamic aperture in the presence of strong space charge

Space charge introduces nonlinearities in the transverse fields both directly and through the effects of image charges. "Rules of thumb" have been established over the years for how much of the aperture can be used, avoiding the introduction of excessive nonlinear image effects and accounting for coherent oscillations. The aperture radius b needed is often assumed to be given by $b = 1.25\, a + c$, where a is the maximum beam radius and c is an additive constant of order 1 cm, to account for misalignments and beam nonuniformities (a smaller constant may be used at very low energy, before errors accumulate). Recent experimental investigations of breakdown limits for actual quadrupoles have established improved scaling laws for transportable current.[16] As pointed out by Laslett, for electrostatic quadrupole focusing systems using four round cylindrical rods per element, a rod:aperture radius ratio of 8:7 leads to cancellation of the unwanted dodecapole term, in the infinite-rod limit.[41] Other prescriptions take into account image effects.[42] This work should be updated. Also, progress in machining and aligning small elements suggests that future systems will optimize at a larger number of smaller beams passing in parallel through a given set of induction cores.

B. Halo causes and avoidance

Minimization of beam halo (stray particles) is of importance because of the desire to preserve the practicality of hands-on maintenance of the driver, while keeping the bore radius small enough that the driver cost is attractive. For space-charge-dominated beams, the profile is closer to a "top hat" than to a Gaussian, and so (as in most astrophysical contexts) the halo particles constitute a distinct population rather than the tails of the main distribution.

Recent work on core-halo models has shown that a "breathing" core is capable of parametrically driving particles to large excursions and so into the halo. Minimization of mismatch is a clearly identified means of minimizing halo formation. Other measures involve avoidance of "hollow" beams and of beams with over-focusing at the edges, out of the injector. Longitudinal mismatches as well as waves launched by imperfect "ears" can in principle contribute to a transverse beam halo as well.

C. Halo scraping

In some contexts it may be attractive to periodically remove stray particles before they hit the wall in an unpredictable manner. The introduction of apertures at appropriate stations along the machine may effectively remove the halo. This also may ensure that the collisions with solid matter which occur when particles are lost are at near-normal incidence, thereby minimizing desorption from the wall (which is generally greater when an ion strikes at grazing incidence). In general it may be necessary to place multiple scrapers so that halo particles at all phases of their betatron oscillations are removed. The question of whether and how quickly the halo reforms must be answered in the context of each particular machine.

D. Beam combining

Transverse beam combining or merging, as described earlier, necessarily involves the entrainment of empty phase space. Much progress has been made in using simulations and experiments to understand the process, and further experiments are underway, but the optimum designs have not yet been identified.

E. "Recovery" of beam when perturbed

After the beam has been perturbed by a change in system parameters leading to a mismatch, by scraping, or by some other process, it relaxes toward a new equilibrium. However, in the absence of collisions the relaxation occurs by other means and is typically incomplete. Many relaxation processes operate on the time scale of a plasma period, while others (such as those which damp centroid oscillations) can be much slower.

F. Beam "steering" vs. precise alignment

In a linear system it may be possible to precisely align all elements and only "steer" in a rudimentary way; however, true steering may afford simpler construction with relaxed tolerances and overall cost savings. Steering of the beam requires adjustment, on the basis of measurements of the beam centroid motion, of existing dipole fields, and/or of auxiliary fields. Such steering is essential to multi-lap operation of the small recirculator, and of driver-scale recirculators as well.

The principal method of beam steering would use beam centroid information obtained on the previous shot (linac), or on the corresponding lap of the previous shot (recirculator). Because of the expected good shot-to-shot repeatability of the system, this information should provide sufficiently reliable input to accomplish this task. This inter-shot (so-called "shot-to-shot") steering can most simply be carried out by setting the steering fields to a value which does not change as the beam passes by. Such "time-independent" steering should provide the bulk of the necessary correction.

In addition, it may be highly desirable to include a small amount of time-dependent steering occasionally. This will serve to remove the small amount of accumulated centroid position error which arises from the head-to-tail energy and current variation of the beam.

The ultimate method of steering would rely on information from the current shot. The beam centroid position will be measured, and steering corrections applied at a location downstream from the measuring station. This "feed-forward steering" would have the advantage of removing random, non-repeatable errors, as well as those which remain systematic from shot to shot.

G. Emittance growth in curved systems

Emittance growth can result from the non-uniform distribution of beam space charge resulting from the action of centrifugal forces. One identified mechanism is a consequence of the beam's thermal longitudinal velocity spread; particles of differing energies naturally ride around the bend at differing radii, and the resulting transverse "smearing" of the space charge leads to nonlinear forces. As revealed in particle simulations using the WARP3d code[26] and confirmed by analytic theory,[25] the effect occurs at changes in the accelerator's curvature where the distribution of beam particles must relax toward a new equilibrium. Thus, a circular recirculator is to be preferred over one with a "racetrack" shape.

H. Insertion and extraction into and out of rings

In order to switch the beam into and out of a ring, it is necessary to apply dipole bending fields that vary in time. In a driver-scale recirculator, insertion and extraction are carried out over a number of half lattice periods. In the small recirculator it is possible to insert or extract the beam in a single half lattice period. However, even for the small ring, extraction will require real-time adjustment of the dipole immediately upstream of the extraction section as well as the dipole within that section. An additional requirement is that transverse confinement of the beam against its own defocusing space-charge forces must be carried out continuously, even during the insertion or extraction process, or else the beam would grow excessively in size. Furthermore, the optical quality of the magnetic quadrupole lenses which provide this transverse confinement must be high, or aberrations will distort the beam over multiple laps. To this end, a large-aperture quadrupole element must be employed; its inside dimension in the vertical plane need be no larger than that of a normal quadrupole, but its dimension in the plane of the ring must be larger by a factor of approximately 2.5. For the small ring we plan to use a permanent-magnet quadrupole with an expanded aperture. For a driver, a set of special large-aperture electromagnets ("Panofsky quadrupoles") will be employed. Initially, the extraction process should be just a mirror-image of insertion; however, at higher energies extraction becomes more challenging because bending the beam becomes more difficult.

IV. Longitudinal dynamics

A. Wave propagation

Waves if excited will propagate along the beam, with a restoring force associated with both space-charge and (usually to a lesser degree) thermal pressure. Much is known about the propagation of such waves, their mode structure and dispersion, both theoretically[30,31,32] and experimentally[29]. In some cases a "longitudinal instability" analogous to the microwave instability seen in storage rings is driven by the accelerating module impedance; recent work has elucidated the circumstances under which growth is significant.[32] Remaining questions concern the impedances of real accelerating structures, the utility of feed-forward stabilization, and the properties of various inherent stabilization mechanisms including longitudinal thermal spread and transverse-longitudinal interactions.

B. Beam-end effects

The shape and timing of the "ear" pulses is critical if the beam tips are to be quiescently confined. These are in general provided using separate waveform generators, and/or by shaping the initial rise of the accelerating pulses. Questions include: How to minimize the waves launched due to periodic application of ear fields? Can they be eliminated somehow? What are effects of various kinds of errors in ear fields: strength, timing, shape, etc.? How do waves reflect at bunch ends for confined, expanding beams?[29]

C. Acceleration

One issue is how best to shape the initial accelerating waveforms so as to perturb the beam minimally during early acceleration? One scenario is termed "load-and-fire," wherein no acceleration takes place until the entire pulse has been injected. Others include use of triangular ramps to start the pulse compressing, with a transition to flat-top pulses. In a short machine, optimum use of the available volt-seconds is important.

A technology which needs to be fully explored is feedback correction of accelerating errors.

In a practical accelerator, tradeoffs between pulse compression, accelerative cooling (II.C), and "equilibration" (VI) must be taken into account

D. Pulse compression & shaping

In a driver, the beam current is increased (via a reduction of the pulse duration) by orders of magnitude. Some of this increase is due to the increase in the beam's velocity. The remainder is due to an imposed axial compression of the beam—it is made physically shorter—as it is accelerated. The design of the driver must take the pulse compression into account.

Current concepts assume that the beam will be compressed by an additional factor of 10-25 after it leaves the driver, in a "drift-compression" process, to achieve a shaped pulse with a duration of order 10 ns (for the main part). Optimization of this process, including an assessment of the relative merits of driven- vs. drift-compression, or of drift-compression with a "mid-course correction," remains to be done. Detailed scenarios for setting up the beam line-charge density and velocity for drift compression have yet to be worked out.

The ultimate limitations on the pulse shape are as yet unknown; this is an active area of study. While not all shapes are practical (and very short pulses are challenging), sufficient flexibility appears to be available that the needs of the fusion application can be met.

As the beam compresses axially its radial extent becomes larger. It thereby samples more of the non-linear fields from the accelerator lattice elements, creating more opportunity for emittance growth and possible beam loss. These effects must be taken into account.

V. 3-D dynamics

3-D particle simulations of HIF-relevant beams are used to study effects introduced by the full geometry, and to do "integrated" simulations of full beams. Most beam physics can be successfully treated in lower dimensional approximation, but some cannot. For example, 4-D or 5-D "slice" models of infinite beams are often useful, but extra information is needed to know "what is in each slice" for real, finite-length beams with longitudinal particle motions in the beam frame. Axisymmetric 5-D models are useful but do not capture effects of A-G systems or of beam bending, not to mention merging and other strongly geometry-dependent processes.

A. Transverse-longitudinal coupling in bent systems

A bent system naturally couples the longitudinal and transverse in-plane motions. For example, in the small recirculator designed at LLNL, particles are slowed or sped longitudinally as they enter the region between the electrostatic dipole plates which serve to bend the beam, with a strong dependence upon their individual transverse position. This "energy effect" is analogous to the one calculated for and observed in the LBNL ESQ injector. If the whole beam centroid moves off the design orbit, the overall timing of the beam's passage around the ring is affected. Also, particles of differing energy naturally follow different orbits (III.G).

B. Other dynamical coupling (due to fringe fields, etc.)

Fringe fields and imperfections can also couple the various degrees of freedom. For example, particles at different transverse positions can be slowed or sped differently as they enter or exit a lattice element.

C. Time-dependent offsets and "corkscrew"

In a beam of non-constant energy, misalignments can lead to a time-dependent centroid displacement (or "corkscrew", as it appears in a system using solenoids). This is further complicated by space-charge image effects.

D. BBU instability (nature and correction)

The beam-breakup instability endemic to relativistic electron beams in induction accelerators is much milder in the context of HIF ion accelerators, but must still be kept in mind as a part of practical machine design.

E. Halo formation due to imperfect beam end confinement

Beam end confinement against electrostatic repulsive forces is effected by time-dependent fields. Particles which happen to pass far into the tip before being driven back toward the main bunch may be imparted a large peculiar velocity, leading to poor matching and large transverse excursions.

VI. Equilibration and relaxation

A. Collective transfer of thermal energy from x,y to z

Simulations [43,44] in both 3-D and r,z geometry have shown the existence of a rapid collective transfer of thermal energy from transverse to longitudinal degrees of freedom, when the transverse temperature $T_\perp$ is significantly warmer than the longitudinal temperature $T_\parallel$. This appears to be a betatron-motion analog to the "Harris mode" in magnetoplasma, which is driven by cyclotron motion. In no instances have these simulations shown a similar transfer in the other direction, when $T_\parallel > T_\perp$. An analytic theory [45] is under development. The nature of the seeds for growth, and the consequences for real systems design, need to be understood. The process may amount to a useful cooling mechanism for transverse motions, or may represent a constraint on fusion driver design. In injectors and elsewhere it may partially offset the "accelerative cooling" process described in (II.C).

B. Existence or nonexistence of rapid transfer from z to x,y?

Such a transfer might be induced by coupling due to fringe fields in external elements, or by the so-called "energy effect." It appears likely in short bunches, where the beam's fields provide a strong geometrical coupling.

C. Existence of mechanisms to drive beam toward an equilibrium

Mechanisms which drive the beam toward a quiescent state may be valuable, as noted in (I.A). Thermalization of mismatch oscillations typically brings with it emittance growth, but the beam may be well-centered and better-behaved as a result. Active cooling measures may be useful in some cases, though those identified to date are in general too slow for linac applications. Careful tailoring of accelerating and confining fields, possibly through a feedback or feed-forward process, can to some degree push the beam toward a state with quiescent macroscopic moments.

CONCLUDING REMARKS

The research program outlined here is, in the author's opinion, broadly based and well-positioned to explore the most important aspects of HIF driver physics and technology. It is complementary to the European program of RF driver development. At present no particular heavy-ion driver concept is clearly best. Each requires further development before its full potential can be assessed, and improvements to each appear both possible and likely. More work is needed integrating target designs, beam transport options, and driver optimization.

The categorization of beam dynamics issues outlined in the second part of this paper may provide a useful framework for future investigations.

The ongoing experimental program, in close conjunction with theory and modeling, will lead to confident predictions of driver behavior. The progression to full-scale beams and the requisite technology to accelerate and control them is a critical step in the development of heavy-ion fusion as an energy source.

ACKNOWLEDGMENTS

The work described in this paper is the result of a collaborative effort of the LBNL and LLNL HIF groups, along with colleagues at other institutions. The author thanks Roger Bangerter, John Barnard, Andy Faltens, Bill Herrmannsfeldt, and Grant Logan for useful comments on the manuscript.

*This work was performed under the auspices of the U.S. Department of Energy by LLNL under contract W-7405-ENG-48.

REFERENCES

The most recent International Symposium on Heavy Ion Inertial Fusion was held at Princeton Plasma Physics Laboratory on September 6-9, 1995. The Proceedings of that meeting will be published in *Fusion Engineering and Design* (Elsevier, Amsterdam), in late 1996.

Il Nuovo Cimento **106 A**, N. 11-12, (1994), *Proceedings of the International Symposium on Heavy Ion Inertial Fusion*, Frascati, Italy, May 25-28, 1993, is also a good general reference.

[1] R. W. Moir et. al., "HYLIFE-II: A Molten-Salt Inertial Fusion Energy Power Plant Design—Final Report," *Fusion Technology* **25**, 5-25 (1994).
[2] W. R. Meier et. al., "OSIRIS and SOMBRERO Inertial Confinement Fusion Power Plant Designs Final Report," DOE/ER/54100-1 (1992).
[3] L. M. Waganer et. al., "Inertial Fusion Energy Reactor Design Studies Prometheus-L & Prometheus-H," DOE/ER/54101 (1992).
[4] W. M. Sharp, "A Comparison of Driver Concepts for Heavy-Ion Fusion," *Proceedings of the 11th Workshop on Laser Interaction and Related Plasma Phenomena*, Monterey, CA, 25-29 October 1993, *AIP Conference Proceedings 318*, AIP Press (1994), G. Miley, Ed.; also *ICF Quarterly Report*, LLNL Report UCRL-LR-105821-94-2, Pp. 70-7 (1994).
[5] R. O. Bangerter, "The Induction Approach to Heavy-Ion Inertial Fusion: Accelerator and Target Considerations," *Il Nuovo Cimento* **106 A** 11, 1445 (1994).
[6] J. J Barnard, F. J. Deadrick, A. Friedman, D. P. Grote, L. V. Griffith, H. C. Kirbie, V. K. Neil, M. A. Newton, A. C. Paul, W. M. Sharp, H. D. Shay, R. O. Bangerter, A. Faltens, C. G. Fong, D. L. Judd, E. P. Lee, L. L. Reginato, S. S. Yu, and T. F Godlove, "Recirculating Induction Accelerators as Drivers for Heavy-Ion Fusion," *Phys. Fluids B* **5**, 2698 (1993).
[7] E. P. Lee, R. O. Bangerter, C. F. Chan, A. Faltens, J. Kwan, E. Henestroza, K. Hahn, and P. Seidl, "Physics Design and Scaling of Elise," *Proc. Int. Sympos. on Heavy Ion Inertial Fusion, 1995*.
[8] T. J. Fessenden et. al., *Plasma Phys. & Cont. Nucl. Fus. Res.* **3**, 89 (1992).
[9] A. Friedman "Recirculating Induction Accelerators for Inertial Fusion: Prospects and Status," *Proc. Int. Sympos. on Heavy Ion Inertial Fusion, 1995*.
[10] D. D-M. Ho, I. Haber, K. R. Crandall, and S. T. Brandon, *Part. Accel.* **36**, 141 (1991).
[11] D. A. Callahan and A. B. Langdon, "Transport of a Partially-Neutralized Ion Beam in a Heavy-Ion Fusion Reactor Chamber," *Proc. 1995 IEEE/APS Part. Accel. Conf.*, Dallas TX, May 1995.
[12] D. A. Callahan, "Chamber Propagation Physics for Heavy Ion Fusion," *Proc. Int. Sympos. on Heavy Ion Inertial Fusion, 1995*.
[13] C. L. Olson et.al., "Physics of Gas Breakdown for Ion Beam Transport in Gas," *Il Nuovo Cimento* **106 A**, N. 11, 1705 (1994)
[14] A. Tauschwitz, E. Boggasch, D. H. H. Hoffmann, M. de Magistris, U. Neuner, M. Stetter, R. Tkotz, T. Wagner, W. Seelig, and H. Wetzler, *Il Nuovo Cimento* **106 A**, N. 11, 1733 (1994).

[15] C. M. Celata, W. Chupp, A. Faltens, W. M. Fawley, W. Ghiorso, H. Hahn, E. Henestroza, C. Peters, and P. A. Seidl, "Transverse Combining of 4 Beams in MBE-4," *Proc. 1995 Part. Accel. Conf.,* Dallas TX, May 1995
[16] P. Seidl and A. Faltens, "Electrostatic Quadrupoles for Heavy-Ion Fusion," *Proc. 1993 Part. Accel. Conf.,* IEEE Cat. No. 93CH3279-7 p. 721 (1993).
[17] H. C. Kirbie, W. R. Cravey, S. A. Hawkins, M. A. Newton and C. W. Ollis, "A FET-Switched Induction Accelerator Cell," *Proc. Ninth Int. Pulsed Power Conf.,* IEEE Cat. No. 93CH3350-6 p. 415 (1993)
[18] M. G. Tiefenback and D. Keefe, "Measurement of Stability Limits for a Space-Charge-Dominated Ion Beam in a Long A.G. Transport Channel," *Proc. 1985 Part. Accel. Conf.,* IEEE Trans. Nucl. Sci. **NS-32**, p. 2483 (1985).
[19] T. Garvey, S. Eylon, T. J. Fessenden, and E. Henestroza, "Beam Acceleration Experiments on a Heavy Ion Linear Induction Accelerator," *Part. Accel.* **37-8**, 241 (1992).
[20] S. S. Yu, "Driver-Scale Ion Injector Experiments," *Proc. Int. Sympos. on Heavy Ion Inertial Fusion, 1995.*
[21] J. J. Barnard, M. D. Cable, D. A. Callahan, T. J. Fessenden, A. Friedman, D. P. Grote, D. L. Judd, S. M. Lund, M. A. Newton, W. M. Sharp, and S. S. Yu, "Physics Design and Scaling of Recirculating Induction Accelerators: From Benchtop Prototypes to Drivers," *Proc. Int. Sympos. on Heavy Ion Inertial Fusion, 1995.*
[22] T. J. Fessenden, J. J. Barnard, M. D. Cable, F. J. Deadrick, M. B. Nelson, S. Eylon, and H. S. Hopkins, "Intense Heavy-Ion Beam Transport with Electric and Magnetic Quadrupoles," *Proc. Int. Sympos. on Heavy Ion Inertial Fusion, 1995.*
[23] F. J. Deadrick, J. J. Barnard, T. J. Fessenden, J. Meredith, and J. Rintamaki, "Development of Beam Position Monitors for Heavy Ion Recirculators," *Proc. 1995 Part. Accel. Conf.,* Dallas, TX, May 1–5, 1995 (to be published).
[24] J. Klabunde *et.al., IEEE Trans. Nucl. Sci* **NS-30** No. 4, 2543 (1983).
[25] J. J. Barnard, H. D. Shay, S. S. Yu, A. Friedman, and D. P. Grote, "Emittance Growth in Heavy-Ion Recirculators," *Proc. 1992 Linear Accelerator Conf,* August 24-28, Ottawa, Ontario, Canada, C.R. Hoffman, ed.; AECL 10728 (AECL Research, Chalk River, Canada) p. 229.
[26] A. Friedman, D. P. Grote, and I. Haber, "Three-Dimensional Particle Simulation of Heavy-Ion Fusion Beams," *Phys. Fluids B* **4**, 2203 (1992)
[27] D. P. Grote, A. Friedman, I. Haber, and S. S. Yu, "Three-Dimensional Simulations of High-Current Beams in Induction Accelerators with WARP3d," *Proc. Int. Sympos. on Heavy Ion Inertial Fusion, 1995.*
[28] M. Stuart, A. Faltens, W. M. Fawley, C. Peters, and M. C. Vella, "Design and Construction of a Large Aperture, Quadrupole Electromagnet prototype for ILSE," *Proc. 1995 IEEE/APS Part. Accel. Conf.,* Dallas TX, May 1995; see also W. M. Fawley, L. J. Laslett, C. M. Celata, A. Faltens, and I. Haber, "Simulation Studies of Space-Charge-Dominated Beam Transport in Large-Aperture Ratio Quadrupoles," *Il Nuovo Cimento* **106 A 11**, 1637 (1994).

[29] J. G. Wang, D. X. Wang, D. Kehne, M. Reiser, and H. Suk, "Studies of Space-Charge Waves due to Localized Perturbations in an Electron Beam," *Il Nuovo Cimento* **106 A**, N. 11, 1745 (1994)
[30] E. P. Lee, "Longitudinal Instability in Heavy-Ion Fusion Induction Linacs," *Il Nuovo Cimento* **106 A**, N. 11, 1679 (1994)
[31] D. A. Callahan, A. B. Langdon, A. Friedman, and I. Haber, "Longitudinal Beam Dynamics for Heavy-Ion Fusion," *Proc. 1993 IEEE/APS Part. Accel. Conf.*, IEEE Cat. No. 93CH3279-7 p. 730 (1993).
[32] D. A. Callahan, "Simulations of Longitudinal Beam Dynamics of Space-Charge Dominated Beams for Heavy Ion Fusion," Ph.D. thesis, University of CA, Davis, 1994; available as LLNL Report UCRL-LR-119364, Dec. 1994.
[33] I. Haber and H. Rudd, "Numerical Limits on P.I.C. Simulation of Low Emittance Transport," *Proc. Conf. on Linear Accel. and Beam Optics Codes*, La Jolla Inst., C. R. Eminhizer, Ed., *AIP Conf. Proc.* **177**, AIP, NY, 1988, p. 161.
[34] R. C. Davidson and Q. Qian, *Phys. Plasmas* **1**, 1328 ff; 3104 ff (1994).
[35] C. Chen and R. C. Davidson, "Nonlinear Properties of the Kapchinskij-Vladimirskij Equilibrium and Envelope Equation in a Periodic Focusing Field," *Phys. Rev.* **E49**, 5679 (1994).
[36] W. M. Sharp, J. J. Barnard, D. P. Grote, S. M. Lund, and S. S. Yu, "Envelope Model of Beam Transport in ILSE," *Proc. 1993 Computational Accelerator Physics Conference*, Pleasanton, CA, Feb. 22-26 1993, pp. 540-548.
[37] D. D-M. Ho, J. A. Harte, and M. Tabak, "Radiation-Driven Targets for Heavy-Ion Fusion," *Proc. 15th Int. Conf. on Plasma Phys. and Controlled Nucl. Fusion Research,* IAEA, Seville, Sept. 26-Oct. 1, 1994 (IAEA-CN-60-B-P-13).
[38] R. Petzoldt, "Inertial Fusion Energy Target Injection, Tracking, and Beam Pointing," Ph.D. thesis, University of CA, Davis, March 7, 1995; available as LLNL Report UCRL-LR-120192.
[39] K. B. Wilson and P. Peterson, "HYLIFE-II Inertial Confinement Fusion Reactor Oscillating Sheet Jet Experiment Design," Master of Science Project report, U.C. Berkeley Dept. of Nuclear Engineering, 1995 (unpubl.); M. Longeot and P. Peterson, abstract sub. to 12th Topical Meeting on the Technology of Fusion Energy, June 16-20, 1996, Reno NV, proc. to be publ. in *Fusion Technology.*
[40] M. Reiser, *Theory and Design of Charged Particle Beams*, Wiley, New York, 1994, p. 393 ff.
[41] L. J. Laslett, in *Selected Works of L. Jackson Laslett*, Lawrence Berkeley Laboratory Pub-616, V. III, chap. 6 (1987).
[42] C. M. Celata, *Proc. 1987 Part. Accel. Conf.,* p. 996 (1987).
[43] A. Friedman, D. A. Callahan, D. P. Grote, A. B. Langdon and I. Haber, *Bull. Am. Phys. Soc.* **35**, 9, 2121 (1990).
[44] I. Haber, D. A. Callahan, A. Friedman, D. P. Grote, and A. B. Langdon, "Transverse-Longitudinal Energy Equilibration in a Long Uniform Beam," *Proc. 1995 Part. Accel. Conf.*
[45] S. M. Lund, private communication.
[46] J. Struckmeier and I. Hofmann, "The Problem of Self-Consistent Particle Phase Space Distributions for Periodic Focusing Channels," *Part. Accel.* **39**, 219-249 (1992)

HEAVY ION FUSION EXPERIMENTS AT LLNL

J.J. Barnard[1], M.D. Cable[1], D. A. Callahan[1], F. J. Deadrick[1], S. Eylon[4],
T.J. Fessenden[2], A. Friedman[1], D.P. Grote[1], K.A. Holm[1], H.A. Hopkins[1],
D.L. Judd[2], R. L. Hanks[1], S.A. Hawkins[1], H.C. Kirbie[1], B.G. Logan[1],
S.M. Lund[1], L.A. Nattrass[1], D. Longinotti[3], M. B. Nelson[1],
M.A. Newton[1], C.W. Ollis[1], T.C. Sangster[1], and W.M. Sharp[1]

[1]Lawrence Livermore National Laboratory
[2]Lawrence Berkeley National Laboratory
[3]EG& G, [4]Participating Guest at LLNL

ABSTRACT

We review the status of the experimental campaign being carried out at Lawrence Livermore National Laboratory, involving scaled investigations of the acceleration and transport of space-charge dominated heavy ion beams. The ultimate goal of these experiments is to help lay the groundwork for a larger scale ion driven inertial fusion reactor, the purpose of which is to produce inexpensive and clean electric power.

1. Introduction

Because of their high efficiency, high repetition rate, and relatively simple target chamber geometries, heavy-ion accelerators are attractive candidates as drivers for inertial fusion energy power plants. Recurring induction accelerators have been studied[1,2] as potential lower cost alternatives to linear induction machines. The cost advantage is achieved, as in any circular accelerator, by the reuse of the accelerating and focusing components. However, the introduction of bends and the higher repetition frequency of the induction core pulsers introduces new challenges, as will be discussed below. A series of experiments leading to the development of a small (2m radius) scaled recirculator at LLNL is underway. The ion source, matching section and a short linear transport section have been constructed and the beam has been characterized. A design for the half-lattice periods of the ring has been developed as well. A complete description of the experimental apparatus and results through October, 1995 have recently been presented in the literature[3]. The purpose of this paper is to summarize these results as well as present subsequent developments.

The ultimate specifications for a heavy ion driver are determined by the power and energy density requirements of the target, and the associated high beam quality requirements needed to ensure that the beam can be focused onto a small spot at the target, necessary for high energy density. Typical requirements[4] set by target include a pulse energy of ~4 MJ, a main pulse duration of ~10 ns, and an ion range in the radiation converter material of ~0.1-0.2 g/cm^2. For the heaviest ion mass of about 200 amu, the total current is minimized (at ~40 kA), and the individual ion energy required is ~10 GeV. The required beam spot radius is in the range ~1.5-3 mm. Achieving this small spot, assuming unneutralized ballistic propagation through a final set of quadrupole lenses, requires that the normalized transverse emittance be less than ~ 10 mm-mrad, and the longitudinal fractional momentum spread $\delta p/p$ be less than $\sim 1\times 10^{-3}$. In contrast, the planned recirculator experiments will have a final current (~ 8 mA), energy (~ 320 keV), and normalized emittance (~ 0.2 mm-mrad) all much lower than the driver. Nevertheless, the relevant dimensionless parameters describing beam propagation and physics, (for example, undepressed phase advance σ_0, depressed phase advance σ, perveance K, and tune ν) are in regimes which are either comparable to or more demanding than in the driver recirculator[6].

2. Recirculator Plans

The goal of the small recirculator experiment is to test nearly all of the beam dynamics issues and many of the technology issues facing a driver. When successful, the experiment will have demonstrated:
- transport of a space charge dominated beam around multiple laps of a circular accelerator,
- longitudinal and transverse beam control in a recirculator,
- emittance control (less than a factor of 2 growth is the goal),
- freedom from resonance instabilities,
- injection and extraction into and out of a ring,
- energy gain (by a factor of 4),
- bunch length compression (by a factor of 2).

The new technology demonstrated will include:
- high repetition rate pulse power with variable pulse format, including acceleration and longitudinal confinement fields,
- coordination of the temporally increasing dipole field with the variable firing of the induction cores, as the beam speed increases,
- beam steering over a muliple laps.

The experiment will not examine:
- stability of background gas against beam desorption,
- high current instabilities,
- manipulations of multiple beams.

The physics design of the small recirculator required a number of parametric choices that are more completely motivated elsewhere[5,6]. Simplicity, low cost, and relevance to a driver-scale machine provided general guidance in the design. The dipoles are ramped electric rather than ramped magnetic for simplicity and cost. The quadrupoles will be permanent magnets, simulating the driver designs which use superconducting magnets (which are also fixed in strength). The insertion/extraction section consists of a permanent magnet elliptical quadrupole, with electric dipoles serving the dual purposes of kickers on insertion and extraction, as well as bending the beam on normal laps. The diagnostics for the recirculator must be non-invasive. Capacitive probes have been selected for determining the location of the beam centroid, providing sufficient information to steer the beam. The vacuum must be $<10^{-8}$ torr, so metallic seals are being used to minimize outgassing. This is still a large residual gas density compared to a driver design[1], which requires pressures less than $\sim 10^{-10}$ torr. The maximum repetition rate of the induction core pulsers is ~ 100 kHz, equal to the lap time around the recirculator. A prototype pulser has been developed which uses solid state switches to achieve the variable repetition rate and pulse format required for recirculator operation[10]. A list of some of the design parameters of the recirculator is given in table 1, and a diagram illustrating the layout of the experiment is given elsewhere in these proceedings[7].

3. Linear Magnetic Transport Experiment

Presently, our experiments are employing a straight magnetic transport section in preparation for bent beam studies. The apparatus consists of an injector, matching section, and magnetic transport section. The 80 keV diode injector has a hot plate source which produces singly charged potassium ions. The injector is followed by seven electric quadrupoles which act as a matching section for converting the circular beam emerging from the diode into the elliptical shape needed for alternating gradient transport. This matching section was adapted from one of the transport sections of the LBNL Single Beam Transport Experiment[9]. Two of the original quadrupoles have been electrically separated so that dipole voltages may be applied for small steering corrections to the beam centroid. The matching section is followed by a drift section consisting of seven permanent magnet quadrupoles, each with pole tip fields of approximately 0.3 T (at a clear aperture radius of 3.5 cm) and an integrated field gradient $(= \int B' dz)$ of 0.95 T, where B' is the quadrupole gradient and the integral is carried out along the quadrupole magnetic axis. Results to date are encouraging. No anomalies have been observed in magnetic transport of a space-charge-dominated beam. The goals of these experiments are to characterize the beam prior to injection into the ring, to develop necessary diagnostics for recirculator operation, and to assess the role that stray electrons play in beam transport. Some examples of data obtained from our initial experiments are briefly summarized below.

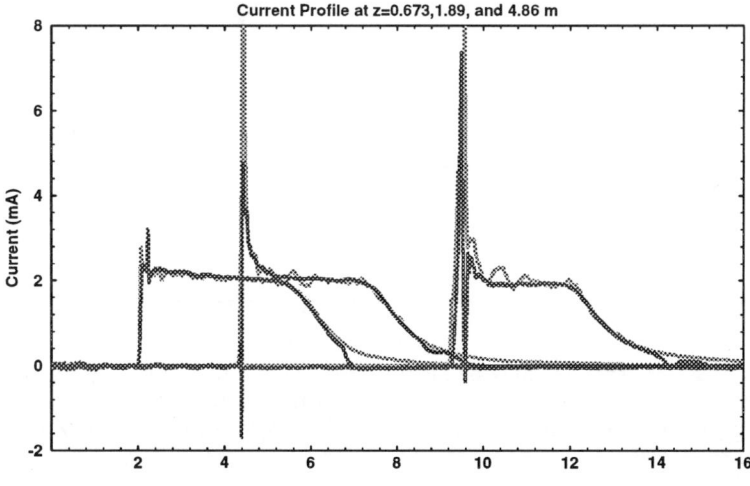

Figure 1. Current measurements (dark gray) and 1D simulations (light gray) at Faraday Cups located 0.67, 1.89, and 4.86 m downstream of the diode source.

Faraday cup current profiles: Figure 1 shows the time dependent current profile at three stages within the matching and transport section.

The 2 mA flattop current is proceeded by a sharp current spike in the beam head. The spike arises because the rise time of the injector pulser is longer than optimal. Ions emitted at the beginning of the pulse are overtaken by faster, higher energy ions emitted slightly later. A new, faster pulser is in the process of being installed in the machine. The shorter rise time of this new pulser will produce a waveform closer to the ideal risetime[11], so that the particles emitted during the rise will get a boost from the space charge field of the particles emitted later, so as to arrive at the diode exit with an energy and current nearly equal to those of the main body of the pulse. The new pulser should eliminate problems associated with the non-ideal waveform. Nevertheless the good agreement between simulations and experiment with the non-ideal pulse has given us confidence in the modeling effort and has served to enhance our understanding of the machine.

Energy analyzer data: An electrostatic energy analyzer in which the beam is bent by a transverse electric field which segregates particles based on kinetic energy, has been used as an independent energy check. Results from this diagnostic located 1.85 m from the source are shown in figure 2. The energy flatness is about ± 3% over 3μs. 1D code comparisons using the inferred injector waveform as input to the code agree with the absolute measure of the energy within 2% (essentially using time of flight as the energy measure), and show virtually identical energy flatness.

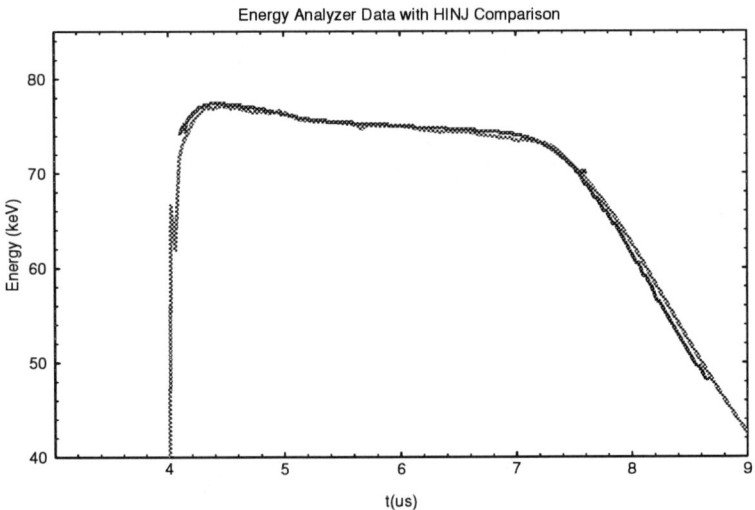

Figure 2. Peak energy signal (dark gray) and 1D simulations (light gray) vs. time at energy analyzer detector located 1.75 m downstream of the source. (The energy normalization and time zero point were adjusted for best fit of simulation to data).

Two-slit emittance scanner: In this dianostic, the beam impinges upon an aperture slit of width 0.1 mm (compared to a typical beam radius of 1 cm), and 21 cm later passes through a second slit with the same dimension. These scanners may be placed at four different locations within the matching and magnetic transport sections. The transmitted beam current is measured. The position of both slits is varied and rms angles can be calculated. The results indicate a normalized emittance which grows from 0.025 mm mrad (just inside the matching section) to approximately 0.040 mm mrad, in about 4.7 m (at the end of of the magnetic transport section). (This compares to a normalized emittance at the beam source estimated to be approximately 0.020 mm mrad, based on its temperature and radius). The origin of this emittance growth is unknown, but our current suspicion is that mismatch oscillations are providing a source for beam thermalization. Although the emittance growth needs explanation, there is still an expectation of meeting our design goal of 0.10 mm mrad after insertion into the recirculator. Work on better beam matches that should alleviate much of this growth will be carried out after the new pulser is installed.

Capacitive probe: Capacitive probes (C-probes) are placed within the bores of the permanent magnet quadrupoles for a non-interceptive measurement of the beam centroid. These C-probes consist of cylindrical insulators with interiors that are copper plated on four equal sectors which run parallel to the longitudinal axis. The voltage on each segment can be measured, and the difference in voltage between the sum of the two left segments and the sum of the two right segments (for example) gives a measure which is nearly proportional to the offset of the beam centroid in the horizontal direction. A similar combination of voltages yields the position in the vertical direction. Preliminary measurements indicate that signals obtained from the probe in the presence of the beam are easily measured and have acceptable signal to noise ratio. It is anticipated that beam centroid position can be measured to 0.1 mm with these C-probes.

Rotating wire scanner: Another method of determining beam position and, in addition, of measuring the cross sectional profile of the beam has been explored using a rotating wire scanner. In this method, a 1 mm diameter wire intercepts a small portion of the beam, and either the current created by the ion striking the wire (and associated ejected electrons) is measured directly, or the electrons which are ejected from the wire by the beam impact are collected and measured. As an initial setup we have adapted a commercially available scanner (a National Electrostatics Corp. Model BPM82) which is shaped as a 45 degree helix. The helix can rotate at a rate on the order of one Hz and the beam is injected when the helix reaches a preset phase. The phase can then be varied, and thus a sequence of projections of the beam profile can be formed. (The advantage of using a helix is that with one wire two orthogonal directions can be measured.) These will provide checks for the C-probes and can be parked outside the beam in normal operation. Initial results with this device have shown promise for direct and reproducible profile and centroid measurements and are being used as a check on the calibration of the C-probes. Position uncertainties of ~ 0.1 mm are expected in these measurements. Slit scanner measurements downstream of the rotating wire scanner have shown negligible changes in the emittance. (In fact, some measurements show a small decrease in the emittance). Hence, these probes are being considered as "minimally" intercepting diagnostic for use in the main recirculating ring.

Gated beam imager: A final diagnostic under development is a gated beam imaging system. This system consists of an optical, time-gated CCD camera, together with a pepper-pot mask and a scintillating screen for the purpose of obtaining rapid measurements of the complete four dimensional phase space of the beam. As in the case of the slit scanner, after the beam passes through the array of circular apertures on the pepperpot mask, it expands, and the intensity, ellipticity and size of each miniature ellipse on the scintillating screen gives information about the detailed transverse phase space distribution function of the beam. Present experiments are examining saturation properties of the scintillating screen to establish the reliability and reproducibility of these measurements.

Spurious electron effects: Free electrons within the accelerator can be produced by stripping of the residual gas by the ion beam or the impact of stray beam particles against the structure, for example. In the electric focusing section, the electric fields of the electrodes readily separate the electrons from the ions, so that an accumulation of electrons in the potential well of the ions is unlikely to occur. In the magnetic transport section electrons could accumulate in the potential well of the ion beam between focusing magnets. (Within the quadrupoles, the small Larmor radius of the electrons would presumably prevent them from copropagating with the ions). In a previous study of space-charge dominated ion beams[8], evidence for beam neutralization was presented at long pulse durations (~ 100 μs). By varying both the bias on the capacitive probes and the residual gas density, evidence for electron effects has been found in our experiment as well. At densities above about $\sim 10^{-6}$ torr there is an apparent decrease in the beam current over the duration of the pulse, particularly when the bias voltage on the C-probe exceeds the beam potential. This is indicative of electrons being collected during the pulse. This effect increases as the residual gas densities increase. At ($\sim 10^{-7}$ torr (ten times the nominal operating residual gas pressure) these effects are minimal and would have minimal impact on beam propagation. We are continuing to refine these experiments to assess the effects of these stray electrons. Near term experiments are expected to have a more ideal injector waveform that will minimize overtaking and thereby minimize particle loss due to a mismatched beam head. This will enable a more concrete assessment of electron production due to particle stripping and the accompanying effects.

4. Conclusion

Previous studies have suggested that recirculating induction accelerators provide a potential lower cost approach for a heavy-ion fusion driver[1]. The LLNL experiments provide a small-scale test of most of the beam dynamics issues and many of the technology issues of high repetition rate induction acceleration relevant to a heavy ion driver for Inertial Fusion Energy. This experiment is being assembled as a series of experiments that will lead to a fully functioning recirculator. The current set of matching section and magnetic transport experiments have characterized the beam and tested a variety of probes which will be used when the complete ring is assembled. The physics of stray electrons is still being assessed, although they appear to play no significant role at the residual gas densities planned for the completed ring. We have briefly summarized the set of Heavy Ion Fusion beam experiments being carried out at LLNL. A more complete description is given in ref. 3.

References

1. J.J. Barnard, A.L. Brooks, F. Coffield, F. Deadrick, L.V. Griffith, H.C. Kirbie, V.K. Neil, M.A. Newton, A.C. Paul, L.L. Reginato, W.M. Sharp,

J. Wilson, S.S. Yu, and D.L. Judd, "Study of Recirculating Induction Accelerators as Drivers for Heavy Ion Fusion," Lawrence Livermore National Laboratory Report UCRL-LR-108095.

2. J.J. Barnard, F. Deadrick, A. Friedman, D.P. Grote, L.V. Griffith, H.C. Kirbie, V.K. Neil, M.A. Newton, A.C. Paul, W.M. Sharp, H.D. Shay, R.O. Bangerter, A. Faltens, C.G. Fong, D.L. Judd, E.P. Lee, L.L. Reginato, S.S. Yu, and T.F. Godlove, "Recirculating Induction Accelerators as Drivers for Heavy Ion Fusion," Physics of Fluids B: Plasma Physics, **5**, 2698 (1993).

3. T.J. Fessenden, J.J. Barnard, M.D. Cable, F.J. Deadrick, M.B. Neslon, T.C. Sangster, S. Eylon, H.S. Hopkins, "Intense Heavy-Ion Beam Transport with Electric and Magnetic Quadrupoles," submitted to Fusion Engineering and Design (1995).

4. R.O. Bangerter, "The Induction Approach to Heavy-Ion Inertial Fusion: Accelerator and Target Considerations," Il Nuovo Cimento, **106A**, 1445, (1993).

5. A. Friedman, J.J. Barnard, M.D. Cable, D.A. Callahan, F.J. Deadrick, S. Eylon, T.J. Fessenden, D.P. Grote, H.A. Hopkins, V.P. Karpenko, D.L. Judd, H.C. Kirbie, D.B. Longinotti, S.M. Lund, L.A. Nattrass, M.B. Neslon, M.A. Newton, W.M. Sharp, and S.S. Yu, "Recirculating Induction Accelerators for Inertial Fusion: Prospects and Status," submitted to Fusion Engineering and Design (1995).

6. J.J. Barnard, M.D. Cable, D.A. Callahan, T.J. Fessenden, A. Friedman, D.P. Grote, D.L. Judd, S.M. Lund, M.A. Newton, W.M. Sharp, and S.S. Yu, "Physics Design and Scaling of Recirculation Induction Accelerators: From Benchtop Prototypes to Drivers," submitted to Fusion Engineering and Design (1995).

7. A. Friedman, "Induction-Accelerator Heavy Ion Fusion: Status and Beam Physics Issues," these proceedings.

8. J. Klabunde, M. Reiser, A. Schonlein, P. Spadtke, J. Struckmeier, "Studies of Heavy Ion Beam Transport in a Magnetic Quadrupole Channel," IEEE Transactions on Nuclear Science, Vol. NS-30, 2543 (1983).

9. M.G. Tiefenback, "Space-Charge Limits on the Transport of Ion Beams in a Long Alternating Gradient System," Ph. D. Dissertation, University of California, Berkeley, LBL-22465, HIF AN-352.

10. M.A. Newton, F.J. Deadrick, R.L. Hanks, H.C. Kirbie, V.P. Karpenko, L.A. Nattrass, "Engineering Developments for a Small-Scale Recirculator," submitted to Fusion Engineering and Design (1995).

11. M. Lampel and M. Tiefenback, "An Applied Voltage to Eliminate Current Transients in a One-Dimensional Diode," App. Phys. Lett. **43**, 57, (1983).

Longitudinal Dynamics and Stability in Beams for Heavy-Ion Fusion[*]

W. M. Sharp, D. A. Callahan, and D. P. Grote
Lawrence Livermore National Laboratory L-440, Livermore, CA 94550, USA

Abstract

Successful transport of induction-driven beams for heavy-ion fusion requires careful control of the longitudinal space charge. The usual control technique is the periodic application of time-varying longitudinal electric fields, called "ears," that, on the average, balance the space-charge field. This technique is illustrated using a fluid/envelope code CIRCE, and the sensitivity of the method to errors in these ear fields is illustrated. The possibility that periodic ear fields also excite the longitudinal instability is examined.

I. Introduction

Since its inception in the 1970s, the American heavy-ion fusion (HIF) program has favored induction accelerators. [1,2] These devices can accelerate much higher currents than the radio-frequency (rf) accelerators preferred in Europe, and they can compress the beam during acceleration, simplifying the problem of longitudinal compression. However, to be economically competitive, induction accelerators must carry nearly the maximum transportable charge along the entire lattice, so that beam dynamics in such devices is invariably dominated by space charge.

The effects of space charge on transverse dynamics are familiar and will not be discussed here in detail. Strong focusing is needed to balance the space-charge force, and this is usually applied by alternating-gradient (AG) electric or magnetic quadrupoles [3]. Since the space-charge force partially cancels the focusing force, space-charge-dominated beams are often substantially tune depressed [4], and since they maintain a nearly uniform charge density except near the edges, such beams have radially density profiles that are much flatter than the Gaussian profiles typically found in rf accelerators [5]. The space-charge fields for these flatter density profile are more nearly linear in the beam interior than an emittance-dominated beams, reducing the rate that transverse fluctuations thermalize. Finally, the transverse space-charge provides a mechanism for halo formation when a beam is imperfectly matched [6].

There are also two important longitudinal effects of space charge. Unlike radio-frequency accelerators, the accelerating fields of induction accelerators provide no longitudinal focusing, so space charge causes the beam to lengthen

---

[*] The research was performed under the auspices of the U. S. Department of Energy by Lawrence Livermore National Laboratory under Contract No. W-7405-ENG-48.

in the absence of supplemental focusing. Longitudinal control is typically maintained by time-varying electric fields, referred to here as "ears," that are added to the acceleration field in at least some induction cells to balance the space-charge force. The other longitudinal effect of space charge is an interaction, known as the "longitudinal instability," between the beam and its resistively retarded image field. This paper will discuss theoretical work on the generation and application of ear fields, and it will review simulations examining whether the periodic application of ear fields is likely to stimulate the longitudinal instability.

II. Effects of Ear Fields

A. Time Dependence of Ears

For the ion beams considered for heavy-ion fusion (HIF), which are typically meters long and only a few centimeters in radius, the longitudinal space-charge field is a highly non-linear function of position in the beam frame, so the ear fields must vary in time proportionally with the space-charge field as the beam moves through the acceleration gaps. The ideal ear field is related to the longitudinal component E_z of the beam space-charge field at corresponding times by

$$E_{ear} \approx -\eta_{ear}\langle E_z \rangle, \qquad (1)$$

where the ear occupancy η_{ear} is the ratio of the induction-cell gap length to the distance between those cells in which ears are applied, termed "ear cells." The angle brackets on $\langle E_z \rangle$ denote a density-weighted average over the beam cross-section. This average is needed because the impulse an ion receives from an induction cell is approximately independent of the transverse location of the particle, making it impossible for the ear field to nullify the space charge at all points across the beam. Using ears calculated from the averaged space-charge field minimizes this inevitable error in longitudinal confinement.

The average axial space-charge field $\langle E_z \rangle$ in Eq. (1) is found to depend significantly on both the current and radius profiles of the beam. This dependence is illustrated by fields calculated for a quadratically decreasing current at the beam ends in Fig. 1 and for a Gaussian current fall off in Fig. 2. In each case, the beam is presumed to be centered in a perfectly conducting cylindrical pipe, and the parameters are those of a small recirculating induction accelerator [7] being developed at the Lawrence Livermore National Laboratory (LLNL), except that the beam is assumed here to the axisymmetric and the beam midsection has been shortened to highlight field changes near the ends. The maximum current I_{b0} is 2 mA of singly charged potassium ions with an energy of 80 keV, and the maximum radius a_0 of the beam is 0.011 m. For each current profile, radius profiles are shown for several choices of the emittance variation, corresponding to a uniform beam

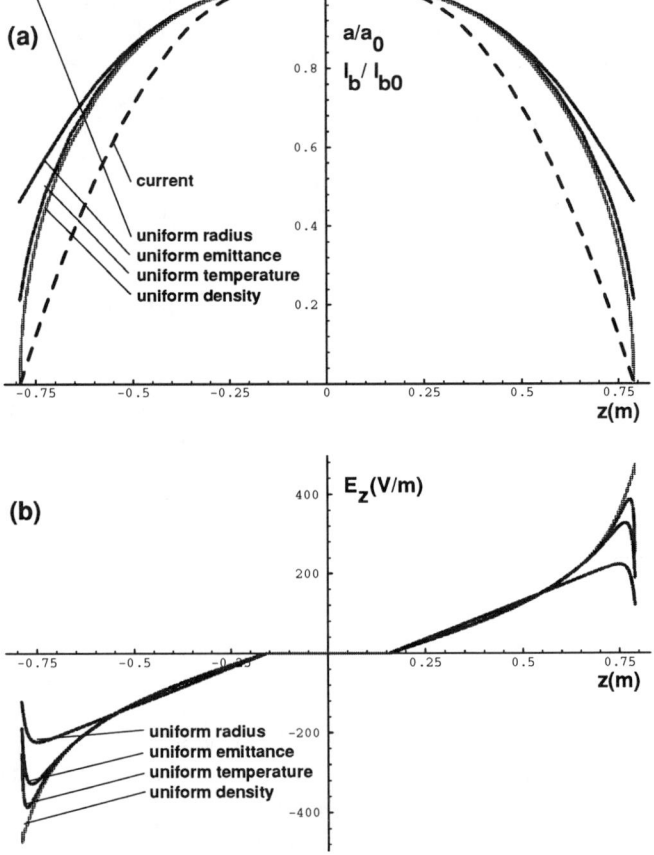

Fig. 1 (a) Radial profiles and (b) radially averaged space-charge fields for axisymmetric beams with quadratically decreasing current I_b near the ends and various choices of emittance variation. Current profile is plotted as a dashed line.

radius a, a uniform transverse emittance $\epsilon_\perp$, a uniform transverse temperature $\epsilon_\perp/a$, and a uniform charge density $I_b/(\pi\beta c a^2)$, where β is the beam axial velocity scaled by the speed of light c. Although these simple profiles are unlikely to match that in an experimental beam, they illustrate the sensitivity of the space-charge field to the beam radial variation. Figs. 1a and 2a show the relative radial profiles of the beams for these various emittance choices, and the corresponding current profiles are plotted as dashed lines.

In the field plot for the "quadratic-end" cases in Fig. 1b, one finds that the peak space-charge field increases for profiles that have smaller end radii, with the field for a uniform-density beam being more than twice that for a uniform-radius beam. It is also evident that $\langle E_z \rangle$ for the uniform-density is

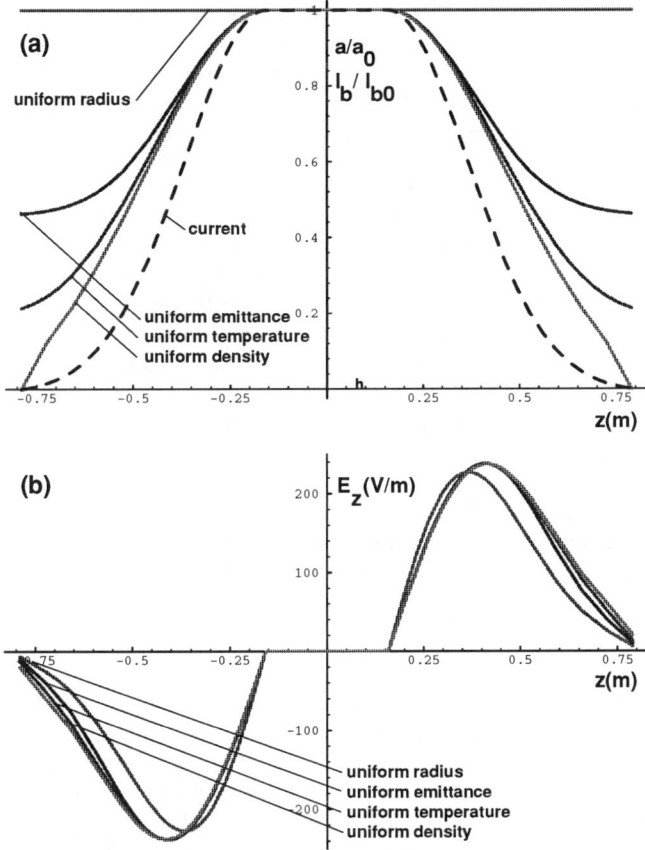

Fig. 2 (a) Radial profiles and (b) radially averaged space-charge fields for axisymmetric beams with Guassian decrease in current near the ends and various choices of emittance variation. Current profile is plotted as a dashed line.

qualitatively different from the others. For the cases with a finite beam-end radius, the field magnitude is seen to drop significantly in a narrow region at the beam end. This region has a characteristic length of $0.415R$ for a pipe with radius R, and within the region, the absence of charge beyond the beam end reduces the axial field. At the endpoints, the field is reduced by approximately half. In contrast, $\langle E_z \rangle$ for the uniform-density case varies monotonically near the end due to the rapidly decreasing radius. The field plots for the "Gaussian-end" cases in Fig. 2b are substantially different. As expected from the analytic expressions for $\langle E_z \rangle$ in Ref. [8], the peak field magnitudes occur near the points of maximum slope in the beam-current, with some shift in magnitude and position arising from the radius variation. However, since there is less difference in the beam radii at these points than in the quadratic-end cases, the peak space-charge field differs by only about 5% between the various emittance profiles. Also, the rapid fall off in field strength near the ends is not seen in these cases. These two cases show that a detailed knowledge of the beam current and radius profiles is important in determining both the waveform and range of variation needed for ear generation.

Despite the assumption of axisymmetry used in calculating the fields in Figs. 1b and 2b, the results are nearly the same both for beams displaced from the accelerator axis and for beams in AG lattices. Using a three-dimensional simulation code WARP3d [9], we find for $a/R = 0.33$ that values of $\langle E_z \rangle$ for a centered beam and one displaced $0.14R$ from the axis differ by no more than 2%. Similarly, the average space-charge field of a beam with a circular cross-section has been compared with that of an elliptical beam with major and minor radii a and b by matching the circular-beam radius to the geometric mean $(ab)^{1/2}$. For $a/b = 1.8$, $\langle E_z \rangle$ differs from the field for the corresponding round beam by at most 4%. Furthermore, since a and b fluctuate in the AG focusing on a period that is typically much smaller than the characteristic longitudinal-expansion time, the effect of beam ellipticity on longitudinal dynamics is substantially less than expected from this static comparison.

B. Calculation of Optimal Ears

The fluid/envelope code CIRCE [10] is used here to assess the effects of space charge on the longitudinal dynamics of heavy-ion beams. Longitudinal dynamics is modeled in the code by treating slices of the beam as Lagrangian fluid elements. The boundaries between slices are assumed to be perpendicular to the beam axis and to respond to the average space-charge field $\langle E_z \rangle$ and to the axial field from acceleration cells. The neglect in this model of radial variation in the space-charge field is justified because this variation remains small so long as the space-charge potential across a beam is large compared with the transverse thermal energy, a condition that characterizes space-charge-dominated beams. The beam can be initialized using any of the emittance profiles discussed in section IIA, and the average space-charge field can optionally be calculated from the analytic expressions in Ref [8].

Optimal ears are generated by a simple procedure. For a particular beam and lattice, CIRCE is run once without ear fields and with the axial space-charge force artificially switched off. This condition is equivalent to having perfect ears that are applied continuously. During the run, the average axial space-charge field $\langle E_z \rangle$ is calculated as the beam enters each ear cell, and Eq. (1) is used to obtain the corresponding ear fields, which are written along with timing data to an external file. For beams with a finite emittance and radius at the ends, $\langle E_z \rangle$ is estimated by the simple expression

$$\langle E_z(z) \rangle \approx -\frac{1}{4\pi\epsilon_0} \left\{ \left[\frac{1}{2} + \ln\left(\frac{R^2}{ab}\right) \right] \frac{\partial \lambda}{\partial z} - \frac{\lambda}{2ab} \frac{\partial(ab)}{\partial z} \right\}, \qquad (2)$$

where $\lambda = I_b/(\beta c)$ is the beam line-charge density, R is the beam-pipe radius, and z is axial position in the beam rest frame. A more complicated Bessel-series form from Ref. [8] is used for beams with uniform charge density. Even though Eq. (2) misses the rapid field variation near the ends seen in Fig. 1b, it is used here because the high frequency response needed to balance the space-charge field near the ends is beyond current pulse-generation technology. On subsequent runs, ears are obtained from the external file, and the space-charge field, including any rapid variation at the ends, is calculated at every step from the appropriate expression from Ref. [8].

C. Dynamics with Optimal Ears

The effects of optimal ears are illustrated here using the lattice and initial beam parameters of the small recirculating induction accelerator being developed at LLNL [7]. As in the field calculations of section IIA above, the initial current of singly charged potassium ions is 2 mA, but the emittance is sufficiently low that at the initial energy of 80 keV, the beam transverse dynamics is space-charge dominated. In the uniform-emittance cases presented here, the beam doubles its energy in 15 laps while compressing by a factor of four in duration and a factor of two in length. A head-to-tail velocity variation or "tilt" of about 35% is imposed during the first lap by triangular accelerating pulses in the first ten active cells, and ear fields are applied in each 36 cm half-lattice period (HLP), except in two sections of three HLPs each that are used for insertion and extraction of the beam. The initial pulse duration is 4 μs, and the current decreases quadratically to zero in 1 μs at both ends.

In the absence of ears, space charge is seen in Fig. 3a to push out the ends, and rarefaction waves move in from each end at the electrostatic wave speed

$$C_s \approx \left(\frac{gqe\lambda}{4\pi\epsilon_0 M}\right)^{1/2} \ll \beta c, \tag{3}$$

where q is the ion charge state, M is the ion mass, and $g = \ln(R^2/a^2)$ is a factor accounting for the wall geometry. The velocity tilt imposed by the first ten cells causes the density at the beam center to increase until, in the fourth lap, the rarefaction waves meet, and after that point the entire beam elongates to about 2.5 times the initial value. This case indicates that a beam of length L_b can be transported without ear fields only over lattice lengths that are short compared with $\beta c L_b/(2C_s)$.

With the axial component of the space-charge field set to zero, the beam current is found to compress in a nearly self-similar manner. Using ears generated by this first run, a second case, including space charge, gives the current history shown in Fig. 3b. On the scale shown, the pulse history is indistinguishable from that for the run without a longitudinal space-charge field, although more detailed plots show some low-amplitude space-charge

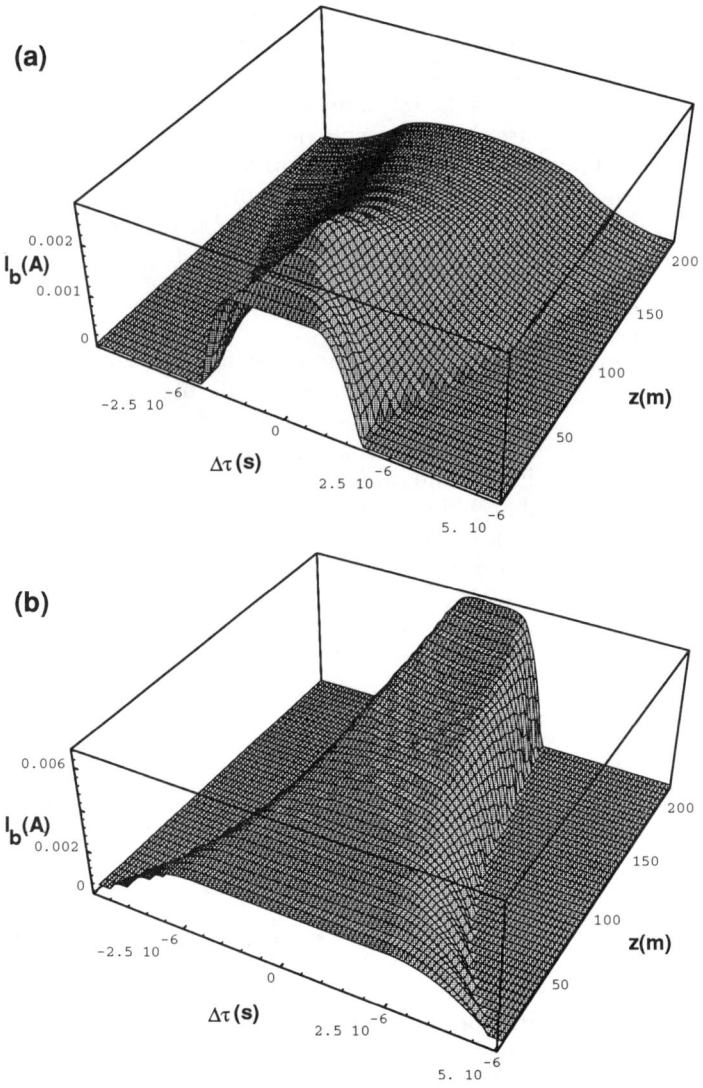

Fig. 3 CIRCE calculation of beam-current histories of a space-charge-dominated beam (a) without "ear" fields and (b) with space charge balanced by optimal ears applied periodically. Current is plotted as a function lattice position z and time $\Delta\tau$ relative to the beam midpoint. Ripples seen in (b) near $z = 0$ are a plotting artifact.

waves launched during the first few laps. These waves result from the initial mismatch near the ends between the monotonically varying ear field and the rapidly varying space-charge field seen in Fig. 1b for a uniform emittance beam. Examination of the beam line-charge indicates, however, that the ends of the beam are quickly compressed, making the space-charge field approximately match the average ear field and effectively ending the generation of space-charge waves. A similar adaptation of the beam to the ear field is found with other beam emittance profiles.

If the optimal ears for a compressing beam are applied but no velocity tilt is imposed by the accelerating voltage, then the beam ends are pushed in according to the compression schedule, but large space-charge waves are generated as the ear fields move into the denser part of the beam. After six laps, the waves collide in a highly non-linear fashion that is not modeled correctly by CIRCE, illustrating that a velocity tilt is necessary for quiescent beam compression, as well as suitable ear fields.

Of necessity, the ear fields are applied periodically, so the ends are repeatedly being pushed in and then expanding. However, provided that this period is short compared with the characteristic time for beam expansion, these kicks are not an important source of noise. This consideration imposes a constraint on the maximum spacing of ear cells, but in practice, the breakdown voltage in induction cells is typically a more stringent limitation.

D. Dynamics with Imperfect Ears

Due to the imprecision in measuring the beam current profile and limitations in waveform generation, the ear fields used in an experiment are certain to be imperfect. A number of runs have been made to test qualitatively the sensitivity of beam dynamics to ear-field errors. For cases in which ear fields generated from one beam profile are applied to a beam with a different profile, the beam is found to be rather insensitive to the discrepancy for any pairing that involves the uniform emittance, density, or transverse temperature profiles discussed in section IIA. The reason for the insensitivity is that mismatches between the average ear fields in these cases and the space-charge field become appreciable only in the low-density region at the ends, where λ is less than about 10% of its mid-pulse value. In such cases, that beam is found to adapt to discrepancies as large as 60% between the ear and space-charge fields. Appreciable space-charge waves are generated, however, in beams with a uniform initial radius controlled by ears from one of the other profiles. As is evident from Fig. 1b, the space charge for a uniform-radius beam differs visibly from that for the other profiles over almost half of the current rise and fall lengths, and the size of the field discrepancy is larger than in the previously cited cases. Both differences are expected to increase wave generation, but the effects have not been isolated and quantified.

A second series of runs was made using current profiles that differ from that used to generate the ear fields. Fig. 4a shows the current history of a beam with 1.2 μs current rise and fall times confined by ears calculated for a 1 μs rise and fall. In this case, the mismatch between the average ear

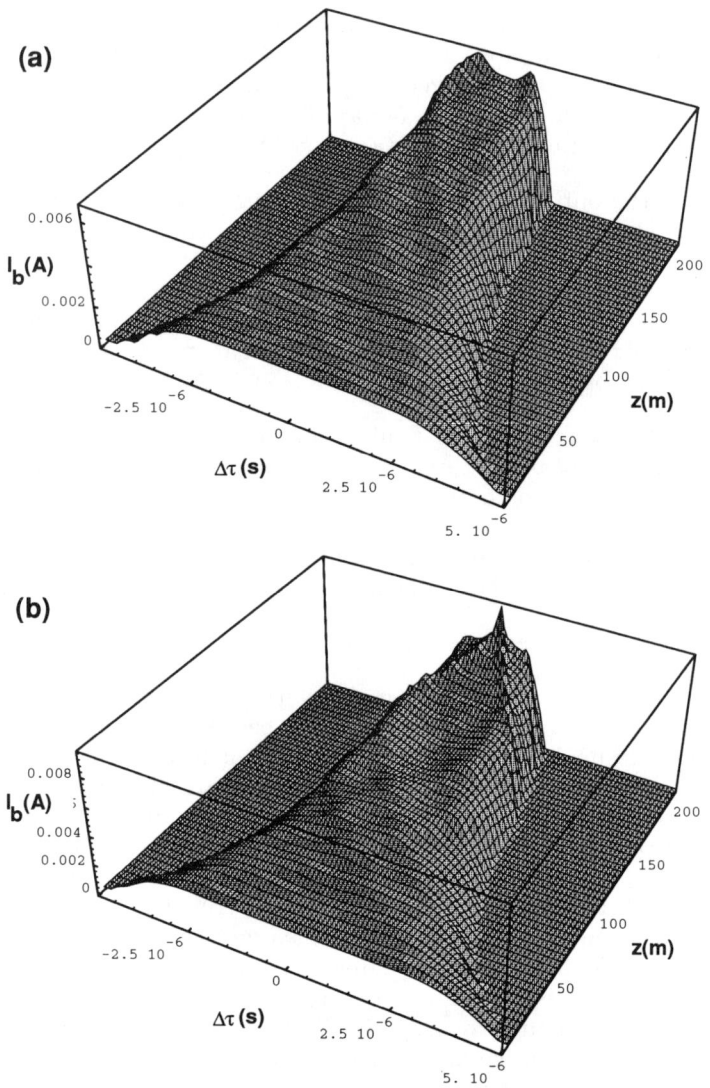

Fig. 4 CIRCE calculation of space-charge waves generated (a) by inappropriate current rise and fall times and (b) by an inappropriate current-profile form

field and the space-charge field occurs near the current flat-top, and sizable linear space-charge waves are generated. The waves are seen to increase in amplitude with z because the fluid model used in CIRCE for longitudinal dynamics has no damping, so the wave energy is trapped in a progressively shorter length as the beam compresses. In an experiment or particle simulation, some damping is expected due to phase mixing and thermalization, and there would be a corresponding growth in longitudinal temperature. A more extreme mismatch is generated by specifying a cubic fall-off in the beam current near the ears but using ears calculated for a quadratic drop-off. The result, shown in Fig. 4b, is that larger linear waves are initially generated and become nonlinear after 15 laps. Both cases in Fig. 4 indicate that the beam is moderately sensitive to errors in the ear field that occur in the higher-density regions of the beam.

III. Longitudinal Instability

A. Mechanism

The longitudinal instability is a concern in HIF drivers because it amplifies small perturbations launched from the beam head. These perturbations may be caused by errors in the acceleration or ear fields, or at a low level, by the periodic application of ear fields. The instability has the same mechanism exploited in "resistive-wall" amplifiers, except that the impedance in HIF accelerators results mainly from the accelerating modules. The instability can be stabilized by a sufficiently large spread in longitudinal momentum, but the required spread exceeds the limit imposed by chromatic aberration in the focusing lens system.

The longitudinal instability can be modeled using a simple fluid model. If we consider an incompressible axisymmetric beam with radius a traveling down a pipe of radius R, one-dimensional (1-D) linear cold-fluid theory shows that two waves will develop: a "fast" wave traveling forward in the beam frame and a "slow" wave traveling backward. These waves propagate with a phase velocity in the beam frame near the electrostatic wave speed C_s, given by Eq. (3). In induction drivers, the resistance of acceleration modules causes the beam image charge to lag behind the beam, and the interaction of perturbations with their retarded images leads to the growth of slow waves, as well as damping of fast waves. In early theoretical work on the instability [10], module impedance was modeled as a distributed resistance per unit length η, and the 1-D cold-fluid model predicted a dangerously high growth rate Γ and a growth length

$$\frac{C_s}{\Gamma} \approx \frac{g}{2\pi\epsilon_0 \eta \beta c} \quad (4)$$

that was much shorter than the beam length. A more sophisticated 1-D model [13], which includes a realistic distributed capacitance in parallel with the resistance, shows a much lower maximum growth rate, occurring for a perturbation wavelength that is large compared with the pipe radius R.

B. Simulations

An r-z branch of the WARP code was developed to model the longitudinal dynamics of HIF drivers [14]. This code is a 2 1/2-dimensional, cylindrically symmetric particle-in-cell code. Field calculation is done in a window that moves with the beam center of mass. In this computational window, the fields are treated as purely electrostatic, since magnetic fields arise only from longitudinal temperature and any velocity tilt, and they affect dynamics only like the square of these relative velocities scaled by c^2. To allow study of the longitudinal instability, a wall model with distributed resistance and capacitance in parallel is used. This approximation for the induction modules contains the relevant physics, and also corresponds well with recent analytic work [13]. The resistive-wall contribution to the electric field is calculated using a Poisson solve at the boundary because this approach is smoother and more physical than using the explicit beam current.

Simulations with a purely resistive wall and perturbations launched from the beam head have shown growth of the backward traveling wave, as expected. The measured growth rate is lower than the prediction of cold-beam fluid theory by about 15%. This discrepancy is probably due to the effects of finite transverse temperature, but the matter requires further research. The perturbation reflects off the beam tail, and some steepening of the perturbation is seen during reflection. This steepening appears to be a nonlinear effect since it is more evident in larger perturbations than in small ones. This nonlinear pulse narrowing is also accompanied by dispersion, so that the wave lengthens and decreases in amplitude due to the velocity spread between different frequency components as the perturbation travels from the beam tail to the head.

When a capacitive component is included on the impedance, WARPrz simulations show both the lower growth rates and the longer wavelength of the fastest growing mode predicted by cold-beam fluid theory. For wavelengths that are comparable with the beam length perturbations are excited, but little growth is seen. These perturbations move back and forth from beam head to tail with little change in amplitude.

For a realistic appraisal of the effects of the longitudinal instability, finite-amplitude sources of perturbations need to be considered. One source of such perturbations is the intermittent application of the ear fields. In most of WARPrz simulations, ear fields are applied at each time step and are designed to keep the beam from expanding or contracting. In an experiment, these fields will be applied at fixed locations along the accelerator and the beam will contract and then expand between ear cells. CIRCE simulations with driver parameters show that a train of low-level perturbations is launched from the beam ends, and with a non-zero impedance, these perturbations will be amplified by the longitudinal-instability mechanism.

In the simulations reviewed here, each application of the ears consists of the following steps: (1) The beam is allowed to expand for 0.48 μs (48 m at 0.33c). (2) The expansion velocity is reversed by applying ear fields at the ends for 0.0875 μs. (3) The beam is allowed to expand for 0.48 μs. After these operations, the beam should be back to its original length.

In the absence of errors, the intermittent application of ears is found to be benign. Perturbations on the beam are minimized when an electric ear field is applied at each end that is proportional to the average particle velocity in the beam frame as a function of z after the first expansion. The proportionality constant is varied until the beam is close to its original state just before the next application, and since the beam is being held at constant length, the same ear fields are used for each application. The simulation shows that these ear fields can be applied more than 20 times without significant perturbations developing on the beam, even in the presence of a 100 ohms/meter resistive wall with no capacitance.[15]

Systematic ear-field errors are also found not to excite the instability. In a typical case, a error field with the form of half the period of a sine wave and a magnitude equaling 5% of the correct ear field was added to the . The same error was added at each application, making the ear fields systematically too large. Although such non-canceling ear errors were expected to be pathological, the beam adjusted quickly to the mismatched ears. The simulation was run for 1.4 km, with 25 applications of the intermittent ears, but perturbations were excited only during the first few applications. The beam remained in its new equilibrium for the remainder of the simulation. Similar adjustment of the beam to mismatched ear fields has also been seen in longitudinal-control experiments done on the Single-Beam Test Experiment (SBTE) at LBNL[16]. In these experiments, ear fields of the form $E_{ear} \sim [1 - \exp(-\alpha(t - t_0)]$ were applied near the beam tail, where t_0 is the time at which the beam current began to decrease in an ear cell, and α is an adjustable constant. No effort was made to match this ear waveform to the beam profile. The initial mismatch caused waves to be launched from the beam tail in the first few acceleration modules, but the beam soon reached a new steady-state configuration.

In contrast to the cases with systematic errors, random error are found to excite perturbations at each ear cell, with no adaptation of the beam profile. As an illustration, ear errors were applied that had the same half-sine form used previously, but the error magnitude varied randomly between being 5% too large to 5% too small. In this case, a train of perturbations was launched from the beam head throughout the 1.4 km run, growing as expected as they convected toward the tail. The perturbation widths were found to approximately match the wavelength of the fastest-growing mode predicted by the cold-fluid analysis. Figure 5 shows the electrostatic potential on axis vs z after 15 applications of the ear fields with 100 ohms/meter wall resistance. Although the waves seen in Fig. 5 are sizable, it should be remembered that capacitance was deliberately omitted in this case to exaggerate the instability. A much smaller wave amplitude is found in more realistic cases.

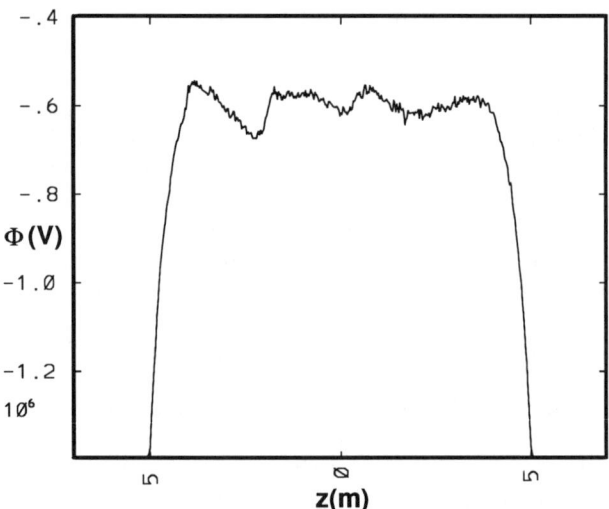

Fig. 5 WARPrz calculation of on-axis electrostatic potential of a driver-like beam after 1.4 km transport with random ear-field errors. Perturbations at the most unstable wavelength are excited.

IV. Conclusions

The CIRCE simulations shown here illustrate that beams for heavy-ion fusion are moderately sensitive to errors in ear fields. Mismatches up to 50% between the average ear field and the beam space-charge field appears to be acceptable at the beam ends due to the low beam density there. However, near the high-density part of the beam, mismatches should not exceed about 20%. The CIRCE results probably overestimate the wave amplitude due to mismatches due to the absence of phase-mix damping, but preliminary comparisons with the 3-D particle simulation WARP3d indicate fair agreement nonetheless.

Simulations with an r-z version of WARP indicate that the longitudinal instability is not a major problem for heavy-ion beams. The inclusion of a realistic induction-module capacitance gives a much lower growth rate than originally predicted from purely resistive models, and the intermittently applied ears do not normally seed the instability. Even the worst case, with random 5% errors in the absence of module capacitance, fails to excite destructive waves after 1.5 km of propagation. The modest growth of the longitudinal instability seen in realistic simulations can probably be stabilized by feed-forward correction, in which perturbations are detected at selected points along the accelerator, and correcting fields are applied downstream of the detectors.

References

[1] J. Hovingh, V. O. Brady, A. Faltens, D. Keefe, and E. P. Lee, *Fusion Technol.* '**3**, 255 (1988).

[2] R. O. Bangerter, *Nuovo Cimento* **106A**, 1445 (1993).

[3] M. Reiser, *Theory and Design of Charged Particle Beams* (John Wiley & Sons, New York, 1994), pp 162-165.

[4] E. P. Lee, T. J. Fessenden, and L. J. Laslett, *IEEE Trans. Nucl. Science* **NS-32**, 2489 (1985).

[5] M. Reiser and N. Brown, *Phys. Rev. Lett.* **71**, 2911 (1993).

[6] R. L. Gluckstern, *Phys. Rev. Lett.* **73**, 1247 (1994).

[7] A. Friedman, J. J. Barnard, M. D. Cable, D. A. Callahan, F. J. Deadrick, S. Eylon, T. J. Fessenden, D. P. Grote, H. A. Hopkins, V. P. Karpenko, D. L. Judd, H. C. Kirbie, D. B. Longinotti, S. M. Lund, L. A. Nattrass, M. B. Nelson, M. A. Newton, T. C. Sangster, and W. M. Sharp, "Recirculating Induction Accelerators for Inertial Fusion: Prospects and Status," to be published in *Fusion Eng. and Design*.

[8] W. M. Sharp, A. Friedman, and D. P. Grote, "Effects of Longitudinal Space Charge in Beams for Heavy-Ion Fusion," to be published in *Fusion Eng. and Design*.

[9] A. Friedman, D. P. Grote, and I. Haber, *Phys. Fluids B* **4**, 2203 (1992).

[10] W. M. Sharp, J. J. Barnard, D. P. Grote, S. M. Lund, and S. S. Yu, "Envelope Model of Beam Transport in ILSE" in *AIP Conference Proceeding 297* (AIP Press, Woodbury, NY, 1994), p. 540.

[11] C. K. Birdsall, G. R. Brewer, and A. V. Haeff, *Proc. of the I. R. E.* **41** (1953).

[12] E. P. Lee, Proc. 1981 Linear Accel. Conf., Los Alamos Repost LA-9234-C, p. 263.

[13] E. P. Lee, *Nuovo Cimento* **106A**, 1679 (1993).

[14] D. A. Callahan, "Simulations of Longitudinal Beam Dynamics of Space-Charge Dominated Beams for Heavy Ion Fusion," PhD. Thesis, Lawrence Livermore National Laboratory Report UCRL-LR-119364 (1994).

[15] D. A. Callahan, A. B. Langdon, A. Friedman, and I. Haber, "Longitudinal Beam Dynamics for Heavy Ion Fusion Using WARPrz" in *AIP Conference Proceeding 297* (AIP Press, Woodbury, NY, 1994), p. 221.

[16] A. Faltens, LBL Half-Year Report LBL-19501, June 1985.

HIGH BRIGHTNESS ELECTRON BEAMS

Collective Effects in Isochronous Storage Rings

A. W. Chao[1]

Stanford Linear Accelerator, Stanford University, Stanford, CA, USA

K.-J. Kim[2]

Lawrence Berkeley National Laboratory, Berkeley, CA 94720, USA

We have studied the collective instabilities in isochronous storage rings with a linac-type analysis. Simple criteria for avoiding the longitudinal and transverse instabilities are developed by employing a two particle model. Numerical examples show that these conditions do not impose serious performance restrictions for currently proposed isochronous storage rings.

I. INTRODUCTION

It has been suggested that ultra-short electron beam bunches can be stored in a quasi-isochronous storage ring whose momentum slip factor η is designed to be very small [1]. Since the peak current is high, the collective instabilities are one of the limiting factors in the operation of the quasi-isochronous storage rings.

In discussing the collective instabilities, we need to distinguish different regimes according to the relative magnitudes of the period of the synchrotron oscillation τ_{syn} and the radiation damping time τ_{rad}. These are given by [2]

$$\tau_{\text{syn}} = 2\pi \sqrt{\frac{2\pi R E_0}{\eta c e \omega_{\text{rf}} V_{\text{rf}}}}, \tag{1}$$

where $2\pi R$ is the storage ring circumference, E_0 is the design particle energy, c is the speed of light, e is the electron charge, ω_{rf} and V_{rf} are the angular frequency and integrated voltage per revolution of the rf cavities, and

$$\tau_{\text{rad}} = \frac{4\pi}{cC_\gamma E_0^3 <G^2>}, \tag{2}$$

---

[1] Work supported by the U.S. Department of Energy under Contract No. DE-AC03-76SF00515.

[2] Work supported by the U.S. Department of Energy under Contract No. DE-AC03-76SF00098.

where $C_\gamma = 8.85 \times 10^{-5}$ m-GeV^{-3}, $G = 1/\rho$, ρ is the bend radius, and the angular brackets imply taking the average over the ring circumference.

In the regime where

$$\tau_{\text{syn}} \ll \tau_{\text{rad}}, \qquad (3)$$

the instability mechanisms are the usual microwave instabilities, discussed extensively in the literature (for a review, see [3]). A storage ring for which the inquality (3) is satisfied will be referred to as the conventional storage ring in this paper.

The synchrotron oscillation period is proportional to $1/\sqrt{\eta}$. When η becomes sufficiently small, therefore, we have the opposite regime

$$\tau_{\text{syn}} \gg \tau_{\text{rad}}. \qquad (4)$$

In this regime the internal longitudinal motion of particles in the bunch can be neglected. A storage ring for which the inquality (4) is satisfied will be referred to as the isochronous storage ring in this paper. In an isochronous storage ring, the usual analysis of microwave instabilities breaks down. The instability mechanisms are now replaced by the linac collective effects and the analyses have also to be replaced.

In this note, we consider the collective instabilities in an isochronous storage ring for the case when the wake fields are short-ranged so that we have to worry only about single-bunch, single-turn wake fields. We further simplify the analysis by employing a two-particle model.

The damping time is different for the longitudinal and transverse oscillation. The radiation damping time τ_{rad} in the above should therefore be interpreted as the energy damping time $\tau_{\text{rad,E}} = \tau_{\text{rad}}/J_E$ in discussing the longitudinal synchrotron oscillation and as the betatron (horizontal or vertical) damping time $\tau_{\text{rad},\beta} = \tau_{\text{rad}}/J_\beta$ in discussing the transverse betatron oscillations. Here the quantities Js are the damping partition numbers given by $J_E \approx 2$ and $J_\beta \approx 1$.

The longitudinal and transverse collective effects for isochronous storage rings are discussed in section II. Two numerical examples, one for a Φ-factory and another for a FEL, are included in Section III.

II. ANALYSIS IN ISOCHRONOUS STORAGE RING

We consider first the collective effects in an isochronous storage ring, where the inequality (4) is valid.

A. Head-Tail Energy Split

The main longitudinal collective effect in a linac, and in an isochronous storage ring, is to cause a head-tail energy split. We consider a two-particle model. Let the bunch head and the bunch tail have longitudinal coordinates

$z = \frac{1}{2}\ell_z$ and $z = -\frac{1}{2}\ell_z$ relative to the bunch center. The equation for the energy error of the bunch head is

$$\dot{\delta}_{\text{head}} = -\frac{2}{\tau_{\text{rad,E}}}\delta_{\text{head}} + \frac{eV_{\text{rf}}\omega_{\text{rf}}\ell_z}{4\pi R E_0}. \tag{5}$$

The energy distribution of the stored beam reaches an equilibrium by a balance between radiation damping and the energy gain from the rf cavities. The bunch head then has an equilibrium energy error

$$\delta_{\text{head}} = \frac{eV_{\text{rf}}\omega_{\text{rf}}\ell_z \tau_{\text{rad,E}}}{8\pi R E_0} \tag{6}$$

Similarly, the equation for the bunch tail energy error is [3]

$$\dot{\delta}_{\text{tail}} = -\frac{2}{\tau_{\text{rad,E}}}\delta_{\text{tail}} - \frac{eV_{\text{rf}}\omega_{\text{rf}}\ell_z}{4\pi R E_0} - \frac{Nr_0 c W_0'}{4\pi R \gamma} \tag{7}$$

where N is the number of electrons in the bunch, r_0 is the electron classical radius, W_0' is the longitudinal wake function (created by the bunch head and seen by the bunch tail) integrated over the storage ring circumference, and γ is the Lorentz energy factor of the stored electrons. In equilibrium, we have

$$\delta_{\text{tail}} = -\frac{eV_{\text{rf}}\omega_{\text{rf}}\ell_z \tau_{\text{rad,E}}}{8\pi R E_0} - \frac{Nr_0 c W_0' \tau_{\text{rad,E}}}{8\pi R \gamma} \tag{8}$$

The head-tail energy split is therefore, using Eqs.(6) and (8),

$$\Delta\delta = \delta_{\text{head}} - \delta_{\text{tail}} = \Delta\delta_{\text{rf}} + \Delta\delta_Z, \tag{9}$$

where

$$\Delta\delta_{\text{rf}} = \frac{eV_{\text{rf}}\omega_{\text{rf}}\ell_z \tau_{\text{rad,E}}}{4\pi R E_0}, \tag{10}$$

and

$$\Delta\delta_Z = \frac{Nr_0 c W_0' \tau_{\text{rad,E}}}{8\pi R \gamma} \tag{11}$$

The first term, $\Delta\delta_{\text{rf}}$, is independent of the wake fields, and is there to maintain an equilibrium energy distribution of the bunch. The second term, $\Delta\delta_Z$, is wake dependent. In the design of the isochronous storage ring, one needs to make sure $\Delta\delta$ is within tolerance.

Equation(10) can be written as

$$\eta c \tau_{\text{rad,E}} \Delta\delta_{\text{rf}} = 2\pi^2 (\tau_{\text{rad,E}}/\tau_{\text{syn}})^2 \ell_z. \tag{12}$$

The LHS is the distance a particle with an energy error $\Delta\delta_{\text{rf}}$ travels in one damping time, which is much smaller than the bunch length ℓ_z in the present case where Eq.(4) is valid. Thus the relative motion of the particles in the bunch can indeed be neglected. It remains however necessary to design the isochronous storage ring in such a way that $\Delta\delta_{\rm rf}$ is within the momentum aperture of the ring.

Assuming the longitudinal wake is due to a longitudinal impedance $Z_0^\parallel/n$, then

$$W_0' \approx \frac{cR}{b^2}\frac{Z_0^\parallel}{n} \qquad (13)$$

where b is the storage ring beam pipe radius, and n is the impedance frequency in units of revolution frequency. The contribution to the head-tail energy split due to the wakefield then reads

$$\Delta\delta_Z \approx \frac{Nr_0 c\tau_{\rm rad,E}}{2b^2\gamma}\frac{1}{Z_0}\frac{Z_0^\parallel}{n} \qquad (14)$$

where $Z_0 = 377\ \Omega$.

B. Beam Break-Up Instability

The main transverse collective effect in a linac is to cause a beam break-up instability. In the beam-break-up instability, the bunch head executes a simple betatron oscillation without being affected by the wake fields. The bunch tail sees the wake field left behind by the bunch head, and is driven resonantly by it. Using a two-particle model, the betatron oscillation of the bunch tail grows by a factor of Υ per turn, where [3]

$$\Upsilon = -\frac{Nr_0 W_1 \beta_Z}{4\gamma} \qquad (15)$$

with W_1 the transverse wake function integrated over the storage ring circumference, and β_Z the β-function at the location of the impedance.

For stability, the growth of the bunch tail must be suppressed by radiation damping. This leads to the stability criterion

$$\frac{T_0}{\tau_{\rm rad,\beta}} > \Upsilon \qquad (16)$$

where T_0 is the revolution period.

We assume the wake fields are produced by a transverse impedance $Z_1^\perp$. For a short bunch with $\ell_z \ll b$, the wake field seen by the bunch tail (which trails the bunch head by a distance ℓ_z) is proportional to ℓ_z. The transverse wake function W_1 seen by the bunch tail is approximately

$$W_1 \approx -\frac{c\ell_z}{b^2}Z_1^\perp \qquad (17)$$

For our purpose, it is more convenient to relate $Z_1^\perp$ to the longitudinal impedance $Z_0^\parallel/n$ by the approximate relation

$$Z_1^\perp = \frac{2R}{b^2} \frac{Z_0^\parallel}{n} \tag{18}$$

Combining Eqs.(16), (17) and (18) then gives

$$\frac{Z_0^\parallel}{n} < Z_0 \frac{\gamma b^4}{cNr_0 \beta_z \ell_z \tau_{\mathrm{rad},\beta}} \tag{19}$$

It should be mentioned that the short length of the bunch helps reducing the beam break-up effect. This, as seen in Eq.(17), is due to the fact that the transverse wake field is smaller for short bunches.

III. NUMERICAL EXAMPLES

A. UCLA Φ Factory

A design of a high luminosity Φ factory based on a small value of η is described in [4]. The ring consists of four cells, each containing two 47 degree beding sections of $\rho = 0.425$ m and one -4 degree inverted bending section of $\rho = -1.7$ m. By controlling the dispersion in the inverted bending section, the momentum slip factor η is variable between -0.005 and 0.008. Other relevent parameters are $2\pi R = 32.7$ m, rf frequency=499 MHz, $eV_{\mathrm{rf}} = 0.1$ MV, and $E_0 = 0.51$ GeV. With six bunches in the ring, each with $N = 1.33 \times 10^{11}$ electrons, and $\beta^* = \sigma_z = 0.4$ cm, where β^* is the beta function at the interaction point, the luminosity becomes 1.6×10^{33} /cm^2/s.

We find $\tau_{\mathrm{rad,E}} \approx 3.7$ ms. If the ring were conventional, one would calculate the energy spread according to

$$\sigma_\delta = \sqrt{C_q \frac{<G^3> \gamma^2}{J_{\mathrm{E}} <G^2>}}, \tag{20}$$

where $C_q = 3.84 \times 10^{-13}$ m. One then obtains $\sigma_\delta = 6.7 \times 10^{-4}$. The momentum slip factor necessary for $\sigma_z = 0.4$ cm is, using the expression

$$\sigma_z = \frac{\eta c \tau_{\mathrm{syn}}}{2\pi} \sigma_\delta = \sqrt{\frac{2\pi R \eta c E_0}{e \omega_{\mathrm{rf}} V_{\mathrm{rf}}}} \sigma_\delta. \tag{21}$$

is found to be $\eta = 2.2 \times 10^{-3}$. Inserting this into Eq.(1), we obtain $\tau_{\mathrm{syn}} = 56\mu$s. Therefore the ring in this parameter regime is conventional although the bunch length σ_z is likely to be much shorter than the pipe radius b.

For a conventional ring with very short bunches, it is conceivable that one may obtain the microwave instability criterion by using the result for an isochronous ring, Eq.(19), but with $\tau_{\mathrm{rad},\beta}$ replaced by τ_{syn}. One then obtains the threhold of the transverse microwave instability given by

$$\frac{Z_0^{\parallel}}{n} < Z_0 \frac{\gamma b^4}{cNr_0\beta_Z\sigma_z\tau_{\text{syn}}} \qquad (22)$$

With similar substitution to Eq.(14), we obtain the contribution to the energy spread due to the impedance effect in a conventional storage ring:

$$\Delta\delta_Z \approx \frac{Nr_0c\tau_{\text{syn}}}{2b^2\gamma} \frac{1}{Z_0} \frac{Z_0^{\parallel}}{n} \qquad (23)$$

We assume $Z_0^{\parallel}/n \approx 0.2\Omega$, which is the value obtained in the ALS. Taking $b = 2$ cm, we obtain from Eq.(23) that $\Delta\delta_Z = 4.2 \times 10^{-3}$. Thus the impedance contribution to the energy spread is about seven times larger than the natural energy spread, implying that the bunch length will be seven times longer than the zero current value of 0.4 cm. On the other hand, the RHS of Eq.(22) using $\sigma_z = 2.8$ cm is about 7 Ω; the transverse microwave effect is not important although strictly speaking, Eq.(22) no longer applies because σ_z is now comparable to b.

B. A Proposed Quasi-Isochronous Ring at ETL

An experimental ring, possibly for FEL application, to be built at Electrotechnical Laboratory in Japan was proposed recently [5]. The general idea of the ring is similar to that discussed in the above, each cell containg two 49 degree degree beding sections of $\rho = 1.5$ m and one -8 degree inverted bending section of $\rho = -10$ m. By controlling the dispersion in the inverted bending section, the momentum slip factor as small as $\eta = 2. \times 10^{-7}$ is contemplated. Other parameters are $2\pi R = 82.4$ m, rf frequency=502 MHz, $eV_{\text{rf}} = 1$ MV, and E_0 up to 1.5 GeV. If we use the formulae for conventional ring, one obtains $\sigma_z = 51\mu$m. However, we also obtain $\tau_{\text{syn}} = 5.1$ ms which is longer than $\tau_{\text{rad,E}} \approx 1.25$ ms and $\tau_{\text{rad},\beta} \approx 2.5$ ms. Thus, the ring is in the isochronous regime in which case the short bunch length is not the result of the radiation damping. It must be injected from the beginning.

The energy acceptance of the ring is about 1%. Taking $\ell_z \approx 2\sigma_z$, we find from Eq.(10) that $\Delta\delta_{\text{rf}} = 1.6 \times 10^{-3}$ is negligible. Assuming again that $b = 2$ cm and $Z_0^{\parallel}/n \approx 0.2\Omega$, we compute from Eq.(14) the number of electrons per bunch N corresponding to the case $\Delta\delta_Z$ equals the enegy acceptance 1%, and find $N = 4.2 \times 10^{10}$. This would be more than what would be required for most applications. With this value of N, the inequality (24) becomes $0.2\Omega < 19.6\Omega/\beta_Z[\text{m}]$. Since the average value of the horizontal or vertical β function of the ring is less than 20 m, this inequality is also easily satisfied, provided $Z_o^{\parallel}/n$ is controlled to 0.2 Ω.

Acknowledgements

We thank D. Robin and H. Ohgaki for discussions on parameters of quasi-isochronous rings.

REFERENCES

1. Claudio Pellegrini and David Robin, Nucl. Instr. Meth. Phys. Res. A301, 27 (1991).
2. See, for example, M. Sands,"*The Physics of Electron Storage Rings, An Introduction*", SLAC preprint, SLAC-121 (1970).
3. Alexander W. Chao, "*Physics of Collective Beam Instabilities in High Energy Accelerators*", John Wiley, New York, 1993.
4. A. Amiry, C. Pellegrini, E. Forest, and D. Robin, Particle Accelerators, Vol-44, 65 (1994)
5. H. Ohgaki, D. Robin, and T. Yamazaki," Quasi- Isochronous Storage Ring for Enhanced FEL Performance", paper submitted to the FEL95, New York, N.Y., August(1995)

Dynamics of Microbeams from a Point Emitter and its Array

S.Kawasaki
Faculty of Science, Saitama University, 255 Shimo-ohkubo, Urawa 338 Japan

H.Ishizuka
Fukuoka Institute of Technology, Higashi-ku, Fukuoka, 811-02 Japan

M.Shiho
Japan Atomic Energy Research Institute, Naka, Ibaraki 311-01 Japan

K.Yokoo
Research Institute of Electrical Communication, Tohhoku Univ. 2-1-1 Katahira, Aoba-ku, Sendai 980 Japan

(Presented by S.Kawasaki)

ABSTRACT

Electron Beams emitted from a point field emitter and its array are proposed to be a promising particle source for the advanced use of the beam where the brightness is required to be as large as possible. Described are the problems to be studied related to the generation and propagation of the beam such as the tunneling from the inside of the cathode material to vacuum, beam dynamics including space charge effect associate to the high current intensity on the emitting surface and redistribution of the beam distribution function, and interaction between the beamlets coming from the array of the emitters. Results of the preliminary measurement of the intrinsic emittance of the beam are also given.

1. INTRODUCTION

In most of the cases in advanced use of particle beams the qualities of the beam in its collective behaviour take a decisive role in their operation. Linear collider[1] concept asks the beam extremely small emittance as well as a high peak intensity in order to obtain a brightness of the beam enough to cause reasonable counts on rare events. Similarly severe requirements are suggested to the qualities of the electron beams for FEL[2] and the heavy ions beams for inertial fusion program[3]. Especially the evolution of the (normalized) emittances as phase space area occupied by the beam particles has a crucial feature since it is not reversible anyway and necessarily tend to be deteriorated to grow up in the course of the beam handling beyond the initial value which the beams had at the particle sources[4].

Beam cooling[5] has given a considerable remedy to improve the situation by reducing the emittance of ion beams through the interaction with the external electron beam or feed-back circuit, while no useful technology of cooling for electrons has not been developed so far. Then, in the study of the operation of the particle beams the two main projects should be oriented:

(1) search and study of a particle source with small intrinsic eminence, and

(2) how to keep the emittance growth minimal (possibly zero) in the course of the sequence of the beam handling including the acceleration, deflection, bunching or debunching, splitting, etc.

Intensive studies on physics issues related to the problem (2) cited above, have been carried out recently[4] and detailed discussions on the mechanisms to cause the emittance growth have been presented in this workshop. We wish to present here a new concept of an electron source with an intrinsic high quality, promising for the injector of high energy accelerator or other advanced uses.

A field emission electron source is an old device, well known in the history of the study of the electron microscope. Recent rapid progress of a point emitter fabricated by Spindt[6] in 1968 opened a new scope of the technology with many novel aspects:

(1) it could be operated with the driving voltage between the cathode and the extracting electrode as low as 30 V, and with the emission intensity as high as 50 µA/emitter.

(2) the cathode is to be processed with the modern semiconductor technology with a very high accuracy of the fabrication, so that each cathode can be well regulated to show the uniform characteristics. Then simply a two dimensional array of the cathodes could easily multiply the current intensity many times without disturbing the original beam quality,

(3) various materials other than metals (semiconductors and even insulating materials) can be used as the emitter on the cathode tip, and it would enable to enhance the emission efficiency considerably,

(4) experimental observations have shown that the emission area is to be of the order of an atomic scale ($\sim 10 \text{Å}^2$) and therefore the intrinsic emittance should have an extremely small value. On the other hand, the fact causes the current density as high as 10^{12} A/cm^2 on the surface and the associated collective phenomena as the diffraction of the electron wave and plasma oscillations in the region close to the cathode, and finally

(5) when the array of the emitter is used to enhance the total current intensity, we will have many beamlets, where the spacial distance between the neighbors is the order of micrometer. The beam-beam interactions are of quite new features in the experimental and theoretical beam dynamics.

The typical form of the Spindt type cathode is shown in Fig. 1[7], where the single tip of the cathode and gate electrode is given as well as the array of the emitters and the electric circuit for extracting the beam.

From the early period of the history of the electron microscope a sharp point metal (especially molybdenum) cathode has been used frequently for the advantage of its small emitting area, where the electrons are extracted from the interior of the cathode by applying 10^3-10^4 V between the cathode and the gate electrode located far from the emitter. The extracted current is the order of tens nA. The fact that the operation of the cathode of this kind is very sensitive to the applied electric field prevents the use of multiple cathodes to increase the emission current.

The invention of Spindt cathode changed the situation completely. With the modern technology of the fabrication of the semiconductor device in submicron tolerance, the working voltage of the field emission diode now has been reduced down to 30 V, and tip to tip variation of the emission characteristics was suppressed to minimal level, so that an array of the emitters where they could be

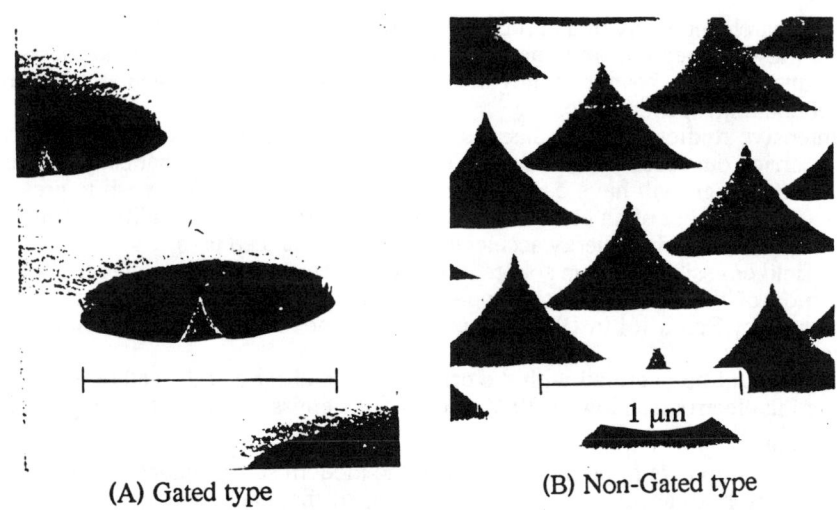

(A) Gated type   (B) Non-Gated type

Fig.1 (a) Electron microscope photograph of the Spindt type emitters

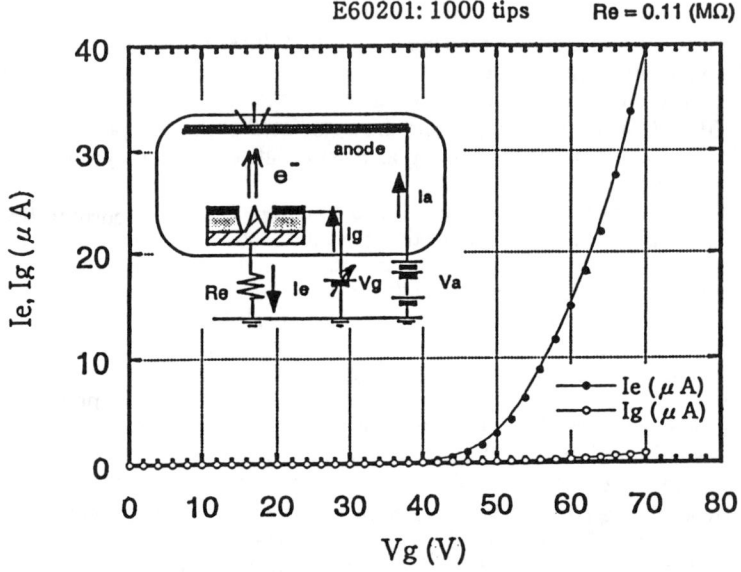

Fig.1 (b) I-V curve of the Spindt type emitter and the driving circuit

arranged in any desired pattern would bring a huge enhancement of the emission current. The technology also enabled to use various materials (pure or composite) other then metals as the emitter and it implies the possibility of getting much larger current density per single emitter. So far the (new) point emitters have been discussed mostly on the applications to as integrated microwave device, displays, ion sources and so on [8]. We would like to point out its promising use as the electron guns for the advanced use of the particle accelerators, such as the FELs, linear collider and tools for studying the internal dynamics of intense particle beams.

In the latter section we will describe these factors briefly and present results of a preliminary experiment carried out in our group.

2. ELECTRON EMISSION WITH TUNNEL EFFECT

When the electrons are emitted from a metal or semiconductor material, thay are generally assumed to be transferred from the conduction band structure inside them into outer vacuum region though a potential barrier formed at the surface, by a tunnel effect under the action of an intense electric field. The derivation of the tunnel current was calculated long ago with one dimensional geometry by Fowler and Nordheim[9]. The formula was revised and applied to various emitters with a quite rich variety of the parameters involved: emitter geometry, emitting materials, and processes of the fabrication and surface treatment. There remain, however, some factors which can not be determined with direct experimental observations such as the exact scale of emitters and emitting area, so that the formula might be somewhat empirical and its validity in analysing the experiments is rather limited. Many topics on the details of the tunneling mechanism are still unsettled and intensively discussed related to the new generation of the electron microscope (STM, FIM, TEM, etc.)

Since the emitting area of the point emitter is inferred to be ~10 $Å^2$, or nearly the site of one or a few atoms by fitting the simple F-N formula to the experiments, it is clear that the theory of the tunneling of the electrons should be reconsidered in a 3-D geometry and with a full quantum mechanical treatment of the problem other than WKB approximation generally used. Recent studies of the tunneling have confirmed that the tunneling current is to be really limited in a small region of a few $Å^2$ on the top of the emitter with an emitting angle of few degree[10]. In addition to the 3-D effect, an image potential produced by the emitted electrons themselves should be taken into consideration in the vacuum side. 3-D calculation of the image potential has shown that the curvature of the emitting point should reduce drastically the image effect compared with 1-D geometry, so that the intensity of the tunneling current could be much higher than expected by the simple F-N equation[11]. When the geometry and atomic composition of the emitter and the anode are determined, the remaining parameters dominating the emitter operation are the work function (or electron affinity in the semiconductor emitter) and the working temperature of the emitter.

Experiments have shown a variety of F-N characteristics depending on the experimental conditions, suggesting that there might be still unknown parameters. The details of the mechanisms of field emission from the point emitter are then to be investigated further. Excellent review written from the point of view of vacuum microelectronics including the history of the development, physics involved, fabrication technology, and applications envisaged was given by Brodie and Spindt[8].

3. PHYSICS ISSUES ON THE INTERFACE OF INTERIOR OF THE EMITTER AND OUTSIDE VACUUM

As stated above the F-N equation is derived with several assumptions based on a simple physical picture and gives a rather approximate formula. Many complex factors intervene therein and make difficult the exact comparison between the theory and experiments. Our concern is mainly on the use of the emitter as an electron gun with a very small emittance, or an extremely high brightness, implying that the beam should be transferred into the region far away from the emitter without being deteriorated in its quality. Therefore the problems to be studied are concentrated on (1) the limiting value of the intrinsic emittance of the beam and (2) how to reduce the growth to minimal in the course of the beam transport.

Tunnel effect is an essentially quantum mechanical phenomenon, and the dynamics and orbits of the electrons would be derived in the framework of the wave dynamics. The de Broglie wave length, however, gives the value much shorter than the atomic scale (<1Å) inside the surface, whereas it is much lager (>4000Å) in the vacuum side. The electrons will behave as a "single particle" on going through the potential barrier without any diffraction effect and coherence between the "particles", while they are to be considered as a wave of rather long wavelength in the outside of the surface, where naturally multi-particle coherence should be taken into account in the region of cathode-extracting electrode at least, to determine the "initial condition" of the beam. No detailed discussion on this subject was given so far.

4. BEAM DYNAMICS IN VACUUM SPACE

When the effect of the wave dynamics of the electrons in neglected, or can be evaluated somehow, the orbits in the vacuum side of the emitter surface could be derived with a classical treatment. We have to determine the initial distributions in phase space at first, and then trace their time and spatial evolution. The analysis is in the theoretical framework of the space charge dominated particle beam, which has been a target of intensive studies in the beam physics. Both analytical and numerical approaches are useful and various computer codes have been developed to apply for a variety of the combinations of accelerated beams and their environments: CW and pulsed beam in steady, periodic, or rf focusing fields. Many papers were presented on this workshop. Nevertheless, so far as we know, no situation similar to that of our concerns has not been treated. A computer code[12] is arrange to calculate the electron orbits near the cathode in a grid spacing as small as 0.5Å, the validity of the code to derivation of the actual particle orbits in our cases, however, is to be investigated further. For the "initial condition" of the emitting electrons, naturally the beam would not be in a K-V or other equilibrium distribution state, and the redistribution of the charge would lead to an inevitable growth of the emittance. Experimental study of the evolution of the beam state near the cathode is required to estimate the limiting value of the intrinsic emittance or brightness, while it seems very difficult.

5. EXPERIMENTAL STUDY OF EMITTANCE MEASUREMENT

Instead of diagnosing the evolution of the beam qualities in the emitter region, we have tried to measure the emittance of the beam for the first step, after it is extracted from the anode into free space and settled in an equilibrium state. A group of multiple (three) beamlets coming from an array of the emitters is injected into an electro-static quadrupole field. Before entering the quadrupole electrons pass through a solenoidal magnetic field to control the focal length and the radial size of the beamlets so as to be "matched". The schematic figure of the experimental device is shown in Fig. 2. The orbits and radial spreads of the beamlets are traced in various positions along the longitudinal axis of the device by observing the images of the beamlets on a fluorescent screen, with the parameter of the quadrupole field. Then the distribution of the beamlets in a phase space are obtained from the dependence of the orbits on the Q field intensity and the evolution of the image size, and thus these emittances are estimated. The measured (normalized) emittance is the order of 10^{-9} mrad. Minimum size of the images on the screen is of the same order with the distance between the images, suggesting that considerable emittance growth took place in the course of the beam propagation from the emitters to the observation point. Intrinsic emittance should be much smaller than the value measured. This technique of the measurement of the emittance was developed by H. Ishizuka et al[13]. Details of the process of the experimental analysis will be presented elsewhere.

6. DYNAMICS OF MULTIPLE BEAMLETS EMITTED FROM EMITTER ARRAY

Two experiments to measure the emittance or charge distribution of the beam in a phase space have been carried out. One is at LBL where Alkali-ions are transported in a series of periodic electro-static Q_s over several tens meters and the evolution of the beam envelope is observed to get the phase advance σ, and compared with σ_0 in the limit of small current intensity[14]. This approach is specifically useful and can give an exact measure of the beam quality when the beam is "matched" to at the entrance of the focusing system and the beam has been settled in an equilibrium. Changing the parameters of the focusing Q_s, σ will be estimated move exquisitely.

Another experiment was executed at Maryland University. Electron beams are introduced into an axially symmetric magnetic field produced by a long series of intermittent solenoidal coils and the envelope evolution of the beam was diagnosed with a fluorescent screen in various locations in the course of the beam propagation. A slit system with multiple circular holes is placed in front of the thermoionic cathode, and the effect of the spatial charge distribution on the emittance growth was measured, comparing with the computer simulation[15].

These kind of approaches could be combined to study the behaviour of the beam in our geometry. An array of the emitter would supply with various types of the spatial distribution as the initial condition, since the beam is composed of many beamlets naturally produced. A new experiment using these arrangements to compliment the results of the preliminary experiment cited in the previous section is being planned. An energy analyser with the resolution of 10^{-4} is being designed to measure the energy spread of the particles in longitudinal direction, with which

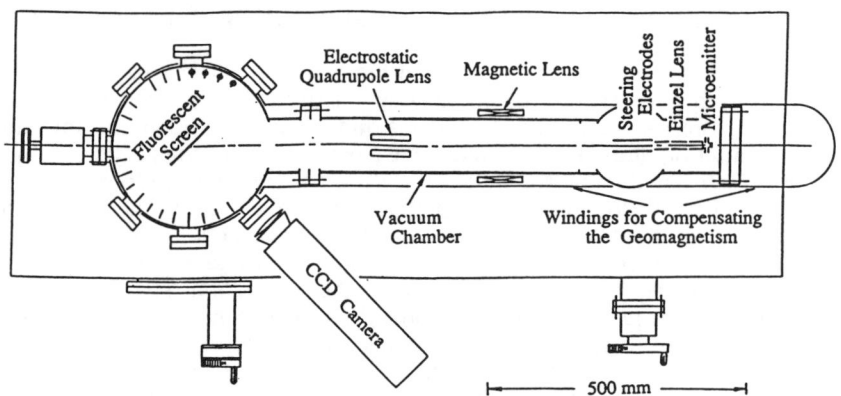

Fig.2 (a) Schematic figure of the experimental apparatus

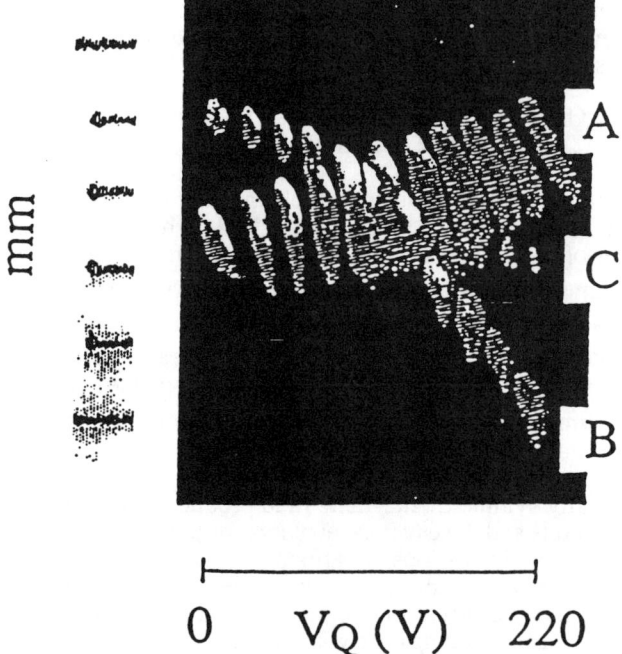

Fig.2 (b) Images of the beamelets on the screen with various field of electrostatic Q

the space charge effect in the point field emitter including the Boersch broadening[16] causing the emittance growth in the longitudinal direction would be investigated. We expect also to get the information (stability, equilibrium state, beam qualities and so on) on the interaction between the beamlets where the distance of the beamlet to beamlet is in the range of micrometer to millimeter.

7. CONCLUDING REMARKS AND FUTURE PLANS

(1) The field emitter fabricated with a modern semiconductor processing technology offers considerably attracting characteristics although it is still at the developing stage presently. It can generate an electron beam with a sufficiently long life and a good reliability, and with an extremely small emittance and quite high brightness.

(2) The emittance and brightness measured at the location far from the emitter after it is extracted through the gate are the order of 10^{-9} m-rad and $10^{12} A/m(m\text{-}rad)^2$. The intrinsic emittance could be much smaller. The emitters are to be promising for the use as the electron source for high energy accelerator in its advanced uses.

(3) Use of an array of the emitter with multiple electrodes would lead to their operation in a mode of much higher (total) intensity and improved stability. The bean intensity from the array can already compete with the level of any ordinary electron emitting cathode. The beamlets from the array has a regular spatial structure and it might profit the interaction with environment if the system of the transport and interaction (wiggler for example) can be arranged to be matched to the beamlets. Superlattice or multiple wigglers would have a spatial structure to accept the beamlets for producing intense channeling radiation or being a micro table-top FEL[17].

(4) The emitters present an ideal experimental tool for studying the theory on the internal dynamics of the intense charged particle beam. The beam diverges from anatomic to a sub micron scale in the emitter region and finally has a radius of a few millimeters. The separation between the beamlets also evolves in the range of the scale. No such situation has been investigated numerically or experimentally.

(5) The emitters could be involved in a plenty of interesting topics: applications to channeling radiation and Smith-Purcell effect, study of surface physics, improvement of the electron microscope, etc., where the beam quality should have the first impotence.

ACKNOWLEDGMENTS

A part o this study is supported by the collaboration program of Research Institute of Electrical Communication, Tohoku Univ. 2-1-1 Katahira, Aoba-ku, Sendai 980 Japan. The point emitters used in the experiments were fabricated in the Institute and supplied in the program.

REFERENCES

[1] On the recent liner collider concept see the followings and the papers cited therein. U. Amaldi, "Introduction to the Next Generation of Liner Colliders", CERN-EP 87-169, in: Frontiers of particle beams, eds. M. Month and S. Tuner, (Springer-Verlag, Berlin, 1988) p341.

P. B. Wilson, in Proc. UCLA Workshop on liner collider $B\overline{B}$ factory conceptual design, ed.. D. H. Stork (World Scientific, Singapore) p.373, 1990.

[2] Application of the field emitter to FEL was discussed by us:
H. Ishizuka, Y. Nakahara, S.Kawasaki, K. Sakamoto, A. Watanabe, N. Ogiwara and M. Shiho, "Ultrahigh-brightness microbeams: Consideration for Their Generation and Relevance to FEL", Proc. of 1993 Particle Accelerator Conference, IEEE p. 1556.

[3] C. Rubbia, Nucl. Instrum. Methods A **278**, 253 (1989).

[4] See the excellent text book:
M. Reiser, Theory and design of charged particle beams, John Wiley & Sons, 1994, Chap.6.

[5] G. I. Budker, Proc. International Symposium on Electron and Positron Storage Rings, Saclay, France, 1966, p II-1-1.
D. Mohl, G. Petrucci, L. Thorndahl, and S. van der Meer, "Physics and Technique of Stochastic Cooling", Phys. Rep. **58** (2) (1980).

[6] C. A. Spindt, "A Thin Film Field Emission Cathode", J. Appl. Phys. **39**, 3504 (1968).

[7] K. Yokoo, M. Arai, M. Mori, J. Gsuck and S. Ono, "Active Control of the Emission Current of Field Emitter Arrays", J. Vac. Sci. Tech. B13 (2) (1995) 491.
K. Yokoo, H. Shimawaki and S. Ono, "Proposal of A High Efficiency Microwave Power Source using A Field Emission Array", Tech. Digest of 6th Int. Vac. Microel. Conf. (1993) 153.

[8] See an excellent review on the development of the field emitter and vacuum microelectronics up to 1992:
I. Brodie and C. A. Spindt, "Vacuum Microelectronics", Advances in Electronics and Electron Physics, ed. by P. W. Hawkes, AP, Vol. 83, pp1-106, 1992.

[9] R. H. Fowler and L. W. Noldheim, "Electron Emission in Intense Fields", Proc. R. Soc. London A **119**, 173 (1928).

[10] B. Das and J. Mahanty, Phys. Rev. Lett., **B36**, 898 (1987).

[11] K. L. Jensen and E. G Zaidman, J. Vac. Sci. Tech, **B12**, 749 (1994).

[12] W. B. Hermannsfeldt, R. Becker, I. Brodie, A. Rosegreen and C. A. Spindt, High-Resolution Simulation of Field Emission, SLAC-PUB-5217 (1990).

[13] H. Ishizuka, A. Watanabe, M. Shiho, S. Kawasaki, J. Itoh, K. Yokoo, M. Arai, H. Shimawaki, "Beam Extraction Experiment with Field-Emission Arrays", Wel-15, 17th. Int. FEL conf. August 21-25, 1995, New York

[14] A. Faltens, D. Keefe, C. Kim, S. Rosenblum, M. Tiefenback, and A. Warwick, Proc. of the 1984 Liner Accelerator Conference, GSI-84-11 (Darmstadt, W. Germany), p312.
On the theoretical background also see J. Struckmeier and M. Reiser, Part. Accel. **14**, 227 (1988).

[15] M. Reiser, C. R. Chang, D. Kehne, K. Low, and T. Shea, Phys. Rev. Lett. **61**, 2933 (1988).

[16] H. Boersch, Z. Phys. **139**, 115 (1954).

[17] J. U. Andersem, E. Bonderup, and R. H. Pantell, Ann. Rev. Nucl. Part. Sci. **33**, 453 (1983).
V. V. Beloshitsky and F. F. Komarov, Phys. Rep. **93**, 117 (1982).

C. M. Tang, A. C. Ting and T. Swyden, Nucl. Instr. and Meth. **A 318**, 353 (1992).

Space Charge Effects in the Injector and Driver Accelerator for a UV FEL at CEBAF

H. Liu and D. Neuffer

CEBAF, 12000 Jefferson Ave., Newport News, VA 23606

Abstract A high-intensity CW photocathode electron injector test stand has been designed and is being built for a 200 MeV two-pass recirculating CW superconducting accelerator. The accelerator is being designed to drive a high average power (1 kW) UV free-electron laser (FEL). This FEL requires beam emittances as small as 11 mm mrad for the transverse normalized rms emittance and 30 deg-keV for the longitudinal rms emittance. The electrons contained in a 135 pC charged bunch are compressed from an initial bunch length of 90 ps (6σ) at the photocathode down to 1.2 ps (a factor of 75 compression) at the wiggler. In this paper, we present our studies on the space charge effects that determine the magnitude of beam emittance growth during the various stages of beam transport from the injector through the driver accelerator to the wiggler.

INTRODUCTION

A high-intensity CW photocathode electron injector test stand has been designed [1] and is being built for a 200 MeV two-pass recirculating CW superconducting accelerator, which is being designed [2] as a driver for a high average power (1 kW) UV free-electron laser (FEL) [3]. The overall layout of this machine is given in Fig. 1. Table 1 shows the stringent emittance requirements on the injector and driver accelerator.

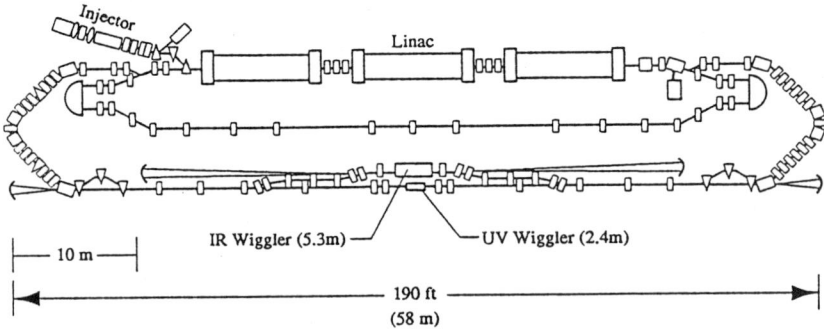

Fig. 1 Overall layout of the CEBAF UV FEL

© 1996 American Institute of Physics

Table 1 Requirements on the CEBAF UV FEL injector and driver accelerator

Beam parameters	@ injector exit	@ wiggler	Units
Momentum (p)	10	200	MeV/c
Charge/bunch (Q)	135	135	pC
Repetition rate (f_p)	37.43	37.43	MHz
Tran. norm. emittance (ε_{nrms})	8	11	mm mrad
Longitudinal emittance ($\varepsilon_{\phi rms}$)	20	30	deg-keV
Bunch length (σ_t)	1.5	0.2	ps

We present a brief description of the beam propagation in this machine. A CW train of electron bunches originate from a 500 kV DC electron gun with a photoemission cathode. They are compressed by a 1497 MHz room-temperature buncher, and then accelerated to ~ 10 MeV by a high-gradient CEBAF SRF cryounit containing two CEBAF-type SRF cavities. Two solenoid lenses in the 500 kV beam line region match the beam from the gun exit into the cryounit. An injection line consisting of a 4-quad zoom lens and a 3-dipole achromatic bending system match and deliver the beam from the cryounit exit into the driver accelerator. In the driver accelerator, the electron beam is accelerated from 10 MeV to 200 MeV using two passes of a 96 MeV linac consisting of three CEBAF cryomodules. The beam is then compressed and matched into the wiggler, where it loses energy (~0.5%) and greatly increases its energy spread. Then it is decelerated for energy recovery through two passes down to ~ 10 MeV, where the spent beam is transported into a dump.

One of the central design issues we have addressed is to preserve the beam emittances as much as possible while delivering the highest possible peak current to the wiggler by circumventing the space charge effects during the various stages of beam transport from the gun exit all the way to the wiggler. In this paper, we summarize our studies on this issue.

SPACE CHARGE EFFECTS IN THE DC LASER GUN

The high-voltage DC laser gun [4] is one of the most critical elements of this machine in the sense that it should produce inherently low emittance CW beams. This gun is designed for the nominal operating parameters of 500 kV voltage, 10 MV/m DC field gradient at the cathode, 135 pC charge/bunch, 90 ps (6σ) initial bunch length, and 5 mA average current. Simulations [1, 5, 6] have shown that it can produce beams with a scaled normalized transverse rms emittance of 1.5 mm mr/100 pC. The initial thermal emittance is about 0.3 mm mr for a small laser spot of 2 mm (diameter) at the cathode [1]. This factor-of-five emittance growth due to space charge indicates that space charge effects dominate the beam dynamics in the gun region.

Fig. 2 shows the dynamic properties of the beam in the nominal design of the DC laser gun as obtained in an 8000-particle PARMELA run simulating the first 10 cm of transport from the cathode [1]. All the symbols in the figure have their usual meaning except G_0 (geometric shape factor), k_p (plasma wave number), T_z/T_x (partition parameter), and $t_{dx,z}$ (space charge tune depressions [7], defined by $R_{x,z}$, the ratios of space charge to emittance in the x- and z-envelope equations [8]).

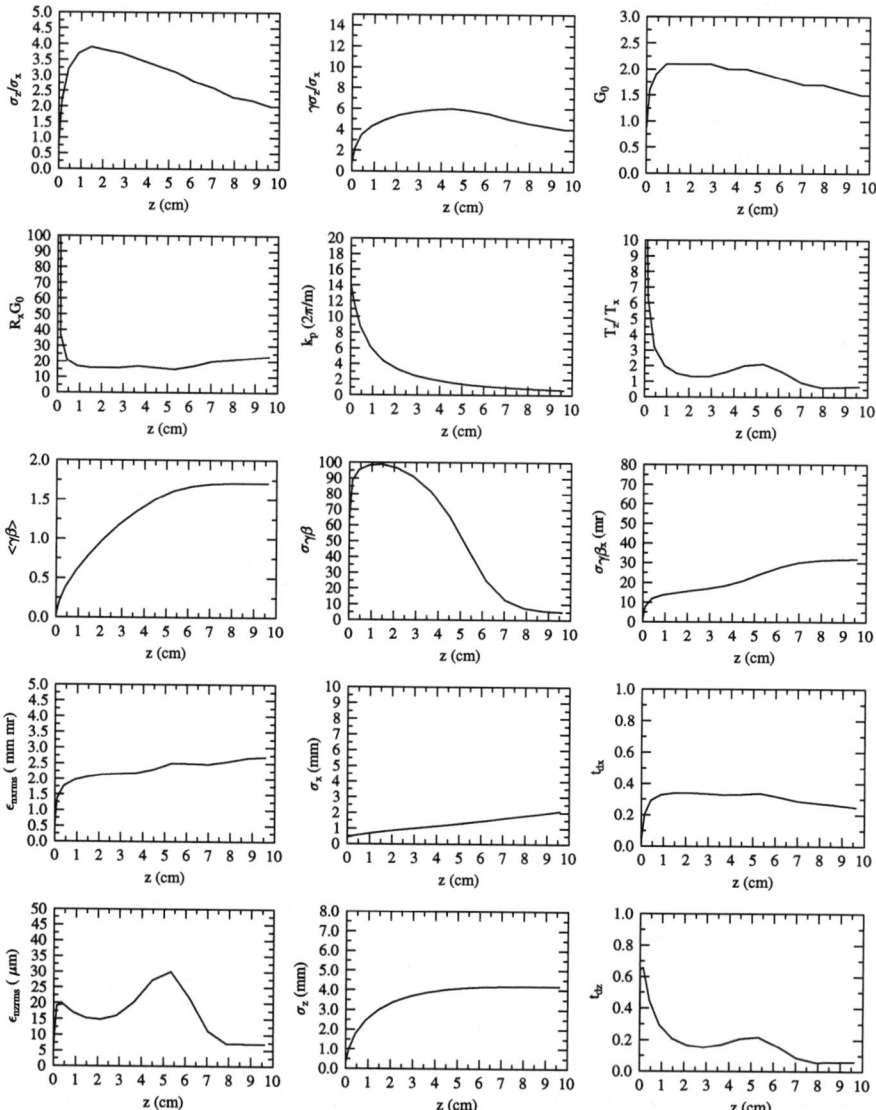

Fig. 2 Beam dynamic properties in the nominal design of the DC laser gun.

The emittance growth of the beam in the gun region may be estimated using the ballistic theory for RF laser guns [9] and the traditional free field energy theory for heavy ion and proton machines [10]. However, these two types of theories assume that the beam sizes are constant during beam propagation. In addition, the free field energy theory assumes constant energy. These assumptions may not be satisfied in our DC laser gun case, as is shown explicitly in Fig. 2. Therefore, we must use caution when applying these theories.

Examining the dynamic properties of the beam as shown in Fig. 2, we found that there is an emittance explosion within the first 5 mm, namely, from $z = 0$ to 5 mm, where z is the axial position of the beam centroid from the cathode surface. The same interesting phenomenon was shown by Lehrman et al. in modeling an RF laser gun using MAGIC [11]. It appears that we can divide the rapidly-varying beam dynamics in the gun region into an initial transient and a subsequent steady-state. We define the constants of motion assumed in the ballistic theory and the free field energy theory as those in the steady state.

A formula for calculating the steady-state transverse rms beam size has been derived [12]

$$\sigma_{xf}^3 = \sigma_{x0}^3 + \frac{6Nr_c}{8(E_0/0.511)^2}\left(1 - \frac{1}{\gamma_f}\right), \tag{1}$$

where σ_{x0} is the initial rms beam size at the cathode, N is the number of electrons in the bunch, $r_c = e^2/4\pi\varepsilon_0 mc^2 = 2.82\times 10^{-15}$ m is the classic electron radius, E_0 is the DC field gradient at the cathode in MV/m, and γ_f is the relativistic beam energy factor at the gun exit. Eq. (1) indicates that a high field gradient at the cathode surface causes less divergence for the beam moving out of the gun.

We generalized the ballistic theory for emittance growth in an RF laser gun to the DC laser gun case by ignoring the RF effect, which does not exist in a DC laser gun, as follows [12]

$$\varepsilon_{nx,\,zrms} = 30\,[\,\mathrm{acos}\,(1/\gamma_f)\,]\,(I/E_0)\,\mu_{x,\,z}(A), \quad (mm\ mrad), \tag{2}$$

where $I = Q/\sqrt{2\pi}\sigma_t$ is the peak current in A, E_0 the DC field gradient at the cathode in MV/m, $A = \sigma_{xf}/\sigma_{zf}$ the aspect ratio of the beam at the gun exit, and $\mu_x^{-1}(A) = 3A + 5$, $\mu_z^{-1}(A) = 1 + 4.5A + 2.9A^2$ for Gaussian distributions. Here we emphasize that σ_{xf}, σ_{zf}, and γ_f refer to the values at the gun exit.

The free energy theory says that when initial beam distributions are nonuniform, the excess energy carried within the beam will be converted into thermal energy (or emittance) through a rapid charge redistribution process (in one quarter plasma period) driven by nonlinear space charge fields, with the beam density distribution becoming uniform ultimately when all the free field energy has been consumed. According to this theory [10]

$$\varepsilon_{nxrms}^{(f)} = \sqrt{\left(\varepsilon_{nxrms}^{(i)}\right)^2 \left(1 + \frac{P_i - P_f}{2 + P_f}\right) + \frac{1}{2 + P_f}\frac{Nr_c\sigma_x^2 G_0}{5\sqrt{5}\gamma\sigma_z}\frac{U}{W_1}}, \quad (3)$$

$$\varepsilon_{nzrms}^{(f)} = \sqrt{\left(\varepsilon_{nzrms}^{(i)}\right)^2 \left(1 - \frac{2(P_i - P_f)}{P_i(2 + P_f)}\right) + \frac{P_f}{2 + P_f}\frac{Nr_c\sigma_z\gamma^3 G_0}{5\sqrt{5}}\frac{U}{W_1}}, \quad (4)$$

where U/W_1 is the normalized free field energy, and is equal to 0.308 for a Gaussian distribution and to 0.0368 for a parabolic distribution of spherical bunches,

$$G_0 = 1 + f(\sigma_z^2/\sigma_x^2 - 1), \quad (5)$$

is the shape factor of the bunch [8] where

$$f = \frac{1}{1-p^2} - \frac{p\arccos(p)}{(1-p^2)^{3/2}}, \quad p = (\sigma_z/\sigma_x) < 1 \quad (6)$$

for an oblate spheroidal bunch, or

$$f = \frac{1}{1-p^2} + \frac{p\,\mathrm{acosh}(p)}{(p^2-1)^{3/2}}, \quad p > 1 \quad (7)$$

for a prolate spheroidal bunch. In the near-spherical limit ($0.8 \leq p \leq 5$), $f \approx 1/3p$ is a good approximation. $G_0 = 1$ for a sphere. And

$$P = \frac{T_z}{T_x} = \frac{\sigma_x}{\varepsilon_{nx}\gamma\sigma_z}\varepsilon_{nz}, \quad (8)$$

is the so-called partition parameter.

It is interesting to note that the ballistic theory and the free field energy theory predict different dependences of the emittances on the total charge contained in a bunch. In the former case, the emittances are proportional to the peak current, whereas in the latter case, there are no simple general relationships between the emittances and the charge.

At the nominal design parameters for our DC laser gun [$N = 8.4\times10^8$ (135 pC), $E_0 = 10$ MV/m, $\gamma_f = 2$, $\sigma_{x0} = 0.5$ mm, and $\sigma_z = 4.5$ mm ($\sigma_t = 15$ ps)], we find from Eq. (1) that the beam size in the steady-state is $\sigma_{xf} = 1.33$ mm. We calculated the transverse and longitudinal emittances using both the ballistic theory and free field energy theory for which the assumption was made that the partition parameter of 0.5 is conserved from the cathode to the gun exit. The results are shown in Table 2. It is seen that both theories agree reasonably well with the simulation.

Table 2 Emittances at the gun exit from theory using $\sigma_{xf}= 1.33$ mm

Emittance	ballistic theory	free field energy theory	modeling
ε_{nxrms}(mm mr)	1.9	3.1	2.3
ε_{nzrms}(μm)	4.0	10.4	6.6

As is seen in Table 2, both the transverse emittance and longitudinal emittance predicted from the free field energy theory are larger than those from the simulation. In addition, if we use Eq. (4) instead of (8), the longitudinal emittance from the free field energy would be 30 μm instead of 10.4 in Table 2. To understand this, we checked the radial charge density distribution at the exit of the gun. As is shown in Fig. 3, the radial charge density distribution from the space charge run is fairly uniform, but not completely uniform, which means not all the free field energy in the radial direction has been consumed. The axial charge density distribution is slightly distorted from an initial Gaussian distribution [12], which means only a little amount of free field energy in the longitudinal direction is consumed in the gun region. We point out that the agreement between the free field energy theory and simulation depends on the amount of free field energy actually consumed in each direction, which generally is unknown analytically.

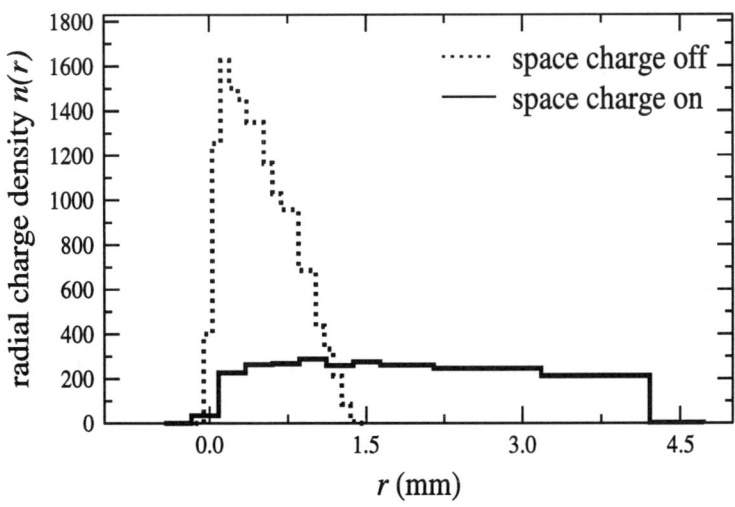

Fig. 3 Radial distributions of 8000 macroparticles at the gun exit.

PHASE SPACE BIFURCATIONS DRIVEN BY SPACE CHARGE

After the beam is extracted from the gun, the emittance may further increase due to space charge. In the straight beam line region from the gun exit to the exit of the cryounit, phase space bifurcation or filamentation is the major mechanism responsible for emittance growth [13]. Phase space bifurcation can be explained analytically [13, 14]. With the normalized spatial coordinates,

$$\bar{x}(s) = x(s)/r_0(s), \bar{y}(s) = y(s)/r_0(s), \bar{z}(s) = z(s)/l_b(s), \qquad (9)$$

where $s=vt$, v is the average axial velocity of the electrons, and $r_0(s)$ and $l_b(s)$ are the beam radius and bunch length at some instant. The equations of motion for betatron oscillation of an electron under linear external forces and arbitrary beam self-forces, including the acceleration terms, are

$$\gamma\beta^2\bar{x}'' + \bar{x}'\gamma' + k_x^2(s)\bar{x} - k_s^2(s)e_x = 0, \qquad (10.1)$$

$$\gamma\beta^2\bar{y}'' + \bar{y}'\gamma' + k_y^2(s)\bar{y} - k_s^2(s)e_y = 0, \qquad (10.2)$$

$$\gamma\beta^2\bar{z}'' + \bar{z}'\gamma' + k_z^2(s)\bar{z} - k_s^2(s)\alpha^{-2}e_z = 0, \qquad (10.3)$$

where all the derivatives are taken with respect to s, $\alpha = l_b/r_0$ is the beam aspect ratio, e_x, e_y and e_z are the space charge fields in the beam frame, normalized with respect to $E_0 = Q/2\pi r_0^2 \varepsilon_0$, and k_s is the plasma wave number defined by

$$k_s = \frac{d\phi_p}{ds} = \sqrt{\frac{2}{3}}\frac{2\pi}{\gamma\lambda_{p0}}, \qquad (11)$$

$$\lambda_{p0} = 2\pi\sqrt{r_0^3/3Nr_c} = 2.16\sqrt{r_0^3(mm)/N(10^6)}, \quad (m) \qquad (12)$$

where r_0 is the beam radius in mm, N the number of electrons in units of 10^6 and λ_{p0} the plasma period in m. Eq. (10.3) implies that for long beams ($\alpha \to \infty$) one can ignore the space charge effect in the longitudinal motion, whereas for slice or "pancake" beams ($\alpha \to 0$), the longitudinal space charge effect may outweigh the transverse space charge effect.

With a general charge distribution function, $\rho(r, z) = qNn_1(z)n_2(r)$, the transverse space charge field can be written

$$e_r = -\frac{1}{2}r\frac{\partial e_z}{\partial z} - 2\pi r_0^2 n_1(z) n_r(r), \tag{13}$$

where $n_r(r) = \frac{1}{r}\int_0^r n_2(r') r' dr'$ is the peripheral charge density distribution function which is equal to $n_2(0)r/2$ around the beam axis. It can be seen that the radial defocussing strength from the collective space charge fields differs from one longitudinal slice to another of the beam, and so does the betatron period. This is the underlying mechanism for phase space bifurcation.

Examining e_r/r in Eq. (13), we see that those electrons at a lower local axial density $n_1(z)$ and a smaller $\partial e_z/\partial z$ will be less affected by the space charge forces. Such electrons usually are those at the wings of the initial axial beam profile. According to Eq. (10), the motion of these electrons will be determined mainly by the external focussing, which means that their phase space distribution will always be linear; whereas the electrons residing in the main part of the beam experience nonlinear space charge forces and their phase space distribution will be determined by both external and internal forces and end up with a S-shaped signature. These separate groups of electrons will branch away from each other along the central linear part of the S-shaped phase space distribution, leading to phase space bifurcation. The bifurcated phase space distributions observed in our simulations are explained by these arguments [13]. A general prescription for reducing the degree of phase space bifurcation is to minimize the number of crossovers along a system (or the number of beam waists), and this has been used as one of the strategies in designing our injector [12].

SPACE CHARGE EFFECTS IN THE 10 MEV INJECTION LINE

A bending system (horizontal) introduces x-z correlations in the electron beam, resulting in a large growth of the effective emittance in the symmetry plane [15]

$$\Delta\varepsilon_{nxrms} = (\eta_c/f_x)\,\sigma_{\gamma\beta}(0)\,\sigma_x(0), \tag{14}$$

where η_c is the maximum dispersion at the symmetry plane, f_x the focal length of the first half of the system, $\sigma_{\gamma\beta}(0)$ the initial rms momentum spread which can be large when beam compression is desired through the bending system, and $\sigma_x(0)$ the initial rms beam size in the x-plane. This effective emittance growth is removed completely in an achromatic bending system. However, space charge may deteriorate the achromaticity by modifying the energy of the electrons, resulting in an incomplete cancellation of the x-z correlation, and a residual emittance growth may develop [15]. Therefore, it is desirable to minimize the dispersion, beam momentum spread, and beam size in order to ease the emittance cancellation through the achromaticity of the system.

In Ref. [15], we discussed some general principles for designing high-current beam injection lines for minimum beam quality degradation. Three versions of injection lines that can possibly be used for the driver accelerator were examined carefully and compared with each other. Simulations have shown that the current 3-dipole injection line design shown in Fig. 1 has the best emittance performance with the required bunching capability.

SPACE CHARGE EFFECTS IN THE FIRST SECTION OF THE LINAC

Our simulations show that the space charge and skew quad effects are important in the first few SRF cavities of the first cryomodule. The skew quad effect originates from the asymmetry of the accelerator structure. It can be minimized by minimizing the beam size, or be compensated by using a skew quad before the accelerating section. However, this compensation is incomplete in the presence of space charge.

Table 3 shows the skew quad and space charge effects on the emittance of the beam passing through the first cryomodule of the linac in the first pass, with the nominal injection beam conditions $\beta_x = \beta_y = 30$ m and $\alpha_x = \alpha_y = 0$ prior to the driver accelerator. It is shown that there is a large projected emittance growth due to the skew quad effect, even with space charge turned off. The emittance change can be compensated completely by a 15-cm long correction skew quad with a gradient optimized at 1 G/cm. However, with space charge turned on, the compensation is incomplete, leaving a net emittance growth of 1 mm mrad in the beam out of the first cryomodule in the first pass. This amount of emittance growth is significant in view of the fact that we have a very tight emittance budget from the end of the injector to the wiggler, and work is in progress toward optimizing the initial injection beam conditions before the driver accelerator in the hope that we can minimize the emittance growth in this part of the machine.

Table 3 Space charge and skew quad effects on emittance in the first cryomodule

ε_{nrms} @ entry	ε_{nrms} @ exit	correction skew quad	space charge
4.6 (mm mr)	7.5	0 (G/cm)	off
4.6 (mm mr)	4.7	1 (G/cm)	off
4.6 (mm mr)	5.5	1 (G/cm)	on

SPACE CHARGE EFFECTS IN THE ARCS

One potential source for additional emittance growth in this machine may come from those arcs in the first and second circulating passes. The possible physical effects may include space charge, wakefield and coherent synchrotron radiation, etc. A program has been set up to investigate these effects theoretically, numerically and experimentally [16].

ACKNOWLEDGMENT

We wish to thank J. Bisognano, Z. Li and C. Sinclair for helpful discussions. This work was supported by the Virginia Center for Innovative Technology and DOE Contract # DE-AC05-84ER40150.

REFERENCES

[1] H. Liu et al., "Design of a high charge cw photocathode injector test stand at CEBAF," presented at PAC-95., Dallas, Texas, May 1-5, 1995.
[2] D. Neuffer et al., "Accelerator design for the high-power industrial FEL," ibid.
[3] S. Benson et al., "A status report on the development of a high power UV and IR FEL at CEBAF," ibid.
[4] C. Sinclair, "A 500 kV photoemission electron gun for the CEBAF FEL," *Nucl. Instru. Methods* **A 318**, 410 (1992).
[5] H. Liu et al, "Numerical investigation of a laser gun injector at CEBAF," *Nucl. Instru. Methods* **A 339**, 415 (1994).
[6] H. Liu et al., "Modeling of space charge dominated performance of the CEBAF FEL injector," *Nucl. Instru. Methods* **A 358**, 475 (1995).
[7] M. Reiser, *Theory and Design of Charged Particle Beams,* (John Wiley & Sons, New York), 1994.
[8] I. Hofmann and J. Struckmeier, "Generalized three-dimensional equations for the emittance and field energy of high-current beams in periodic focusing structures," *Part. Accel.* **21**, 69(1987).
[9] K. Kim, "Rf and space charge effects in laser-driven RF electron guns," *Nucl. Instru. Methods* **A 275**, 201 (1989).
[10] T. Wangler and F. Guy, "The influence of equipartitioning on the emittance of intense charged-particle beams," *Proc. 1986 Linear Accel. Conf.*, SLAC Report 303, p. 340.
[11] I. Lehrman et al., "Design of a high-brightness, high-duty factor photocathode electron gun," *Nucl. Instru. Methods* **A 318**, 247 (1992).
[12] H. Liu, "Strategies for minimizing emittance growths in high intensity CW FEL injectors," to be published.
[13] H. Liu, "Phase space bifurcation observed in modeling a CW photocathode

FEL injector at CEBAF," *Micro-bunches workshop,* BNL, Sept. 28-30, 1995.
[14] O. Anderson, "Internal dynamics and emittance growth in space-charge-dominated beams," *Part. Accel.* **21,** 197 (1987).
[15] H. Liu and D. Neuffer, "Design principles for high current beam injection lines," presented at PAC-95
, Dallas, Texas, May 1-5, 1995.
[16] C. Bohn, "A program to research emittance growths in bends," *Micro-bunches workshop,* BNL, Sept. 28-30, 1995.

High Voltage High Brightness Electron Accelerator with MITL Voltage Adder Coupled to Foilless Diode

Michael G. Mazarakis, J. W. Poukey, D. Rovang, S. Cordova, P. Pankuch, R. Wavrik, D. L. Smith, J. E. Maenchen, L. Bennett, K. Shimp, and K. Law

Sandia National Laboratories
P. O. Box 5800
Albuquerque, NM 87185-1193

Abstract

The design and analysis of a high brightness electron beam experiment under construction at Sandia National Laboratory is presented. The beam energy is 12 MeV, the current 35-40 kA, the rms radius 0.5 mm, and the pulse duration FWHM 40 ns. The accelerator is SABRE [J. Corley, J. A. Alexander, P. J. Pankuch, C. E. Heath, D. L. Johnson, J. J. Ramirez, and G. J. Denison, in *Proceedings of the Eighth International IEEE Pulsed Power Conference*, San Diego, CA, 1991 (IEEE, New York, 1991), p. 920], a pulsed inductive voltage adder, and the electron source is a magnetically immersed foilless diode. This experiment has as its goal to stretch the technology to the edge and produce the highest possible electron current in a submillimeter radius beam.

Introduction

During the last 15 years, Sandia National Laboratories dedicated a considerable effort toward developing ultra high current and high brightness electron beams. The energy range was between 4 MeV and 20 MeV and the current was 30-100 kA. The accelerators utilized were single pulse devices, which can be divided into two groups: the single stage Blumlein-type accelerators such as Hermes II[1] and IBEX,[2] and the multistage devices such as RADLAC I,[3] RADLAC II,[4] RADLAC II/SMILE[5], and MABE.[6]

The electron source of choice was the magnetically immersed foilless diode.[7] This diode is ideally suited to produce high current and high brightness beams. The beams are generated and propagated in a strong axial magnetic field; thus, large amounts of current can be contained and tightly focused in small radius cross sections with relatively very small transverse velocities.

The present experiments were motivated by the success of converting RADLAC II into an inductive voltage adder fitted with a magnetically immersed foilless diode (RADLAC II/SMILE). The RADLAC II accelerator in its original configuration was an electron induction linear accelerator designed to produce ~ 100 kA, 16 MeV electron beams. The beam was produced by a magnetically immersed foilless diode located at the lower voltage end of the device (4-MV injector). It was then transported magnetically and further accelerated through the remaining six postaccelerating gaps of 2-MV accelerating voltage each. The SMILE modification converted RADLAC II into an inductive voltage adder similar to Hermes III[8] and SABRE.[9] The RADLAC II/SMILE configuration had a higher, 150 ohm, impedance which matched the original linac impedance. It proved that an inductive voltage adder can successfully be coupled to a magnetically immersed foilless diode to produce high quality electron beams. Annular beams of 50-100 kA, 1 cm radius were produced with very sharply defined 3-mm thick annulus and low transverse velocities ($\beta_\perp \approx 0.05$).

The beam experiments described here utilize the SABRE accelerator modified to a higher impedance voltage adder (~ 120 ohm) and fitted with a foilless diode immersed in a very strong (23 Tesla) solenoidal magnetic field.

In the next sections, the SABRE accelerator and the design which modifies it to a higher impedance voltage adder, the magnetically immersed foilless diode, and numerical simulation analysis of the produced e⁻ beam properties are presented.

SABRE Modifications

The SABRE accelerator (Fig. 1) design is based on the successful Hermes-III technology developed at Sandia during the last ten years in collaboration with Pulse Sciences Inc. this technology is fairly simple and couples the self-magnetically insulated transmission line (MITL)[10] principle with that of the induction linac[11] to generate a new family of linear induction accelerators, which we call linear inductive voltage adders. In these accelerators, the particle beam which drifts through the multiple cavities of conventional induction linacs is replaced by a metal conductor which extends along the entire length of the device and effectuates the voltage addition of the accelerating cavities. These devices can operate in either polarity to produce negative or positive voltage pulses. In a negative polarity voltage adder (Fig. 2), the center conductor is negatively charged relative to the outer conductor which is interrupted at regular intervals by the cavity gaps. SABRE in its original configuration is a relatively low impedance inductive voltage adder (Fig. 3). It was designed to operate in negative polarity at 40 ohm maximum output impedance and in positive polarity at approximately 20 ohm. It has 10 inductively insulated cavities each rated to maximum voltage of 1.2-1.3 MV. Ideally SABRE should be able to produce a 12-MV, 300-kA output in negative polarity and ~ 8 MV, 400 kA in positive polarity. Because of higher than expected energy losses in the pulse forming network, the operating input cavity voltage is of the order of 800 kV which limits the total output voltage to ~ 8 MV for negative polarity and 6 MV for positive polarity.

The modifications proposed here aim to increase the output voltage in both polarities and reduce the current proportionally since we do not plan to increase the total energy stored in the device. SABRE's main energy storage section consists of a Marx generator, two intermediate store capacitors and two energy transfer switches. Each switch controls the charging of ten pulse forming lines. Each cavity is fed by two 7.8-ohm pulse forming and transmission lines. The only major modification of the pulse forming network is the reduction of the total number of pulse forming and transmission lines into half (from 20 to 10). Thus, each cavity is fed by only one pulse forming line.

In addition, because each intermediate store will now feed power to only five pulse forming lines, the intermediate store-pulse forming line ringing ratio will substantially increase providing to the transmission lines and to the cavities a voltage pulse of ~ 50% higher amplitude (~ 1.2 MV). With this modification we have increased the impedance and the output voltage of each cavity. To take advantage of and maintain the voltage increase all the way to the output of the voltage adder, we must also increase the voltage adder impedance. To that effect a new smaller diameter cathode electrode was designed and constructed. Normally the impedance of the voltage adder is matched to that of the cavities; however, here to maintain an additional capability of further increasing the voltage output, the impedance of the voltage adder is 40% higher. To avoid exceeding 1.2 MV per cavity the Marx generator must be charged to lower voltage (85 kV instead of the 95 kV presently used) and the transfer switches must be triggered to close at a lower voltage level, 2-2.2 mV, instead of the usual 2.6 to 2.8 MV.

Figure 1. Photograph of SABRE accelerator

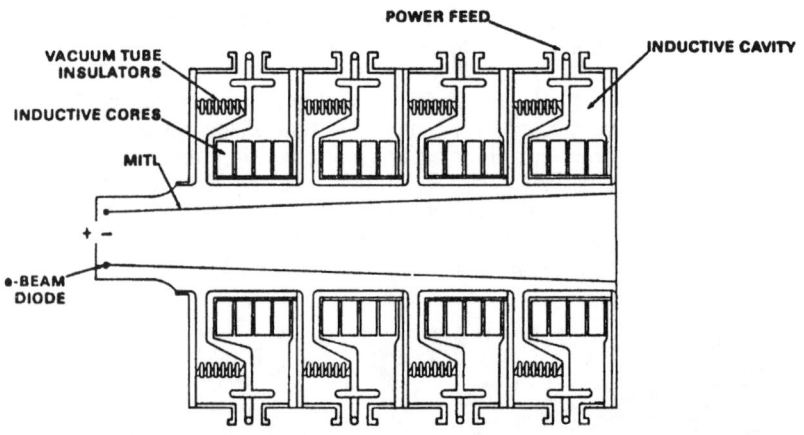

Figure 2. A simple negative polarity voltage adder of the SABRE type

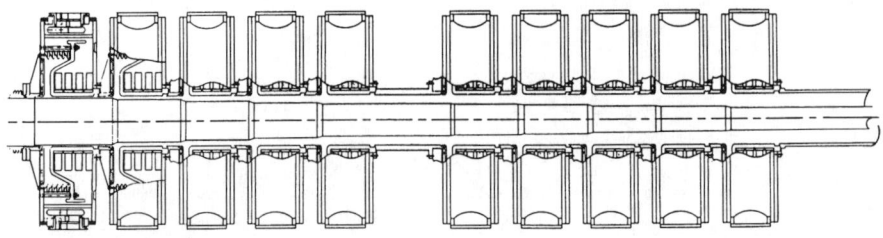

Figure 3. SABRE original low impedance voltage adder

The High Impedance MITL Design

The design of the high impedance voltage adder was done utilizing Creedon formalism.[10] It is based on a pulse forming line-fed self-magnetically insulated transmission line system which performs the series addition of voltage pulses from 10 cavity gaps (feeds). The cathode geometry is shown in Fig. 4. It is preferred over a continuous taper for the following reasons: it is easier and cheaper to manufacture, the constant radius segments provide constant vacuum impedance along each MITL segment, and the impedance increases gradually at each successive voltage feed with a rate of increase which follows the voltage axial gradient along the feed. The latter assures constant current flow over the entire length of SABRE.

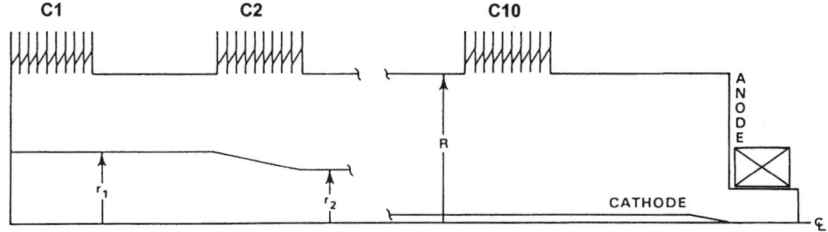

Figure 4. Schematic cross section of a high impedance voltage adder illustrating key design parameters

The vacuum impedance Z_i of each section i depends only on the dimensions of Fig. 4 and can be easily calculated from the following expression:

$$Z_i = 60 \ \ln(R/r_i)[\Omega]; \ i = 1, 2, \ldots 8 \tag{1}$$

where R = 19 cm and is the anode inner radius and r_i is the radius of the i-th cathode segment. The selection of the radius r_i of each cylindrical section was done in a fashion to provide the same operating load impedance for all of the SABRE cavity feeds. Some small variations were allowed for mechanical and construction reasons (see Table I for actual cathode radius).

The point design is for 110 kA and assumes equal 1.2 MV voltages at each cavity feed. Because of the relatively short voltage pulse (40-ns FWHM) of each feed, the current flow is self limited and, to a considerable extent, independent of the diode impedance conditions. However, in our design we stayed as close as possible to the constant current conditions all the way to the end of the cathode electrode. Near the upstream edge of the solenoid (Fig. 5) the cathode electrode smoothly tapers to accept the 1-mm cathode electrode of the immersed diode. Here the self-magnetic insulation is adiabatically replaced by the externally applied solenoidal field insulation. The anode inner radius is constant (19 cm) and defined by the existing SABRE anode cylinder. At first we determined the desired degree of overmatch of the MITL to the cavity feeds (~ 1.44). This in turn defined the operating impedance of each segment and the constant current I_ℓ of the voltage adder. With these initial parameters and Creedon equations for minimum current flow, I_ℓ, to establish self-limited magnetic insulation:

$$I_\ell = 8500 \ g\gamma_\ell^3 \ln\left[\gamma_\ell + (\gamma_\ell^2 - 1)^{1/2}\right],$$

$$\gamma_i = \gamma_\ell + (\gamma_\ell^2 - 1)^{3/2} \ln\left[\gamma_\ell + (\gamma_\ell^2 - 1)^{1/2}\right], \tag{2}$$

$$g = [\ln R/r_i]^{-1} \ \text{and} \ \gamma_i = V_i[MV]/mc^2 + 1 \ ,$$

we estimate the cathode radii and operating impedances of the entire new voltage adder for SABRE.

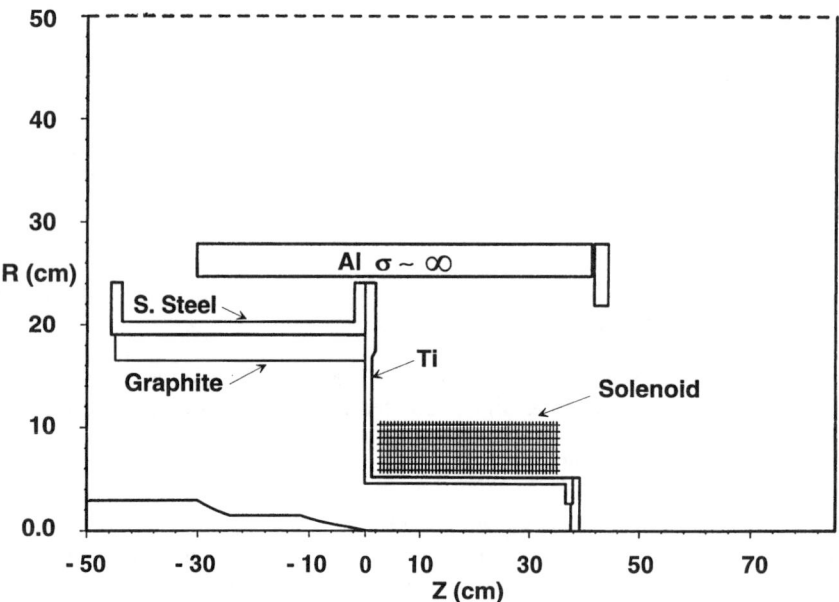

Figure 5. Schematic diagram of diode design and transition region

Our main concern was to keep the currents I_ℓ the same. The relativistic factor γ_ℓ is for electrons at the outer boundary of the electron sheath in the minimum current case. It can be approximated by the following formula which is tested to be correct[12] for up to 20 MV adders:

$$\gamma_\ell = \frac{12\ \gamma_i^{1/3}}{12 + \ell n \left[\dfrac{\gamma_i}{5.9314}\right]} \qquad (3)$$

Table I summarizes dimensions and design parameters. The cathode electrode (~ 9.2 m) includes the voltage adder section (6 m long) and a constant radius (2.2 cm) extension section (3.2 m long) and is cantilevered from the low voltage end of the accelerator. It starts with a 14-cm radius cylinder at the cathode end plate and tapers off to 2.2-cm radius after the 10th cavity gap (Fig. 6). Nine conical tapers were utilized along with 10 cylindrical sections and 11 flex-adjusting, double washer sections. The outer shell (anode cylinder) is formed by ten insulating stacks (feeds) alternating with ten cylinders, plus the final anode extension cylinder. The cathode electrode (Fig. 7) was preloaded before insertion into the anode cylinder to compensate for gravitational droop. The final adjustment was made in situ. Because of the large difference in radius between anode and cathode shank, precise alignment and centering of the cathode stock inside the anode cylinder is not very critical since the electrical potential is a logarithmic function of the radii.

Table I

Distance from cathode plate Z (cm)	Segment i (0 - 12)	Segment Voltage V_i (MV)	Cathode Radius R_c (cm)	Vacuum Impedance Z_i (Ω)	Operating Impedance Z_i (Ω)
0 - 33	0	0	13.65	20	11.6
37 - 80	1	1.2	13.65	20	11.6
84 - 127	2	2.4	10.95	33.2	22.6
131 - 174	3	3.6	8.41	49.0	35.3
178 - 221	4	4.8	7.14	58.9	44.4
225 - 366	5	6.0	5.715	72.2	55.5
370 - 413	6	7.2	4.76	83.2	65.0
417 - 460	7	8.4	3.81	96.6	76.6
464 - 507	8	9.6	3.175	107.5	86.3
511 - 554	9	10.8	2.54	120.9	98.1
558 - 900	10	12.0	2.22	129.0	105.5

R_{anode} = 19.05 cm

Figure 6. Line drawing of the new high impedance cathode electrode

Figure 7. Photograph of the cathode electrode. It is composed of 12 segments preloaded to compensate for the gravitation droop which is of the order of 20 cm.

Diode Design

The magnetically immersed foilless diode is similar to those of RADLAC II/SMILE and IBEX.[13] However, the impedance and solenoidal magnetic field are much higher. To generate beams of millimeter sizes the diode must be immersed in solenoidal fields of ~ 20 Tesla. A schematic diagram of the diode design including the solenoidal magnet is shown in Fig. 5. The anode and cathode electrodes are made of titanium because it is not magnetic material and has large resistivity, allowing the pulsed magnetic field to penetrate without appreciable losses.

The pulsed solenoids were designed and constructed in-house by our magnet team. They draw upon the large experience we acquired over many years of research and development for the high field coils of the inertial confinement fusion ion diodes. We have constructed four of them for redundancy since we do not know yet their life time. These solenoids are among the strongest ever built. The total magnetic energy stored is ~ 1.8 MJ. The inductance is 8 mH, the bore 12 cm, and length is 30 cm. The shape of the fringe field is tailored by a 2-cm thick aluminum cylinder of 25-cm inner radius coaxially enclosing the entire diode assembly. The diode assembly and solenoids have been tested successfully to 23 Tesla. The magnetic field profile and strength (Fig. 8) are in good agreement with numerical simulation predictions done with our magnet code "Atheta." [14] The agreement is within the estimated measurement errors of 10%.

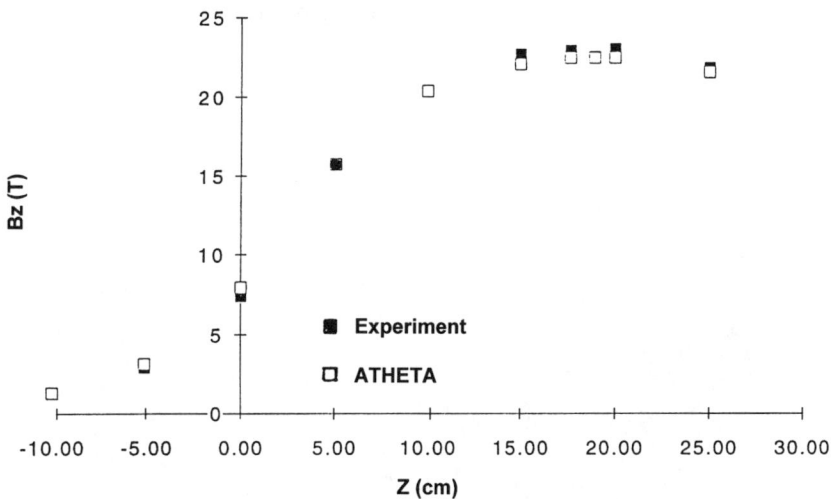

Figure 8. Measured and calculated magnetic field profile of the diode solenoid.

Simulation Results

The design of the MITL voltage adder (Table I) and the foilless diode (Fig. 5) were validated with a large number of TWOQUICK[15] particle-in-cell code simulations. Because of the large range in space and time scales, it was necessary to divide the entire design into three parts: voltage adder, from cavity feeds through the extension MITL to a (more or less) self-limited diode load; transition region where the coupling from MITL to immersed diode was studied; and finally, immersed diode where the beam generation and beam parameters were analyzed in fine detail. The optimum results are shown in Figures 9, 10, and 11. They go from the large scale of the entire SABRE voltage adder to the small scale of the immersed foilless diode and merge smoothly into one another.

Figure 9 shows an electron map at 60 ns following the arrival of the voltage pulse at the first cavity (t = 0). The line is magnetically insulated with the self-field (B_θ) of the current flowing along the voltage adder. Electron maps at earlier times (t = 20-30 ns) show some electron losses to the anode electrode. This is to be expected since self-limited magnetic insulated flow is established by driving some electron current to the anode wall during the rise time of the voltage pulse as it travels along the voltage adder. The anode-cathode gap of the planar diode at the end of the MITL is large, 20 cm, to allow operation in the self-limited mode. No applied magnetic field is assumed here. In this simulation, the cavity input voltage is a trapezoidal pulse of 1.2-MV peak value and 40-ns flat top (Fig. 9). Because the voltage adder is overmatched to the cavity impedance (12 ohm versus 7.8 ohm), the actual operating voltage pulse at the cavity gaps is higher, ~ 1.4 MV. These pulses combine and travel down the voltage adder toward the slightly undermatched diode. The final voltage is 13 MV and the total current is 135 kA. The estimated operating impedance of 96 ohms agrees with the parapotential theory and pressure balance theory within 10%.

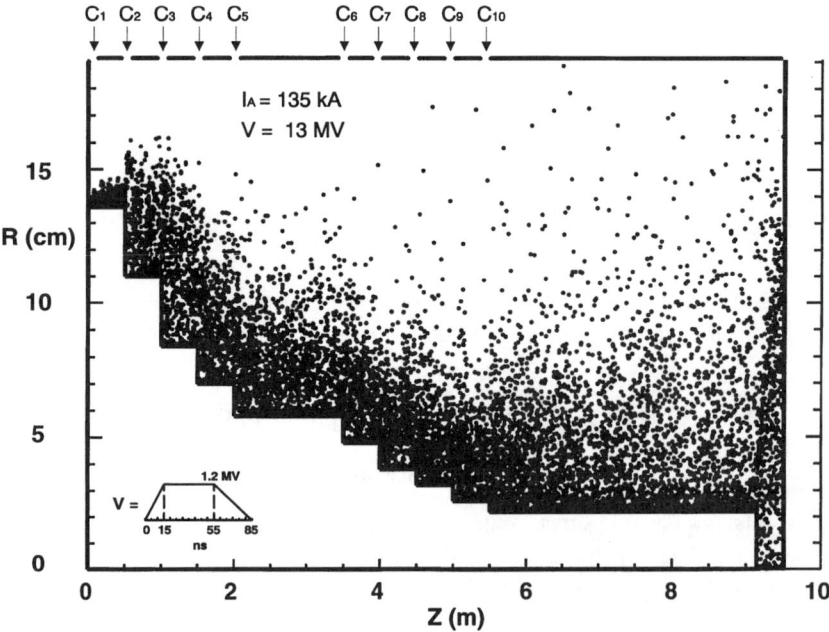

Figure 9. Electron map for the high impedance cathode electrode obtained with TWOQUICK PIC code at 60 ns following the arrival of the voltage pulse at the first cavity (t = 0).

Figure 10 is a simulation of the transition region. In this region all the sheath electrons are lost to the anode. The sum of the loss current and beam current is equal to the total current flowing along the voltage adder. Thus the transition region is the effective load for the SABRE MITL with an impedance equal to the MITL operating impedance (matched load). The location of the taper relative to the shape and strength of the solenoidal fringe field is very critical. Figure 10 shows the configuration that produces the highest brightness and lowest emittance beam. The magnetic field lines are also shown. The losses near the conically tapered section are due to the radial component B_r of the applied solenoidal magnetic field. They occur at the point where the self field B_θ becomes equal to the B_r component of the applied field. The applied B field was calculated for the actual coil configuration using the Atheta code which includes diffusion into the various materials based on their electrical conductivities. As seen in Fig. 10, the current splits, with about 70% striking the anode wall which is lined with a graphite insert to prevent activation of the stainless steel cylinder. The remaining 36 kA form a pencil-like beam of about 0.6-mm radius. The resolution of the simulation is not fine enough to give the precise beam parameters.

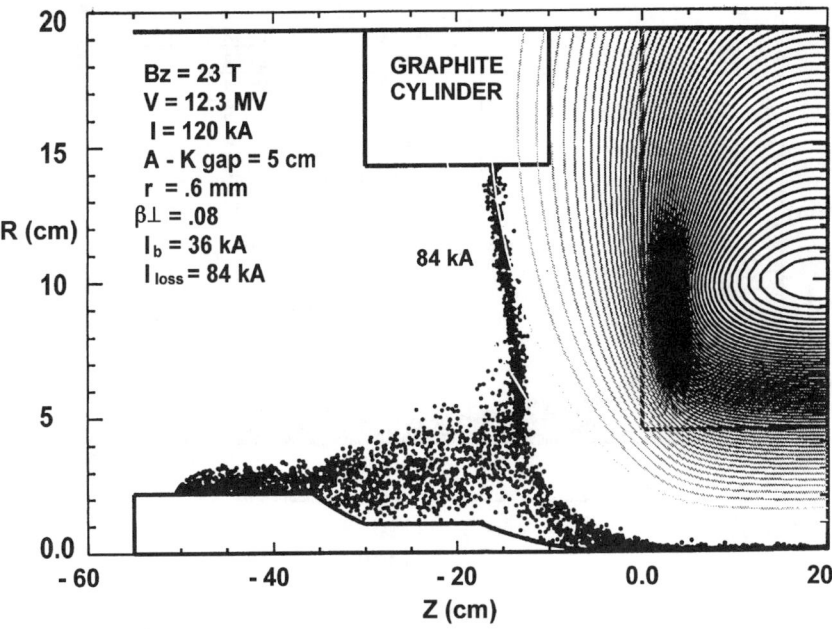

Figure 10. TWOQUICK simulation of the transition region. The losses near the conically tapered section are due to the radial component B_r of the applied solenoidal magnetic field.

The simulation of Fig. 11 was done with the above concern in mind, so only the immersed diode was included. The applied magnetic field B is the actual one, and the anode-cathode voltage is 12 MV. A beam of 36 kA with 0.44 mm rms radius is produced. This beam is the ultimate goal of the proposed experiments. This simulation represents an ideal situation assuming no cathode plasma radial expansion, perfect cylindrical symmetry without instabilities, and negligible beam perturbation due to possible beam-stop plasma blow offs. Some inconclusive scoping studies of those problems have been done up to now; however, only the experiments will tell how "ideal" is the simulated beam. According to the simulations, we should produce a beam with emittance $\varepsilon = \pi . r_b . \beta_\perp = 5 \; 10^{-5}$ m.rad. The brightness $B = 3.10^{13}$ A / (m.rad)^2 is calculated from the expression:

$$B = 2I / \varepsilon^2, \qquad (4)$$

where I is the beam current and ε the geometric emittance.

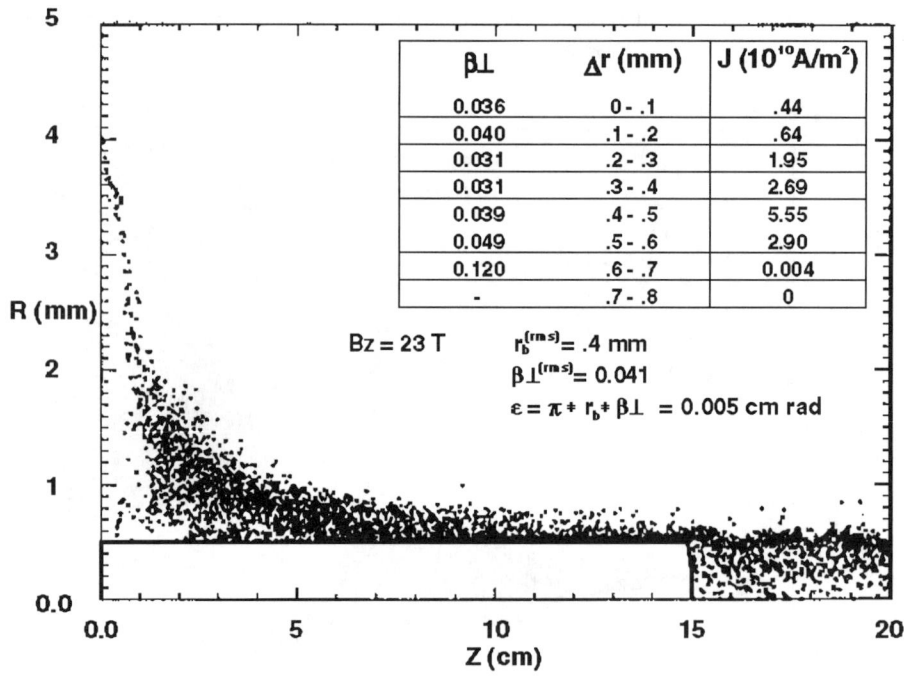

Figure 11. High resolution diode simulation evaluating the beam parameters.

Figure 12 is the v_θ, Z phase space. The maximum rotation velocity is close to the speed of light. Figure 13 provides an interesting insight and comparison between the electrons self-field and the applied field. The electrons exhibit a strong rotation at the transition region and around the cathode electrode with a net diamagnetic component; however, the beam rotates much slower, and its diamagnetism is smaller and becomes zero at the beam stop. The phase space of Fig. 12 corresponds to the motion of the electrons of Fig. 10: the electrons escaping radially from the cathode electrode initially have a strong rotation because they cross the magnetic field lines while accelerated by the radial electric field. At Z = +20 cm (end of cathode tip), the electric field becomes axial, the electrons accelerate forward and their v_θ decreases.

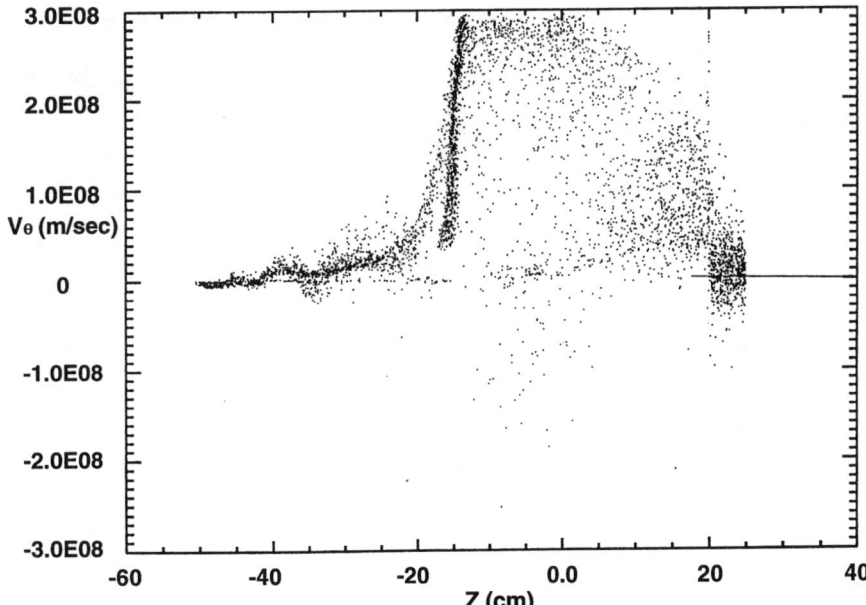

Figure 12. v_θ, z phase space of diode and transition region. It corresponds to the electron map of Figure 10. In the transition region and around the cathode tip the v_θ can be as high as the speed of light. However, in the beam (z = 20-25 cm) the electrons rotate much slower $\sim v_\theta = 0.10 \; v_c$.

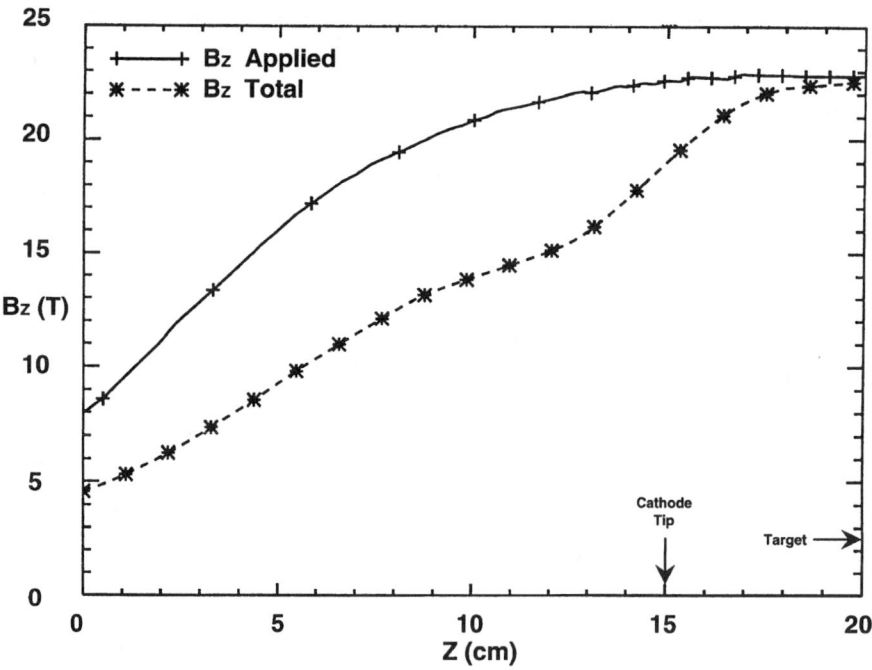

Figure 13. Applied and total B_z field at r = 0.5 mm (cathode tip radius).

The simulated beams are very tightly focused, depositing on the target very impressive amounts of energy per unit surface area. For instance, the current density, power density, and energy density are, respectively, 5.5 10^6 A/cm^2, 6.0 10^{13} W/cm^2, 2.4 10^6 J/cm^2. Also the energy absorbed by the surface of a heavy metal target is of the order of 6 x 10^5 J/gr. This makes our beams ideal for beam-target-plasma interaction studies.

Summary

We have designed and constructed an immersed diode and a high impedance voltage adder for SABRE which, if they perform as expected, should produce a very intense high brightness electron beam of millimeter size. Extensive numerical simulations and previous experience with RADLAC II/SMILE suggests that these beams should be achievable. Halo, radial plasma expansion at the cathode and instabilities may limit the minimum possible beam radius. Experimental verification of the design and numerical simulation predictions is planned for the beginning of 1996. Experimental results will be reported in future publication.

Acknowledgment

This work is supported by the U. S. Department of Energy under Contract No. DE-AC04-94AL85000.

References

1. T. M. Martin, *IEEE Trans. Nucl. Sci.* **NS-16**, 59 (1969).

2. J. J. Ramirez, J. P. Corley, and M. G. Mazarakis, *Proceedings of the 5th International Conference on High-Power Particle Beams*, San Francisco (Physics International, San Leandro, CA, 1983), p. 256.

3. R. B. Miller, K. R. Prestwich, J. W. Poukey, B. G. Epstein, J. R. Freeman, A. W. Sharpe, W. K. Tucker, and S. L. Shope, *J. Appl. Phys.* **52**, 1184 (1981).

4. M. G. Mazarakis, G. T. Leifeste, R. S. Clark, C. A. Ekdahl, C. A. Frost, D. E. Hasti, D. L. Johnson, R. B. Miller, J. W. Poukey, K. R. Prestwich, S. L. Shope, and D. L. Smith, *Proceedings of the 1987 IEEE Particle Accelerator Conference*, Washington, DC (IEEE, New York, 1987), p. 908.

5. M. G. Mazarakis, J. W. Poukey, S. L. Shope, C. A. Frost, B. N. Turman, J. J. Ramirez, and K. R. Prestwich, *Proceedings of the 8th International IEEE Pulsed Power Conference*, San Diego, CA, (IEEE, New York, 1991), p. 86.

6. D. E. Hasti, J. J. Ramirez, P. D. Coleman, C. W. Huddle, A. W. Sharpe, L. L. Torrison, *Proceedings of the 5th International IEEE Pulsed Power Conference*, Arlington, VA (IEEE, New York, 1985), p. 147.

7. L. M. Friedman and M. Ury, *Rev. Sci. Instrum.* **41**, 1334 (1970).

8. J. J. Ramirez, K. R. Prestwich, E. L. Burgess, J. P. Furaus, R. A. Hamil, D. L. Johnson, T. W. L. Sanford, L. E. Seamons, L. X. Schneider, and G. A. Zawadzkas, *Proceedings of the 6th International IEEE Pulsed Power Conference*, Arlington, VA (IEEE, New York, 1987), p. 294.

9. J. Corley, J. A. Alexander, P. J. Pankuch, C. E. Heath, D. L. Johnson, J. J. Ramirez, and G. J. Denison, *Proceedings of the 8th International IEEE Pulsed Power Conference*, San Diego, CA (IEEE, New York, 1991), p. 920.

10. J. H. Creedon, *J. Appl. Phys.* **48** (3), 1070 (1977).

11. N. Christophilos, R. E. Hester, W. A. S. Lamb, D. D. Reagan, W. A. Sherwood, and R. E. Wright, *Rev. Sci. Instrum.* **35** (7), 886 (1964).

12. T. W. L. Sanford, J. W. Poukey, T. P. Wright, J. Bailey, C. E. Heath, and R. Mock, *J. Appl. Phys.* **63** (3), 681 (1988).

13. M. G. Mazarakis, R. B. Miller, and J. W. Poukey, *J. Appl. Physics* **62** (10), 4024 (1987).

14. R. S. Coats, ATHETA/DATHETA User's Guide (Sandia National Laboratory document, Albuquerque, NM, 1995).

15. D. B. Seidel, M. L. Kiefer, R. S. Coats, T. D. Pointon, J. P. Quintenz, and W. A. Johnson, *Proceedings of the CP90 Europhysics Conference on Computational Physics*, Singapore (World Scientific, Singapore, 1991), p. 475.

On minimum emittance lattices[1]

S.Y. Lee

Department of Physics, Indiana University, Bloomington, IN 47405

General conditions for minimum emittance double bend achromat, single dipole minimum emittance, and the three bend achromat lattices are derived. A condition for the minimum emittance three bend (or multi-bend) achromat lattice is derived. The condition shows that the minimum emittance achievable is the same as that of the double bend achromat.

I. INTRODUCTION

The amplitudes of the betatron and synchrotron oscillations are determined by the equilibrium of the quantum excitation arising from the emission of photons and the rf acceleration fields used in compensating the energy loss of the synchrotron radiation. The horizontal emittance for an isomagnetic ring (constant bending radius) is given by [1]

$$\epsilon_x = C_q \gamma^2 \frac{\langle \mathcal{H} \rangle_{dipole}}{J_x \rho}, \tag{1}$$

where $C_q = 3.84 \cdot 10^{-13}$ m, γ is the relativistic energy factor, ρ is the bending radius, J_x is the damping partition number, and $\langle \mathcal{H} \rangle_{dipole}$ is the average of the dispersion action in dipoles with

$$\mathcal{H} = \frac{1}{\beta_x}[D^2 + (-\frac{\beta_x'}{2}D + \beta_x D')^2]. \tag{2}$$

Here β_x is the Courant-Snyder [2] betatron amplitude functions, D, D' are the dispersion function and its derivative. Since $\mathcal{H}$ is proportional to $L\theta^2$, where L is the length of the dipole and θ is the bending angle, the average of the $\mathcal{H}$ function obeys a scaling law:

$$\langle \mathcal{H} \rangle = \rho \theta^3 \mathcal{F}, \tag{3}$$

where the scaling factor $\mathcal{F}$ depends on the storage ring lattice arrangement. The design of low emittance optics is to minimize $\langle \mathcal{H} \rangle / J_x$ in dipoles.

For separate function (dipoles without field gradient) storage rings, the dispersion function in the dipole region is given by

$$D = \rho(1 - \cos\phi) + D_0 \cos\phi + \rho D_0' \sin\phi, \quad D' = (1 - \frac{D_0}{\rho})\sin\phi + D_0' \cos\phi,$$

---

[1] Work supported by grants from NSF PHY-9221402 and the US DOE DE-FG02-93ER40801

where $s = 0$ corresponds to the entrance of the dipole with $\phi = s/\rho$, D_0 and D_0' are respectively the values of the dispersion function and its derivative at $s = 0$. For the double bend achromat (or called the Chasman-Green Lattice), we have $D_0 = 0$ and $D_0' = 0$ to attain the achromatic condition.

The evolution of $\mathcal{H}$-function in a dipole is given by [3]

$$\mathcal{H}(\phi) = \mathcal{H}_0 + 2(\alpha_0 D_0 + \beta_0 D_0')\sin\phi - 2(\gamma_0 D_0 + \alpha_0 D_0')\rho(1-\cos\phi)$$
$$+ \beta_0 \sin^2\phi + \gamma_0 \rho^2(1-\cos\phi)^2 - 2\alpha_0 \rho \sin\phi(1-\cos\phi), \quad (4)$$

where $\mathcal{H}_0 = \gamma_0 D_0^2 + 2\alpha_0 D_0 D_0' + \beta_0 D_0'^2$, α_0, β_0, and γ_0 are Courant Snyder parameters at $s = 0$, and $\phi = s/\rho$. Averaging the $\mathcal{H}$-function in the dipole, one obtains

$$\langle \mathcal{H} \rangle = \mathcal{H}_0 + (\alpha_0 D_0 + \beta_0 D_0')\theta^2 E(\theta) - \frac{1}{3}(\gamma_0 D_0 + \alpha_0 D_0')\rho\theta^2 F(\theta)$$
$$+ \frac{\beta_0}{3}\theta^2 A(\theta) - \frac{\alpha_0}{4}\rho\theta^3 B(\theta) + \frac{\gamma_0}{20}\rho^2\theta^4 C(\theta). \quad (5)$$

Here θ is the bending angle of the dipole and

$$E(\theta) = \frac{2(1-\cos\theta)}{\theta^2}, \quad F(\theta) = \frac{6(\theta - \sin\theta)}{\theta^3}, \quad A(\theta) = \frac{6\theta - 3\sin 2\theta}{4\theta^3},$$
$$B(\theta) = \frac{6 - 8\cos\theta + 2\cos 2\theta}{\theta^4}, \quad C(\theta) = \frac{30\theta - 40\sin\theta + 5\sin 2\theta}{\theta^5}.$$

In the small angle limit, $A \to 1, B \to 1, C \to 1, E \to 1$, and $F \to 1$. Using the normalized scaling parameters:

$$d_0 = \frac{D_0}{L\theta}, \quad d_0' = \frac{D_0'}{\theta}, \quad \tilde{\beta}_0 = \frac{\beta_0}{L}, \quad \tilde{\gamma}_0 = \gamma_0 L, \quad \tilde{\alpha}_0 = \alpha_0, \quad (6)$$

where $L = \rho\theta$ is the length of the dipole, the averaged $\mathcal{H}$-function is given by

$$\langle \mathcal{H} \rangle = \rho\theta^3 \{\tilde{\gamma}_0 d_0^2 + 2\tilde{\alpha}_0 d_0 d_0' + \tilde{\beta}_0 d_0'^2 + (\tilde{\alpha}_0 E - \frac{\tilde{\gamma}_0}{3}F)d_0$$
$$+ (\tilde{\beta}_0 E - \frac{\tilde{\alpha}_0}{3}F)d_0' + \frac{\tilde{\beta}_0}{3}A - \frac{\tilde{\alpha}_0}{4}B + \frac{\tilde{\gamma}_0}{20}C\}. \quad (7)$$

II. THE DOUBLE BEND ACHROMAT

In a special case with achromat condition: $d_0 = 0$, and $d_0' = 0$, the average $\mathcal{H}$-function is given by

$$\langle \mathcal{H} \rangle = \rho\theta^3 \{\frac{\tilde{\beta}_0}{3}A - \frac{\tilde{\alpha}_0}{4}B + \frac{\tilde{\gamma}_0}{20}C\}. \quad (8)$$

Using the condition $\tilde{\beta}_0 \tilde{\gamma}_0 = (1 + \tilde{\alpha}_0^2)$, the minimum of $\langle \mathcal{H} \rangle$ is given by

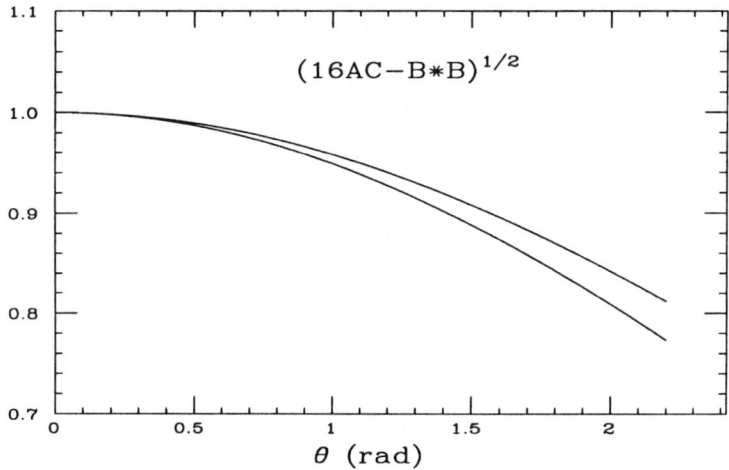

FIG. 1. The minimum $\langle \mathcal{H} \rangle$ factors $\sqrt{16AC - 15B^2}$ for the DBA (the lower curve) and $\sqrt{16\tilde{A}\tilde{C} - 15\tilde{B}^2}$ for the ME (the upper curve) lattices are plotted as a function of the bending angle θ. Note that $\langle \mathcal{H} \rangle$ is slightly smaller in long dipoles

$$\langle \mathcal{H} \rangle_{\text{MEDBA}} = \frac{G}{4\sqrt{15}} \rho \theta^3, \tag{9}$$

where $G = \sqrt{16AC - 15B^2}$. The corresponding betatron amplitude functions are

$$\tilde{\beta}_0 = \frac{6C}{\sqrt{15}G}, \quad \tilde{\alpha}_0 = \frac{\sqrt{15}B}{G}, \quad \tilde{\gamma}_0 = \frac{8\sqrt{5}A}{\sqrt{3}G}. \tag{10}$$

The factor $G = \sqrt{16AC - 15B^2}$ depends slowly on the dipole bending angle θ shown in Fig. 1, where $\langle \mathcal{H} \rangle$ is slightly smaller due to the horizontal focusing of the bending radius. In the small angle approximation, one obtains easily $\langle \mathcal{H} \rangle_{\text{MEDBA}} = \frac{1}{4\sqrt{15}} \rho \theta^3$ with $\beta^*_{\text{MEDBA}} = \frac{3}{4\sqrt{60}} L$, $s^*_{\text{MEDBA}} = \frac{3}{8} L$. The corresponding minimum emittance is

$$\epsilon_{\text{MEDBA}} = \frac{C_q \gamma^2 \theta^3}{4\sqrt{15} J_x}.$$

III. MINIMUM $\langle \mathcal{H} \rangle$-FUNCTION LATTICE

Without the achromat constraint, the lattice can be consider a single dipole lattice. Thus the minimum $\langle \mathcal{H} \rangle$ will be symmetric with respect to the center of the dipole. Therefore, the dispersion function and the betatron amplitude functions are symmetric with respect to the center of the dipole. The minimization procedure can be achieved through the following steps. First, the $\langle \mathcal{H} \rangle$

can be minimized by finding the optimal dispersion functions with

$$\frac{\partial \langle \mathcal{H} \rangle}{\partial d_0} = 0, \quad \frac{\partial \langle \mathcal{H} \rangle}{\partial d'_0} = 0.$$

The solution is given by $d_{0,min} = \frac{1}{6}F$, $d'_{0,min} = -\frac{1}{2}E$, with

$$\langle \mathcal{H} \rangle = \frac{1}{12}\rho\theta^3(\tilde{\beta}_0\tilde{A} - \tilde{\alpha}_0\tilde{B} + \frac{4\tilde{\gamma}_0}{15}\tilde{C}), \quad (11)$$

where $\tilde{A} = 4A - 3E^2, \tilde{B} = 3B - 2EF, \tilde{C} = \frac{9}{4}C - \frac{5}{4}F^2$. Using the relation that $\tilde{\beta}_0\tilde{\gamma}_0 = 1 + \tilde{\alpha}_0^2$, we obtain

$$\langle \mathcal{H} \rangle_{\text{ME}} = \frac{\tilde{G}}{12\sqrt{15}}\rho\theta^3, \quad (12)$$

where $\tilde{G} = \sqrt{16\tilde{A}\tilde{C} - 15\tilde{B}^2}$ which is also shown in Fig. 1. Thus the minimum $\langle \mathcal{H} \rangle$ without achromatic constraint is a factor of 3 smaller than that with the achromat condition. The minimum condition corresponds to

$$\tilde{\beta}_0 = \frac{8\tilde{C}}{\sqrt{15}\tilde{G}}, \quad \tilde{\alpha}_0 = \frac{\sqrt{15}\tilde{B}}{\tilde{G}}, \quad \tilde{\gamma}_0 = \frac{2\sqrt{15}\tilde{A}}{\tilde{G}}. \quad (13)$$

In the small angle approximation with $\theta \ll 1$, where $\tilde{A} \to 1, \tilde{B} \to 1, \tilde{C} \to 1$, and $\tilde{G} \to 1$. The waist of the optimal betatron amplitude function for the minimum $\langle \mathcal{H} \rangle$ is located at the middle of the dipole, i.e. $s^*_{\text{ME}} = L/2$. The corresponding minimum betatron amplitude function at the waist location is $\beta^*_{\text{ME}} = L/\sqrt{60}$, i.e. the required minimum betatron amplitude function is

$$\beta^*_{\text{ME}} = \frac{4}{3}\beta^*_{\text{MEDBA}}.$$

The attainable emittance is

$$\epsilon_{\text{ME}} = \frac{C_q\gamma^2\theta^3}{12\sqrt{15}J_x}.$$

The values of the dispersion $\mathcal{H}$-function at both sides of the dipole are important to determine the beam size in straight sections, where insertion devices such as the undulators are located. At the ME condition, we have

$$\mathcal{H}(0) = \mathcal{H}(\theta) = \frac{1}{3\sqrt{15}}\rho\theta^3\{6\tilde{C}F^2 - \frac{15}{2}\tilde{B}EF + \frac{5}{2}\tilde{A}E^2\}\tilde{G}^{-1}. \quad (14)$$

In the small bending angle approximation, we have $\mathcal{H}(\theta) = \frac{1}{3\sqrt{15}}\rho\theta^3$.

The brilliance of the photon beam from the undulator depends on the electron beam width. From Eq. (12), the minimum betatron emittance is given by

$$\epsilon_\beta = \frac{1}{12\sqrt{15}} \frac{C_q \gamma^2 \theta^3}{J_x}. \tag{15}$$

Now we define the dispersion emittance as

$$\epsilon_d \equiv \gamma_x (D\delta)^2 - \beta'_x (D\delta)(D'\delta) + \beta_x (D'\delta)^2 = \mathcal{H}(0)\delta^2, \tag{16}$$

where $\delta^2 = (\frac{\sigma_E}{E})^2 = C_q \frac{\gamma^2}{J_E \rho}$ is the equilibrium energy spread in the beam, J_E is the damping partition in synchrotron phase space. Because the $\mathcal{H}$-function is invariant in the straight section, ϵ_d is invariant in the straight section. Substituting $\mathcal{H}(0)$ of Eq. (14) into Eq.(16), we obtain

$$\epsilon_d = \frac{1}{3\sqrt{15}} \frac{C_q \gamma^2 \theta^3}{J_E}. \tag{17}$$

For a separated function lattice, $J_E \approx 2$, $J_x \approx 1$ or $J_E \approx 2J_x$. The total emittance for a bi-Gaussian distribution is given by

$$\epsilon = \epsilon_\beta + \epsilon_d = \frac{1}{4\sqrt{15}} \frac{C_q \gamma^2 \theta^3}{J_x} = \epsilon_{\text{MEDBA}}. \tag{18}$$

Thus the decrease in the betatron emittance is consumed by the dispersion beam size. The brilliance of the photon beam (namely the size of the electron beam in the "dispersion free" straight section) is not affected by the dispersion introduced to minimize the betatron emittance. The total electron beam size in the straight section remains unchanged. Thus the minimization procedure does not impair the beam brilliance of undulators.

IV. MINIMUM EMITTANCE IN COMBINED FUNCTION DBA

We have only discussed the minimum $\langle \mathcal{H} \rangle$ for sector dipoles. To simplify the design of DBA in synchrotron storage rings, combined function dipoles have often been used, e.g. the ELETTRA at Trieste and UVU and XRay rings at NSLS. The combined function defocusing dipole has the advantage of having a minimum β_x inside the dipole. Thus the lattice design would become slightly easier. Computer codes such as the Methodological Accelerator Design (MAD) [4] or the SYNCH can be used to optimize $\langle \mathcal{H} \rangle$. However, the analytic solution of the minimum emittance can serve as a guidance for realistic lattice design.

The dispersion function in the combined function dipole satisfies

$$D'' - KD = \frac{1}{\rho},$$

where K is the effective defocusing strength function. For the DBA, the dispersion function is given by

$$D(s) = \frac{1}{\rho K}(\cosh \sqrt{K}s - 1), \tag{19}$$

and the $\langle \mathcal{H} \rangle$ is given by

$$\langle \mathcal{H} \rangle = \rho \theta^3 [\frac{\tilde{\beta}_0}{3} A(q) - \frac{\tilde{\alpha}_0}{4} B(q) + \frac{\tilde{\gamma}_0}{20} C(q)],$$

where

$$A(q) = \frac{3(\sinh 2q - 2q)}{4q^3}, \quad B(q) = \frac{6 - 8\cosh q + 2\cosh 2q}{q^4},$$

$$C(q) = \frac{30q - 40\sinh q + 5\sinh 2q}{q^5},$$

with $q = \sqrt{K}L$, $\tilde{\beta}_0 = \beta_0/L$, $\tilde{\alpha}_0 = \alpha_0$, and $\tilde{\gamma}_0 = \gamma_0 L$. Thus the minimum of $\langle \mathcal{H} \rangle$ is given by

$$\langle \mathcal{H} \rangle_{min} = \frac{\sqrt{16AC - 15B^2}}{4\sqrt{15}} \rho \theta^3.$$

Figure 2 shows $\sqrt{16AC - 15B^2}$ as a function of the quadrupole strength q. As expected, we find $\sqrt{16AC - 15B^2} \geq 1$, i.e. the combined function DBA gives rise to a larger $\langle \mathcal{H} \rangle$. However, emittance is also affected by the damping partition given by

$$J_x = 1 + 2\frac{\sinh q - q}{q} - \frac{\alpha R}{\rho}, \tag{20}$$

where α is the momentum compaction factor, R is the average radius, and ρ is the bending radius. Figure 2 shows also $\sqrt{16AC - 15B^2}/J_x$ as a function of the quadrupole strength q. Depending on the focusing strength, the emittance can be reduced accordingly. For the ELETTRA lattice, the q parameter is about 0.94 and the emittance can be decreased by about 20% shown in Fig. 2.

V. THREE BENDS ACHROMAT

In recent years, three bend achromat (TBA) lattices have often been used in synchrotron radiation sources. The TBA is a combination of DBA lattice with a single dipole cell at the center. If all dipoles have equal length, the minimum emittance is usually quoted to be

$$\epsilon_{METBA} = \frac{2}{3}\epsilon_{MEDBA} + \frac{1}{3}\epsilon_{ME}. \tag{21}$$

The emittance is about 7/9 of the minimum DBA emittance while retaining the dispersion free sections for rf cavities and insertion devices. One can argue that the minimum emittance of QBA would be 1/2 of that of the minimum emittance DBA lattice. Is the minimum emittance in TBA attainable?

To simplify our discussion, we use small angle approximation. The normalized dispersion coordinates for the minimum emittance DBA and minimum emittance dipole lattices are given respectively by

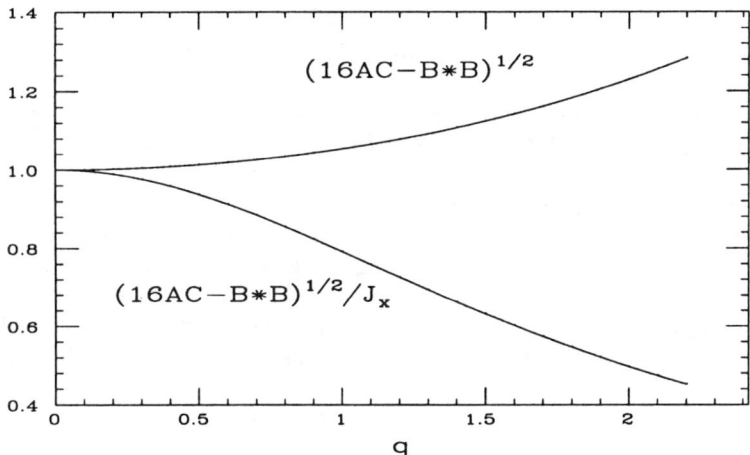

FIG. 2. The factor $\sqrt{16AC - B^2}$ for the combined function DBA lattice is shown as a function of the quadrupole strength $q = \sqrt{K}\ell$. The emittance factor $\sqrt{16AC - B^2}/J_x$ is also shown. Note that the combined function DBA can achieve lower emittance due to the damping partition

$$X_{\text{MEDBA}} = \frac{D}{\sqrt{\beta}} = \frac{(15)^{1/4}}{8} \frac{\ell_1^{3/2}}{\rho_1} \qquad (22)$$

$$P_{\text{MEDBA}} = \frac{\alpha D + \beta D'}{\sqrt{\beta}} = \pm \frac{7}{8(15)^{1/4}} \frac{\ell_1^{3/2}}{\rho_1}, \qquad (23)$$

at the dispersive ends of the dipoles in the MEDBA lattice, and

$$X_{\text{ME}} = \frac{D}{\sqrt{\beta}} = \frac{\sqrt{2}(15)^{1/4}}{24} \frac{\ell_2^{3/2}}{\rho_2} \qquad (24)$$

$$P_{\text{ME}} = \frac{\alpha D + \beta D'}{\sqrt{\beta}} = -\frac{3}{4\sqrt{2}(15)^{1/4}} \frac{\ell_2^{3/2}}{\rho_2}, \qquad (25)$$

at the entrance and exit locations of the dipole in the ME lattice, where ρ_1 and ℓ_1 are the bending radius and the length of the DBA dipoles, and ρ_2 and ℓ_2 are the bending radius and the length of the ME dipoles.

The optic Matching between the MEDBA module and the ME single dipole module is accomplished with quadrupoles, where the normalized dispersion functions is transformed by coordinate rotations, i.e.

$$\begin{pmatrix} X_{\text{ME}} \\ P_{\text{ME}} \end{pmatrix} = \begin{pmatrix} \cos\Phi & \sin\Phi \\ -\sin\Phi & \cos\Phi \end{pmatrix} \begin{pmatrix} X_{\text{MEDBA}} \\ P_{\text{MEDBA}} \end{pmatrix}, \qquad (26)$$

where Φ is the betatron phase advance. The necessary and sufficient condition for achieving betatron phase space matching is

$$\frac{\ell_2^3}{\rho_2^2} = 3\frac{\ell_1^3}{\rho_1^2} \tag{27}$$

with a corresponding phase advance $\Phi = 127.76°$. The necessary betatron matching condition for Eq. (27) is $L_2 = 3^{1/3} L_1$ for isomagnetic TBA lattice, or $\rho_1 = \sqrt{3}\rho_2$ for TBA lattice with equal length dipoles.

Thus we have proved a theorem stating that the TBA module with equal length dipoles can *not* be optically matched to attain the advertised minimum emittance. For an isomagnetic storage ring, we find that the center dipole for the TBA should be $3^{1/3}$ longer than those of outer dipoles in order to achieve betatron matching. In this case, one can prove an interesting trivial theorem: The emittance of the matched minimum TBA (QBA etc.) lattices is equal to

$$\epsilon_{\text{METBA}} = \frac{1}{4\sqrt{15}} \frac{C_q \gamma^2 \theta_1^3}{J_x}, \tag{28}$$

where θ_1 is the dipole bending angle of the DBA module. The minimum emittance is identical to that of the MEDBA.

VI. CONCLUSION

In conclusion, we have derived general formula for the emittance of electron storage rings. We show that the combined function DBA attains its smaller possible emittance through the damping partition number. We also show that the minimum emittance TBA is not attainable with equal dipole length. The necessary condition for achieving a minimum emittance in the TBA lattice is to change the length of the middle dipole. The resulting emittance is identical to that of DBA lattice. This logic can be used to prove that the QBA has also equal emittance formula as that of the MEDBA. Further reduction in emittance can only be achieved by varying the damping partition number. These analytic formula for minimum emittance should be helpful for a realistic low emittance lattice design.

REFERENCES

1. M. Sands, in *Physics with intersecting storage rings*, edited by B. Touschek, pp.257-411, (Academic Press, N.Y. 1971)
2. E.D. Courant and H.S. Snyder, Ann. Phys. **3**, 1 (1958).
3. R.H. Helm, M.J. Lee, and P.L. Morton, IEEE Trans. Act. NS**20**, 900 (1973).
4. H. Grote and F.C. Iselin, CERN report CERN/SL/90-13(AP) (CERN, Geneva, 1993).

PANEL DISCUSSIONS

Issues presented at the 8th ICFA beam dynamics workshop panel discussions

Panel Chair: R.L. Gluckstern (rlg@quark.umd.edu)

Panel members:

Alex Chao (achao@slac.stanford.edu)
Kohji Hirata (hirata@kekvax.kek.jp)
Steve Holmes (holmes@fnal.gov)
David F. Sutter (hep-tech@oer.doe.gov)
W.T. Weng (weng@bnldag.bnl.gov)
Alex Friedman (af@llnl.gov)
Ingo Hofmann (I.Hofmann@gsi.de)
Kwang-Je Kim (K_Kim@lbl.gov)
T. Wangler (twangler@lanl.gov)

Introductory Comments to the Panel Discussion on SCBD (Session VIII)

The panel members are drawn from several different communities:

- Those with interest in high current ion linacs (APT, spallation sources, etc.)
- Those with interest in heavy ion fusion
- Those with interest in high current electron linacs
- DOE perspective
- ICFA perspective

We shall hear briefly from each, followed by questions first from other panel members, then from members of the audience.

Comments from Panel members

Tom Wangler (LANL)
Beam physics needed in proton LINAC applications

A new generation of proton linacs that can deliver high beam power are being considered for a number of important applications throughout the world. The beam physics is important because of the need to control beam losses, which create radioactivation of the accelerator. Except for applications requiring circular machines, such spallation neutron sources, emittance growth itself is not the main concern, but rather the outer part of the beam distribution known as the beam halo. I will present what I believe are the beam-physics areas needing more study and development that directly relate to the needs of the proton-linac applications.

Beam Halo

Transverse mismatch of the beam has been well established as an important mechanism for producing halo. Progress in this area has resulted not only from numerical simulations, but also from an analytic model of the interaction of individual particles with the core, which provided understanding of the physics. The particle-core model has been applied to both uniform and periodic-focusing channels, and it has shown that a resonance between the core breathing oscillation, and the individual particle motion is a main cause of particles acquiring large amplitudes. It is found that there is a maximum amplitude for the resonantly driven particles, which provides some guidance for choosing the aperture sizes to avoid beam loss. Although there still seems to be a lack of a complete consensus about whether matched beams in quadrupole channels can produce halo, these seems to be good agreement about most other features of halo produced by the transverse dynamics. The important unanswered questions, include a) Is there still a maximum halo amplitude when many modes are excited by a beam mismatch? b) What are the characteristics of halo produced by longitudinal mismatch? This is a very important problem because the longitudinal beam-measurements needed to ensure good longitudinal matching are much more difficult. Very little progress has been made so far in the study of halo from longitudinally mismatched beams. c) How important is chaos in the halo formation? Is it important for beams of moderate tune depression? d) Can halo scraping be effective, and how should it be implemented?

Equipartitioning

More systematic study is needed to extend Hofmann's work on the x-y problem, beginning with the r-z problem. Most of the following questions were already identified in Hofmann's talk. What are the thermal-asymmetry thresholds needed to produce energy transfer between planes? What time scales are involved? Does energy transfer proceed until equipartitioning is established? Is halo formed as a result, and is there a maximum amplitude?

Space-Charge Codes

Space-charge-code development is important if we are going to be able to predict fractional beam losses at the level of 10-5 fractional losses, or about 10-8/m. Furthermore we would like to carry out beam simulations with at least 107 particles to have adequate statistical confidence that the total beam losses are sufficiently low. For a perspective, the real number of particles per bunch is about 109. Because of the difficulty of benchmarking the codes against experiment, it may be necessary to carry out such computations with more than one space-charge code to provide some checks.

Experimental Measurements

Measurements are necessary for benchmarking the codes. Ideally, this means measurements of transverse and longitudinal beam halo are necessary. Such measurements are not easy because the beam diagnostics must be capable of detection of the beam distribution over a large range of intensity.

Matching into RFQ's

Beam matching of a high-intensity dc beam into an RFQ has not been easy. High-intensity RFQ's have ended up with less RFQ output current than was predicted by the simulation codes, and this has generally been attributed to poorly matched beams at injection. This is not a problem for radioactivation, because of the low energy of the lost beam. The issue is the ability to deliver the design value of the beam current. Both experimental and computational studies may be required to address this problem.

These five areas are some of the most important. They will most likely require experiment, theory, and computational methods to make maximum progress.

Discussion after Wangler presentation

- Gluckstern: Referring to the particle-core model, to avoid particle losses is it best to choose the apertures outside the outer separatrix of the resonance region, or is the better strategy to relax the aperture requirement and concentrate on controlling the mismatch?
- Wangler: The effective threshold for generating large particle amplitudes and beam halo from mismatch is very low; a mismatch of only 10pretty large amplitudes. It will be difficult to guarantee that you can match even this well, especially longitudinally. Therefore, until we know more, I believe it is prudent to keep the apertures outside the outer separatrix. Ultimately the goal should also be to control the mismatch, but I think we are not yet in a position to claim we can do it.
- Friedman: We are the ideas for making 10^7 particle simulations of real linacs?
- Wangler: We don't have good answers for this. This is what we think we need to do, but at present our large runs have about 10^5 particles.

Ingo Hofmann (GSI)
Remarks concerning the issue of halo & loss:

The present understanding of halo origin seems to focus on the concept of an oscillating nonlinearity due to mismatch (core oscillations) + particles (in general assumed outside the core, i.e. in the "halo region") driven into nonlinear resonance by the core oscillations. It may be necessary to extend the analysis beyond the presently studied envelope mismatch modes for the core and include the effect of higher order mismatch modes as well. Can we ignore small initial amplitudes of such modes? This raises also question of self-consistency: does the model predict how particles get from a realistic initial distribution into the halo region?

There is still a big gap between linac design codes and "modelled simulations" looking at specific modes and resonances. The latter presently focus on infinitely long beams, which are also symmetric in the two transverse degrees of freedom. Hence, one parameter (i.e. the tune depression) seems enough to describe high phase space density. By going to ellipsoidal bunches (symmetric in the two transverse degrees of freedom) three parameters are required to describe the bunch adequately. These can be, for instance, the transverse rsp. Longitudinal tune depressions and the ratio of semi-axi or - equivalently - "temperatures". The latter leads to the question of "equipartitioning". A warning seems in place here: on the short time scales considered in linacs collisions play a negligible role and it may be more appropriate to discuss the question of temperature exchange in terms of collective instabilities and nonlinear resonances rather than stressing too much thermodynamics.

Alex Friedmann (LLNL)
Some thoughts on Beam Dynamics issues in Induction-Accelerator Based Heavy Ion Fusion

In general, Heavy-Ion Fusion driver development involves a simultaneous optimization of two budgets, one describing the beam emittance and the other the driver cost. The fusion target gain (energy multiplication) increases as the beam spot size is reduced (there is less radiation converter mass to heat), and this leads to a reduced cost of electricity. If HIF is to provide cost-effective electric power, the driver must produce a set of beams with an appropriate energy per ion, a large enough current, and a small enough emittance to hit the converters, and must do so at an attractive price. Other issues such as lifetime and availability are important as well, but can be folded into a life-cycle cost.

$$\text{emittance} \longleftrightarrow \text{cost}$$

As part of the author's formal talk earlier in the meeting, a "taxonomy" of the beam dynamics issues associated with induction HIF was presented. The comments made by the author as a panelist supplemented that material and are reproduced below.

Experiments

Beam dynamics research for HIF can and does make good use of scaled experiments to explore the critical issues. These include:

- The 4-into-1 beam combiner experiment about to be fielded at LBNL
- The small, scaled ESQ injector used as a test-bed for the recently-completed full-scale 2 MeV injector at LBNL
- The planned ILSE facility and program of experiments, which will employ beams with driver-scale radius and line-charge density but having reduced ion mass and energy
- Experiments in magnetic transport, bending and recirculation at LLNL, leading up to an operational prototype Small Recirculator with key dimensionless parameters similar to those of a driver
- Electron beam experiments at the University of Maryland exploring a variety of issues in transverse and longitudinal beam dynamics

Researchers in the field are fortunate in that (1) beam dynamics is readily scalable, and key dimensionless parameters are readily identified, in marked contrast with systems governed by atomic physics, wall interactions, etc.; (2) shot-to-shot repeatability is excellent, implying that a detailed picture of the beam particle distribution function can be built up over many shots. Also, small changes can be measured; thus it is often unnecessary to follow a beam over a long distance, and a relatively short and inexpensive experiment suffices; and (3) planned ring experiments at LLNL and the University of Maryland will facilitate the study of those phenomena that require a long path length for proper study, such as growth of unstable waves on the beam, wave reflection, the effects of imperfect longitudinal confinement, and "slow" equilibration mechanisms.

Theory and Simulations

In contrast with some fields of study, theory and simulations for Heavy Ion Fusion (and for space-charge dominated beam applications in general) are typically kept relevant and well-coupled to experiments, except in those cases (driver-scale specific issues) where it is necessary to study systems which do not yet exist. A hierarchy of tools is available, ranging from simple envelope models, through single-particle tracking and fluid models, to full discrete-particle (usually particle-in-cell, or PIC) simulations, including detailed 3-D calculations. In many ways the accelerator is a simpler system to model than (e.g.) a magnetic fusion device, since the beam resides in an induction linac for only 100 plasma periods. Furthermore, the physics does not generally require that a wide range of space- or time-scales be captured for a useful description. However, in some cases it is necessary to resolve the thin Debye length-scale sheath at the edge of the beam, and/or short axial wavelengths (the beam radius) which arise during collective equilibration. There are good opportunities to benchmark the theoretical and numerical tools versus experiments, and this is important since those same tools are used to predict the behavior of planned and proposed machines. True source-to-target modeling is a real possibility. Full PIC simulations of the beam in an induction linac appear to be within reach (after further code development and optimization), and particle simulations of LLNL's planned Small Recirculator over the full 15 (nominal) laps have been carried out using the WARP3d code. However, full PIC simulation of a driver-scale recirculator, with its long path length, will remain out of reach for a long time, requiring that a hierarchy of tools be used. Some calculations are inherently challenging. The production of halo particles often involves "rare" events, and capturing such events with good statistics requires that a very large number of simulation particles be used. Increased understanding has been gained through the combination of analytic theory, clever model calculations, and large-scale simulations.

Personal Views and Priorities

The use of self-consistent models should be expanded. Detailed study which relies on an unrealistic beam state may yield misleading results.

In general, "realistic" beam models should be used, and K-V equilibria should be employed with great care, and avoided when possible. Researchers have been misled by instabilities in K-V beams that do not arise in beams with realistic, smooth particle distributions. Ideally, the full 6-D particle distribution of the beam at a fixed time (or fixed plane) might be obtained from experiment and used to specify the beam at the beginning of a simulation. However, this in general will not be possible, and the use in simulations of beams formed "by injection" (that is, by simulating the injection process) is recommended when possible.

It will be important for practitioners to learn from each other, both across sub-fields and among those practicing theory, experiment, and simulation. In HIF it is also necessary to work both "forward from the source" and "backward from the target."

Technology development hasn't been discussed much at this meeting, but it serves to define the realm of possibilities. It is important for experimenters and theorists to communicate well with those involved in technology development, and to promote promising activities.

Three "feedback loops" ought to be encouraged and kept healthy:

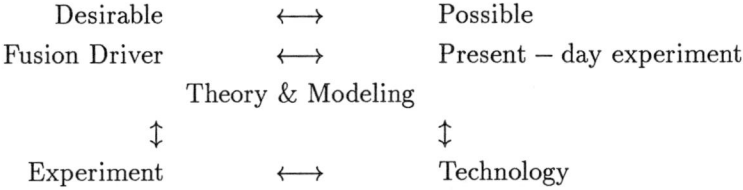

In general, HIF researchers in the U.S. and Europe appear to be doing well at identifying the important issues and making progress on them, within resource constraints. Cooperation with other researchers who are also concerned with space-charge effects is leading to real benefits.

W.T. Weng (BNL)
Comments on issues related to pulsed spallation neutron source

The typical design parameters for a 5 MW pulsed spallation neutron source are a proton synchrotron with an energy of a few GeV, an intensity of about 3×10^{14} ppp, and a repetition rate of around 50 Hz. Such a facility exceeds the current best performing proton synchrotron (for example ISIS – 800 MeV, 2×10^{13} ppp, 50 Hz, and the AGS – 28 GeV, 6×10^{13} ppp, 1 Hz) by a factor of 20 to 50 in total proton flux per second. Therefore the biggest challenge in the design of the next generation PNS is the particle loss in the injection and acceleration process. The typical loss of a few percent in today's accelerators has to be reduced to less than a tenth of a percent to keep the radiation effects to an acceptable level. To achieve that goal, the following areas need more R&D.

1. Charge distribution and halo formation in the linac and synchrotron.
2. Space charge effect at injection and methods of multi-turn injection to minimize particle losses. Issues such as painting, resonance corrections, second harmonic cavity, rf feedback, $\cdots$ etc., naturally come in here.
3. Coherent instabilities and their prevention and cures.
4. Reliable H^- source capable of delivering more than 100 mA.
5. Prototype chopper and stripping foil design.
6. Scraping, collimation and radiation shielding of beam.
7. Accelerator machine studies on critical processes.

Kwang-Je Kim (LBNL)
Beam Dynamics of Ultrashort, High-Brightness Electron Beams

Ultrashort, high-brightness electron beams would have important applications in the future, for example, in TeV linear colliders, x-ray free electron lasers, and short pulse synchrotron radiation generation. The beam parameters required for these applications are very demanding–the bunch length less than a picosecond, the energy spread less than 0.1 less than a few mm-mrad, and the pulse charge of about one nC. There are also more speculative applications, such as in developing compact, high energy accelerators based on the laser acceleration schemes, for which the requirements are even tighter.

Prerequisite to generating such beams is a thorough understanding of the beam generated electromagnetic field, and its effects on the beam qualities. Although the topic is not new for longer pulse beams, some new effects have been identified recently which could have a significant impact on the design of the accelerators producing ultrashort, high-brightness beams. The main goal of the Ultrashort Bunch Working Group was to understand and evaluate these effects in various accelerator schemes.

For linacs, two important components for generation of ultrashort, high brightness beams are the laser driven RF photocathode gun invented recently at LANL and the the bunching section. The space charge effect plays important role in both of these components: It is important in the RF photocathode gun because the beam at the beginning is non-relativistic. In a bunching section, where the orbit has a bend, the space charge effect turns out to be important even at relativistic energies. For storage rings, an isochronous design has been proposed for storing ultrashort beams. The collective instabilities in such a ring behave similar to the linac instabilities, and must be analyzed accordingly. In the following we summarize each of these topics separately:

The beam dynamics in laser driven RF photocathode gun have been investigated by several authors. The original study based on the analysis of the individual particle trajectory resulted in a simple analytic estimate of the emittance growth as the beam travels from the photocathode surface to the cavity exit. However, the growth is due primarily to the fact that the phase space ellipses at different locations along the bunch are not aligned with each other because of the variation in the space charge force. It is important for the operation of the laser driven RF photocathode gun as a source of ultrashort, high brightness electron beam that the phase space ellipses be realigned by means of a focusing element in the gun. This so called emittance compensation scheme has been modeled by detailed numerical simulation and tested conclusively by experiment. However, a clear analytical understanding of the operation of the scheme has been lacking, which could provide guidelines for practical designs. The Working Group discussed two recent approaches, which appear to be promising toward obtaining such an understanding. One is based on an envelope equation, and explains the emittance compensation by the phenomena that the solution of the envelope equation approaches to an invariant envelope. Another is based on an emittance evolution equation, and describes the emittance growth in two parts; a reversible part driven by the free energy stored in the electrostatic field, for which the emittance compensation is applicable, and an irreversible part arising from the entropy change. One should note that both approaches are based on the concepts and the formalisms developed by the ion beam community. The interaction between the electron beam and the ion beam communities during the workshop may hopefully result in a deeper understanding of the beam dynamics in both fields.

Space charge effects are normally negligible for highly relativistic beams moving in a straight pass because of the cancelation of the electric and the magnetic forces. However, two effects are identified recently which could be significant for relativistic, bunched beams moving in a curved pass, as in a bunching section. One effect arises from an incomplete cancelation of the electric and magneto forces for beams on a curved orbit. For unbunched beams, it was established sometimes ago that the effect of the residual force, referred to as the centrifugal space charge force, is canceled by the potential depression in the beam. Recent analysis shows that, for a bunched beam, the centrifugal space charge force remains uncanceled. The second effect is the phenomena that the coherent synchrotron radiation emitted by the trailing part overtakes and exerts force on the leading part of the bunch. Both the centrifugal space charge effect and the coherent synchrotron effect could lead to an energy change of particles within an achromatic bend, which in turn could lead to an emittance growth. Recent estimates indicate that the effects could be large, placing severe constraints on the design of the bunching section. During the working group discussion, different methods to evaluate these effects were discussed, with which a more accurate calculation could hopefully be made.

Recently, there have been some theoretical and experimental studies on the storage ring operation with a small momentum compaction factor for storing short bunches. In the limit of the vanishing momentum compaction, the relative longitudinal motion of the particles in the bunch stops. It was pointed out during the Working Group session that the collective instabilities then behave more like those in an extended linac rather than in a conventional storage ring. The main collective effects in a linac– and therefore in an isochronous storage ring–are the head-tail energy split in the longitudinal dimension and the beam break-up instability in the transverse dimension. The implication of these effects on the operation of isochronous storage rings are being evaluated.

Comments from the panel:

- *T. Wangler:* In principle the physics of space-charge-induced ion emittance growth also apply for electrons. The difference is that electrons leave the dc injector moving at about half the speed of light, and are accelerated to near the speed of light very quickly. Unlike ions, the electrons avoid the low-beta region, where space-charge effects are most important, and spend little time in the medium beta region.

Alex Chao (SLAC)
Emittance of Linear Colliders

Linear colliders require beams with high intensities and low emittances. For example, the 250 GeV NLC design requires $N = 0.7 \times 10^{10}$, $\epsilon_{Nx} = 5$ mm-mrad, $\epsilon_{Ny} = 0.05$ mm-mrad at the collision point, to yield a luminosity of 6×10^{33} cm^{-2}s^{-1}. The figure of merit for a linear collider (luminosity), however, is somewhat different from that for, e.g., an FEL (brightness). The luminosity scales as $\frac{N^2}{\epsilon_N}$, while the brightness scales as $N/\epsilon_N^2 \ell_z$, where ℓ_z is the bunch length. The optimization of a linear collider therefore is not identical to optimizing the brightness. In particular, in a linear collider, a large N is more useful than a small ϵ_N, while the opposite is true if one tries to optimize the brightness. Also, bunch length does not matter when one optimizes luminosity.

Almost all issues concerning low emittances discussed in the workshop are relevant to a linear collider. Space charge effect is critical at the gun. Wake fields, halos, misalignments, and jitters are important emittance spoilers in the linac. Nonlinearities, halos are important in the final focus. Among these issues, due to its long linacs, the wake field effects carry a heavier weight than many other applications, while due to its high beam energy, the importance of the space charge effects is restricted only to the gun area.

In the NLC design, damping rings are required to damp the emittances before the beams are injected into the long linacs. One question one may ask is whether the beam emittances right after the gun can be made small enough that these damping rings can be avoided. To inject the beams into the linacs directly from the gun, we must consider round beams. For the NLC, this means we should consider $\epsilon_{Nx} = \epsilon_{Ny} = 0.5$ mm-mrad. Such a small emittance seems achievable (within a factor of 2) by the state-of-art rf photocathode guns, but only for electrons. For positrons, a damping ring seems necessary.

One may also ask if NLC can be built as a laser accelerator. This means replacing the klystrons by lasers as power source. Let f be the ratio of the rf wavelength to the laser wavelength ($f \gg 1$). For acceleration, we then imagine miniaturizing the size of the rf structures by a factor f. The problem with this laser accelerator is that the number of particles per bunch N must scale down by a factor f in order to keep beam loading effects in check. To avoiding losing luminosity, we need then to increase the collision rate by a factor of f and to reduce the emittance by a factor of f. The increase of collision rate by

f is a difficult technical challenge. The reduction of emittance by a factor of f is in principle achievable for electrons by collimating the emission area at the cathode. If $\lambda_{\rm rf} = 2.6$ cm and $\lambda_{\rm laser} = 10\mu$m, the resulting emittance would then be $\epsilon_N = 1.2 \times 10^{-4}$ mm-mrad. This incredibly small emittance is required for a weak beam bunch with $N = 2.7 \times 10^6$. It fortunately does not (yet) violate the fundamental uncertainty principle limit of $\epsilon_N = \frac{\hbar}{2mc} = 1.9 \times 10^{-7}$ mm-mrad. There remains the question whether it is possible to reduce the positron emittance by a factor of f, even with weak beam bunches.

S. Holmes (Fermilab)
Round Table Comments on Beam Intensity Issues in Hadron Colliders

Comments and Observations

- Lower emittance in the hadron collider complex is (almost) always desirable.
- In hadron facilities the problem isn't at the front end. H^- injection and coalescing allow increase in the beam intensity and/or phase space density. Emittance preservation is much bigger issue than creation of low emittance.
- Mechanism(s) for emittance dilution due to space-charge in low energy synchrotrons are not well understood.
- Raising the injection energy has proven effective in improving performance significantly in low energy synchrotrons (for example the FNAL linac upgrade and the BNL Booster).
- Beam-beam interactions can be a limitation in TeV range proton-antiproton colliders. The long-range becomes important with more bunches and in high energy proton-proton machines with parasitic crossings.
- High energy proton-proton machines will place ever higher premium on low emittance (because of the impact of synchrotron radiation).
- Intrabeam scattering will be significant in colliders operating in the year 2000 (i.e. the Tevatron and RHIC).

Areas Requiring Attention

- Effort should be invested into developing a better understanding of the role of space-charge in low energy synchrotrons
- Improved diagnostics, dampers, kickers are required to minimize dilution during transfers.

- Bunched beam stochastic cooling with cooling times of a few hours would be extremely beneficial in combating emittance growth due to intrabeam scattering, power supply noise, rf noise, etc.
- Medium energy electron cooling presents an opportunity for creating low emittance beams beyond the space-charge regime. This technology should be developed.
- A better understanding of long-rang beam-beam effects will be required in future hadron colliders.

In the end the solution may be to make the proton beam energy high enough that beam is synchrotron radiation damped.

Discussions after Holmes' presentation

- Wangler: Given what we now know about controlling emittance in hadron colliders, how difficult would it have been to achieve the design luminosity for SSC, had it been built.
- Holmes: It would have been a challenge, but I think it was doable.

Kohji Hirata (ICFA Beam dynamics panel chairman, KEK)
What ICFA can do for us?

ICFA is the organization for high energy physics. For this purpose, it has the beam dynamics panel. The mission of the panel is to encourage and promote the international collaboration for the present and future accelerators. There, however, does not exist "the beam dynamics for high energy accelerators". All the knowledge obtained from all the accelerators is useful for all the accelerators. There is no limitation for the issues treated by the beam dynamics panel. The panel might be a good place to construct the beam dynamics society, like other societies in physics, high energy society for example. The panel encourages the more collaboration within the space-charge community, between different communities in accelerator physics, and between other communities in physics.

R.L. Gluckstern (Univ. of Maryland)
Closing Comments

The perspectives presented by the panel members have been extremely interesting. Moreover, the significant occurrence at this workshop is that the different communities have come together to consider important problems of overlapping interest. We hope this continues.

I would like to make some comments about the work on high intensity ion beam transport, in which I am also a "player".

- It seems likely that we have identified the primary source of halos: a resonance between a coherent oscillation of the beam (breathing mode?) and the non-linear motion of individual particles in the beam.
- From analytic and simulation studies of a K-V beam one can predict the halo boundaries. The time frame for halo growth is not yet well understood. In fact one needs another mechanism (instability?) to get the ions to the region where they can generate the halo.
- Further studies are needed on non-KV beams
- Further studies are needed on 3-D beam bunches
- Further studies are needed on AG focusing and the role of chaos.
- Further studies are needed on other beam modes and other resonances
- Much is being learned from envelope studies
- Further study is needed to determine the role played by thermal effects and equipartitioning
- Further experiments on collective beam effects in space charge dominated beams will be very useful
- The work on the Fokker-Planck equation seems interesting, but it is directed toward understanding how a mismatched beam seeks a new equilibrium. For almost matched beams, it may not be of direct use.

Discussion after Gluckstern's talk

- Wangler: My interpretation of the particle-core model is that the K-V core part of the model, represents the high-current central density of a typical particle distribution in a linac, which is peaked in the center and falls off at the edge. The K-V core in the model gives us approximate fields that are seen by the few particles in the tails. Given this interpretation, there is no problem answering the question how the particles got outside the core. There were already there as part of the original particle distribution. What the model then shows is that the mismatched beam core drives the outer particles in resonance and their amplitudes become very large, forming a halo.

Comments from workshop participants

S.Y. Lee (Indiana University)
On the systematic experimental measurements and theoretical analyses of high intensity beam parameters

Halo formation has been extensively studied by numerical simulations. It has been shown to be related to nonlinear parametric resonances, particularly overlapping resonances, resulting from the mismatched space charge dominated beam in the LINAC structure. These nonlinear resonances are characteristically similar to those of circular synchrotrons, where nonlinear resonances have been studied experimentally and theoretically. To better understand the physics of halo formation, better theory has to be developed. This means that we should gain knowledge of the parametric resonances in the global parametric space.

Experimental measurements of the beam properties in the global parametric space, e.g. *the space perveance, the phase advance, and the mismatch parameters* are essential to understand the validity of the simplified KV model vs the thermal beam distribution model. Since the resonance driving term for the space charge dominated beam is basically a universal function (Coulomb force) of the beam property. Thus systematic experimental measurements of space charge dominated beams are urgently needed. These systematic measurements can be used to gain confidence of our model studies.

M. Reiser (Univ. of Maryland)

I have two questions and some comments. The first is a question to Alex Chao:

- M. Reiser: "You mentioned the two options for achieving the luminosity in linear colliders - higher beam current or lower emittance - and you pointed out that increasing the beam current would be the preferred option. But in view of the high average power in a future linear collider, which is in the 100 MW range, this choice is not so obvious to me and lowering the emittance to the extent that this is feasible would appear to be preferable."
- A. Chao : I agree that increasing the current is the preferred choice only as long as other considerations such as average power do not impose a limit.
- M. Reiser: The second contribution is a comment to the statements by Tom Wangler and Bob Gluckstern. "Tom Wangler and Bob Gluckstern discussed the importance of the halo problem in high-current linacs and what has been learned in theoretical and computer simulation studies. I

would like to point out that these studies as well as our electron beam experiments on halo formation in mismatched beams at Maryland are concerned with long or continuous beams. In future work the much more difficult problem of halo formation in bunched beams must be addressed. Here, the coupling between longitudinal and transverse motion and the associated equipartitioning via space charge forces and the combined effects of transverse and longitudinal beam mismatch must be taken into account. This is a very challenging task for the theoreticians and experimentalists. As we pointed out in our talk (M. Reiser and N. Brown), a design where the beam is in thermodynamic equilibrium -throughout the entire rf linac or at least in the more critical low-energy region - appears to be an attractive way of avoiding or minimizing particle losses via halo formation."

C. Chen (MIT)

I'd like to add one area to Tom Wangler's list of areas of research on beam dynamics in high-current accelerators, namely, the role of charge density inhomogeneities in halo production. Indeed, recent studies by Qian Qian, Ron Davidson and myself [Phys. Rev. E51, 5216 (1995)] have shown that nonlinear space-charge forces due to beam density nonuniformities not only can induce chaotic particle motion, but also can cause a small fraction of particles to escape from the beam interior to form a halo.

Tom Katsouleas (USC)

After 3 days of interesting talks, many involving PIC simulations, I was struck by the observation that almost all of the PIC simulation output was in the form of particle plots. PIC codes are powerful tools, full of useful information. However, it appeared that field diagnostics were rarely turned on or presented. This is something like chopping down trees with a chainsaw with its engine turned off. Turn on the field and energy plots and the ω and k spectra of the collective modes that we think are responsible for the isotropization of the beam temperature. Then perhaps we will learn what modes are.

A similar problem of anisotropic temperature relaxation was recently encountered in laser-ionized gases (of the type used for plasma accelerators). The transverse laser field results in a high $T_\perp$ and cold $T_\parallel$ plasma. Subsequent thermalization is observed on a time scale much faster than a collision period. Through PIC simulations the thermalization mechanism was found to be the Weibel instability [W. Leemans, et al., Phys. Rev. A **46**, 1091 (1992)] – a zero frequency mode with a characteristic wave number. Perhaps a similar mechanism is responsible for the isotropization observed in heavy ion beams.

Shane Koscielniaki (TRIUMF)
Comment after panel discussion ICFA workshop on S-C dominated beams

Numerical methods play an important part in increasing our understanding of complex systems, because of the facility to trivially alter initial and boundary conditions and the force law, and because of the exact repeatability of computer experiments. The problem of the time evolution of a distribution of N ions mutually interacting by Coulombic repulsion (in the beam rest frame) leads to an $N \times N$ step algorithm for the calculation of the forces on the ions when direct methods are employed. Often, the N^2 algorithm is replaced by the PIC or particle-mesh algorithm: charges are assigned to the mesh, fields calculated on the mesh, and forces interpolated to the ion locations; which is a $2N$ algorithm plus some overhead for solving the fields on the mesh. In either case, N^2 or $2N$ we have no method to predict the minimum N to give realistic results; and for the $2N$ algorithm, we have no procedure to relate the mesh dimensions (i.e. number of cells) to the minimum number of particles. It is my opinion that there is essential work to be done here, so that reliable ensemble sizes can be estimated in advance of the detailed numerical simulations.

K. Bongardt (KFA)

For high intensity proton linacs, the high energy transfer line, either to a target station or to a following ring, suffers from space charge forces. In a longitudinal 'drift' the rms energy spread increases due to longitudinal space charge forces, even for energies above 1 GeV (=1000 MeV). As a consequence of this rms energy spread increase the design of an achromatic bending section gets complicated. The space charge forces cause also filmentation of the longitudinal phase space area. One way to reduce this filamentation is to keep the beam bunched which can be in conflict for the application of a bunch-rotator in order to reduce the energy spread. As the bunch length is increasing in the longitudinal 'drift', image forces cannot be neglected as the bunch length is comparable to the beam pipe diameter.

Another important effect that you didn't mention that affects short-pulse spallation neutron sources is in the transfer line between the linac and the ring. The beam debunches, and becomes very space-charge dominated in the longitudinal dimension. Thus it can become very difficult to control the longitudinal emittance. This is a very serious problem.

Author Index

B

Barnard, J., 427
Batygin, Y. K., 290
Bennett, L., 479
Billen, J. H., 60
Blaskiewicz, M., 283
Bohn, C. L., 322
Bongardt, K., 343, 354
Brown, N., 329
Bruhwiler, D. L., 219

C

Cable, M. D., 427
Callahan, D. A., 105, 244, 427, 434
Celata, C. M., 244
Chan, K. C. D., 60
Chao, A. W., 451
Chen, C., 169
Cho, Y., 363, 375
Cordova, S., 479
Crosbie, E., 260

D

Deadrick, F. J., 427

E

Eylon, S., 134, 427

F

Fawley, W. M., 244
Fessenden, T. J., 54, 427
Friedman, A., 105, 244, 401, 427

G

Garnett, R. W., 60
Genzlinger, R., 60

Gluckstern, R. L., 505
Gray, E. R., 60
Grote, D. P., 105, 134, 244, 427, 434
Guharay, S. K., 124
Guignard, G., 74

H

Haber, I., 105, 244
Hanks, R. L., 427
Harkay, K., 375
Hawkins, S. A., 427
Henestroza, E., 134
Holm, K. A., 427
Holmes, S. D., 42
Hopkins, H. A., 427

I

Ishizuka, H., 458

J

Judd, D. L., 427

K

Kawasaki, S., 458
Kehne, D., 105
Kim, K. J., 451
Kirbie, H. C., 427

L

Lagniel, J.-M., 206
Langdon, A. B., 105, 244
Law, K., 479
Lee, S. Y., 187, 234, 495
Lessner, E., 363, 375
Letchford, A. P., 343, 354
Liu, H., 468

Logan, B. G., 427
Longinotti, D., 427
Lund, S. M., 427

M

Machida, S., 160
Maenchen, J. E., 479
Mazarakis, M. G., 479

N

Nath, S., 60
Nattrass, L. A., 427
Nelson, M. B., 427
Neuffer, D., 468
Newton, M. A., 427
Ng, K. Y., 234

O

Ollis, C. W., 427
O'Shea, P. G., 309

P

Pabst, M., 343, 354
Pankuch, P., 479
Poukey, J. W., 479
Prior, C. R., 391

R

Rees, G. H., 23
Reiser, M., 95, 105, 329
Riabko, A., 187

Rovang, D., 479
Rudd, H., 105
Rusnak, B., 60

S

Sangster, T. C., 427
Schrage, D. L., 60
Sharp, W. M., 427, 434
Shiho, M., 458
Shimp, K., 479
Shoji, Y., 160
Smith, D. L., 479
Stovall, J. E., 60
Suk, H., 105
Symon, K., 260, 375

T

Takeda, H., 60

W

Wang, D. X., 105
Wang, J. G., 95, 105
Wangler, T. P., 3, 60
Wavrik, R., 479
Weiss, M., 32
Weng, W. T., 145
Wood, R., 60

Y

Yokoo, K., 458
Young, L. M., 60
Yu, S., 134

AIP Conference Proceedings

	Title	L.C. Number	ISBN
No. 343	High Energy Spin Physics Eleventh International Symposium (Bloomington, IN 1994)	95-78431	1-56396-374-4
No. 344	Nonlinear Dynamics in Particle Accelerators: Theory and Experiments (Arcidosso, Italy 1994)	95-78135	1-56396-446-5
No. 345	International Conference on Plasma Physics ICPP 1994 (Foz do Iguaçu, Brazil 1994)	95-78438	1-56396-496-1
No. 346	International Conference on Accelerator-Driven Transmutation Technologies and Applications (Las Vegas, NV 1994)	95-78691	1-56396-505-4
No. 347	Atomic Collisions: A Symposium in Honor of Christopher Bottcher (1945–1993) (Oak Ridge, TN 1994)	95-78689	1-56396-322-1
No. 348	Unveiling the Cosmic Infrared Background (College Park, MD, 1995)	95-83477	1-56396-508-9
No. 349	Workshop on the Tau/Charm Factory (Argonne, IL, 1995)	95-81467	1-56396-523-2
No. 350	International Symposium on Vector Boson Self-Interactions (Los Angeles, CA 1995)	95-79865	1-56396-520-8
No. 351	The Physics of Beams Andrew Sessler Symposium (Los Angeles, CA 1993)	95-80479	1-56396-376-0
No. 352	Physics Potential and Development of $\mu^+\mu^-$ Colliders: Second Workshop (Sausalito, CA 1994)	95-81413	1-56396-506-2
No. 353	13th NREL Photovoltaic Program Review (Lakewood, CO 1995)	95-80662	1-56396-510-0
No. 354	Organic Coatings (Paris, France, 1995)	96-83019	1-56396-535-6
No. 355	Eleventh Topical Conference on Radio Frequency Power in Plasmas (Palm Springs, CA 1995)	95-80867	1-56396-536-4
No. 356	The Future of Accelerator Physics (Austin, TX 1994)	96-83292	1-56396-541-0
No. 357	10th Topical Workshop on Proton-Antiproton Collider Physics (Batavia, IL 1995)	95-83078	1-56396-543-7
No. 358	The Second NREL Conference on Thermophotovoltaic Generation of Electricity	95-83335	1-56396-509-7

	Title	L.C. Number	ISBN
No. 360	The Physics of Electronic and Atomic Collisions XIX International Conference (Whistler, Canada, 1995)	95-83671	1-56396-440-6
No. 361	Space Technology and Applications International Forum (Albuquerque, NM 1996)	95-83440	1-56396-568-2
No. 362	Two-Cneter Effects in Ion-Atom Collisions (Lincoln, NE 1994)	96-83379	1-56396-342-6
No. 363	Phenomena in Ionized Gases XXII ICPIG (Hoboken, NJ, 1995)	96-83294	1-56396-550-X
No. 364	Fast Elementary Processes in Chemical and Biological Systems (Villeneuve d'Ascq, France, 1995)	96-83624	1-56396-564-X
No. 365	Latin-American School of Physics XXX ELAF Group Theory and Its Applications (México City, México, 1995)	96-83489	1-56396-567-4
No. 366	High Velocity Neutron Stars and Gamma-Ray Bursts (La Jolla, CA 1995)	96-84067	1-56396-593-3
No. 367	Micro Bunches Workshop (Upton, NY, 1995)	96-83482	1-56396-555-0
No. 368	Acoustic Particle Velocity Sensors: Design, Performance and Applications (Mystic, CT, 1995)	96-83548	1-56396-549-6
No. 369	Laser Interaction and Related Plasma Phenomena (Osaka, Japan 1995)	96-85009	1-56396-445-7
No. 370	Shock Compression of Condensed Matter-1995 (Seattle, WA 1995)	96-84595	1-56396-566-6
No. 371	Sixth Quantum 1/f Noise and Other Low Frequency Fluctuations in Electronic Devices Symposium (St. Louis, MO, 1994)	96-84200	1-56396-410-4
No. 372	Beam Dynamics and Technology Issues for + - Colliders 9th Advanced ICFA Beam Dynamics Workshop (Montauk, NY, 1995)	96-84189	1-56396-554-2
No. 373	Stress-Induced Phenomena in Metallization (Palo Alto, CA 1995)	96-84949	1-56396-439-2
No. 374	High Energy Solar Physics (Greenbelt, MD 1995)	96-84513	1-56396-542-9
No. 376	Chaos and the Changing Nature of Science and Medicine: An Introduction (Mobile, AL 1995)	96-85220	1-56396-442-2
No. 377	Space Charge Dominated Beams and Applications of High Brightness Beams (Bloomington, IN 1995)	96-85165	1-56396-625-7